GENES, ENVIRONMENT AND ALZHEIMER'S DISEASE

GENES, ENVIRONMENT AND ALZHEIMER'S DISEASE

Edited by

ORLY LAZAROV
University of Illinois at Chicago, Chicago, IL, USA

GIUSEPPINA TESCO
Tufts University School of Medicine, Boston, MA, USA

AMSTERDAM • BOSTON • HEIDELBERG • LONDON
NEW YORK • OXFORD • PARIS • SAN DIEGO
SAN FRANCISCO • SINGAPORE • SYDNEY • TOKYO

Academic Press is an imprint of Elsevier

Academic Press is an imprint of Elsevier
125 London Wall, London EC2Y 5AS, UK
525 B Street, Suite 1800, San Diego, CA 92101-4495, USA
50 Hampshire Street, 5th Floor, Cambridge, MA 02139, USA
The Boulevard, Langford Lane, Kidlington, Oxford OX5 1GB, UK

Notices
Knowledge and best practice in this field are constantly changing. As new research
and experience broaden our understanding, changes in research methods, professional
practices, or medical treatment may become necessary.

Practitioners and researchers must always rely on their own experience and knowledge
in evaluating and using any information, methods, compounds, or experiments described
herein. In using such information or methods they should be mindful of their own safety
and the safety of others, including parties for whom they have a professional responsibility.

To the fullest extent of the law, neither the Publisher nor the authors, contributors, or
editors, assume any liability for any injury and/or damage to persons or property as a
matter of products liability, negligence or otherwise, or from any use or operation of any
methods, products, instructions, or ideas contained in the material herein.

ISBN: 978-0-12-802851-3

British Library Cataloguing-in-Publication Data
A catalogue record for this book is available from the British Library

Library of Congress Cataloging-in-Publication Data
A catalog record for this book is available from the Library of Congress

For information on all Academic Press publications
visit our website at http://store.elsevier.com/

Typeset by MPS Limited, Chennai, India
www.adi-mps.com

Printed and bound in the United States of America

Working together
to grow libraries in
developing countries

www.elsevier.com • www.bookaid.org

Publisher: Mara Conner
Acquisition Editor: Mara Conner
Editorial Project Manager: Kristi Anderson
Production Project Manager: Chris Wortley
Designer: Mark Rogers

Cover image: The three-dimensional rendering was constructed using Pixlogic's ZBrush
software on the framework of a voxel human model for electromagnetic dosimetry,
'TARO', which was created from a set of 2mm interval MRI images of a brain.
Image provided by David Gate

Contents

List of Contributors   xi
Foreword   xv
Preface   xvii
Introduction   xxi

1. Molecular Mechanisms of Learning and Memory
DANIELA PUZZO, JOLE FIORITO, ROSITA PURGATORIO, WALTER GULISANO,
AGOSTINO PALMERI, OTTAVIO ARANCIO AND RUSSELL NICHOLLS

From Plato's Wax Tablet to Genes   2
Memory Processes   2
Brain Structures Underlying Memory Processes   5
Cellular Basis of Memory   6
The Role of CREB in Memory   8
Pathways Activating CREB   9
HDACs/HATs Role in Gene Transcription   13
References   17

2. When Cognitive Decline Becomes Pathology:
From Normal Aging to Alzheimer's Disease
ELIEZER MASLIAH AND DAVID P. SALMON

Models of Cognitive Decline in Healthy Aging   30
Distinguishing AD-Associated Cognitive Decline from Normal Aging   31
Distinguishing Preclinical AD from Normal Aging with Event-Related
     Potentials (ERPs)   35
Memory Deficits in Preclinical AD   37
Asymmetric Cognitive Decline in Preclinical AD   42
Summary and Conclusions   43
References   44

3. Adult Neurogenesis and Cognitive Function: Relevance
for Disorders Associated with Human Aging
KERI MARTINOWICH AND ROBERT J. SCHLOESSER

Introduction   52
Identification and Characterization of Adult Neurogenesis   54

Development and Maturation of Adult-Born Neurons   55
Adult Mammalian Hippocampal Circuitry   58
Factors Affecting Adult Neurogenesis   59
Proposed Functions of Adult Neurogenesis   67
Neurogenesis, Hippocampal Dysfunction, and Alzheimer's Disease   73
Methodologies for Studying Adult Neurogenesis   75
Conclusion   78
References   78

I

HALLMARKS AND GENETIC FORMS OF AD

4. The Amyloid β Precursor Protein and Cognitive Function in Alzheimer's Disease

ROBERT A. MARR

Introduction   98
The APP Gene and Its Homologs   99
The APP Protein, Its Domains, and Its Proteolytic Fragments   100
Intracellular Trafficking and Processing of APP   102
Cholesterol and APP Processing   103
APP-Interacting Proteins: Signal Transduction and Processing   104
Functions of APP   106
Does Aβ Have a Function?   111
APP Knockout Mice   113
APP and AD   113
References   117

5. Molecular Pathways in Alzheimer's Disease and Cognitive Function: New Insights into Pathobiology of Tau

XU CHEN, MEREDITH C. REICHERT AND LI GAN

Introduction   136
Normal Production and Function of Tau   136
Pathogenic Tau   138
Tau-Mediated Neuronal Deficits   143
Targeting Tau in AD: Therapeutic Implications   154
References   156

II

GENETIC AND ENVIRONMENTAL RISK FACTORS

6. Apolipoprotein E and Amyloid-β-Independent Mechanisms in Alzheimer's Disease

TAKAHISA KANEKIYO AND GUOJUN BU

Introduction	172
APOE Genotypes and Cognitive Functions in Nondemented Individuals	173
APOE and Cholesterol Metabolism in Synaptic Functions	174
APOE, Brain Glucose Metabolism, and Insulin Signaling	176
APOE, Mitochondria Dysfunction, and Tau Phosphorylation	177
APOE, ApoE Receptors, and Synaptic Functions	179
APOE and Cerebrovascular Functions	181
APOE and Inflammatory Response	183
Summary and Perspectives	185
Acknowledgments	187
References	188

7. Lifestyle and Alzheimer's Disease: The Role of Environmental Factors in Disease Development

NANCY BARTOLOTTI AND ORLY LAZAROV

Introduction	198
Epidemiological Studies	201
The Benefits of Cognitive Complexity Following the Onset of Dementia	207
Physical Activity and Exercise	209
Chemical Exposure and AD Risk	211
Metals	212
Air Pollution and Tobacco Smoke	213
Nutrition and the Microbiome	214
Sleep and Circadian Rhythm	218
Socialization	219
Conclusion	219
References	220

8. Role of BACE1 in Cognitive Function, from Alzheimer's Disease to Traumatic Brain Injury

SYLVIA LOMBARDO AND GIUSEPPINA TESCO

Introduction	240
BACE1	241
References	258

9. Traumatic Brain Injury and Rationale for a Neuropsychological Diagnosis of Diffuse Axonal Injury

AMANDA R. RABINOWITZ AND DOUGLAS H. SMITH

Introduction	268
Common TBI-Related Cognitive Deficits	271
Mechanisms of Cognitive Disturbance Following TBI	273
Diffuse Axonal Injury	274
DAI as a Mechanism of Cognitive Dysfunction Following TBI	274
Neurochemical and Neurometabolic Changes	277
Neurometabolic Changes as a Mechanism of Cognitive Dysfunction Following TBI	277
Focal Injuries	279
Focal Brain Injury as a Mechanism of Cognitive Dysfunction Following TBI	279
Neurodegenerative Processes	281
Neurodegenerative Processes as a Mechanism of Cognitive Dysfunction Following TBI	282
Conclusions	283
References	285

10. Alzheimer's Disease and the Sleep–Wake Cycle

ADAM W. BERO AND LI-HUEI TSAI

Introduction	295
To Sleep, Perchance to Learn	297
Sleep Disturbances and Cognitive Decline	300
Sleep and Aβ Pathology: A Pathogenic Loop	301
Summary and Closing Remarks	308
References	309

11. Stroke, Cognitive Function, and Alzheimer's Disease

KATHERINE A. JACKMAN, TOBY CUMMING AND ALYSON A. MILLER

Introduction	320
Stroke: An Overview	320
Cognitive Impairment and Dementia: AD, VaD, and VCI	324
The Role of the NVU in AD and VCI	327
VCI After Stroke	333
AD and Stroke	341
Prospects for Treatment and Prevention	344
Conclusions	347
References	347

12. Cerebral Innate Immunity: A New Conceptual Framework for Alzheimer's Disease

DAVID GATE AND TERRENCE TOWN

Cerebral Innate Immunity in Alzheimer Pathoetiology	362
An Historical Perspective: Innate Immunity in AD	364

Aβ Immunotherapy and the Role of Mononuclear Phagocytes in Amyloid
    Plaque Clearance   365
Novel Strategies for Targeting Peripheral Macrophages Versus
    Brain-Resident Microglia   367
Targeting Cardinal Anti-inflammatory Cytokines to Restrict Cerebral Amyloidosis   369
Chemokines Recruit Monocytes to Aβ Plaques and Dying Neurons   371
Blocking Inflammatory ILs 12 and 23 Prevents Plaque Buildup in Transgenic Mice   372
PPARγ Agonists Reduce Inflammation While Boosting Microglial Aβ Uptake   373
Inflammasome Activation in AD   375
Beclin 1 Regulates Microglial Phagocytosis and May Be Impaired in AD   375
A New Generation of AD Pharmacotherapeutics Targeting Innate Immunity   376
Concluding Remarks   378
Acknowledgements   380
References   380

13. Type 2 Diabetes Mellitus as a Risk Factor for Alzheimer's Disease

JACQUELINE A. BONDS, PETER C. HART, RICHARD D. MINSHALL, ORLY LAZAROV,
JACOB M. HAUS AND MARCELO G. BONINI

Epidemiology   388
T2DM: Clinical Description   390
Mouse Models of Type 2 Diabetes   391
Mechanism and Pathways   394
Cerebrovascular Complications in T2DM: Implications for AD Development   400
Alzheimer's Disease: When Inflammation and Vascular Dysfunction
    Get to the Brain   404
Conclusion   407
References   407

Index   **415**

List of Contributors

Ottavio Arancio Department of Pathology and Cell Biology, Taub Institute for Research on Alzheimer's Disease and the Aging Brain, Columbia University, New York, NY, United States

Nancy Bartolotti Department of Anatomy and Cell Biology, College of Medicine, University of Illinois at Chicago, Chicago, IL, United States

Adam W. Bero Department of Brain and Cognitive Sciences, Picower Institute for Learning and Memory, Massachusetts Institute of Technology, Cambridge, MA, United States; Department of Neuroscience, Merck Research Laboratories, Boston, MA, United States

Jacqueline A. Bonds Graduate Program in Neuroscience and Department of Anatomy and Cell Biology, College of Medicine, University of Illinois at Chicago, Chicago, IL, United States

Marcelo G. Bonini Department of Pathology, University of Illinois at Chicago, Chicago, IL, United States; Department of Medicine, University of Illinois at Chicago, Chicago, IL, United States

√ **Guojun Bu** Department of Neuroscience, Mayo Clinic Jacksonville, Jacksonville, FL, United States

Xu Chen Department of Neurology, Gladstone Institute of Neurological Disease, University of California, San Francisco, CA, United States

Toby Cumming Melbourne Brain Centre (Austin Campus), The Florey Institute of Neuroscience and Mental Health, The University of Melbourne, Heidelberg, Victoria, Australia

Jole Fiorito Department of Pathology and Cell Biology, Taub Institute for Research on Alzheimer's Disease and the Aging Brain, Columbia University, New York, NY, United States

Li Gan Department of Neurology, Gladstone Institute of Neurological Disease, University of California, San Francisco, CA, United States

David Gate Department of Physiology and Biophysics, Zilkha Neurogenetic Institute, Keck School of Medicine, University of Southern California, Los Angeles, CA, United States

Walter Gulisano Department of Biomedical and Biotechnological Sciences, Section of Physiology, University of Catania, Catania, Italy

Peter C. Hart Department of Pathology, University of Illinois at Chicago, Chicago, IL, United States

Jacob M. Haus Department of Kinesiology and Nutrition, University of Illinois at Chicago, Chicago, IL, United States

Katherine A. Jackman Melbourne Brain Centre (Austin Campus), The Florey Institute of Neuroscience and Mental Health, The University of Melbourne, Heidelberg, Victoria, Australia

✓ **Takahisa Kanekiyo** Department of Neuroscience, Mayo Clinic Jacksonville, Jacksonville, FL, United States

Orly Lazarov Department of Anatomy and Cell Biology, College of Medicine, University of Illinois at Chicago, Chicago, IL, United States

Sylvia Lombardo Alzheimer's Disease Research Laboratory, Department of Neuroscience, Tufts University School of Medicine, Boston, MA, United States

Robert A. Marr Department of Neuroscience, Rosalind Franklin University of Medicine and Science, North Chicago, IL, United States

Keri Martinowich Lieber Institute for Brain Development, Baltimore, MD, United States; Department of Psychiatry and Behavioral Sciences, Johns Hopkins School of Medicine, Baltimore, MD, United States; Department of Neuroscience, Johns Hopkins School of Medicine, Baltimore, MD, United States

Eliezer Masliah Department of Neurosciences, University of California, San Diego, CA, United States; Department of Pathology, University of California, San Diego, CA, United States

Alyson A. Miller School of Medical Sciences, Health Innovations Research Institute, RMIT University, Melbourne, Victoria, Australia

Richard D. Minshall Department of Anesthesiology, University of Illinois at Chicago, Chicago, IL, United States; Department of Pharmacology, University of Illinois at Chicago, Chicago, IL, United States

Russell Nicholls Department of Pathology and Cell Biology, Taub Institute for Research on Alzheimer's Disease and the Aging Brain, Columbia University, New York, NY, United States

Agostino Palmeri Department of Biomedical and Biotechnological Sciences, Section of Physiology, University of Catania, Catania, Italy

Rosita Purgatorio Department of Pathology and Cell Biology, Taub Institute for Research on Alzheimer's Disease and the Aging Brain, Columbia University, New York, NY, United States

Daniela Puzzo Department of Biomedical and Biotechnological Sciences, Section of Physiology, University of Catania, Catania, Italy

Amanda R. Rabinowitz Center for Brain Injury and Repair, Department of Neurosurgery, University of Pennsylvania, Philadelphia, PA, United States

Meredith C. Reichert Department of Neurology, Gladstone Institute of Neurological Disease, University of California, San Francisco, CA, United States

David P. Salmon Department of Neurosciences, University of California, San Diego, CA, United States

Robert J. Schloesser Sheppard Pratt-Lieber Research Institute, Baltimore, MD, United States

Douglas H. Smith Center for Brain Injury and Repair, Department of Neurosurgery, University of Pennsylvania, Philadelphia, PA, United States

Giuseppina Tesco Alzheimer's Disease Research Laboratory, Department of Neuroscience, Tufts University School of Medicine, Boston, MA, United States

Terrence Town Department of Physiology and Biophysics, Zilkha Neurogenetic Institute, Keck School of Medicine, University of Southern California, Los Angeles, CA, United States

Li-Huei Tsai Department of Brain and Cognitive Sciences, Picower Institute for Learning and Memory, Massachusetts Institute of Technology, Cambridge, MA, United States; Broad Institute of Harvard University and Massachusetts Institute of Technology, Cambridge, MA, United States

Foreword

There is a great awareness of the contribution of lifestyle to cardio-vascular disease, diabetes, fitness, and general health. However, it is not well known that, to some degree, lifestyle can affect the development of Alzheimer's disease. This book is the first to summarize the most advanced information based on state-of-the-art experimental and epide-miological evidence concerning the interplay of genes and environment in the development of Alzheimer's disease.

Preface

It is only during the last decade that neuroscientists discovered that environmental factors play a role in the development of Alzheimer's disease. Until then, research was focused on the genetic forms of familial Alzheimer's disease, simply because an adequate experimental model for the sporadic disease was not available. The knowledge we have gained on the molecular signaling of familial Alzheimer's disease is invaluable; however, the epidemiology shows that more than 95% of the people affected have the sporadic, late onset form of the disease. In fact, the greatest risk factor for the disease is aging. The likelihood of developing Alzheimer's disease doubles about every 5 years after age 65. After age 85, the risk reaches nearly 50%. The fact that there is not a genetic cause for the late onset form of the disease and the strong relationship of age to the development of the disease suggest that it is a gene-environment interplay that supports or defers the development of the disease.

The discovery that the environment affects brain function, and in particular learning and memory, was made by the neuropsychologist Donald Hebb, who showed that pet rats, raised in an open environment in his house, were cognitively superior to caged lab rats. This was the first evidence that behavior modulates brain function. This seminal discovery was received with nothing but skepticism. Nearly 80 years later, it is now accepted that behavior has a profound effect on brain structure and function. There is strong evidence to suggest that the previous dogma that the brain is immutable is far from accurate. On the contrary, the brain exhibits a great deal of plasticity. It changes in response to stimuli, such as stress, learning, and exercise. An increasing body of evidence suggests that factors such as education, cognitive activity, and physical exercise enhance memory, while stress, interrupted sleep, and social isolation compromise it.

When we discovered that letting our Alzheimer's mouse model experience an environment with toys, running wheels, and their mates' company led to reduced amyloid pathology, we were skeptical but intrigued. It was clear that this observation is a game changer. During the last 10 years we and others have established that, for our Alzheimer's mice, an enriched environment has numerous positive effects on the brain, including enhanced synaptic plasticity, increased neurogenesis, rescue of memory deficits, and reduced tau pathology.

This observation, that lifestyle affects brain dysfunction and pathology, suggests that we might be able to control our own destiny and attenuate, delay, or prevent pathology. This book is meant to bring into focus this important development in Alzheimer's research, increase public awareness, and discuss the possibility that our daily actions maybe accountable, at least in part, for the health of the central nervous system. Accumulative and summative effects will play a role, in concert with our genetic material, in determining the degree of our brain's function as we age.

What lifestyle factors should we consider in relation to brain function and Alzheimer's disease? The book will discuss evidence suggesting that level of education, cognitive stimulation, and physical activity play a role in the development of the disease. Interestingly, cholesterol, glucose, and insulin metabolism are implicated in Alzheimer's disease. Type-2 diabetes is a known risk factor for Alzheimer's disease. These metabolic pathways are linked to dietary habits, obesity, and age, and are thought to be preventable or controllable by lifestyle. Another important aspect is brain injury and trauma to the central nervous system. This issue has been brought into focus in relation to football players who suffer cognitive deficits and personality change, as well as in relation to soldiers who suffer traumatic brain injury (TBI). The book will discuss evidence suggesting that molecular components of familial Alzheimer's disease play a role in the response to stroke and TBI.

Research in our labs seeks to elucidate the molecular mechanisms underlying environmental enrichment-induced brain plasticity and the attenuation of Alzheimer's disease pathology. It is our hope that future basic and epidemiological research will provide specific insights into the factors that are critical for the prevention of Alzheimer's disease, and their effective, therapeutic dose.

The large brain, like large government, may not be able to do simple things in a simple way.

Donald O. Hebb

Introduction

Alzheimer's disease was described for the first time by Alois Alzheimer, a Bavarian psychiatrist, in a conference in 1907. He presented the case of a woman in her fifties who exhibited memory impairments, disorientation in her own dwelling, and paranoia. In a postmortem autopsy of her brain he found two types of lesions that are considered as the hallmarks of the disease: amyloid deposits (plaques) and neurofibrillary tangles. Since then, substantial research has been conducted aiming at elucidating the cause for the disease and finding ways to cure it. Currently, there is no effective medication for curing or attenuating the disease.

Currently, more than 35 million individuals suffer from Alzheimer's disease worldwide. About 5 million Americans are affected with the disease, and it is estimated that this number will triple in the next decade. Patients may live up to 20 years after the initial diagnosis, although the average survival is 5–10 years. The quality of life of an Alzheimer's patient is poor. Coupled with the advancement of life-prolonging technologies and increased life expectancy, an effective therapeutic approach is a high priority for researchers.

Alzheimer's disease is characterized by progressive loss of memory and cognitive decline. A large proportion of these individuals will develop dementia. Forgetting recently learned information is one of the most common early signs of dementia. Other signs are difficulty planning or completing everyday tasks, forgetting words, disorientation to place and time, and deterioration of mental thinking and executive decisions. Sometimes, change in personality or behavior becomes apparent.

Neuropathologically, the disease is characterized by progressive loss of neurons in critical areas of learning and memory, such as the hippocampus and cortex. Nevertheless, neuronal pathology is thought to start decades before neuronal loss takes place, possibly in the form of synaptic loss and alterations in dendrites. Nevertheless, the mechanism underlying cognitive decline is not fully elucidated.

For several decades the dominant theory in the field has been the amyloid hypothesis, which stated that cognitive deficits are caused by malformation of β-amyloid peptides, the precursor of amyloid deposits. Recently, this hypothesis has been strongly challenged mostly in light of the failure of recent medications that were developed based on this theory. The amyloid hypothesis has evolved based on the molecular pathways of the familial mutant forms that cause early onset of Alzheimer's disease, namely, amyloid precursor protein and presenilin 1, 2. While autosomal dominant mutations in these genes are extremely rare, these molecular signals are thought to play a role in both the familial and sporadic forms of the disease. In this book we will consider the evidence that these signals and other genetic risks play a role, along with environmental factors, in the development of Alzheimer's disease.

CHAPTER

1

Molecular Mechanisms of Learning and Memory[*]

Daniela Puzzo[1], Jole Fiorito[2], Rosita Purgatorio[2],
Walter Gulisano[1], Agostino Palmeri[1],
Ottavio Arancio[2] and Russell Nicholls[2]

[1]Department of Biomedical and Biotechnological Sciences, Section
of Physiology, University of Catania, Catania, Italy
[2]Department of Pathology and Cell Biology,
Taub Institute for Research on Alzheimer's Disease and the Aging Brain,
Columbia University, New York, NY, United States

OUTLINE

From Plato's Wax Tablet to Genes	2
Memory Processes	2
Brain Structures Underlying Memory Processes	5
Cellular Basis of Memory	6
The Role of CREB in Memory	8
Pathways Activating CREB	9
HDACs/HATs Role in Gene Transcription	13
References	17

[*] The authors declare no competing financial interests.

Genes, Environment and Alzheimer's Disease.
DOI: http://dx.doi.org/10.1016/B978-0-12-802851-3.00001-2

1

FROM PLATO'S WAX TABLET TO GENES

Learning and memory are two closely related phenomena that allow living beings to acquire new knowledge and to retain this information. Undoubtedly, they are ubiquitous characteristics of several species, ranging from invertebrates to mammals, probably because of the need to adapt behavior to changing environmental conditions, thus aiding survival.

The higher complexity of the human brain makes them even more fascinating, because memory is the book, or better the Plato's wax tablet, where we impress our thoughts, our feelings, and images of ourselves and the world around us. Apart from the various philosophical, theological, and sociocultural implications, as neuroscientists we have been interested in understanding the neurobiological mechanisms underlying memory. What makes our brain to be able to "memorize" new information? What are memories made of? To answer these questions, we will outline the cellular, electrophysiological, and molecular processes underlying hippocampus-dependent declarative memory with a focus on the role of the transcription factor, cAMP responsive element-binding protein (CREB), and histone acetylation in this process.

MEMORY PROCESSES

Memory can be divided into three stages: encoding, storage, and retrieval. Encoding refers to the initial processing of the information that is captured and converted within a certain construct to be stored. Storage is the process that allows maintaining a copy or a permanent recording of the coded information. And retrieval refers to the subsequent use of information filed in memory to perform a certain task.

In 1949, Hebb understood the complexity of this process and suggested that short-term and long-term memory were underpinned by different mechanisms (Hebb, 1949). For example, short-term memory for verbal material is likely to be based on a specific phonological encoding related to the sounds of words, whereas long-term verbal memory relies upon semantic codes; that is, the meaning of words rather than their sound. One of the first models proposed to explain the relationship between short- and long-term was the so-called "modal model" by Atkinson & Shiffrin (1968), according to which information enters three separate compartments: (i) a series of sensory registers; (ii) a short-term store; (iii) a long-term store (Figure 1.1). To ensure long-term memory it was sufficient to store for a longer time the information in the short-term system, a theory that raised several criticisms (Craik & Watkins, 1973; Shallice & Warrington, 1970). Baddeley & Hitch (1974) abandoned this

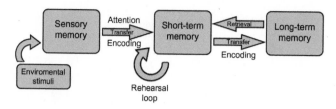

FIGURE 1.1   Atkinson–Shiffrin memory model.

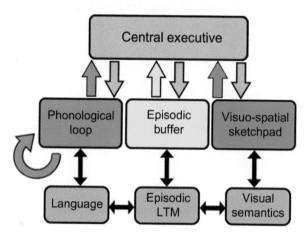

FIGURE 1.2   Baddeley's model of working memory.

concept of short-term memory as a unitary system and introduced the term "working memory" to indicate a memory able to manipulate information rather than passively maintain it. Working memory was organized as a multicomponent model consisting of several subsystems: two "slave systems" responsible for maintaining information, and a central executive responsible for supervision, integration, and coordination between slaves systems. Slave systems consisted of the "phonological loop," which stores phonological information (e.g., sounds of language) and prevents the decay by recurrence (e.g., repetition of a telephone number); and the "visuo-spatial sketchpad," which processes and preserves visuo-spatial information (Figure 1.2). The latter can be used, for example, to construct and manipulate visual images and for the representation of mental maps. Later, a fourth component, named "episodic buffer," was added to this model to maintain representations that integrate phonological and visuo-spatial information within time sequencing (Figure 1.2).

According to Cowan (Cowan, 1995, 2005), working memory is organized into two integrated levels: one activated by long-term memory

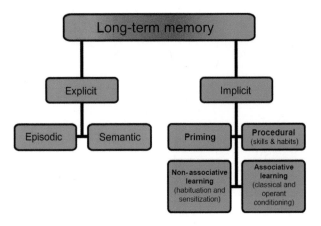

FIGURE 1.3   Different types of long-term memory.

representations, and one that allows focusing attention on specific memory blocks (chunks) that we want to retain. This was based on the classic study, "The magic number 7 plus or minus two" (Miller, 1956), in which the author argued that we are capable of processing and transmitting about seven units of information at a time. However, the primary information can be grouped together using other knowledge or information already learned.

With regard to long-term memory, it is commonly classified into two different types: declarative or explicit memory and nondeclarative or implicit memory (Figure 1.3). Declarative memory is related to autobiographical knowledge of past events, depending upon cognitive processes such as evaluation, comparison, and inference, and directly accessible to consciousness since it can be described in words. Declarative memory can be categorized as either episodic or semantic memory. Episodic memory evokes particular experiences or episodes and is influenced by attention and organization, whereas semantic memory is conditioned by education, perceptual knowledge of the physical world around us, and includes language and knowledge of specialized skills. Semantic memory is based on the summary of information built up from many episodes, and reflects our ability to comprehensively evaluate these episodes, highlighting the characteristics that are common to a group of different episodes, whereas episodic memory is the ability to extract and recover a single event (Tulving, 1985). Nondeclarative memory, also identified as procedural memory, is not directly accessible to consciousness and cannot be described in words in terms of facts, specific data, and events localized in time and space. It is a memory formed in an automatic way by repeated motor or perceptual practices. This memory includes several forms of learning such as priming, procedural learning, associative conditioning, and evaluative conditioning.

BRAIN STRUCTURES UNDERLYING MEMORY PROCESSES

Medial temporal lobes and medial diencephalic structures are thought to play a crucial role in declarative memory. They serve as a temporary deposit of information, the final storage being at neocortex level. Thus, declarative memory encoding-storage-retrieval processes are ensured by the interaction between medial temporal lobes and neocortex (i.e., prefrontal cortex). The neural circuit responsible for long-term storage of procedural memory is less clear, but possible candidates are represented by basal ganglia, cortico-striatal connections, cerebellar cortex, and the cerebellar nuclei.

Multiple areas of the brain have been shown to play a role in different forms of learning and memory. Among these, the hippocampus, a structure within the temporal lobe, has been recognized as fundamental in the formation of declarative memory, in both semantic and episodic aspects. In 1957, Scolville and Milner observed that patient H.M., after the bilateral removal the hippocampus for severe epilepsy, manifested anterograde amnesia (Scolville & Milner, 1957). Since then, several studies have been performed on other patients with hippocampal damage or in lesioned animal models, confirming the role of hippocampus and temporal lobes in memory formation (Squire, 1992). The hippocampal capability to fix memories related to a particular context correlates with its role in helping to create a map of space, a sort of internal GPS that allows us to remember the environment and forms the basis of spatial memory and navigation (Smith & Mizumori, 2006).

From an anatomical point of view, the hippocampus is folded and curved in a complex three-dimensional shape organized in three layers (archi-cortex). The entorhinal cortex possesses most of the afferent and efferent neurons. It transmits information to all hippocampal areas, and, in turn, receives signals back from them in a loop-circuit—the so-called trisynaptic organization of the hippocampus. This circuit begins with granule cells in the dentate gyrus that receive information from the entorhinal cortex via axons of the perforant pathway. These cells then make axonal projections via the mossy fiber pathway and synapse on pyramidal cells in the CA3 region. CA3 pyramidal cells then project via the Schaffer collateral pathway to synapse on pyramidal cells in the CA1 region, and these cells then project to the subiculum (Figure 1.4). All these fields are also interconnected with one another and also to the entorhinal cortex, which makes projections to both the prefrontal cortex and hypothalamus.

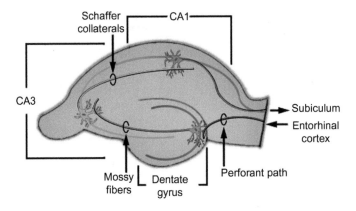

FIGURE 1.4   Schematic representation of the hippocampal trisynaptic circuit.

CELLULAR BASIS OF MEMORY

One of the ways in which information is stored in the brain is in the form of activity-dependent changes in the efficacy of synaptic transmission (Takeuchi, Duszkiewicz, & Morris, 2014). Learning elicits activity in a neural network that results in persistent changes in the strength of the synaptic connections among the neurons in that network. Information is then stored in the network in the form of these synaptic alterations, and this pattern of alterations is thought to constitute the memory trace. These activity-dependent changes that underlie memory share some similarities with, but are distinct from, activity dependent changes that help refine the pattern of synaptic connections during development. They also work in concert with mechanisms that mediate homeostatic plasticity—a process that also alters the efficacy of synaptic transmission within a neuron to maintain its firing within a normal range (Schacher & Hu, 2014; Siddoway, Hou, & Xia, 2014).

Beginning with the pioneering experiments of Bliss and Lomo in rabbit hippocampus (Bliss & Lomo, 1973), a number of experimental manipulations have been developed that elicit activity-dependent increases or decreases in the efficacy of synaptic transmission at many types of excitatory and inhibitory synapses in a variety of organisms. In general, increases in synaptic efficacy are referred to as long-term potentiation (LTP) and are produced by high frequency stimulation; decreases in synaptic efficacy are referred to as long-term depression (LTD) and are produced by low-frequency stimulation. Activity-dependent changes in synaptic efficacy can also be produced by more complex, and arguably more physiological protocols that manipulate the temporal relationship between presynaptic action potentials and postsynaptic depolarization—so-called spike-timing dependent plasticity. While experimentally induced changes in synaptic efficacy are clearly different in important ways from changes in synaptic efficacy that occur *in vivo* during learning, there is accumulating evidence suggesting that a mechanism similar to experimentally observed LTP plays an important role in memory (reviewed in Takeuchi et al., 2014).

Multiple neurotransmitter receptors have been implicated in mediating or modulating synaptic plasticity (e.g., metabotropic glutamate receptors, dopamine receptors, kainite receptors, acetylcholine receptors, cannabinoid receptors) and their involvement depends on multiple factors including synapse type, developmental stage, stimulation protocol, and recording conditions. By far the most studied of these is the NMDA receptor (NMDAR) and its role in mediating NMDAR-dependent forms of synaptic plasticity in the hippocampus. NMDARs are calcium permeable ion channels that are activated by the excitatory neurotransmitter, glutamate. NMDARs have a number of distinctive characteristics, including their sensitivity to contemporaneous presynaptic glutamate release and postsynaptic depolarization. This sensitivity derives from the fact that at negative membrane potentials, the NMDARs are blocked by the presence of Mg^{++} in the pore, and this block is relieved by postsynaptic depolarization that allows calcium influx in the presence of extracellular glutamate. These receptors therefore function only when both criteria are met, conferring upon them the ability to act as coincidence detectors for pre- and postsynaptic activity. Upon NMDAR activation, the influx of calcium is thought to activate a number of intracellular signaling pathways that ultimately lead to changes in synaptic efficacy.

In principle, changes in synaptic efficacy can result from either an increase in the amount of neurotransmitter present at a synapse following an action potential, or an increase in the responsiveness of the postsynaptic cell to a fixed concentration of neurotransmitter. The degree to which these two mechanisms contribute to particular forms of synaptic plasticity has been a topic of intense debate (Kullmann, 2012). While evidence exists to support both mechanisms, even within the same synapse at the same developmental stage (Bliss & Collingridge, 2013), a great deal of work has focused on the role of AMPA receptor trafficking in this process (Anggono & Huganir, 2012). AMPA receptors are ligand gated ion channels that respond to glutamate, and their presence at the cell surface has been found to be regulated both *in vitro* by protocols that induce changes in synaptic efficacy, and *in vivo* by learning. Increases in the number of AMPA receptors at a synapse increase the postsynaptic response of a neuron to presynaptic glutamate release and increase synaptic efficacy while decreases in synaptic AMPA receptor numbers reduce synaptic efficacy.

The duration of changes in efficacy can vary according to the details of the experiment, but are frequently classified as either short-term or long-term. Classically, short-term changes are thought to result primarily from posttranslational modifications to existing proteins such as neurotransmitter receptor or ion channel phosphorylation, while long-term changes are thought to require transcription and synthesis of new proteins and structural changes. While generally valid, this distinction has been complicated by findings that show the involvement of local protein synthesis, and significant synaptic structural reorganization at early time points.

To identify newly synthesized proteins responsible for the long-term changes in synaptic efficacy, several screens for genes regulated by neuronal activity have been carried out (reviewed in Leslie & Nedivi, 2011). These screens identified hundreds of genes, and while progress has been made in understanding the mechanisms by which many of these contribute to synaptic plasticity, a complete understanding of the process is still emerging. Some of the best characterized activity regulated genes include transcription factors such as *c-fos* and *Zif268*, genes that regulate AMPA receptor trafficking such as *arg/arg3.1*, genes that regulate the interaction of proteins within the postsynaptic density such as *homer1a*, genes that regulate the extracellular matrix such as tPA, and neurotrophic factors such as *bdnf*.

While there is clear evidence to support a role for cell-wide gene expression changes in synaptic efficacy, these changes are known to be synapse specific—not all the synapses that a given neuron makes are potentiated by learning or LTP induction. The basis for this synapse specificity is thought to be due to a process known as synaptic tagging, whereby the products of transcription within the nucleus are recruited selectively to particular synapses that are molecularly marked by synaptic activity (Redondo & Morris, 2011). While there is substantial electrophysiological evidence to support the existence of synaptic tagging, the molecular mechanism or mechanisms underlying this process are still poorly understood. Candidate mechanisms include processes whereby structural changes within dendritic spines increase their ability to recruit receptors to the cell surface as well as mechanisms whereby changes in RNA binding proteins and proteins that regulate translation alter the local translation of mRNAs trafficked from the nucleus.

THE ROLE OF CREB IN MEMORY

One of the key proteins responsible for regulating gene expression in response to neuronal activity is the nuclear transcription factor cAMP-responsive element binding protein, CREB. CREB is broadly expressed in a variety of tissues and developmental stages where it regulates the expression of genes involved in development, survival, cell proliferation and differentiation, circadian clock, cardiovascular functions, synaptic plasticity, and memory (Lonze & Ginty, 2002). CREB was originally purified in 1987 as a 43 kD protein capable of binding selectively to the cAMP responsive element (CRE) in the somatostatin gene (Montminy & Bilezikjian, 1987). CREB activity is regulated, in part, through phosphorylation at serine residue 133 (Ser133). Binding of phosphorylated CREB to a CRE sequence (TGACGTCA) of a target gene recruits CREB binding protein (CBP), a transcriptional coactivator with intrinsic histone acetyltransferase (HAT) activity that activates gene transcription (Mayr & Montminy, 2001). In addition, mechanisms apart from CREB phosphorylation at Ser133 may be involved in CREB activation. For instance, the cofactors CREB-regulated transcriptional co-activator (CRTC) and the transducer of regulated CREB activity have also been demonstrated to be involved in hippocampal synaptic plasticity and memory (Ch'ng et al., 2012; Kovács et al., 2007; Nonaka et al., 2014; Sekeres et al., 2012).

Some of the first evidence implicating CREB in long-term plasticity was obtained in the laboratory of Eric Kandel through work on the marine slug, Aplysia (Dash, Hochner, & Kandel, 1990). These authors showed that microinjection of the CRE sequence into the nucleus, which prevented CREB to exert its function, blocked the long-term increase in synaptic strength without affecting short-term facilitation. Later, CREB was implicated in long-term memory in mice and flies, where it was found that expression of a dominant-negative CREB transgene blocked long-term memory in Drosophila (Yin et al., 1994; Dubnau & Tully, 1998), and disruption of CREB genes in mice impaired long-term memory as well as late-LTP (Bourtchuladze et al., 1994). Additional support for the involvement of CREB in LTP and memory has been obtained in the intervening years by multiple genetic and pharmacological manipulations that targeted CREB and its upstream and downstream pathways (Barco & Marie, 2011; Kandel, 2012).

PATHWAYS ACTIVATING CREB

Hundreds of stimuli (hormones, peptides, neurotransmitters, cytokines, microorganisms, stress factors, etc.) lead to CREB activation through different transduction signals (Johannese et al., 2004). The first pathway implicated in CREB activation is the cAMP/PKA pathway from which CREB gets its name. cAMP, discovered in 1958 by Earl Sutherland, is a second messenger able to transfer at intracellular level the information from proteins (first messenger) that cannot cross the plasma membrane (Sutherland, 1992). The role of cAMP in regulating synaptic transmission was first underlined in Aplysia sensory neurons where serotonin-induced presynaptic facilitation was mediated by cAMP (Brunelli, Castellucci, &

Kandel, 1976). Serotonin is thought to act on a G-protein coupled receptor. Activation of the G-protein is then thought to activate adenylate cyclase, leading to an increase in the intracellular concentration of cAMP. cAMP then activates PKA. Specifically, the binding of cAMP with the regulatory subunits of PKA leads to the activation of its catalytic subunits, which, in turn, catalyzes the phosphorylation of several substrates including CREB, histones, and several enzymes involved in metabolic processes (Corbin and Krebs, 1969). Multiple studies have found that cAMP/PKA pathway stimulates CREB phosphorylation (for a review see Montminy, 1997), as well as other CREB family members named ATF-1 and cAMP response element modulator (Foulkes, Borrelli, & Sassone-Corsi, 1991). This PKA capability to phosphorylate transcription factors and to activate genes has been linked to long-term synaptic plasticity and, consequently, long-term memory (Kandel, 2012).

Actually, the cAMP/PKA pathway plays a role in both short- and long-term memory, involving different substrates, whereas transcription or translation of genes seems to be required only for long-term storage (Kandel, 2001). Although it is now widely accepted that inhibition of CREB disrupts long-term memory whereas its stimulation improves it, recent studies have also proposed a role for CREB in short-term memory by the modulation of genes involved in BDNF expression (Suzuki et al., 2011).

CREB phosphorylation at Ser133 is also triggered following calcium entry into the cell (Sheng, Thompson, & Greenberg, 1991), as demonstrated by the activation of dihydropyridine-sensitive calcium channels (Sheng et al., 1991; Sun, Lou, & Maurer, 1996; Enslen et al., 1994; Matthews et al., 1994). Interestingly, calcium might induce CREB activation through the cAMP/PKA pathway (Ginty, Glowacka, Bader, Hidaka, & Wagner, 1991; Impey, Wayman, Wu, & Storm, 1994) or via Ca^{2+}-calmodulin-dependent protein kinases (CaMK). The family of CaMK includes CaMKI, CaMKII, and CaMKIV, widely expressed in the CNS. While CaMKII and CaMKIV are localized in both cytoplasm and nucleus, CaMKI is exclusively cytoplasmic (Picciotto, Zoli, Bertuzzi, & Nairn, 1995) and it is still discussed whether it is able to directly or indirectly phosphorylate CREB. CaMKII phosphorylation at Ser133 seems to not induce a CRE-mediated transcription (Matthews et al., 1994; Sun, Enslen, Myung, & Maurer, 1994); moreover, CaMKII can phosphorylate CREB at Ser142 (Sun et al., 1994) to negatively modulate CREB-dependent transcription by inducing a decreased binding affinity of CREB for CBP. CaMKII and CaMKIV are relevant in glutamatergic transmission (Hu, Chrivia, & Ghosh, 1999) and have been involved in LTP and memory mechanisms. CaMKII is highly concentrated in the postsynaptic density, where it responds to the calcium entry mediated by NMDARs during LTP (Lisman, Yasuda, & Raghavachari, 2012). Indeed, the use of pharmacological inhibitors or mutant mice for CaMKII induced a block of hippocampal LTP (Ito, Hidaka, & Sugiyama, 1991;

Malenka et al., 1989; Malinow, Schulman, & Tsien, 1989; Silva, Stevens, Tonegawa, & Wang, 1992) and spatial memory (Silva, Paylor, Wehner, & Tonegawa, 1992), confirming that CaMKII has a role in these processes. CaMKII has also been demonstrated to have a presynaptic function during synaptic plasticity in hippocampal cultured neurons, since its presynaptic inhibition blocked synaptic plasticity induced by tetanus, glutamate, or NO/cGMP pathway activation (Ninan & Arancio, 2004).

Changes of CaMK IV activity in the nuclei of CA1 pyramidal neurons were also observed during LTP induced by a high frequency stimulation (Miyamoto, 2006); this is consistent with studies indicating that CaMKIV is one of the main kinase inducing CREB activation after a stimulation in cultured rat hippocampal neurons (Kasahara, Fukunaga, & Miyamoto, 1999; Kasahara, Fukunaga, & Miyamoto, 2000). Also, inhibition of CaMKIV or genetic deletion (i.e., CaMKIV-/-mice) induced synaptic plasticity and memory deficits (Ho et al., 2000; Kang et al., 2001; Wei et al., 2002), whereas its up-regulation led to an improvement of memory in contextual fear conditioning (FC), passive avoidance, and social recognition (Fukushima et al., 2008).

It is interesting to notice that CaMKII activity is related to nitric oxide (NO) during learning and memory. Indeed, CaMKII is involved in both pre- and postsynaptic function mediated by the NO/cGMP/protein kinase G (PKG) pathway. Neurotransmitter release induced by stimulating the NO/cGMP/PKG signaling is inhibited by injections of a CaMKII inhibitor at a presynaptic level (Feil and Kleppisch, 2008), and recent studies suggest that NO activates CaMKII during memory (Harooni, Naghdi, Sepehri, & Rohani, 2009). On the other hand, physiological production of NO from L-arginine is catalyzed by Ca^{2+}/calmodulin-sensitive NO synthases (NOS), which are highly expressed at the hippocampus (Hardingham, Dachtler, & Fox, 2013; Vincent & Kimura, 1992). In particular, NO production by neural NOS has been involved in pre- and postsynaptic mechanisms underlying hippocampal LTP (Arancio, Kiebler, et al., 1996; Böhme, Bon, Stutzmann, Doble, & Blanchard, 1991, 1993) and memory (Arancio, Lev-Ram, Tsien, Kandel, & Hawkins, 1996; Bernabeu, de Stein, Fin, Izquierdo, & Medina, 1995; Bohme et al., 1993), such as inhibitory avoidance (Fin et al., 1995; Hölscher & Rose, 1993; Khavandgar, Homayoun, & Zarrindast, 2003; Qiang, Chen, Wang, Wu, & Qiao, 1997; Telegdy and Kokavszky, 1997; Zarrindast, Shendy, & Ahmadi, 2007), object recognition (Blokland, Prickaerts, Honig, & de Vente, 1998), and spatial memory (Choopani, Moosavi, & Naghdi, 2008; Ingram, Spangler, Kametani, Meyer, & London, 1998; Kendrick et al., 1997; Majlessi, Kadkhodaee, Parviz, & Naghdi, 2003; Yamada et al., 1996) in different species.

NO mainly exerts its physiological functions activating its soluble receptor guanylate cyclase leading to the production of the second messenger cGMP, which, in turn, is able to activate PKG (Boulton, Southam, & Garthwaite, 1995; Doyle, Hölscher, Rowan, & Anwyl, 1996; Edwards,

Rickard, & Ng, 2002), whose activity maintains CREB phosphorylation at Ser133 (Wong, Bathina, & Fiscus, 2012). NO also mediates CREB-DNA binding via a Ser133-independent mechanism by the S-nitrosylation of nuclear proteins associated with CREB target genes (Riccio et al., 2006). The NO/cGMP/PKG has been widely demonstrated to intervene in memory processes and various studies have suggested that it cooperates with the cAMP/PKA cascade to ensure early- and late-phase LTP converging on CREB (Bernabeu, Schmitz, Faillace, Izquierdo, & Medina, 1996, 1997; Izquierdo et al., 2000; Lu, Kandel, & Hawkins, 1999). Thus, cAMP and cGMP cascades underpin memory acquisition and consolidation (Bach et al., 1999; Bernabeu et al., 1996; Bourtchouladze et al., 1998; Frey, Huang, & Kandel, 1993; Lu et al., 1999; Prickaerts et al., 2002; Son et al., 1998). In particular, cGMP seems to be involved in early consolidation, whereas cAMP is involved in late consolidation (Izquierdo et al., 2006; Rutten et al., 2007). A recent study further supports this differential time-dependent role of cAMP and cGMP synaptic plasticity and memory consolidation by showing that for late-phase LTP and long-term memory the cAMP-PKA pathway is required after cGMP signaling (Bollen et al., 2014).

Another pathway leading to CREB activation is represented by mitogen-activated protein kinases (MAPKs), a group of kinases divided into four main groups: MAPKs ERK1/2, p38 MAPKs, JNKs, and atypical MAPKs. ERK1/2 and p38 MAPKs are known to phosphorylate CREB at Ser133 not directly but through downstream kinases such as RSKs, MK2, and MSKs (Anjum & Blenis, 2008; Dümmler et al., 2005; Roux and Blenis, 2004; Roux, Richards, & Blenis, 2003; Xing, Ginty, & Greenberg, 1996). In particular, ERK1/2 is activated by an initial increase of Ras-GTP levels followed by activation of protein kinase Raf and, in turn, the phosphorylation of MAPK/ERK kinase (MEK). This latter phosphorylates and activates ERK1/2, a serine/threonine kinase whose targets are downstream kinases such as RSK and MSK. ERK signaling has been found to be required for several forms of hippocampal LTP, in CA1 and DG areas (Atkins, Selcher, Petraitis, Trzaskos, & Sweatt, 1998; Bolshakov, Carboni, Cobb, Siegelbaum, & Belardetti, 2000; Coogan, O'Leary, & O'Connor, 1999; English & Sweatt, 1996, 1997; Impey et al., 1998; Kanterewicz et al., 2000; Ohno, Frankland, Chen, Costa, & Silva, 2001; Patterson et al., 2001; Selcher et al., 2003; Zhu, Qin, Zhao, Van Aelst, & Malinow, 2002), but also for LTP in other brain areas such as the amygdala (Huang, Martin, & Kandel, 2000; Schafe et al., 2000) and cortex (Di Cristo et al., 2001), as well as cerebellar LTP in Purkinje cells (Kawasaki et al., 1999). The use of MEK inhibitors has also shown that ERK is required for both spatial learning and FC (Atkins et al., 1998; Blum, Moore, Adams, & Dash, 1999; Schafe et al., 2000; Selcher, Atkins, Trzaskos, Paylor, & Sweatt, 1999; Hebert and Dash, 2002).

p38 is a MAPK family member that is also highly expressed in the brain and has been implicated in synaptic plasticity and memory (Lee, Park, Che,

Han, & Lee, 2000). p38 MAPK is required for the induction of some forms of synaptic plasticity (Collingridge, Peineau, Howland, & Wang, 2010; Thomas and Huganir, 2004), including some forms of mGluR-dependent and NMDA-dependent LTD (Huang, You, Wu, & Hsu, 2004; Moult, Corrêa, Collingridge, Fitzjohn, & Bashir, 2008), as well as short- and long-term memory for inhibitory avoidance (Alonso, Bevilaqua, Izquierdo, Medina, & Cammarota, 2003), spatial memory (Ashabi et al., 2012; Yang et al., 2013), and eye-blink conditioning (Zhen, Du, Romano, Friedman, & Harvey, 2001).

In summary, multiple signaling pathways involved in synaptic plasticity activate CREB through phosphorylation (Figure 1.5), and once activated, CREB activates transcription of genes involved in synaptic plasticity and memory (Bourtchouladze, 2002), for example, by remodeling chromatin and influencing histone acetylation/deacetylation, as discussed in the next paragraph.

HDACs/HATs ROLE IN GENE TRANSCRIPTION

In the last decade several studies have supported the role of epigenetics in neuronal memory (Levenson et al., 2004; Levenson and Sweatt, 2005; Lipsky, 2013). Epigenetics is defined as the modification in gene transcription without changing DNA sequence (Gräff, Kim, Dobbin, & Tsai, 2011; Russo, Martienssen, & Riggs, 1996). Such a phenomenon is mediated by two major chemical modifications: (i) methylation of DNA at its cytosines positioned adjacent to guanine nucleobases (CpG) by methyltransferase enzymes (DNMTs) (Bird, 2002) and (ii) covalent posttranslation modifications (PTMs) of histone proteins, such as methylation, acetylation, phosphorylation, SUMOylation, and ubiquitination (Peterson & Laniel, 2004). Based on data obtained from different animal models, both DNA methylation and PTMs are found to be implicated in regulation of memory functions (Castro-Gomez et al., 2013; Chwang, O'Riordan, Levenson, & Sweatt, 2006; Gupta et al., 2010; Gupta-Agarwal et al., 2012; Miller & Sweatt, 2007). Here, we will discuss the critical role of histone acetylation during learning and memory formation.

Histone PTMs influence the degree of the DNA packing inside the nucleus. Indeed, DNA is coiled around a repeating unit of histone proteins, called nucleosome, to form the chromatin structure. The nucleosome consists of two pairs of histone H2A-H2B dimers and an H3-H4 histone tetramer (Felsenfeld & Groudine, 2003; Kornberg, 1974; Quina, Buschbeck, & Di Croce, 2006), and it is stabilized by a linker histone H1 (Happel & Doenecke, 2009). The acetylation of lysine (Lys, K) residues located in the histone tails is one of the general mechanisms for regulating chromatin structure and subsequently gene transcription (Felsenfeld & Groudine, 2003).

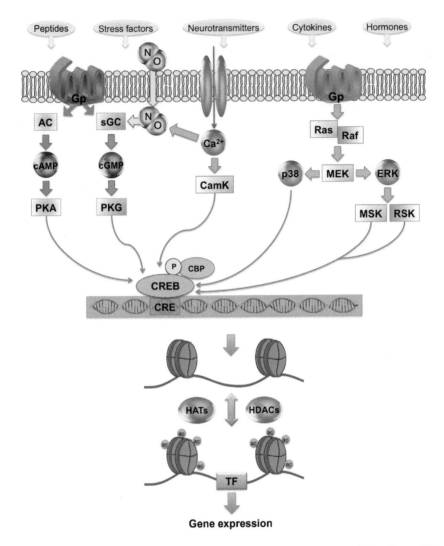

FIGURE 1.5 Pathways activating CREB phosphorylation. AC, adenylyl cyclase; sGC, soluble guanylyl cyclase; NO, nitric oxide; PKA, protein kinase A; PKG, protein kinase G; CamK, calcium-calmodulin kinase; CREB, cAMP-responsive element binding protein; CRE, cAMP responsive element; CBP, CREB binding protein; P, phosphate; ERK, extracellular signal-regulated kinase; MEK, MAPK/ERK kinase; HDAC, histone deacetylase; HAT, histone acetyltransferase; ac, acetylation; TF, transcription factor.

Histone acetylation is a dynamic process that has been associated with memory formation (Chwang, Arthur, Schumacher, & Sweatt, 2007; Chwang et al., 2006; Guan et al., 2002; Levenson et al., 2004; Vecsey et al., 2007). This posttranslational mechanism is catalyzed by two different families of enzymes: HATs, which add acetyl groups to histones, and histone deacetylase (HDAC), which perform the removal of the acetyl groups (Legube & Trouche, 2003; Marmorstein and Roth, 2001; Thiagalingam et al., 2003) (Figure 1.5). HATs are divided into five major groups: GNAT (including Gcn5 and PCAF), MYST, CBP/p300, transcription factor-related HATs, and nuclear receptor-associated HATs (Schneider et al., 2013), whereas the HDAC family comprises 18 members, grouped into four classes: HDAC I (HDAC1-3 and HDAC8); HDAC II, which is divided in IIA (HDAC4, HDAC5, HDAC7 and HDAC9) and IIB (HDAC6 and HDAC10); HDAC III (Sirt1-7); and HDAC IV (HDAC11) (Bolden, Peart, & Johnstone, 2006).

The activity of these two families controls the status of chromatin. Specifically, transferring acetyl groups on histone lysine residues leads to the opening of chromatin (so-called euchromatin) and thus enhances the DNA accessibility to transcription factors. Instead, removing acetyl groups facilitates the interaction between histones and DNA, promoting a closed chromatin structure (so-called heterochromatin) and repressing gene transcription (Felsenfeld & Groudine, 2003; Muhlbacher, Schiessel, & Holm, 2006; Quina et al., 2006). The balance between HATs and HDACs activity is crucial for regulating a dynamic change in gene transcription, which establishes appropriate neurophysiological conditions underlying neuronal response outputs, such as synaptic plasticity and memory (Saha & Pahan, 2006).

Growing evidence suggests that dysregulation of neuronal acetylation levels leads to an impairment of memory formation and that an increase of histone acetylation levels occurs in rodents after different hippocampus-dependent tasks (e.g., contextual FC, Morris water maze (MWM), and environmental enrichment (EE)) (Chwang et al., 2007; Chwang et al., 2006; Fischer, Sananbenesi, Wang, Dobbin, & Tsai, 2007; Fontán-Lozano et al., 2008; Koshibu et al., 2009; Levenson et al., 2004; Peleg et al., 2010), and this is thought to result from disregulation of genes that mediate synaptic plasticity and memory. Levenson and colleagues demonstrated that H3 (but not H4) acetylation levels increase in young adult rat hippocampus 1h after FC (Levenson et al., 2004). Interestingly, the increase of acetylated H3 fails to occur following the administration of a DNMT inhibitor, demonstrating that epigenetic mechanisms are not isolated but, rather, work in concert (Miller & Sweatt, 2007; Miller, Campbell, & Sweatt, 2008). Furthermore, in a study conducted on the APP/PS1 mouse model of Alzheimer's disease, H4 acetylation levels in hippocampus 1h after FC training are decreased with respect to wild-type (WT) mice (Francis et al., 2009). The increase of H4 acetylation levels was also observed in a latent inhibition paradigm (Levenson et al., 2004).

Numerous studies have explored the correlation between memory formation and acetylation of specific Lys residues. A recent work by Peleg et al. (2010) shows increased acetylation of K9 and K14 of H3 and K5, K8, and K12 of H4 in 3-month-old mice 1h after FC, whereas 16-month-old mice failed to up-regulate H4K12 acetylation. This finding revealed that memory impairment is associated with a deficit in learning-induced H4K12 acetylation. WT mice exposed to EE showed an increase in H3 (K9 and K14) and H4 (K5, K8, and K12) acetylation in the hippocampus upon enhancement of associative and spatial learning (Fischer et al., 2007). Histone acetylation levels were also measured after MWM training in young rats. An increase in H2B (K5, K12, K15, and K20) and H4K12 acetylation in the rat dorsal hippocampus during spatial memory consolidation (Bousiges et al., 2010) was found. Thus, although it has been reported that neuronal activation induces epigenetic changes in chromatin, the modified histone sites are different during various memory tasks, suggesting the specificity of epigenetic regulation during memory processes.

Interestingly, HATs (CBP, p300, and PCAF) activity and protein expression in dorsal hippocampus are found to be up-regulated during the spatial memory formation and therefore the increase of histone acetylation is likely to be a direct consequence of this up-regulating event (Bousiges et al., 2010). In particular, CBP is identified as the major contributor to memory formation and, indeed, in a cognitively impaired rat model neither mRNA nor protein levels of CBP are up-regulated upon spatial learning. This down-regulation of CBP-histone acetylation dependent pathway is correlated to a lack of acetylated-H2B levels at promoters for memory/plasticity-related genes *BDNF* and *zif268* with consequent reduction of their expression (Bousiges et al., 2010). Additionally, modulation of CBP levels in hippocampus via viral delivery was shown to rescue learning and memory deficits in a mouse model of AD (Caccamo, Maldonado, Bokov, Majumder, & Oddo, 2010). Accordingly, several studies have supported the important role of HATs in memory formation by using HAT activity deficient mutant mouse models of CBP (Alarcón et al., 2004; Korzus, Rosenfeld, & Mayford, 2004) (reviewed by Barrett & Wood, 2008), p300 (Oliveira, Wood, McDonough, & Abel, 2007), and PCAF (Duclot, Meffre, Jacquet, Gongora, & Maurice, 2010; Maurice et al., 2008). In light of these discoveries, HATs have been considered drug targets for improving memory consolidation processes necessary for long-term memory. To this end, HAT activators are under investigation for treating neurodegenerative diseases (Chatterjee et al., 2013; Selvi, Cassel, Kundu, & Boutillier, 2010; Wei et al., 2012).

On the other side, targeting HDACs has represented a well-established approach for enhancing histone acetylation levels during memory formation. Pharmacological interventions through HDAC inhibitors have been successfully used for improving synaptic plasticity and learning, demonstrating the molecular mechanisms underlying memory functions (Alarcón et al., 2004; Bredy & Barad, 2008; Bredy et al., 2007; Fischer et al., 2007; Korzus et al., 2004; Levenson et al., 2004; Ricobaraza et al., 2009; Vecsey et al., 2007). Guan et al. (2009) examined the specific HDACs involved in memory formation. By using a transgenic mouse in which HDAC2 is overexpressed, they found that HDAC2 impairs spatial working memory and synaptic plasticity. To further investigate the role of HDAC2 in associative learning, they generated a second transgenic mouse lacking HDAC2, which showed an enhancement of memory functions. These results together confirm that HDAC2 has a deleterious effect in regulating memory formation (Guan et al., 2009). Moreover, the expression of HDAC2 was measured in aged male mice, which showed declined recognition memory consolidation, and it was found that HDAC2 mRNA and protein levels were increased (Singh & Thakur, 2014). Recent studies have found that other HDACs are associated with memory formation (Gao et al., 2010; Kim et al., 2012; McQuown et al., 2011; Michan et al., 2010). Similar to HDAC2, both the deletion and the inhibition of HDAC3 significantly enhance long-term memory and elevate expression of *Nr4a2* gene (Hawk & Abel, 2011), demonstrating that HDAC3 plays a critical role in gene expression programs regulating memory formation (McQuown et al., 2011). Conversely, the knockout of HDAC4 and SIRT1 (but not HDAC5) leads to memory deficits (Gao et al., 2010; Kim et al., 2012; Michan et al., 2010). Therefore, it can be concluded that HDAC regulation on memory is type-specific, suggesting unique roles in brain for individual HDACs.

Chromatin structure remodeling has been established as a required mechanism for memory. The epigenetic modulation of memory-related gene expression is extremely dynamic; it occurs during the ongoing phase of memory formation. This is in agreement with the evidence that transient increase of gene expression leads to newly acquired memory dependent on hippocampus. Taken together, these findings have explained new molecular processes underlying memory and proposed innovative areas of therapeutic intervention for neurodegenerative diseases.

References

Alarcón, J. M., Malleret, G., Touzani, K., Vronskaya, S., Ishii, S., Kandel, E. R., et al. (2004). Chromatin acetylation, memory, and LTP are impaired in CBP+/− mice: A model for the cognitive deficit in Rubinstein–Taybi syndrome and its amelioration. *Neuron, 42*, 947–959.

Alonso, M., Bevilaqua, L. R., Izquierdo, I., Medina, J. H., & Cammarota, M. (2003). Memory formation requires p38MAPK activity in the rat hippocampus. *Neuroreport, 14*, 1989–1992.

Anggono, V., & Huganir, R. L. (2012). Regulation of AMPA receptor trafficking and synaptic plasticity. *Current Opinion in Neurobiology, 22*, 461–469.

Anjum, R., & Blenis, J. (2008). The RSK family of kinases: Emerging roles in cellular signalling. *Nature Reviews Molecular Cell Biology, 9*, 747–758.

Arancio, O., Kiebler, M., Lee, C. J., Lev-Ram, V., Tsien, R. Y., Kandel, E. R., et al. (1996). Nitric oxide acts directly in the presynaptic neuron to produce long-term potentiation in cultured hippocampal neurons. *Cell, 87*, 1025–1035.

Arancio, O., Lev-Ram, V., Tsien, R. Y., Kandel, E. R., & Hawkins, R. D. (1996). Nitric oxide acts as a retrograde messenger during long-term potentiation in cultured hippocampal neurons. *Journal of Physiology, Paris, 90*, 321–322.

Ashabi, G., Ramin, M., Azizi, P., Taslimi, Z., Alamdary, S. Z., Haghparast, A., et al. (2012). ERK and p38 inhibitors attenuate memory deficits and increase CREB phosphorylation and PGC-1α levels in Aβ-injected rats. *Behavioural Brain Research, 232,* 165–173.

Atkins, C. M., Selcher, J. C., Petraitis, J. J., Trzaskos, J. M., & Sweatt, J. D. (1998). The MAPK cascade is required for mammalian associative learning. *Nature Neuroscience, 1,* 602–609.

Atkinson, R. C., & Shiffrin, R. M. (1968). Human memory: A proposed system and its control processes. In K. W. Spence (Ed.), *The psychology of learning and motivation: Advances in research and theory* (Vol. 2). New York: Academic Press.

Bach, M. E., Barad, M., Son, H., Zhuo, M., Lu, Y. F., Shih, R., et al. (1999). Age-related defects in spatial memory are correlated with defects in the late phase of hippocampal long-term potentiation *in vitro* and are attenuated by drugs that enhance the cAMP signaling pathway. *Proceedings of the National Academy of Sciences of the United States of America, 96,* 5280–5285.

Baddeley, A. D., & Hitch, G. J. (1974). Working memory. In G. A. Bower (Ed.), *The psychology of learning and motivation: Advances in research and theory* (Vol. 8, pp. 47–89). New York: Academic Press.

Barco, A., & Marie, H. (2011). Genetic approaches to investigate the role of CREB in neuronal plasticity and memory. *Molecular neurobiology, 44,* 330–349.

Barrett, R. M., & Wood, M. A. (2008). Beyond transcription factors: The role of chromatin modifying enzymes in regulating transcription required for memory. *Learning & Memory, 15,* 460–467.

Bernabeu, R., de Stein, M. L., Fin, C., Izquierdo, I., & Medina, J. H. (1995). Role of hippocampal NO in the acquisition and consolidation of inhibitory avoidance learning. *Neuroreport, 6,* 1498–1500.

Bernabeu, R., Schmitz, P., Faillace, M. P., Izquierdo, I., & Medina, J. H. (1996). Hippocampal cGMP and cAMP are differentially involved in memory processing of inhibitory avoidance learning. *Neuroreport, 7,* 585–588.

Bernabeu, R., Schroder, N., Quevedo, J., Cammarota, M., Izquierdo, I., & Medina, J. H. (1997). Further evidence for the involvement of a hippocampal cGMP/cGMP-dependent protein kinase cascade in memory consolidation. *Neuroreport, 8,* 2221–2224.

Bird, A. (2002). DNA methylation patterns and epigenetic memory. *Genes and Development, 16,* 6–21.

Bliss, T. V., & Collingridge, G. L. (2013). Expression of NMDA receptor-dependent LTP in the hippocampus: Bridging the divide. *Molecular Brain, 6,* 5.

Bliss, T. V., & Lomo, T. (1973). Long-lasting potentiation of synaptic transmission in the dentate area of the anaesthetized rabbit following stimulation of the perforant path. *The Journal of Physiology, 232,* 331–356.

Blokland, A., Prickaerts, J., Honig, W., & de Vente, J. (1998). State-dependent impairment in object recognition after hippocampal NOS inhibition. *Neuroreport, 9,* 4205–4208.

Blum, S., Moore, A. N., Adams, F., & Dash, P. K. (1999). A mitogen-activated protein kinase cascade in the CA1/CA2 subfield of the dorsal hippocampus is essential for long-term spatial memory. *The Journal of Neuroscience, 19,* 3535–3544.

Böhme, G. A., Bon, C., Lemaire, M., Reibaud, M., Piot, O., Stutzmann, J. M., et al. (1993). Altered synaptic plasticity and memory formation in nitric oxide synthase inhibitor-treated rats. *Proceedings of the National Academy of Sciences of the United States of America, 90,* 9191–9194.

Böhme, G. A., Bon, C., Stutzmann, J. M., Doble, A., & Blanchard, J. C. (1991). Possible involvement of nitric oxide in long-term potentiation. *European Journal of Pharmacology, 199,* 379–381.

Bolden, J. E., Peart, M. J., & Johnstone, R. W. (2006). Anticancer activities of histone deacetylase inhibitors. *Nature Reviews Drug Discovery, 5,* 769–784.

Bollen, E., Puzzo, D., Rutten, K., Privitera, L., De Vry, J., Vanmierlo, T., et al. (2014). Improved long-term memory via enhancing cGMP-PKG signaling requires cAMP-PKA signaling. *Neuropsychopharmacology, 39,* 2497–2505.

GENES, ENVIRONMENT AND ALZHEIMER'S DISEASE

Bolshakov, V. Y., Carboni, L., Cobb, M. H., Siegelbaum, S. A., & Belardetti, F. (2000). Dual MAP kinase pathways mediate opposing forms of long-term plasticity at CA3-CA1 synapses. *Nature Neuroscience, 3*, 1107–1112.

Boulton, C. L., Southam, E., & Garthwaite, J. (1995). Nitric oxide-dependent long-term potentiation is blocked by a specific inhibitor of soluble guanylyl cyclase. *Neuroscience, 69*, 699–703.

Bourtchouladze, R. (2002). *Memories are made of this: How memory works in humans and animals.* New York: Columbia University Press.

Bourtchouladze, R., Abel, T., Berman, N., Gordon, R., Lapidus, K., & Kandel, E. R. (1998). Different training procedures recruit either one or two critical periods for contextual memory consolidation, each of which requires protein synthesis and PKA. *Learning & Memory, 5*, 365–374.

Bourtchuladze, R., Frenguelli, B., Blendy, J., Cioffi, D., Schutz, G., & Silva, A. J. (1994). Deficient long-term memory in mice with a targeted mutation of the cAMP-responsive element-binding protein. *Cell, 79*, 59–68.

Bousiges, O., Vasconcelos, A. P., Neidl, R., Cosquer, B., Herbeaux, K., Panteleeva, I., et al. (2010). Spatial memory consolidation is associated with induction of several lysine-acetyltransferase (histone acetyltransferase) expression levels and H2B/H4 acetylation-dependent transcriptional events in the rat hippocampus. *Neuropsychopharmacology, 35*, 2521–2537.

Bredy, T. W., & Barad, M. (2008). The histone deacetylase inhibitor valproic acid enhances acquisition, extinction, and reconsolidation of conditioned fear. *Learning & Memory, 15*, 39–45.

Bredy, T. W., Wu, H., Crego, C., Zellhoefer, J., Sun, Y. E., & Barad, M. (2007). Histone modifications around individual BDNF gene promoters in prefrontal cortex are associated with extinction of conditioned fear. *Learning & Memory, 14*, 268–276.

Brunelli, M., Castellucci, V., & Kandel, E. R. (1976). Synaptic facilitation and behavioral sensitization in Aplysia: Possible role of serotonin and cyclic AMP. *Science, 194*, 1178–1181.

Caccamo, A., Maldonado, M. A., Bokov, A. F., Majumder, S., & Oddo, S. (2010). CBP gene transfer increases BDNF levels and ameliorates learning and memory deficits in a mouse model of Alzheimer's disease. *Proceedings of the National Academy of Sciences of the United States of America, 107*, 22687–22692.

Castro-Gomez, S., Barrera-Ocampo, A., Machado-Rodriguez, G., Castro-Alvarez, J. F., Glatzel, M., Giraldo, M., et al. (2013). Specific de-SUMOylation triggered by acquisition of spatial learning is related to epigenetic changes in the rat hippocampus. *Neuroreport, 24*, 976–981.

Chatterjee, S., Mizar, P., Cassel, R., Neidl, R., Selvi, B. R., Mohankrishna, D. V., et al. (2013). A novel activator of CBP/p300 acetyltransferases promotes neurogenesis and extends memory duration in adult mice. *The Journal of Neuroscience, 33*, 10698–10712.

Ch'ng, T. H., Uzgil, B., Lin, P., Avliyakulov, N. K., O'Dell, T. J., & Martin, K. C. (2012). Activity-dependent transport of the transcriptional coactivator CRTC1 from synapse to nucleus. *Cell, 150*, 207–221.

Choopani, S., Moosavi, M., & Naghdi, N. (2008). Involvement of nitric oxide in insulin induced memory improvement. *Peptides, 29*, 898–903.

Chwang, W. B., Arthur, J. S., Schumacher, A., & Sweatt, J. D. (2007). The nuclear kinase mitogen- and stress-activated protein kinase 1 regulates hippocampal chromatin remodeling in memory formation. *The Journal of Neuroscience, 27*(46), 12732–12742.

Chwang, W. B., O'Riordan, K. J., Levenson, J. M., & Sweatt, J. D. (2006). ERK/MAPK regulates hippocampal histone phosphorylation following contextual fear conditioning. *Learning & Memory, 13*, 322–328.

Collingridge, G. L., Peineau, S., Howland, J. G., & Wang, Y. T. (2010). Long-term depression in the CNS. *Nature Reviews Neuroscience, 11*, 459–473.

Coogan, A. N., O'Leary, D. M., & O'Connor, J. J. (1999). P42/44 MAP kinase inhibitor PD98059 attenuates multiple forms of synaptic plasticity in rat dentate gyrus *in vitro*. *Journal of Neurophysiology, 81*, 103–110.

Corbin, J. D., & Krebs, E. G. (1969). A cyclic AMP--stimulated protein kinase in adipose tissue. *Biochemical and Biophysical Research Communications, 36*, 328–336.

Cowan, N. (1995). *Attention and memory: An integrated framework*. New York: Oxford University Press.

Cowan, N. (2005). *Working memory capacity*. New York: Psychology Press.

Craik, F. I. M., & Watkins, M. J. (1973). The role of rehearsal in short term memory. *Journal of Verbal Learning and Verbal Behavior, 12*, 599–607.

Dash, P. K., Hochner, B., & Kandel, E. R. (1990). Injection of the cAMP-responsive element into the nucleus of Aplysia sensory neurons blocks long-term facilitation. *Nature, 345*, 718–721.

Di Cristo, G., Berardi, N., Cancedda, L., Pizzorusso, T., Putignano, E., Ratto, G. M., et al. (2001). Requirement of ERK activation for visual cortical plasticity. *Science, 292*, 2337–2340.

Doyle, C., Hölscher, C., Rowan, M. J., & Anwyl, R. (1996). The selective neuronal NO synthase inhibitor 7-nitro-indazole blocks both long-term potentiation and depotentiation of field EPSPs in rat hippocampal CA1 *in vivo*. *The Journal of Neuroscience, 16*, 418–424.

Dubnau, J., & Tully, T. (1998). Gene discovery in Drosophila: New insights for learning and memory. *Annual Review of Neuroscience, 21*, 407–444.

Duclot, F., Meffre, J., Jacquet, C., Gongora, C., & Maurice, T. (2010). Mice knock out for the histone acetyltransferase p300/CREB binding protein-associated factor develop a resistance to amyloid toxicity. *Neuroscience, 167*, 850–863.

Dümmler, B. A., Hauge, C., Silber, J., Yntema, H. G., Kruse, L. S., Kofoed, B., et al. (2005). Functional characterization of human RSK4, a new 90-kDa ribosomal S6 kinase, reveals constitutive activation in most cell types. *The Journal of Biological Chemistry, 280*, 13304–13314.

Edwards, T. M., Rickard, N. S., & Ng, K. T. (2002). Inhibition of guanylate cyclase and protein kinase G impairs retention for the passive avoidance task in the day-old chick. *Neurobiology of Learning and Memory, 77*, 313–326.

English, J. D., & Sweatt, J. D. (1996). Activation of p42 mitogen-activated protein kinase in hippocampal long term potentiation. *The Journal of Biological Chemistry, 271*, 24329–24332.

English, J. D., & Sweatt, J. D. (1997). A requirement for the mitogen-activated protein kinase cascade in hippocampal long term potentiation. *The Journal of Biological Chemistry, 272*, 19103–19106.

Enslen, H., Sun, P., Brickey, D., Soderling, S. H., Klamo, E., & Soderling, T. R. (1994). Characterization of Ca2+/calmodulin-dependent protein kinase IV. Role in transcriptional regulation. *The Journal of Biological Chemistry, 269*, 15520–15527.

Feil, R., & Kleppisch, T. (2008). NO/cGMP-dependent modulation of synaptic transmission. *Handbook of Experimental Pharmacology, 184*, 529–560.

Felsenfeld, G., & Groudine, M. (2003). Controlling the double helix. *Nature, 421*, 448–453.

Fin, C., da Cunha, C., Bromberg, E., Schmitz, P. K., Bianchin, M., Medina, J. H., et al. (1995). Experiments suggesting a role for nitric oxide in the hippocampus in memory processes. *Neurobiology of Learning and Memory, 63*, 113–115.

Fischer, A., Sananbenesi, F., Wang, X., Dobbin, M., & Tsai, L. H. (2007). Recovery of learning and memory is associated with chromatin remodelling. *Nature, 447*, 178–182.

Fontán-Lozano, A., Romero-Granados, R., Troncoso, J., Múnera, A., Delgado-García, J. M., & Carrión, A. M. (2008). Histone deacetylase inhibitors improve learning consolidation in young and in KA-induced-neurodegeneration and SAMP-8-mutant mice. *Molecular and Cellular Neurosciences, 39*, 193–201.

GENES, ENVIRONMENT AND ALZHEIMER'S DISEASE

Foulkes, N. S., Borrelli, E., & Sassone-Corsi, P. (1991). CREM gene: Use of alternative DNA-binding domains generates multiple antagonists of cAMP-induced transcription. *Cell, 64*, 739–749.

Francis, Y. I., Fà, M., Ashraf, H., Zhang, H., Staniszewski, A., Latchman, D. S., et al. (2009). Dysregulation of histone acetylation in the APP/PS1 mouse model of Alzheimer's disease. *Journal of Alzheimer's Disease, 18*, 131–139.

Frey, U., Huang, Y. Y., & Kandel, E. R. (1993). Effects of cAMP simulate a late stage of LTP in hippocampal CA1 neurons. *Science, 260*, 1661–1664.

Fukushima, H., Maeda, R., Suzuki, R., Suzuki, A., Nomoto, M., Toyoda, H., et al. (2008). Upregulation of calcium/calmodulin-dependent protein kinase IV improves memory formation and rescues memory loss with aging. *The Journal of Neuroscience, 28*, 9910–9919.

Gao, J., Wang, W. Y., Mao, Y. W., Gräff, J., Guan, J. S., Pan, L., et al. (2010). A novel pathway regulates memory and plasticity via SIRT1 and miR-134. *Nature, 466*, 1105–1109.

Ginty, D. D., Glowacka, D., Bader, D. S., Hidaka, H., & Wagner, J. A. (1991). Induction of immediate early genes by Ca2+ influx requires cAMP-dependent protein kinase in PC12 cells. *The Journal of Biological Chemistry, 266*, 17454–17458.

Gräff, J., Kim, D., Dobbin, M. M., & Tsai, L. H. (2011). Epigenetic regulation of gene expression in physiological and pathological brain processes. *Physiological Reviews, 91*, 603–649.

Guan, J. S., Haggarty, S. J., Giacometti, E., Dannenberg, J. H., Joseph, N., Gao, J., et al. (2009). HDAC2 negatively regulates memory formation and synaptic plasticity. *Nature, 459*, 55–60.

Guan, Z., Giustetto, M., Lomvardas, S., Kim, J. H., Miniaci, M. C., Schwartz, J. H., et al. (2002). Integration of long-term-memory-related synaptic plasticity involves bidirectional regulation of gene expression and chromatin structure. *Cell, 111*, 483–493.

Gupta, S., Kim, S. Y., Artis, S., Molfese, D. L., Schumacher, A., Sweatt, J. D., et al. (2010). Histone methylation regulates memory formation. *The Journal of Neuroscience, 30*, 3589–3599.

Gupta-Agarwal, S., Franklin, A. V., Deramus, T., Wheelock, M., Davis, R. L., McMahon, L. L., et al. (2012). G9a/GLP histone lysine dimethyltransferase complex activity in the hippocampus and the entorhinal cortex is required for gene activation and silencing during memory consolidation. *The Journal of Neuroscience, 32*, 5440–5453.

Happel, N., & Doenecke, D. (2009). Histone H1 and its isoforms: Contribution to chromatin structure and function. *Gene, 431*, 1–12.

Hardingham, N., Dachtler, J., & Fox, K. (2013). The role of nitric oxide in pre-synaptic plasticity and homeostasis. *Frontiers in Cellular Neuroscience, 7*, 190.

Harooni, H. E., Naghdi, N., Sepehri, H., & Rohani, A. H. (2009). The role of hippocampal nitric oxide (NO) on learning and immediate, short- and long-term memory retrieval in inhibitory avoidance task in male adult rats. *Behavioural Brain Research, 201*, 166–172.

Hawk, J. D., & Abel, T. (2011). The role of NR4A transcription factors in memory formation. *Brain Research Bulletin, 85*, 21–29.

Hebb, D. O. (1949). *The organization of behaviour: A neuropsychological theory.* New York: Wiley.

Hebert, A. E., & Dash, P. K. (2002). Extracellular signal-regulated kinase activity in the entorhinal cortex is necessary for long-term spatial memory. *Learning & Memory, 9*, 156–166.

Ho, N., Liauw, J. A., Blaeser, F., Wei, F., Hanissian, S., Muglia, L. M., et al. (2000). Impaired synaptic plasticity and cAMP response element-binding protein activation in Ca2+/ calmodulin-dependent protein kinase type IV/Gr-deficient mice. *The Journal of Neuroscience, 20*, 6459–6472.

Hölscher, C., & Rose, S. P. (1993). Inhibiting synthesis of the putative retrograde messenger nitric oxide results in amnesia in a passive avoidance task in the chick. *Brain Research, 619*, 189–194.

Hu, S. C., Chrivia, J., & Ghosh, A. (1999). Regulation of CBP-mediated transcription by neuronal calcium signaling. *Neuron, 22*, 799–808.

Huang, C. C., You, J. L., Wu, M. Y., & Hsu, K. S. (2004). Rap1-induced p38 mitogen-activated protein kinase activation facilitates AMPA receptor trafficking via the GDI.Rab5 complex. Potential role in (S)-3,5-dihydroxyphenylglycine-induced long term depression. *The Journal of Biological Chemistry, 279*, 12286–12292.

Huang, Y. Y., Martin, K. C., & Kandel, E. R. (2000). Both protein kinase A and mitogen-activated protein kinase are required in the amygdala for the macromolecular synthesis-dependent late phase of long-term potentiation. *The Journal of Neuroscience, 20*, 6317–6325.

Impey, S., Obrietan, K., Wong, S. T., Poser, S., Yano, S., Wayman, G., et al. (1998). Cross talk between ERK and PKA is required for Ca2+ stimulation of CREB-dependent transcription and ERK nuclear translocation. *Neuron, 21*, 869–883.

Impey, S., Wayman, G., Wu, Z., & Storm, D. R. (1994). Type I adenylyl cyclase functions as a coincidence detector for control of cyclic AMP response element-mediated transcription: Synergistic regulation of transcription by Ca2+ and isoproterenol. *Molecular and Cellular Biology, 14*, 8272–8281.

Ingram, D. K., Spangler, E. L., Kametani, H., Meyer, R. C., & London, E. D. (1998). Intracerebroventricular injection of N omega-nitro-L-arginine in rats impairs learning in a 14-unit T-maze. *European Journal of Pharmacology, 341*, 11–16.

Ito, I., Hidaka, H., & Sugiyama, H. (1991). Effects of KN-62, a specific inhibitor of calcium/calmodulin-dependent protein kinase II, on long-term potentiation in the rat hippocampus. *Neuroscience Letters, 121*, 119–121.

Izquierdo, I., Bevilaqua, L. R., Rossato, J. I., Bonini, J. S., Medina, J. H., & Cammarota, M. (2006). Different molecular cascades in different sites of the brain control memory consolidation. *Trends in Neurosciences, 29*, 496–505.

Izquierdo, L. A., Vianna, M., Barros, D. M., Mello e Souza, T., Ardenghi, P., Sant'Anna, M. K., et al. (2000). Short- and long-term memory are differentially affected by metabolic inhibitors given into hippocampus and entorhinal cortex. *Neurobiology of Learning and Memory, 73*, 141–149.

Johannessen, M., Delghandi, M. P., & Moens, U. (2004). What turns CREB on? *Cell Signalling, 16*, 1211–1227.

Kandel, E. R. (2001). The molecular biology of memory storage: A dialogue between genes and synapses. *Science, 294*, 1030–1038.

Kandel, E. R. (2012). The molecular biology of memory: cAMP, PKA, CRE, CREB-1, CREB-2, and CPEB. *Molecular Brain, 5*, 14.

Kang, H., Sun, L. D., Atkins, C. M., Soderling, T. R., Wilson, M. A., & Tonegawa, S. (2001). An important role of neural activity-dependent CaMKIV signaling in the consolidation of long-term memory. *Cell, 106*, 771–783.

Kanterewicz, B. I., Urban, N. N., McMahon, D. B., Norman, E. D., Giffen, L. J., Favata, M. F., et al. (2000). The extracellular signal-regulated kinase cascade is required for NMDA receptor-independent LTP in area CA1 but not area CA3 of the hippocampus. *The Journal of Neuroscience, 20*, 3057–3066.

Kasahara, J., Fukunaga, K., & Miyamoto, E. (1999). Differential effects of a calcineurin inhibitor on glutamate-induced phosphorylation of Ca2+/calmodulin-dependent protein kinases in cultured rat hippocampal neurons. *The Journal of Biological Chemistry, 274*, 9061–9067.

Kasahara, J., Fukunaga, K., & Miyamoto, E. (2000). Activation of CA(2+)/calmodulin-dependent protein kinase IV in cultured rat hippocampal neurons. *Journal of Neuroscience Research, 59*, 594–600.

Kawasaki, H., Fujii, H., Gotoh, Y., Morooka, T., Shimohama, S., Nishida, E., et al. (1999). Requirement for mitogen-activated protein kinase in cerebellar long-term depression. *The Journal of Biological Chemistry, 274*, 13498–13502.

Kendrick, K. M., Guevara-Guzman, R., Zorrilla, J., Hinton, M. R., Broad, K. D., Mimmack, M., et al. (1997). Formation of olfactory memories mediated by nitric oxide. *Nature, 388,* 670–674.

Khavandgar, S., Homayoun, H., & Zarrindast, M. R. (2003). The effect of l-NAME and l-arginine on impairment of memory formation and state-dependent learning induced by morphine in mice. *Psychopharmacology (Berl), 167,* 291–296.

Kim, M. S., Akhtar, M. W., Adachi, M., Mahgoub, M., Bassel-Duby, R., Kavalali, E. T., et al. (2012). An essential role for histone deacetylase 4 in synaptic plasticity and memory formation. *The Journal of Neuroscience, 32,* 10879–10886.

Kornberg, R. D. (1974). Chromatin structure: A repeating unit of histones and DNA. *Science, 184,* 868–871.

Korzus, E., Rosenfeld, M. G., & Mayford, M. (2004). CBP histone acetyltransferase activity is a critical component of memory consolidation. *Neuron, 42,* 961–972.

Koshibu, K., Gräff, J., Beullens, M., Heitz, F. D., Berchtold, D., Russig, H., et al. (2009). Protein phosphatase 1 regulates the histone code for long-term memory. *The Journal of Neuroscience, 29,* 13079–13089.

Kovács, K. A., Steullet, P., Steinmann, M., Do, K. Q., Magistretti, P. J., Halfon, O., et al. (2007). TORC1 is a calcium- and cAMP-sensitive coincidence detector involved in hippocampal long-term synaptic plasticity. *Proceedings of the National Academy of Sciences of the United States of America, 104,* 4700–4705.

Kullmann, D. M. (2012). The Mother of All Battles 20 years on: Is LTP expressed pre- or postsynaptically? *The Journal of Physiology, 590,* 2213–2216.

Lee, S. H., Park, J., Che, Y., Han, P.-L., & Lee, J.-K. (2000). Constitutive activity and differential localization of p38α and p38β MAPKs in adult mouse brain. *Journal of Neuroscience Research, 60,* 623–631.

Legube, G., & Trouche, D. (2003). Regulating histone acetyltransferases and deacetylases. *EMBO Reports, 4,* 944–947.

Leslie, J. H., & Nedivi, E. (2011). Activity-regulated genes as mediators of neural circuit plasticity. *Progress in Neurobiology, 94,* 223–237.

Levenson, J. M., O'Riordan, K. J., Brown, K. D., Trinh, M. A., Molfese, D. L., & Sweatt, J. D. (2004). Regulation of histone acetylation during memory formation in the hippocampus. *The Journal of Biological Chemistry, 279,* 40545–40559.

Levenson, J. M., & Sweatt, J. D. (2005). Epigenetic mechanisms in memory formation. *Nature Reviews Neuroscience, 6,* 108–118.

Lipsky, R. H. (2013). Epigenetic mechanisms regulating learning and long-term memory. *International Journal of Developmental Neuroscience, 31,* 353–358.

Lisman, J., Yasuda, R., & Raghavachari, S. (2012). Mechanisms of CaMKII action in long-term potentiation. *Nature Reviews Neuroscience, 13,* 169–182.

Lonze, B. E., & Ginty, D. D. (2002). Function and regulation of CREB family transcription factors in the nervous system. *Neuron, 35,* 605–623.

Lu, Y. F., Kandel, E. R., & Hawkins, R. D. (1999). Nitric oxide signaling contributes to late-phase LTP and CREB phosphorylation in the hippocampus. *The Journal of Neuroscience, 19,* 10250–10261.

Majlessi, N., Kadkhodaee, M., Parviz, M., & Naghdi, N. (2003). Serotonin depletion in rat hippocampus attenuates l-NAME-induced spatial learning deficits. *Brain Research, 963,* 244–251.

Malenka, R. C., Kauer, J. A., Perkel, D. J., Mauk, M. D., Kelly, P. T., Nicoll, R. A., et al. (1989). An essential role for postsynaptic calmodulin and protein kinase activity in long-term potentiation. *Nature, 340,* 554–557.

Malinow, R., Schulman, H., & Tsien, R. W. (1989). Inhibition of postsynaptic PKC or CaMKII blocks induction but not expression of LTP. *Science, 245,* 862–866.

Marmorstein, R., & Roth, S. Y. (2001). Histone acetyltransferases: Function, structure, and catalysis. *Current Opinion in Genetics & Development, 11,* 155–161.

Matthews, R. P., Guthrie, C. R., Wailes, L. M., Zhao, X., Means, A. R., & McKnight, G. S. (1994). Calcium/calmodulin-dependent protein kinase types II and IV differentially regulate CREB-dependent gene expression. *Molecular Cell Biology, 14*, 6107–6116.

Maurice, T., Duclot, F., Meunier, J., Naert, G., Givalois, L., Meffre, J., et al. (2008). Altered memory capacities and response to stress in p300/CBP-associated factor (PCAF) histone acetylase knockout mice. *Neuropsychopharmacology, 33*, 1584–1602.

Mayr, B., & Montminy, M. (2001). Transcriptional regulation by the phosphorylation-dependent factor CREB. *Nature Reviews Molecular Cell Biology, 2*, 599–609.

McQuown, S. C., Barrett, R. M., Matheos, D. P., Post, R. J., Rogge, G. A., Alenghat, T., et al. (2011). HDAC3 is a critical negative regulator of long-term memory formation. *The Journal of Neuroscience, 31*, 764–774.

Michan, S., Li, Y., Chou, M. M., Parrella, E., Ge, H., Long, J. M., et al. (2010). SIRT1 is essential for normal cognitive function and synaptic plasticity. *The Journal of Neuroscience, 30*, 9695–9707.

Miller, C. A., Campbell, S. L., & Sweatt, J. D. (2008). DNA methylation and histone acetylation work in concert to regulate memory formation and synaptic plasticity. *Neurobiology of Learning and Memory, 89*, 599–603.

Miller, C. A., & Sweatt, J. D. (2007). Covalent modification of DNA regulates memory formation. *Neuron, 53*, 857–869.

Miller, G. A. (1956). The magical number seven, plus or minus two: Some limits on our capacity for processing information. *Psychological Review, 63*, 81–97.

Miyamoto, E. (2006). Molecular mechanism of neuronal plasticity: Induction and maintenance of long-term potentiation in the hippocampus. *Journal of Pharmacological Sciences, 100*, 433–442.

Montminy, M. (1997). Transcriptional regulation by cyclic AMP. *Annual Review of Biochemistry, 66*, 807–822.

Montminy, M. R., & Bilezikjian, L. M. (1987). Binding of a nuclear protein to the cyclic-AMP response element of the somatostatin gene. *Nature, 328*, 175–178.

Moult, P. R., Corrêa, S. A., Collingridge, G. L., Fitzjohn, S. M., & Bashir, Z. I. (2008). Co-activation of p38 mitogen-activated protein kinase and protein tyrosine phosphatase underlies metabotropic glutamate receptor-dependent long-term depression. *Journal of Physiology, 586*, 2499–2510.

Muhlbacher, F., Schiessel, H., & Holm, C. (2006). Tail-induced attraction between nucleosome core particles. *Physical Review. E, Statistical, Nonlinear, and Soft Matter Physics, 74*, 031919.

Ninan, I., & Arancio, O. (2004). Presynaptic CaMKII is necessary for synaptic plasticity in cultured hippocampal neurons. *Neuron, 42*, 129–141.

Nonaka, M., Kim, R., Fukushima, H., Sasaki, K., Suzuki, K., Okamura, M., et al. (2014). Region-specific activation of CRTC1-CREB signaling mediates long-term fear memory. *Neuron, 84*, 92–106.

Ohno, M., Frankland, P. W., Chen, A. P., Costa, R. M., & Silva, A. J. (2001). Inducible, pharmacogenetic approaches to the study of learning and memory. *Nature Neuroscience, 4*, 1238–1243.

Oliveira, A. M., Wood, M. A., McDonough, C. B., & Abel, T. (2007). Transgenic mice expressing an inhibitory truncated form of p300 exhibit long-term memory deficits. *Learning & Memory, 14*, 564–572.

Patterson, S. L., Pittenger, C., Morozov, A., Martin, K. C., Scanlin, H., Drake, C., et al. (2001). Some forms of cAMP-mediated long-lasting potentiation are associated with release of BDNF and nuclear translocation of phospho-MAP kinase. *Neuron, 32*, 123–140.

Peleg, S., Sananbenesi, F., Zovoilis, A., Burkhardt, S., Bahari-Javan, S., Agis-Balboa, R. C., et al. (2010). Altered histone acetylation is associated with age-dependent memory impairment in mice. *Science, 328*, 753–756.

GENES, ENVIRONMENT AND ALZHEIMER'S DISEASE

Peterson, C. L., & Laniel, M. A. (2004). Histones and histone modifications. *Current Biology, 14*, R546–R551.

Picciotto, M. R., Zoli, M., Bertuzzi, G., & Nairn, A. C. (1995). Immunochemical localization of calcium/calmodulin-dependent protein kinase I. *Synapse, 20*, 75–84.

Prickaerts, J., de Vente, J., Honig, W., Steinbusch, H. W., & Blokland, A. (2002). cGMP, but not cAMP, in rat hippocampus is involved in early stages of object memory consolidation. *European Journal of Pharmacology, 436*, 83–87.

Qiang, M., Chen, Y. C., Wang, R., Wu, F. M., & Qiao, J. T. (1997). Nitric oxide is involved in the formation of learning and memory in rats: Studies using passive avoidance response and Morris water maze task. *Behavioural Pharmacology, 8*, 183–187.

Quina, A. S., Buschbeck, M., & Di Croce, L. (2006). Chromatin structure and epigenetics. *Biochemical Pharmacology, 72*, 1563–1569.

Redondo, R. L., & Morris, R. G. (2011). Making memories last: The synaptic tagging and capture hypothesis. *Nature Reviews. Neuroscience, 12*, 17–30.

Riccio, A., Alvania, R. S., Lonze, B. E., Ramanan, N., Kim, T., Huang, Y., et al. (2006). A nitric oxide signaling pathway controls CREB-mediated gene expression in neurons. *Molecular Cell, 21*, 283–294.

Ricobaraza, A., Cuadrado-Tejedor, M., Pérez-Mediavilla, A., Frechilla, D., Del Río, J., & García-Osta, A. (2009). Phenyl butyrate ameliorates cognitive deficit and reduces tau pathology in an Alzheimer's disease mouse model. *Neuropsychopharmacology, 34*, 1721–1732.

Roux, P. P., & Blenis, J. (2004). ERK and p38 MAPK-activated protein kinases: A family of protein kinases with diverse biological functions. *Microbiology and Molecular Biology Reviews, 68*, 320–344.

Roux, P. P., Richards, S. A., & Blenis, J. (2003). Phosphorylation of p90 ribosomal S6 kinase (RSK) regulates extracellular signal-regulated kinase docking and RSK activity. *Molecular and Cellular Biology, 23*, 4796–4804.

Russo, V. E. A., Martienssen, R. A., & Riggs, A. D. (1996). *Epigenetic mechanisms of gene regulation*. Plainview. New York: Cold Spring Harbor Laboratory Press.

Rutten, K., Prickaerts, J., Hendrix, M., van der Staay, F. J., Sik, A., & Blokland, A. (2007). Time-dependent involvement of cAMP and cGMP in consolidation of object memory: Studies using selective phosphodiesterase type 2, 4 and 5 inhibitors. *European Journal of Pharmacology, 58*, 107–112.

Saha, R. N., & Pahan, K. (2006). HATs and HDACs in neurodegeneration: A tale of disconcerted acetylation homeostasis. *Cell Death & Differentiation, 13*, 539–550.

Schacher, S., & Hu, J. Y. (2014). The less things change, the more they are different: Contributions of long-term synaptic plasticity and homeostasis to memory. *Learning & Memory, 21*, 128–134.

Schafe, G. E., Atkins, C. M., Swank, M. W., Bauer, E. P., Sweatt, J. D., & LeDoux, J. E. (2000). Activation of ERK/MAP kinase in the amygdala is required for memory consolidation of Pavlovian fear conditioning. *The Journal of Neuroscience, 20*, 8177–8187.

Schneider, A., Chatterjee, S., Bousiges, O., Selvi, B. R., Swaminathan, A., Cassel, R., et al. (2013). Acetyltransferases (HATs) as targets for neurological therapeutics. *Neurotherapeutics, 10*, 568–588.

Scolville, R. M., & Milner, B. (1957). Loss recent memory after bilateral hippocampal lesions. *Journal of Neurology, Neurosurgery & Psychiatry, 20*, 11–21.

Sekeres, M. J., Mercaldo, V., Richards, B., Sargin, D., Mahadevan, V., Woodin, M. A., et al. (2012). Increasing CRTC1 function in the dentate gyrus during memory formation or reactivation increases memory strength without compromising memory quality. *The Journal of Neuroscience, 32*, 17857–17868.

Selcher, J. C., Atkins, C. M., Trzaskos, J. M., Paylor, R., & Sweatt, J. D. (1999). A necessity for MAP kinase activation in mammalian spatial learning. *Learning & Memory, 6*, 478–490.

Selcher, J. C., Weeber, E. J., Christian, J., Nekrasova, T., Landreth, G. E., & Sweatt, J. D. (2003). A role for ERK MAP kinase in physiologic temporal integration in hippocampal area CA1. *Learning & Memory, 10,* 26–39.

Selvi, B. R., Cassel, J. C., Kundu, T. K., & Boutillier, A. L. (2010). Tuning acetylation levels with HAT activators: Therapeutic strategy in neurodegenerative diseases. *Biochimica et Biophysica Acta, 1799,* 840–853.

Shallice, T., & Warrington, E. K. (1970). Independent functioning of verbal memory stores: A neuropsychological study. *The Quarterly Journal of Experimental Psychology, 22,* 261–273.

Sheng, M., Thompson, M. A., & Greenberg, M. E. (1991). CREB: a Ca(2+)-regulated transcription factor phosphorylated by calmodulin-dependent kinases. *Science, 252,* 1427–1430.

Siddoway, B., Hou, H., & Xia, H. (2014). Molecular mechanisms of homeostatic synaptic downscaling. *Neuropharmacology, 78,* 38–44.

Silva, A. J., Paylor, R., Wehner, J. M., & Tonegawa, S. (1992). Impaired spatial learning in alpha-calcium-calmodulin kinase II mutant mice. *Science, 257,* 206–211.

Silva, A. J., Stevens, C. F., Tonegawa, S., & Wang, Y. (1992). Deficient hippocampal long-term potentiation in alpha-calcium-calmodulin kinase II mutant mice. *Science, 257,* 201–206.

Singh, P., & Thakur, M. K. (2014). Reduced recognition memory is correlated with decrease in DNA methyltransferase1 and increase in histone deacetylase2 protein expression in old male mice. *Biogerontology, 15,* 339–346.

Smith, D. M., & Mizumori, S. J. (2006). Hippocampal place cells, context, and episodic memory. *Hippocampus, 16,* 716–729.

Son, H., Lu, Y. F., Zhuo, M., Arancio, O., Kandel, E. R., & Hawkins, R. D. (1998). The specific role of cGMP in hippocampal LTP. *Learning & Memory, 5,* 231–245.

Squire, L. R. (1992). Memory and the hippocampus: A synthesis from findings with rats, monkeys, and humans. *Psychological Review, 99,* 195–231.

Sun, P., Enslen, H., Myung, P. S., & Maurer, R. A. (1994). Differential activation of CREB by Ca2+/calmodulin-dependent protein kinases involves phosphorylation of a negative regulatory site. *Genes & Development, 8,* 2527–2539.

Sun, P., Lou, L., & Maurer, R. A. (1996). Regulation of activating transcription factor-1 and the cAMP response element-binding protein by Ca2+/calmodulin-dependent protein kinases type I, II, and IV. *The Journal of Biological Chemistry, 271,* 3066–3073.

Sutherland, E. W. (1992). Studies on the mechanism of hormone action. In Lindsten (Ed.), *Nobel lectures in physiology or medicine (1971–1980).* Singapore: World Scientific Publishing Co.

Suzuki, A., Fukushima, H., Mukawa, T., Toyoda, H., Wu, L. J., Zhao, M. G., et al. (2011). Upregulation of CREB-mediated transcription enhances both short- and long-term memory. *The Journal of Neuroscience, 31,* 8786–8802.

Takeuchi, T., Duszkiewicz, A. J., & Morris, R. G. (2014). The synaptic plasticity and memory hypothesis: Encoding, storage and persistence. *Philosophical Transactions of the Royal Society of London Series B, Biological Sciences, 369,* 20130288.

Telegdy, G., & Kokavszky, R. (1997). The role of nitric oxide in passive avoidance learning. *Neuropharmacology, 36,* 1583–1587.

Thiagalingam, S., Cheng, K. H., Lee, H. J., Mineva, N., Thiagalingam, A., & Ponte, J. F. (2003). Histone deacetylases: Unique players in shaping the epigenetic histone code. *Annals of the New York Academy of Sciences, 983,* 84–100.

Thomas, G. M., & Huganir, R. L. (2004). MAPK cascade signalling and synaptic plasticity. *Nature Reviews Neuroscience, 5,* 173–183.

Tulving, E. (1985). How many memory system are there? *American Psychologist, 40,* 385–398.

Vecsey, C. G., Hawk, J. D., Lattal, K. M., Stein, J. M., Fabian, S. A., Attner, M. A., et al. (2007). Histone deacetylase inhibitors enhance memory and synaptic plasticity via CREB:CBP-dependent transcriptional activation. *The Journal of Neuroscience, 27,* 6128–6140.

Vincent, S. R., & Kimura, H. (1992). Histochemical mapping of nitric oxide synthase in the rat brain. *Neuroscience, 46,* 755–784.

Wei, F., Qiu, C. S., Liauw, J., Robinson, D. A., Ho, N., Chatila, T., et al. (2002). Calcium calmodulin-dependent protein kinase IV is required for fear memory. *Nature Neurosciene, 5,* 573–579.

Wei, W., Coelho, C. M., Li, X., Marek, R., Yan, S., Anderson, S., et al. (2012). p300/CBP-associated factor selectively regulates the extinction of conditioned fear. *The Journal of Neuroscience, 32,* 11930–11941.

Wong, J. C., Bathina, M., & Fiscus, R. R. (2012). Cyclic GMP/protein kinase G type-Iα (PKG-Iα) signaling pathway promotes CREB phosphorylation and maintains higher c-IAP1, livin, survivin, and Mcl-1 expression and the inhibition of PKG-Iα kinase activity synergizes with cisplatin in non-small cell lung cancer cells. *Journal of Cellular Biochemistry, 113,* 3587–3598.

Xing, J., Ginty, D. D., & Greenberg, M. E. (1996). Coupling of the RAS-MAPK pathway to gene activation by RSK2, a growth factor-regulated CREB kinase. *Science, 273,* 959–963.

Yamada, K., Noda, Y., Hasegawa, T., Komori, Y., Nikai, T., Sugihara, H., et al. (1996). The role of nitric oxide in dizocilpine-induced impairment of spontaneous alternation behavior in mice. *Journal of Pharmacology and Experimental Therapeutics, 276,* 460–466.

Yang, S., Zhou, G., Liu, H., Zhang, B., Li, J., Cui, R., et al. (2013). Protective effects of p38 MAPK inhibitor SB202190 against hippocampal apoptosis and spatial learning and memory deficits in a rat model of vascular dementia. *BioMed Research International, 2013,* 215798.

Yin, J. C., Wallach, J. S., Del Vecchio, M., Wilder, E. L., Zhou, H., Quinn, W. G., et al. (1994). Induction of a dominant negative CREB transgene specifically blocks long-term memory in drosophila. *Cell, 79,* 49–58.

Zarrindast, M. R., Shendy, M. M., & Ahmadi, S. (2007). Nitric oxide modulates state dependency induced by lithium in an inhibitory avoidance task in mice. *Behavioural Pharmacology, 18,* 289–295.

Zhen, X., Du, W., Romano, A. G., Friedman, E., & Harvey, J. A. (2001). The p38 mitogen-activated protein kinase is involved in associative learning in rabbits. *The Journal of Neuroscience, 21,* 5513–5519.

Zhu, J. J., Qin, Y., Zhao, M., Van Aelst, L., & Malinow, R. (2002). Rasand Rap control AMPA receptor trafficking during synaptic plasticity. *Cell, 110,* 443–455.

When Cognitive Decline Becomes Pathology: From Normal Aging to Alzheimer's Disease

Eliezer Masliah[1,2] *and David P. Salmon*[1]

[1]Department of Neurosciences, University of California, San Diego, CA, United States [2]Department of Pathology, University of California, San Diego, CA, United States

OUTLINE

Models of Cognitive Decline in Healthy Aging   30

Distinguishing AD-Associated Cognitive Decline from Normal Aging   31

Distinguishing Preclinical AD from Normal Aging with
Event-Related Potentials (ERPs)   35

Memory Deficits in Preclinical AD   37
Deficits in Associative Memory   37
Deficits in Pattern Separation   38
Deficits in Recollection and Familiarity   38
Deficits in Prospective Memory   39
Deficits in Remote Memory   40
Deficits in Working Memory   41

Asymmetric Cognitive Decline in Preclinical AD   42

Summary and Conclusions   43

References   44

Genes, Environment and Alzheimer's Disease.
DOI: http://dx.doi.org/10.1016/B978-0-12-802851-3.00002-4

29

Current models of Alzheimer's disease (AD) suggest that neuropathological changes begin years before cognitive and other clinical symptoms of the disease become apparent (e.g., Katzman, 1994). One widely-known model proposes that AD begins with the very early accrual of β-amyloid (Aβ) plaques in the brain followed by accumulation of tau-positive neurofibrillary tangles, neuronal dysfunction, and neurodegeneration (Jack et al., 2010). Neurodegeneration is accompanied by synaptic dysfunction that is correlated with cognitive symptoms (Terry et al., 1991). As these pathologic changes gradually accumulate, a threshold for the initiation of the cognitive symptoms of the disease is eventually reached. The disease continues to progress until the cognitive symptoms evolve into a global dementia syndrome usually characterized by early and prominent amnesia with additional deficits in language and semantic knowledge (i.e., aphasia and agnosia), abstract reasoning, "executive" functions, attention, and constructional (i.e., apraxia) and visuo-spatial abilities (Salmon & Bondi, 2009). It is only when cognitive symptoms progress to a point where they are obvious and have a clear impact on everyday activities that dementia is clinically diagnosed.

It is apparent from this sequence of events that cognitive decline is likely to occur in someone with AD well before the clinical diagnosis can be made with any certainty. However, it is difficult to detect this early AD-related decline because it occurs against the backdrop of relatively subtle cognitive changes that take place during the course of normal aging (for reviews, see Hedden & Gabrieli, 2004; Park, O'Connell, & Thomson, 2003). Age-related cognitive decline is particularly evident in information processing abilities such as effortful encoding of new information, processing speed, inductive reasoning, and working memory. In contrast, little age-related decline is evident in semantic knowledge and vocabulary, autobiographical remote memory, and automatic memory processes (e.g., priming).

MODELS OF COGNITIVE DECLINE IN HEALTHY AGING

The nature of the cognitive decline that occurs with healthy aging has led to several psychological and neurological models to account for these changes (Band, Ridderinkhof, & Segalowitz, 2002; Buckner, 2004; Kramer, Bherer, Colcombe, Dong, & Greenough, 2004). One influential psychological model (Salthouse, 1996) suggests that a general decline in processing speed underlies most of the cognitive decline that occurs with age. According to this model, an age-related decline in information processing speed reduces the ability to efficiently integrate and organize information and causes a decline in memory by reducing the efficiency of information encoding, rehearsal, and retrieval. Similar psychological models suggest that cognitive decline in healthy elderly is caused by a decline in a single factor such as working memory, inhibitory processes, or sensory function (Kramer et al., 2004; Park et al., 2003).

Neurologically-based models suggest that atrophy of prefrontal cortex and loss of frontal white matter integrity occurs as a normal consequence of aging (Greenwood, 2000; West, 1996), which results in age-related declines in working memory, cognitive flexibility, verbal fluency, directed and divided attention, and self-monitoring performance (Grady & Craik, 2000; Raz, 2005; West, 1996). In addition, a general loss of white matter connectivity may lead to an age-related "disconnection" syndrome that causes cognitive changes beyond those that are usually considered frontal lobe functions such as memory and visuo-spatial abilities (Bartzokis, 2004; O'Sullivan et al., 2001; Pfefferbaum, Adalsteinsson, & Sullivan, 2005; Raz, 2005).

DISTINGUISHING AD-ASSOCIATED COGNITIVE DECLINE FROM NORMAL AGING

The quest to distinguish between normal age-related and early AD-related cognitive decline has been bolstered by recent advances in detecting reliable biological markers of AD in cognitively normal elderly people. It is now possible to use positron emission tomography (PET) imaging with Pittsburgh compound-B ($[^{11}C]$-PIB) or similar agents to detect the deposition of Aβ in the brain, or to use biochemical assays of cerebrospinal fluid (CSF) to detect abnormal levels of Aβ and tau proteins that constitute the plaques and tangles of AD (for review, see Jack et al., 2010). Each of these biomarkers has been used to predict the development of AD dementia in nondemented elderly individuals (for review, see Jack & Holtzman, 2013). These discoveries prompted a revision of the diagnostic criteria for AD that now incorporate the presence of a biomarker as supporting evidence of the disease (McKhann et al., 2011). Mild cognitive impairment (MCI) that precedes frank dementia is now considered as early AD with varying degrees of confidence determined by the presence of AD biomarkers (Albert et al., 2011; Dubois et al., 2007), and an even earlier stage of "preclinical AD" is identified by the presence of biomarkers in asymptomatic individuals (Sperling et al., 2011).

Several recent studies have shown that cognitively normal elderly adults with significant amyloid burden evident on PET imaging with PIB or related compounds (Aβ+) show faster cognitive decline than do those without amyloid burden (Aβ−) (Chetelat et al., 2012; Doraiswamy et al., 2012; Lim et al., 2012, 2014). This is particularly true when evidence of neurodegeneration such as decreased hippocampal volume and fluorodeoxyglucose (FDG)-PET hypometabolism is also present (Mormino et al., 2014). The decline that is observed in cognitively normal Aβ+ individuals is most evident in the domains of episodic and working memory. In a study that illustrates this effect, Lim and colleagues (Lim et al., 2014) administered a battery of neuropsychological tests to Aβ+ and Aβ− cognitively normal older adults at baseline and again after 18 months and 36 months. They found that the Aβ+ individuals showed faster decline than the Aβ− individuals on the learning and delayed recall measures of the California Verbal Learning Test-II (CVLT) and on the One-Back working memory test, whereas the groups did not differ in rate of decline on tests of attention or psychomotor function. Consistent with these results, a deficit in face-name paired-associate learning was observed in nondemented elderly Aβ+ individuals but not in those who were Aβ− (Rentz et al., 2011). These latter changes occurred in conjunction with functional MRI evidence of reduced deactivation in entorhinal cortex and reduced connectivity with other regions within the so-called default mode network that consists of brain structures that are active when no specific cognitive activity is being performed (Huijbers et al., 2014).

Evidence of preclinical cognitive decline in patients with AD also comes from studies that examined the performance of nondemented elderly individuals who subsequently developed dementia. In one such study, Mickes and colleagues (2007) retrospectively examined detailed neuropsychological evaluations of cognitively-normal elderly individuals over the course of 3 years up to and including the first year of a nonnormal diagnosis (i.e., MCI or dementia). While performance fell off rapidly in all areas of cognitive functioning, episodic and semantic memory abilities mediated by the medial and lateral temporal lobes were substantially more impaired than "executive" functions dependent upon the frontal lobes (Figure 2.1). This pattern of progression is consistent with the proposed early involvement of medial temporal lobe structures in AD (Braak & Braak, 1991) and with imaging evidence of neurodegeneration in medial temporal cortex, frontal cortex, and anterior cingulate cortex in preclinical AD (Albert, Moss, Tanzi, & Jones, 2001; Andrews-Hanna, Snyder, & Vincent, 2007; Small, Mobly, Laukka, Jones, & Backman, 2003).

Consistent with these results, numerous studies have shown that poor episodic memory performance (particularly poor delayed recall) at the initial neuropsychological evaluation of nondemented elderly individuals predicts those who subsequently developed dementia (e.g., Backman, Small, & Fratiglioni, 2001; Grober & Kawas, 1997; Howieson et al., 1997; Jacobs et al., 1995; Kawas et al., 2003; Linn, Wolf, Bachman, Knoefel, & Cobb, 1995; Small, Fratiglioni, Viitanen, Winblad, & Bäckman, 2000). For example, poor performance by nondemented elderly individuals on recall measures from the Fuld Object Memory Test or the Selective Reminding Test correctly predicted the subsequent development of AD within the

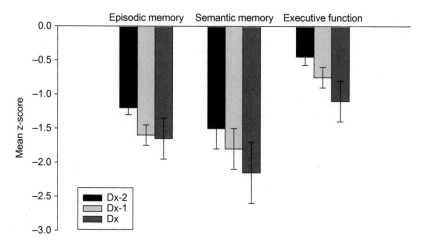

FIGURE 2.1   The mean z-scores (relative to normal control subjects) on composite measures of episodic memory, semantic memory, and executive function achieved by 11 nondemented elderly individuals in the 2-year period immediately preceding a diagnosis of Alzheimer's disease (AD) (i.e., the preclinical AD period). Although episodic memory and semantic memory were more impaired than executive function, significant decline was observed in all three cognitive domains from 2 years prior to diagnosis (Dx-2 years), to 1 year prior to diagnosis (Dx-1 year), to the year of diagnosis (Dx). *Source: Adapted from Mickes et al. (2007).*

next 5 years (Fuld, Masur, Blau, Crystal, & Aronson, 1990). An extensive follow-up to this study showed that delayed recall measures from the Selective Reminding Test and the Fuld Object Memory Test were moderately effective in identifying individuals who later developed AD and provided excellent specificity for identifying individuals who remained free of dementia over a subsequent 11-year period (Masur, Sliwinski, Lipton, Blau, & Crystal, 1994).

There is also evidence that episodic memory deteriorates rapidly in the period immediately preceding the diagnosis of AD dementia (Chen et al., 2001; Lange et al., 2002). Lange et al. (2002) compared the performances of nondemented elderly normal adults who remained cognitively stable for at least two additional years, nondemented older adults who subsequently developed dementia within 1 or 2 years, and patients with mild AD on two tests of episodic memory, the Logical Memory subtest of the Wechsler Memory Scale-Revised and the CVLT. Results showed a precipitous decline in verbal memory abilities 1–2 years prior to the onset of the dementia syndrome (Figure 2.2). These and similar results (e.g., Albert, Blacker, Moss, Tanzi, & McArdle, 2007) suggest that the rate of decline in episodic memory performance in patients with MCI may predict an imminent dementia diagnosis more effectively than absolute

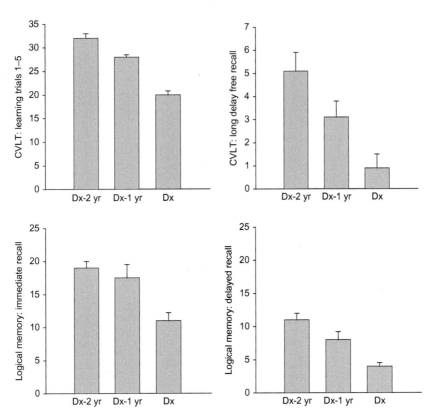

FIGURE 2.2   The mean scores achieved on key measures from the California Verbal Learning Test (CVLT) and the WMS-R Logical Memory Test by nondemented elderly individuals in the 2-year period immediately preceding a diagnosis of Alzheimer's disease (AD) (i.e., the preclinical AD period). Each measure showed significant and relatively linear decline from 2 years prior to diagnosis (Dx-2 years), to 1 year prior to diagnosis (Dx-1 year), to the year of diagnosis (Dx). These results suggest that the rate of decline in episodic memory performance in nondemented elderly may predict an imminent dementia diagnosis more effectively than absolute level of memory performance. *Source: Adapted from Lange et al. (2002).*

level of memory performance. This possibility was supported by a longitudinal study of MCI patients, which showed that delayed recall ability, and particularly 1-year decline in delayed recall, was more effective than neuroimaging (e.g., hippocampal volume) and CSF protein (e.g., Aβ 1–42 and tau) biomarkers in predicting conversion from MCI to AD dementia within the subsequent 4 years (Gomar, Bobes-Bascaran, Conejero-Goldberg, Davies, & Goldberg, 2011; also see Landau et al., 2010).

DISTINGUISHING PRECLINICAL AD FROM NORMAL AGING WITH EVENT-RELATED POTENTIALS (ERPs)

The results of these neuropsychological studies are supported by studies that examine changes in event-related potentials (ERPs) sensitive to abnormal episodic and semantic processing in cognitively normal individuals with preclinical AD. Studies of ERPs in this context are particularly relevant since they are comprised primarily of summed excitatory and inhibitory postsynaptic potentials (Nunez & Srinivasan, 2006) and may provide a noninvasive measure of synaptic dysfunction underlying very early cognitive alternations due to AD. Increasing evidence suggests that synaptic dysfunction is the primary correlate of AD dementia (Klein, 2006; Selkoe, 2002; Terry et al., 1991). These studies employed an ERP paradigm in which target words that were semantically congruous ("hammer") or incongruous ("tree") with a presented category cue (e.g., "tool") were repeated or not repeated after initial presentation. The repetition of a target word (particularly a category congruent target word) elicited a P600 (or "Late Positive") response that is thought to index both memory encoding and retrieval processes (Olichney et al., 2000; Paller & Kutas, 1992; Van Petten & Kutas, 1991). Nonrepeated category incongruous words elicited a N400 response related to semantic processing. In an initial study, Olichney and colleagues (2002) showed that patients with MCI had a reduced P600 response to repeated congruous words but a normal N400 response to category incongruous words. The results were particularly salient when the analyses were limited to MCI patients who subsequently developed AD dementia (Figure 2.3). In a subsequent longitudinal study (Olichney et al., 2008), patients with MCI with normal N400 and P600 repetition effects had an 11–27% likelihood of developing dementia within 3 years while those with abnormal N400 or P600 word repetition effects had an 87–88% likelihood. In a study of preclinical AD, Olichney et al. (2013) used their incidental verbal learning paradigm to compare the P600 and N400 ERPs of cognitively normal elderly individuals with preclinical AD (i.e., they showed subsequent cognitive decline and/or AD pathology at autopsy) and normal elderly individuals who remained cognitively normal. Results showed that the P600 congruous word repetition ERP effects were significantly smaller in the preclinical AD patients than in the normal elderly (mean amplitudes = 0.10 vs 3.28 μV). A P600 amplitude cutoff of ~1.5 μV correctly classified 84% of the elderly individuals as preclinical AD or not. Taken together, the results of these studies suggest that these ERP components may be an important sign of synaptic dysfunction related to episodic and semantic memory deficits in preclinical AD and may be useful biomarkers for the detection and staging of the disease.

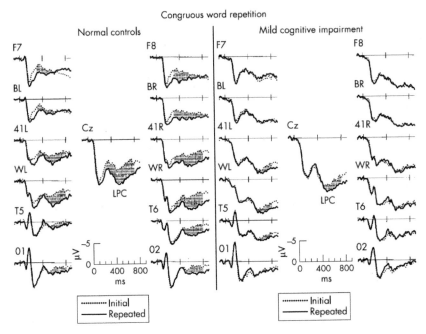

FIGURE 2.3   Grand average event-related potentials (ERPs) for normal control (NC) and mild cognitive impairment (MCI) groups elicited by new and repeated semantically congruous words. The shaded area designates the P600 word repetition effect (between 300 and 800 ms) that reflects greater positivity to new words than to repeated words. This memory effect is present in the NC group but not in the MCI group. *Source: Adapted from Olichney et al. (2002).*

Although episodic memory performance declines precipitously just before the diagnosis of AD dementia can be made, several studies suggests that there may be a plateau in the course of this decline in episodic memory rather than a monotonic or linear decrease (Bäckman et al., 2001; Small et al., 2000; Smith, Pankratz, & Negash, 2007). Functional neuroimaging studies (Bondi, Houston, Eyler, & Brown, 2005; Bookheimer et al., 2000; Han et al., 2007) indicate that this plateau period may result from compensatory brain responses to the development of AD pathology. These studies show that a wider network of activation in medial temporal lobe and other cortical regions is necessary to maintain the same level of memory performance in those who subsequently develop dementia compared to those who do not progress. It is only when these compensatory responses are overwhelmed that memory decline becomes evident. These results suggest that an abrupt decline in memory in an elderly individual might predict the imminent onset of dementia better than poor memory per se.

MEMORY DEFICITS IN PRECLINICAL AD

Given the prominence and preeminence of memory decline in pre-clinical AD, a considerable amount of neuropsychological research has been carried out to characterize the specific aspects of memory that are affected. These studies have focused on the amnestic MCI condition that is usually the precursor to AD dementia. A number of these studies have shown that the episodic memory deficit in amnestic MCI is character-ized, in most cases, by abnormally rapid forgetting on tests of delayed recall (Libon et al., 2011; Manes, Serrano, Calcagno, Cardozo, & Hodges, 2008; Perri, Serra, Carlesimo, & Caltagirone, 2007) and comparable levels of impairment on tests of free recall and recognition (Libon et al., 2011). This pattern is usually attributed to ineffective encoding and consolida-tion of new information and is virtually identical to the pattern shown by patients with circumscribed amnesia arising from bilateral damage to medial temporal lobe structures (for review, see Salmon & Squire, 2009). This is not surprising given that the earliest pathological changes of AD involve these structures (Braak & Braak, 1991).

Deficits in Associative Memory

Several studies indicate that a deficit in associative memory may be a particularly early marker of AD in patients with MCI. Associative mem-ory refers to the ability to remember relationships between two or more items or between an item and its context (e.g., when or where something was seen). This form of memory "binding" is thought to be critically dependent upon the hippocampus and is impaired in patients with cir-cumscribed amnesia (Eichenbaum, 1997). The associative memory deficit in amnestic MCI was recently shown in a study that examined the ability of patients and control subjects to remember six simple geometric forms (memory for items) and their location within a spatial array (associative memory), or to remember nine symbols and the digits with which they were paired (Troyer et al., 2008). The results showed that in both tasks amnestic MCI patients were impaired relative to controls and the impair-ment was greater for associative information than for item information. Similar deficits in associative memory in amnestic MCI patients have been shown using various paired-associate learning tasks that use word or object pairs (de Rover et al., 2011; Pike, Rowe, Moss, & Savage, 2008). These deficits in associative memory have been tied to hippocampus and entorhinal cortex abnormalities through structural and functional imaging (de Rover et al., 2011).

Deficits in Pattern Separation

Just as the hippocampus is important for binding item and context information in memory, it also plays an important role in separating similar representations (or patterns) into distinct representations that can be reliably distinguished from one another (Marr, 1971). Pattern separation and its neural substrates were recently examined in patients with amnestic MCI using high resolution functional MRI during a difficult recognition memory task in which subjects had to distinguish between previously shown stimuli ("old" items), "lure" stimuli very similar to old items (Yassa et al., 2010), and completely novel ("new") items. The results showed that patients with MCI were more likely than controls to report a lure item as previously seen and had an overall reduction in separation bias (i.e., the difference between the probability of calling a lure item similar and the probability of calling a novel item similar). This occurred despite normal performance in discriminating novel items from previously seen items. This selective impairment of pattern separation in patients with amnestic MCI was accompanied by hyperactivity in the left dentate gyrus and CA3 field of the hippocampus. Furthermore, the degree of this hyperactivity was negatively correlated with separation bias scores. These results suggest that poor performance on memory tasks that place heavy demands on pattern separation may be a particularly sensitive marker of early disease.

Deficits in Recollection and Familiarity

Although many studies have shown that recognition memory is impaired in patients with amnestic MCI (for review, see Twamley, Ropacki, & Bondi, 2006), few studies have examined the qualitative nature of this deficit. Current theories of recognition memory propose that the ability to recognize a previously presented stimulus involves relatively independent processes of recollection and familiarity (e.g., Mandler, 1980). Recollection refers to the conscious reexperience of a recent event (or stimulus), while familiarity is the feeling of having previously encountered an event (or stimulus) with no associated contextual information. Westerberg et al. (2006) compared recognition memory under yes/no recognition and forced-choice recognition procedures with the assumption that forced-choice was more likely to be mediated by familiarity. They found that amnestic MCI patients were impaired in the yes/no condition, but normal in the forced-choice condition, suggesting that they have impaired recollection with relatively preserved familiarity (also see Bennett, Golob, Parker, & Starr, 2006). The same conclusion was reached in a study that used a process dissociation procedure to separate the effects of recollection and familiarity in amnestic MCI (Anderson et al.,

2008). In this study, analysis of recognition for words presented either auditorally or visually was tested under two sets of instructions: (1) to identify repeated words regardless of presentation modality or (2) to identify repeated words only if they were repeated in the original modality. By comparing the two conditions, the ability to recognize a word in its original context (recollection) or independent of its context (familiarity) could be contrasted. Patients with amnestic MCI showed impaired recollection and normal familiarity under these conditions (also see Serra et al., 2010). Finally, a study that used a Remember/Know paradigm in which subjects indicated if their recognition of an event (or stimulus) was based (Remember/recollection) or not based (Know/familiarity) on remembering contextual information showed that recollection was impaired and familiarity was spared in patients with amnestic MCI (Hudon, Belleville, & Gauthier, 2009; also see Serra et al., 2010).

The majority of studies of the contributions of recollection and familiarity to recognition memory in patients with amnestic MCI suggest that recollection is impaired and familiarity is spared (but see Algarabel et al., 2009). This conclusion is bolstered by studies that show normal feeling-of-knowing judgments (Anderson & Schmitter-Edgecombe, 2010; but see Perrotin, Belleville, & Isingrini, 2007), susceptibility to false recognition lures (Hudon et al., 2006), and incidental recognition of shallowly encoded words (Mandzia, McAndrews, Grady, Graham, & Black, 2009) in patients with amnestic MCI. Each of these processes is thought to depend upon familiarity to some degree. There are, however, several studies that show impaired familiarity in patients with amnestic MCI using some of the same procedures described above (e.g., Ally, McKeever, Waring, & Budson, 2009; Wolk, Signoff, & Dekosky, 2008). These discrepant results could be related to different degrees of extra-hippocampal pathology related to more advanced disease and suggest the need for further study.

Deficits in Prospective Memory

A number of recent studies have investigated the impact of amnestic MCI on the ability to remember a delayed intention to act at a certain time or when some external event occurs in the future (Costa et al., 2010; Karantzoulis, Troyer, & Rich, 2009; Schmitter-Edgecombe, Woo, & Greeley, 2009; Thompson, Henry, Rendell, Withall, & Brodaty, 2010; Troyer & Murphy, 2007). This form of prospective memory is essential for carrying out activities critical for independent living such as remembering to pay bills on a certain date, remembering to take medication at a certain time of day, or remembering to make a turn when a particular landmark is spotted. Various components of prospective memory are thought to involve separate cognitive functions that could be affected by damage to distinct brain regions. For example, an episodic memory component is involved in remembering the specific act to be performed at the appropriate time, while an executive and attention-related component supports cognitive operations such as monitoring the passage of time, planning sequential activities, and switching from an on-going activity to the intended activity.

Studies of prospective memory have shown that both time-based and event-based tasks are impaired in patients with amnestic MCI (Costa et al., 2010; Karantzoulis et al., 2009; Schmitter-Edgecombe et al., 2009; Thompson et al., 2010; Troyer & Murphy, 2007). Several of these studies also indicate that time-based prospective memory is more impaired than event-based (Karantzoulis et al., 2009; Troyer & Murphy, 2007). Because time-based prospective memory tasks place greater demands on executive function and attention (e.g., self-initiation, time monitoring) than event-based tasks, this pattern of results suggests that deficits in both hippocampus-dependent and frontal cortex-dependent functions contribute to the prospective memory deficit in patients with amnestic MCI. This possibility is supported by a study that showed that patients with MCI were more impaired on recall of the intention to act (an executive-based function) than on execution of the action (a memory-based function), and this discrepancy was greater in amnestic MCI patients with additional executive impairment than in those without additional impairment (Costa et al., 2010). A study that showed a correlation between prospective memory performance and other frontal cortex mediated functions (e.g., temporal order memory, source memory) provides additional support (Schmitter-Edgecombe et al., 2009). Taken together, these findings suggest that time-based prospective memory impairment in patients with amnestic MCI may reflect a more advanced neurodegenerative process that may help to predict imminent development of dementia.

Deficits in Remote Memory

A number of studies indicate that the ability to remember past events that were successfully remembered prior to the onset of the disease occurs in patients with amnestic MCI. Leyhe and colleagues, for example, showed that patients with amnestic MCI were impaired on a test of memory for historic public events that occurred over the past 60 years (Leyhe, Muller, Eschweiler, & Saur, 2010), and on a test of remote autobiographical memory (Leyhe, Muller, Milian, Eschweiler, & Saur, 2009). A temporal gradient was observed for the autobiographical information with older remote memories better retained than newer remote memories. Poettrich and colleagues (2009) found the pattern of activation observed with functional MRI during recall of remote autobiographical events involved a left-lateralized network of frontal, temporal, and parietal areas and the medial frontal cortex in normal control subjects, but also included the symmetrical activation of these areas on the right in patients with MCI.

This spread of activation may indicate a compensatory response to maintain remote memory performance in patients with MCI.

While studies consistently show that remote autobiographical memory is impaired in amnestic MCI, there is less consistency in the evidence for a loss of personal semantics (i.e., one's own name, names of relatives, one's date of birth). Murphy and colleagues (2008) found that personal semantics were preserved in amnestic MCI, but Irish et al. (2010) found that they were impaired. The inconsistency in these findings could be related to degree of spread of AD pathology from the hippocampus, which mediates remote autobiographical memory, to temporal association cortices, which mediate semantic memory. If this is the case, the presence of impairment in both remote autobiographical memory and personal semantics could indicate an increased likelihood of progression to dementia.

Subtle deficits in nonautobiographical semantic memory occur in patients with amnestic MCI and may be an indication of imminent progression to dementia (e.g., Dudas, Clague, Thompson, Graham, & Hodges, 2005; Joubert et al., 2010; Thompson, Graham, Patterson, Sahakian, & Hodges, 2002). Semantic memory refers to our general fund of knowledge, which consists of the meanings and representations of words, concepts, and overlearned facts that are not dependent upon contextual cues for their retrieval. Semantic knowledge is usually assumed to be organized as a complex associative network of concepts stored in a distributed manner in neocortical association areas of the temporal and parietal lobes. Semantic memory is not dependent upon the medial temporal lobe structures that are important for episodic memory. Semantic memory deficits in patients with amnestic MCI are demonstrated by their poor performance on semantically demanding naming tasks that require producing proper nouns such as the names of famous people or buildings (Adlam, Bozeat, Arnold, Watson, & Hodges, 2006; Ahmed, Arnold, Thompson, Graham, & Hodges, 2008; Borg, Thomas-Atherion, Bogey, Davier, & Laurent, 2010; Joubert et al., 2010; Seidenberg et al., 2009) and their impaired ability to generate exemplars from a specific semantic category (e.g., "animals") (Adlam et al., 2006; Biundo et al., 2011; Murphy, Rich, & Troyer, 2006). Patients with amnestic MCI also fail to benefit in a normal fashion from deep semantic encoding in episodic memory tasks (Hudon, Villeneuve, & Belleville, 2011), and they have an enhanced tendency to produce prototypical intrusion errors during free recall (Libon et al., 2011).

Deficits in Working Memory

A deficit in working memory has been observed in patients with preclinical AD (Grober et al., 2008; Rapp & Reischies, 2005) or amnestic MCI (e.g., Darby, Maruff, Collie, & McStephen, 2002; Gagnon & Belleville, 2011; Saunders & Summers, 2011; Sinai, Phillips, Chertkow, & Kabani, 2010).

Working memory refers to a limited capacity memory system in which information that is the immediate focus of attention can be temporarily held in limited-capacity language-based or visual-based buffers while being manipulated through a primary central executive system (Baddeley, 1986). The working memory deficit in MCI is consistent with that seen in early AD (Baddeley, Bressi, Della Sala, Logie, & Spinnler, 1991; Collette, Van der Linden, Bechet, & Salmon, 1999) and is usually mild and limited to disruption of the central executive due to a decline in the control of attention. Consistent with this interpretation, patients with amnestic MCI are impaired on divided attention tasks and this impairment grows during the transition to AD dementia (Bellieville, Chertkow, & Gauthier, 2007; Saunders & Summers, 2011).

Other aspects of "executive" functions (e.g., concurrent mental manipulation of information, concept formation, problem solving, cue-directed behavior) can also be impaired in patients with MCI or preclinical AD (Albert et al., 2007; Brandt et al., 2009; Mickes et al., 2007). Balota et al. (2010) compared the performances of nondemented elderly individuals with or without preclinical AD on the Stroop interference task that timed how quickly they could read words that are the names of colors (i.e., color words), name the color of ink patches (or a series of three x's), and name the color of the ink in which noncongruent color words were printed (i.e., say "red" when the word green is printed in red ink). The results showed that preclinical AD patients made more errors than control subjects on the noncongruent trials and had a larger Stroop effect (i.e., difference between congruent and noncongruent trial reaction times) consistent with a decline in inhibitory control (also see Belanger, Belleville, & Gauthier, 2010). Patients with amnestic MCI also exhibit deficits on other tests of inhibitory control such as the Hayling task, which requires them to complete a sentence with a nondominant response (e.g., He hit the nail with a____.; Belanger & Belleville, 2009), an expectancy violation task in which they must rapidly judge the relatedness of coordinate word pairs (e.g., apple-pear) that occasionally (i.e., unexpectedly) occur in the context of judging category word pairs (e.g., peach-fruit) (Davie et al., 2004), and a digit-number identification cognitive set switching task (Sinai et al., 2010).

ASYMMETRIC COGNITIVE DECLINE IN PRECLINICAL AD

In addition to deficits on tests of episodic, semantic, and working memory, subtle cognitive changes during the preclinical phase of AD can be detected as an asymmetric profile of performance across tests in cognitive domains that remain relatively intact. It has been known for some time that lateralized cognitive deficits (e.g., greater verbal than visuo-spatial deficits or vice versa) are apparent in subgroups of mildly demented AD patients (e.g., Haxby, Duara, Grady, Cutler, & Rapoport, 1985). Based on these findings, Jacobson et al. (2002) postulated that subtle asymmetric cognitive decline due to AD might be detectable during the preclinical period in a subset of individuals, even if their performance in all cognitive domains remained above normal. Subtle deficits might be detectable because the individual would be essentially serving as their own control. Jacobson et al. (2002) compared 20 cognitively normal elderly adults who were in a preclinical phase of AD (i.e., they were diagnosed with AD approximately 1 year later) and 20 age- and education-matched normal control subjects on a number of cognitive tests and derived a score that reflected cognitive asymmetry: the absolute difference between verbal (Boston Naming Test) and visuo-spatial (Block Design Test) standardized scores. The results showed that the two groups performed similarly on individual tests of memory, language, and visuo-spatial ability. In contrast, cognitive asymmetry (in either direction) was significantly greater in the preclinical AD patients than in normal controls. These results support the notion that there is a subgroup of patients with preclinical AD who have asymmetric cognitive decline that is apparent even when the absolute level of performance in all cognitive domains is normal. Consideration of nonmemory asymmetric changes, in addition to consideration of subtle declines in memory, might improve the ability to detect AD in its earliest stages.

SUMMARY AND CONCLUSIONS

In summary, the neuropsychological studies reviewed above clearly indicate that there is a preclinical phase of detectable cognitive decline that precedes the clinical diagnosis of AD. The detection of incipient dementia is most effectively accomplished with sensitive measures of learning and memory. Identification of the initial cognitive changes of AD can help to reliably detect the disease in its earliest stages and provide a target for very early symptomatic treatment. Advances in the neuropsychological detection of preclinical AD have occurred in parallel with advances in detecting reliable biological markers of the disease. It is now possible to use MRI to detect reductions in hippocampal volume and cortical thickness typically associated with AD, to use PET imaging with PIB or similar agents to detect the deposition of $A\beta$ in the brain, and to use biochemical assays of CSF to detect abnormal levels of $A\beta$ and tau proteins. The combined use of neuropsychological assessment and biomarkers of AD may be particularly effective in reliably detecting the disease in its earliest stages (Nordlund et al., 2008).

References

Adlam, A. L., Bozeat, S., Arnold, R., Watson, P., & Hodges, J. R. (2006). Semantic knowledge in mild cognitive impairment and mild Alzheimer's disease. *Cortex, 42*, 675–684.

Ahmed, S., Arnold, R., Thompson, S. A., Graham, K. S., & Hodges, J. R. (2008). Naming of objects, faces and buildings in mild cognitive impairment. *Cortex, 44*, 746–752.

Albert, M. S., Blacker, D., Moss, M. B., Tanzi, R., & McArdle, J. J. (2007). Longitudinal change in cognitive performance among individuals with mild cognitive impairment. *Neuropsychology, 21*, 158–169.

Albert, M. S., Dekosky, S. T., Dickson, D., Dubois, B., Feldman, H. H., Fox, N. C., et al. (2011). The diagnosis of mild cognitive impairment due to Alzheimer's disease: Recommendations from the National Institute on Aging-Alzheimer's Association Workgroups on diagnostic guidelines for Alzheimer's disease. *Alzheimer's & Dementia, 7*, 270–279.

Albert, M. S., Moss, M. B., Tanzi, R., & Jones, K. (2001). Preclinical prediction of AD using neuropsychological tests. *Journal of the International Neuropsychological Society, 7*, 631–639.

Algarabel, S., Escudero, J., Mazon, J. F., Pitarque, A., Fuentes, M., Peset, V., et al. (2009). Familiarity-based recognition in young, healthy elderly, mild cognitive impaired and Alzheimer's patients. *Neuropsychologia, 47*, 2056–2064.

Ally, B. A., McKeever, J. D., Waring, J. D., & Budson, A. E. (2009). Preserved frontal memorial processing for pictures in patients with mild cognitive impairment. *Neuropsychologia, 47*, 2044–2055.

Anderson, J. W., & Schmitter-Edgecombe, M. (2010). Mild cognitive impairment and feeling-of-knowing in episodic memory. *Journal of Clinical and Experimental Neuropsychology, 32*, 505–514.

Anderson, N. D., Ebert, P. L., Jennings, J. M., Grady, C. L., Cabezza, R., & Graham, S. (2008). Recollection- and familiarity-based memory in healthy aging and amnestic mild cognitive impairment. *Neuropsychology, 22*, 177–187.

Andrews-Hanna, J. R., Snyder, A. Z., Vincent, J. L., et al. (2007). Disruption of large-scale brain systems in advanced aging. *Neuron, 56*, 924–935.

Backman, L., Small, B. J., & Fratiglioni, L. (2001). Stability of the preclinical episodic memory deficit in Alzheimer's disease. *Brain, 124*, 96–102.

Baddeley, A. D. (1986). *Working memory*. Oxford: Claredon Press.

Baddeley, A. D., Bressi, S., Della Sala, S., Logie, R., & Spinnler, H. (1991). The decline of working memory in Alzheimer's disease: A longitudinal study. *Brain, 114*, 2521–2542.

Balota, D. A., Tse, C., Hutchison, K. A., Spieler, D. H., Duchek, J. M., & Morris, J. C. (2010). Predicting conversion to dementia of the Alzheimer's type in a healthy control sample: The power of errors in Stroop color naming. *Psychology and Aging, 25*, 208–218.

Band, G. P., Ridderinkhof, K. R., & Segalowitz, S. (2002). Explaining neurocognitive aging: Is one factor enough? *Brain and Cognition, 49*, 259–267.

Bartzokis, G. (2004). Age-related myelin breakdown: A developmental model of cognitive decline and Alzheimer's disease. *Neurobiology of Aging, 25*, 5–18.

Belanger, S., & Belleville, S. (2009). Semantic inhibition impairment in mild cognitive impairment: A distinctive feature of upcoming cognitive decline? *Neuropsychology, 23*, 592–606.

Belanger, S., Belleville, S., & Gauthier, S. (2010). Inhibition impairments in Alzheimer's disease, mild cognitive impairment and healthy aging: Effect of congruency proportion in a Stroop task. *Neuropsychologia, 48*, 581–590.

Bellieville, S., Chertkow, H., & Gauthier, S. (2007). Working memory and control of attention in persons with Alzheimer's disease and mild cognitive impairment. *Neuropsychology, 21*, 458–469.

Bennett, I. J., Golob, E. J., Parker, E. S., & Starr, A. (2006). Memory evaluation in mild cognitive impairment using recall and recognition tests. *Journal of Clinical and Experimental Neuropsychology, 28*, 1408–1422.

Biundo, R., Gardini, S., Caffarra, P., Concari, L., Martorana, D., Neri, T. M., et al. (2011). Influence of APOE status on lexical-semantic skills in mild cognitive impairment. *Journal of the International Neuropsychological Society*, 17, 423–430.

Bondi, M. W., Houston, W. S., Eyler, L. T., & Brown, G. G. (2005). FMRI evidence of compensatory mechanisms in older adults at genetic risk for Alzheimer's disease. *Neurology*, 64, 501–508.

Bookheimer, S. Y., Strojwas, M. H., Cohen, M. S., Saunders, A. M., Pericak-Vance, M. A., Mazziotta, J. C., et al. (2000). Patterns of brain activation in people at risk for Alzheimer's disease. *The New England Journal of Medicine*, 343, 450–456.

Borg, C., Thomas-Atherion, C., Bogey, S., Davier, K., & Laurent, B. (2010). Visual imagery processing and knowledge of famous names in Alzheimer's disease and MCI. *Aging, Neuropsychology and Cognition*, 17, 603–614.

Braak, H., & Braak, E. (1991). Neuropathological staging of Alzheimer-related changes. *Acta Neuropathologica*, 82, 239–259.

Brandt, J., Aretouli, E., Neijstrom, E., Samek, J., Manning, K., Albert, M. S., et al. (2009). Selectivity of executive function deficits in mild cognitive impairment. *Neuropsychology*, 23, 607–618.

Buckner, R. L. (2004). Memory and executive function in aging and AD: Multiple factors that cause decline and reserve factors that compensate. *Neuron*, 44, 195–208.

Chen, P., Ratcliff, G., Belle, S. H., Cauley, J. A., DeKosky, S. T., & Ganguli, M. (2001). Patterns of cognitive decline in presymptomatic Alzheimer disease: A prospective community study. *Archives of General Psychiatry*, 58, 853–858.

Chételat, G., Villemagne, V. L., Villain, N., Jones, G., Ellis, K. A., Ames, D., et al. (2012). Accelerated cortical atrophy in cognitively normal elderly with high β-amyloid deposition. *Neurology*, 78, 477–484.

Collette, F., Van der Linden, M., Bechet, S., & Salmon, E. (1999). Phonological loop and central executive functioning in Alzheimer's disease. *Neuropsychologia*, 37, 905–918.

Costa, A., Perri, R., Serra, L., Barban, F., Gatto, I., Zabberoni, S., et al. (2010). Prospective memory functioning in mild cognitive impairment. *Neuropsychology*, 24, 327–335.

Darby, D., Maruff, P., Collie, A., & McStephen, M. (2002). Mild cognitive impairment can be detected by multiple assessments in a single day. *Neurology*, 59, 1042–1046.

Davie, J. E., Azuma, T., Goldinger, S. D., Connor, D. J., Sabbagh, M. N., & Silverberg, N. B. (2004). Sensitivity to expectancy violations in healthy aging and mild cognitive impairment. *Neuropsychology*, 18, 269–275.

de Rover, M., Pironti, V. A., McCabe, J. A., Acosta-Cabronero, J., Arana, F. S., Morein-Zamir, S., et al. (2011). Hippocampal dysfunction in patients with mild cognitive impairment: A functional neuroimaging study of a visuospatial paired associates learning task. *Neuropsychologia*, 49, 2060–2070.

Doraiswamy, P. M., Sperling, R. A., Coleman, R. E., Johnson, K. A., Reiman, E. M., Davis, M. D., et al. (2012). Amyloid-β assessed by florbetapir F 18 PET and 18-month cognitive decline: A multicenter study. *Neurology*, 79, 1636–1644.

Dubois, B., Feldman, H., Jacova, C., Dekosky, S. T., Barberger-Gateau, P., Cummings, J., et al. (2007). Research criteria for the diagnosis of Alzheimer's disease: Revising the NINCDS-ADRDA criteria. *Lancet Neurology*, 6, 734–746.

Dudas, R. B., Clague, F., Thompson, S. A., Graham, K. S., & Hodges, J. R. (2005). Episodic and semantic memory in mild cognitive impairment. *Neuropsychologia*, 43, 1266–1276.

Eichenbaum, H. (1997). How does the brain organize memories? *Science*, 277, 333–335.

Fuld, P. A., Masur, D. M., Blau, A. D., Crystal, H., & Aronson, M. K. (1990). Object-memory evaluation for prospective detection of dementia in normal functioning elderly: Predictive and normative data. *Journal of Clinical and Experimental Neuropsychology*, 12, 520–528.

Gagnon, L. G., & Belleville, S. (2011). Working memory in mild cognitive impairment and Alzheimer's disease: Contribution of forgetting and predictive value of complex span tasks. *Neuropsychology*, 25, 226–236.

Gomar, J. J., Bobes-Bascaran, M. T., Conejero-Goldberg, C., Davies, P., & Goldberg, T. E. (2011). Utility of combinations of biomarkers, cognitive markers, and risk factors to predict conversion from mild cognitive impairment to Alzheimer disease in patients in the Alzheimer's Disease Neuroimaging Initiative. *Archives of General Psychiatry, 68*, 961–969.

Grady, C. L., & Craik, F. I. (2000). Changes in memory processing with age. *Current Opinion in Neurobiology, 10*, 224–231.

Greenwood, P. M. (2000). The frontal aging hypothesis evaluated. *Journal of the International Neuropsychological Society, 6*, 705–726.

Grober, E., Hall, C. B., Lipton, R. B., Zonderman, A. B., Resnick, S. M., & Kawas, C. (2008). Memory impairment, executive dysfunction, and intellectual decline in preclinical Alzheimer's disease. *Journal of the International Neuropsychological Society, 14*, 266–278.

Grober, E., & Kawas, C. (1997). Learning and retention in preclinical and early Alzheimer's disease. *Psychology and Aging, 12*, 183–188.

Han, S. D., Houston, W. S., Jak, A. J., Eyler, L. T., Nagel, B. J., Fleisher, A. S., et al. (2007). Verbal paired-associate learning by APOE genotype in non-demented older adults: fMRI evidence of a right hemisphere compensatory response. *Neurobiology of Aging, 28*, 238–247.

Haxby, J. V., Duara, R., Grady, C. L., Cutler, N. R., & Rapoport, S. I. (1985). Relations between neuropsychological and cerebral metabolic asymmetries in early Alzheimer's disease. *Journal of Cerebral Blood Flow & Metabolism, 5*, 193–200.

Hedden, T., & Gabrieli, J. D. E. (2004). Insights into the ageing mind: A view from cognitive neuroscience. *Nature Reviews, 5*, 87–97.

Howieson, D. B., Dame, A., Camicioli, R., Sexton, G., Payami, H., & Kaye, J. A. (1997). Cognitive markers preceding Alzheimer's dementia in the healthy oldest old. *Journal of the American Geriatrics Society, 45*, 584–589.

Hudon, C., Belleville, S., & Gauthier, S. (2009). The assessment of recognition memory using the remember/know procedure in amnestic mild cognitive impairment and probable Alzheimer's disease. *Brain and Cognition, 70*, 171–179.

Hudon, C., Belleville, S., Souchay, C., Gely-Nargeot, M., Chertkow, H., & Gauthier, S. (2006). Memory for gist and detail information in Alzheimer's disease and mild cognitive impairment. *Neuropsychology, 20*, 566–577.

Hudon, C., Villeneuve, S., & Belleville, S. (2011). The effect of orientation at encoding on free-recall performance in amnestic mild cognitive impairment and probable Alzheimer's disease. *Journal of Clinical and Experimental Neuropsychology, 33*, 631–638.

Huijbers, W., Mormino, E. C., Wigman, S. E., Ward, A. M., Vannini, P., McLaren, D. G., et al. (2014). Amyloid deposition is linked to aberrant entorhinal activity among cognitively normal older adults. *The Journal of Neuroscience, 34*, 5200–5210.

Irish, M., Lawlor, B. A., O'Mara, S. M., & Coen, R. F. (2010). Exploring the recollective experience during autobiographical memory retrieval in amnestic mild cognitive impairment. *Journal of the International Neuropsychological Society, 16*, 546–555.

Jack, C. R., & Holtzman, D. M. (2013). Biomarker modeling of Alzheimer's disease. *Neuron, 80*, 1347–1358.

Jack, C. R., Knopman, D. S., Jagust, W. J., Shaw, L. M., Aisen, P. S., Weiner, M., et al. (2010). Hypothetical model of dynamic biomarkers of the Alzheimer's pathological cascade. *Lancet Neurology, 9*, 119–128.

Jacobs, D. M., Sano, M., Dooneief, G., Marder, K., Bell, K. L., & Stern, Y. (1995). Neuropsychological detection and characterization of preclinical Alzheimer's disease. *Neurology, 45*, 957–962.

Jacobson, M. W., Delis, D. C., Bondi, M. W., & Salmon, D. P. (2002). Do neuropsychological tests detect preclinical Alzheimer's disease? Individual-test versus cognitive discrepancy analyses. *Neuropsychology, 16*, 132–139.

Joubert, S., Brambati, S. M., Ansado, J., Barbeau, E. J., Felician, O., Didac, M., et al. (2010). The cognitive and neural expression of semantic memory impairment in mild cognitive impairment and early Alzheimer's disease. *Neuropsychologia, 48*, 978–988.

Karantzoulis, S., Troyer, A. K., & Rich, J. B. (2009). Prospective memory in amnestic mild cognitive impairment. *Journal of the International Neuropsychological Society, 15,* 407–415.

Katzman, R. (1994). Apolipoprotein E and Alzheimer's disease. *Current Opinion in Neurobiology, 4,* 703–707.

Kawas, C. H., Corrada, M. M., Brookmeyer, R., Morrison, A., Resnick, S. M., Zonderman, A. B., et al. (2003). Visual memory predicts Alzheimer's disease more than a decade before diagnosis. *Neurology, 60,* 1089–1093.

Klein, W. L. (2006). Synaptic targeting by A beta oligomers (ADDLS) as a basis for memory loss in early Alzheimer's disease. *Alzheimer's & Dementia, 2,* 43–55.

Kramer, A. F., Bherer, L., Colcombe, S. J., Dong, W., & Greenough, W. T. (2004). Environmental influences on cognitive and brain plasticity during aging. *Biological Sciences and Medical Sciences, 59,* 940–957.

Landau, S. M., Harvey, D., Madison, C. M., Reiman, E. M., Foster, N. L., Aisen, P. S., et al. (2010). Comparing predictors of conversion and decline in mild cognitive impairment. *Neurology, 75,* 230–238.

Lange, K. L., Bondi, M. W., Salmon, D. P., Galasko, D., Delis, D. C., Thomas, R. G., et al. (2002). Decline in verbal memory during preclinical Alzheimer's disease: Examination of the effect of APOE genotype. *Journal of the International Neuropsychological Society, 8,* 943–955.

Leyhe, T., Muller, S., Eschweiler, G. W., & Saur, R. (2010). Deterioration of the memory for historic events in patients with mild cognitive impairment and early Alzheimer's disease. *Neuropsychologia, 48,* 4093–4101.

Leyhe, T., Muller, S., Milian, M., Eschweiler, G. W., & Saur, R. (2009). Impairment of episodic and semantic autobiographical memory in patients with mild cognitive impairment and early Alzheimer's disease. *Neuropsychologia, 35,* 547–557.

Libon, D. J., Bondi, M. W., Price, C. C., Lamar, M., Eppig, J., Wambach, D. M., et al. (2011). Verbal serial list learning in mild cognitive impairment: A profile analysis of interference, forgetting, and errors. *Journal of the International Neuropsychological Society, 17,* 905–914.

Lim, Y. Y., Ellis, K. A., Pietrzak, R. H., Ames, D., Darby, D., Harrington, K., et al. (2012). Stronger effect of amyloid load than APOE genotype on cognitive decline in healthy older adults. *Neurology, 79,* 1645–1652.

Lim, Y. Y., Maruff, P., Pietrzak, R. H., Ellis, K. A., Darby, D., Ames, D., et al. (2014). Aβ and cognitive change: Examining the preclinical and prodromal stages of Alzheimer's disease. *Alzheimer's & Dementia, 10,* 743–751.

Linn, R. T., Wolf, P. A., Bachman, D. L., Knoefel, J. E., Cobb, J. L., et al. (1995). The 'preclinical phase' of probable Alzheimer's disease. *Archives of Neurology, 52,* 485–490.

Mandler, G. (1980). Recognizing: The judgment of previous occurrence. *Psychological Review, 87,* 252–271.

Mandzia, J. L., McAndrews, M. P., Grady, C. L., Graham, S. J., & Black, S. E. (2009). Neural correlates of incidental memory in mild cognitive impairment: An fMRI study. *Neurobiology of Aging, 30,* 717–730.

Manes, F., Serrano, C., Calcagno, M. L., Cardozo, J., & Hodges, J. R. (2008). Accelerated forgetting in subjects with memory complaints. *Journal of Neurology, 255,* 1067–1070.

Marr, D. (1971). Simple memory: A theory for archicortex. *Philosophical Transactions of the Royal Society of London. Series B, Biological sciences, 262,* 23–81.

Masur, D. M., Sliwinski, M., Lipton, R. B., Blau, A. D., & Crystal, H. A. (1994). Neuropsychological prediction of dementia and the absence of dementia in healthy elderly persons. *Neurology, 44,* 1427–1432.

McKhann, G. M., Knopman, D. S., Chertkow, H., Hyman, B. T., Jack, C. R., Kawas, C. H., et al. (2011). The diagnosis of dementia due to Alzheimer's disease: Recommendations from the National Institute on Aging-Alzheimer's Association workgroups on diagnostic guidelines for Alzheimer's disease. *Alzheimer's & Dementia, 7,* 263–269.

Mickes, L., Wixted, J. T., Fennema-Notestine, C., Galasko, D., Bondi, M. W., Thal, L. J., et al. (2007). Progressive impairment on neuropsychological tasks in a longitudinal study of preclinical Alzheimer's disease. *Neuropsychology, 21,* 696–705.

Mormino, E. C., Betensky, R. A., Hedden, T., Schultz, A. P., Amariglio, R. E., Rentz, D. M., et al. (2014). Synergistic effect of β-amyloid and neurodegeneration on cognitive decline in clinically normal individuals. *JAMA Neurology, 71*, 1379–1385.

Murphy, K. J., Rich, J. B., & Troyer, A. K. (2006). Verbal fluency patterns in amnestic mild cognitive impairment are characteristic of Alzheimer's type dementia. *Journal of the International Neuropsychological Society, 12*, 570–574.

Murphy, K. J., Troyer, A. K., Levine, B., & Moscovitch, M. (2008). Episodic, but not semantic, autobiographical memory is reduced in amnestic mild cognitive impairment. *Neuropsychologia, 46*, 3116–3123.

Nordlund, A., Rolstad, S., Klang, O., Lind, K., Pedersen, M., Blennow, K., et al. (2008). Episodic memory and speed/attention deficits are associated with Alzheimer-typical CSF abnormalities in MCI. *Journal of the International Neuropsychological Society, 14*, 582–590.

Nunez, P. L., & Srinivasan, R. (2006). *Electric fields of the brain: The neurophysics of EEG* (2nd ed.) (pp. 163–166). New York: Oxford University Press.

Olichney, J., Morris, S., Ochoa, C., Salmon, D., Thal, L., Kutas, M., et al. (2002). Abnormal verbal event related potentials in mild cognitive impairment and incipient AD. *Journal of Neurology, Neurosurgery, and Psychiatry, 73*, 377–384.

Olichney, J. M., Pak, J., Salmon, D. P., Yang, J. C., Gahagan, T., Nowacki, R., et al. (2013). Abnormal P600 word repetition effect in elderly persons with preclinical Alzheimer's disease. *Cognitive Neuroscience, 4*, 143–151.

Olichney, J. M., Taylor, J. R., Gatherwright, J., Salmon, D. P., Bressler, A. J., Kutas, M., et al. (2008). Patients with MCI and N400 or P600 abnormalities are at very high risk for conversion to dementia. *Neurology, 70*, 1763–1770.

Olichney, J. M., Van Petten, C., Paller, K. A., Salmon, D. P., Iragui, V. J., & Kutas, M. (2000). Word repetition in amnesia: Electrophysiological measures of impaired and spared memory. *Brain, 123*, 1948–1963.

O'Sullivan, M., Jones, D. K., Summers, P. E., Morris, R. G., Williams, S. C., & Markus, H. S. (2001). Evidence for cortical "disconnection" as a mechanism of age-related cognitive decline. *Neurology, 57*, 632–638.

Paller, K. A., & Kutas, M. (1992). Brain potentials during memory retrieval provide neurophysiological support for the distinction between conscious recollection and priming. *Journal of Cognitive Neuroscience, 4*, 375–391.

Park, H. L., O'Connell, J. E., & Thomson, R. G. (2003). A systematic review of cognitive decline in the general elderly population. *International Journal of Geriatric Psychiatry, 18*, 1121–1134.

Perri, R., Serra, L., Carlesimo, G. A., & Caltagirone, C. (2007). Amnestic mild cognitive impairment: Difference of memory profile in subjects who converted or did not convert to Alzheimer's disease. *Neuropsychology, 21*, 549–558.

Perrotin, A., Belleville, S., & Isingrini, M. (2007). Metamemory monitoring in mild cognitive impairment: Evidence of a less accurate episodic feeling-of-knowing. *Neuropsychologia, 45*, 2811–2826.

Pfefferbaum, A., Adalsteinsson, E., & Sullivan, E. (2005). Frontal circuitry degradation marks healthy adult aging: Evidence from diffusion tensor imaging. *Neuroimage, 26*, 891–899.

Pike, K. E., Rowe, C. C., Moss, S. A., & Savage, G. (2008). Memory profiling with paired associate learning in Alzheimer's disease, mild cognitive impairment, and healthy aging. *Neuropsychology, 22*, 718–728.

Poettrich, K., Weiss, P. H., Werner, A., Lux, S., Donix, M., Gerber, J., et al. (2009). Altered neural network supporting declarative long-term memory in mild cognitive impairment. *Neurobiology of Aging, 30*, 284–298.

Rapp, M. A., & Reischies, F. M. (2005). Attention and executive control predict Alzheimer's disease in late life: Results from the Berlin aging study (BASE). *The American Journal of Geriatric Psychiatry, 13*, 134–141.

Raz, N. (2005). The aging brain observed *in vivo*: Differential changes and their modifiers. In R. Cabeza, L. Nyberg, & D. Park (Eds.), *Cognitive neuroscience of aging: Linking cognitive and cerebral aging* (pp. 19–57). New York: Oxford University Press.

Rentz, D. M., Amariglio, R. E., Becker, J. A., Frey, M., Olson, L. E., Frishe, K., et al. (2011). Face-name associative memory performance is related to amyloid burden in normal elderly. *Neuropsychologia, 49*, 2776–2783.

Salmon, D. P., & Bondi, M. W. (2009). Neuropsychological assessment of dementia. *Annual Review of Psychology, 60*, 257–282.

Salmon, D. P., & Squire, L. R. (2009). The neuropsychology of memory dysfunction and its assessment. In I. Grant & K. Adams (Eds.), *Neuropsychological assessment of neuropsychiatric & neuromedical disorders* (3rd ed.). New York: Oxford University press.

Salthouse, T. A. (1996). The processing-speed theory of adult age differences in cognition. *Psychological Review, 103*, 403–428.

Saunders, N. L. J., & Summers, M. J. (2011). Longitudinal deficits to attention, executive, and working memory in subtypes of mild cognitive impairment. *Neuropsychology, 25*, 237–248.

Schmitter-Edgecombe, M., Woo, E., & Greeley, D. R. (2009). Characterizing multiple memory deficits and their relation to everyday functioning in individuals with mild cognitive impairment. *Neuropsychology, 23*, 168–177.

Seidenberg, M., Guidotti, L., Nielson, K. A., Woodard, J. L., Durgerian, S., Zhang, Q., et al. (2009). Semantic knowledge for famous names in mild cognitive impairment. *Journal of the International Neuropsychological Society, 15*, 9–18.

Selkoe, D. J. (2002). Alzheimer's disease is a synaptic failure. *Science, 298*, 789–791.

Serra, L., Bozzali, M., Cercignani, M., Perri, R., Fadda, L., Caltagirone, C., et al. (2010). Recollection and familiarity in amnestic mild cognitive impairment. *Neuropsychology, 24*, 316–326.

Sinai, M., Phillips, N. A., Chertkow, H., & Kabani, N. J. (2010). Task switching performance reveals heterogeneity amongst patients with mild cognitive impairment. *Neuropsychology, 24*, 757–774.

Small, B. J., Fratiglioni, L., Viitanen, M., Winblad, B., & Bäckman, L. (2000). The course of cognitive impairment in preclinical Alzheimer disease: Three- and 6-year follow-up of a population-based sample. *Archives of Neurology, 57*, 839–844.

Small, B. J., Mobly, J. L., Laukka, E. J., Jones, S., & Backman, L. (2003). Cognitive deficits in preclinical Alzheimer's disease. *Acta Neurologica Scandinavica Supplementum, 179*, 29–33.

Smith, G. E., Pankratz, V. S., Negash, S., et al. (2007). A plateau in pre-Alzheimer memory decline: Evidence for compensatory mechanisms? *Neurology, 69*, 133–139.

Sperling, R. A., Aisen, P. S., Beckett, L. A., Bennett, D. A., Craft, S., Fagan, A. M., et al. (2011). Toward defining the preclinical stages of Alzheimer's disease: Recommendations from the National Institute on Aging-Alzheimer's Association workgroups on diagnostic guidelines for Alzheimer's disease. *Alzheimer's & Dementia, 7*, 280–292.

Terry, R. D., Masliah, E., Salmon, D. P., Butters, N., DeTeresa, R., Hill, R., et al. (1991). Physical basis of cognitive alterations in Alzheimer's disease: Synapse loss is the major correlate of cognitive impairment. *Annals of Neurology, 30*, 572–580.

Thompson, C., Henry, J. D., Rendell, P. G., Withall, A., & Brodaty, H. (2010). Prospective memory function in mild cognitive impairment. *Journal of the International Neuropsychological Society, 16*, 318–325.

Thompson, S. A., Graham, K. S., Patterson, K., Sahakian, B. J., & Hodges, J. R. (2002). Is knowledge of famous people disproportionately impaired in patients with early and questionable Alzheimer's disease? *Neuropsychology, 16*, 344–358.

Troyer, A. K., & Murphy, K. J. (2007). Memory for intentions in amnestic mild cognitive impairment: Time- and event-based prospective memory. *Journal of the International Neuropsychological Society, 13*, 365–369.

Troyer, A. K., Murphy, K. J., Anderson, N. D., Hayman-Abello, B. A., Craik, F. I. M., & Moscovitch, M. (2008). Item and associative memory in amnestic mild cognitive impairment: Performance on standardized memory tests. *Neuropsychology, 22*, 10–16.

Twamley, E. W., Ropacki, S. A. L., & Bondi, M. W. (2006). Neuropsychological and neuroimaging changes in preclinical Alzheimer's disease. *Journal of the International Neuropsychological Society, 12*, 707–735.

Van Petten, C., & Kutas, M. (1991). Influences of semantic and syntactic context on open- and closed-class words. *Memory & Cognition, 19*, 95–112.

West, R. L. (1996). An application of prefrontal cortex function theory to cognitive aging. *Psychological Bulletin, 120*, 272–292.

Westerberg, C. E., Paller, K. A., Weintraub, S., Mesulam, M., Holdstock, J. S., Mayes, A. R., et al. (2006). When memory does not fail: Familiarity-based recognition in mild cognitive impairment and Alzheimer's disease. *Neuropsychology, 20*, 193–205.

Wolk, D. A., Signoff, E. D., & Dekosky, S. T. (2008). Recollection and familiarity in amnestic mild cognitive impairment: A global decline in recognition memory. *Neuropsychologia, 46*, 1965–1978.

Yassa, M. A., Stark, S. M., Bakker, A., Albert, M. S., Gallagher, M., & Stark, C. E. L. (2010). High-resolution structural and functional MRI of hippocampal CA3 and dentate gyrus in patients with amnestic mild cognitive impairment. *Neuroimage, 51*, 1242–1252.

Adult Neurogenesis and Cognitive Function: Relevance for Disorders Associated with Human Aging

Keri Martinowich[1,2,3] and Robert J. Schloesser[4]

[1]Lieber Institute for Brain Development, Baltimore, MD, United States
[2]Department of Psychiatry and Behavioral Sciences, Johns Hopkins School of Medicine, Baltimore, MD, United States [3]Department of Neuroscience, Johns Hopkins School of Medicine, Baltimore, MD, United States
[4]Sheppard Pratt-Lieber Research Institute, Baltimore, MD, United States

OUTLINE

Introduction   52

Identification and Characterization of Adult Neurogenesis   54

Development and Maturation of Adult-Born Neurons   55

Adult Mammalian Hippocampal Circuitry   58

Factors Affecting Adult Neurogenesis   59
  Neurogenic Niche   60
  Stress   61
  Aging   62
  Environmental Enrichment, Physical Activity, and Diet   64
  Antidepressants and Mood-Stabilizing Treatments   65

Proposed Functions of Adult Neurogenesis   67
  Pattern Separation   68
  Other Aspects of Spatial and Contextual Memory   70
  Regulation of Mood and the HPA-Axis   71

Genes, Environment and Alzheimer's Disease.
DOI: http://dx.doi.org/10.1016/B978-0-12-802851-3.00003-6

51

Neurogenesis, Hippocampal Dysfunction, and Alzheimer's Disease　　73
　Proteins Linked to FAD Regulate Neurogenesis　　74

Methodologies for Studying Adult Neurogenesis　　75
　Labeling Dividing and Maturing Cells　　76
　Manipulating Neurogenesis in Rodents　　76
　Human Technologies　　77

Conclusion　　78

References　　78

INTRODUCTION

To produce behavior, the brain integrates perceptions from incoming sensory stimuli to generate appropriate responses by engaging outgoing synaptic connections that innervate the musculature. This allows the organism to control movement in space, produce language, and coordinate the activities and function of internal organs and glands. The integration and computation of current and prior percepts is referred to as cognition, or "thinking." Cognitive processes use existing knowledge and generate new knowledge by utilizing processes that include attention, learning and memory, problem solving, judgment, and evaluation.

Multiple underlying neurobiological mechanisms must simultaneously function and converge to perform cognitive tasks on demand. At a basic level, these include the generation as well as electrochemical transmission of information via synaptic activity in neurons, which initiates induction of oscillatory activity to regulate large-scale dynamics in neuronal networks. Cognitive processes also depend on the brain's capacity to change the molecular and cellular characteristics within relevant circuits. This latter capacity to change and adapt in response to extrinsic and intrinsic stimuli is termed neuroplasticity, which is crucial for learning new information as well as for carrying out various memory functions. Neuroplasticity encompasses mechanisms that modulate the strength of synaptic connections, including changes in the molecular composition of synapses, the generation or deletion of synaptic connections, and structural changes in the neuronal dendrites and axons. One of the most striking forms of structural plasticity is the *de novo* generation and integration of new neurons into the existing circuitry through a process termed adult neurogenesis (Aimone et al., 2014; Drew, Fusi, & Hen, 2013; Ming & Song, 2011; Vivar & van Praag, 2013).

Originally thought to occur only during embryogenesis, the existence of adult neurogenesis is now documented in almost all mammalian species that have been examined (Lledo, Alonso, & Grubb, 2006; Ming & Song, 2011). Adult neurogenesis exists in only two discreet regions of the mammalian brain. In the subventricular zone (SVZ) of the lateral ventricles, neural progenitors are born and migrate to the olfactory bulb where they primarily differentiate into interneurons (Ernst et al., 2014; Lledo et al., 2006). Second, in the subgranular layer (SGL) of the dentate gyrus of the hippocampus, new granule cells are born, mature, and integrate into the local hippocampal circuitry (Aimone et al., 2014; Drew et al., 2013; Ming & Song, 2011). Young adult rodents generate approximately 9000 new cells each day, which corresponds to approximately 6% of the total granule cell population each month (Cameron & McKay, 2001), and in the adult human brain, it is estimated that the rate of annual turnover is approximately 1.75% (Spalding et al., 2013). Since the adult hippocampus is one of the primary neural structures involved in memory formation (Squire, Stark, & Clark, 2004), the potential contribution of adult neurons in cognition has garnered significant interest. Current thinking hypothesizes that the integration of newborn neurons, which have unique cellular and physiological properties, may significantly contribute to structural and functional plasticity, which regulates specific components of hippocampal function (Aimone et al., 2014; Drew et al., 2013; Ming & Song, 2011).

Since the discovery of adult neurogenesis, significant progress has been made in understanding the generation, differentiation, and integration of adult-born neurons (Aimone et al., 2014; Christian, Song, & Ming, 2014). In particular, substantial knowledge has been acquired regarding the molecular and cellular mechanisms that are used throughout the process of adult neurogenesis—from stem cell proliferation to integration of new, immature granule cells in the circuitry. These studies have allowed for ongoing exploration into the potential contributions of adult neurogenesis to normal brain function as well as toward dysfunction evidenced in neurological and psychiatric disorders and in aging. This chapter will review current knowledge on the functional roles of adult-born neurons and how they may contribute to cognitive function. We will first review the basic neurobiological characteristics of new neurons and the process by which they are generated in the adult brain. We will then examine how factors relevant to cognitive function (e.g., aging, stress, and environmental influences) affect adult neurogenesis. Finally, we will review studies that have investigated the functional roles of newborn neurons to examine current hypotheses regarding the proposed contributions of adult neurogenesis in hippocampal function.

Most research on adult neurogenesis has been conducted in rodent models and accordingly, most of the studies cited in this chapter have utilized such models. Rapidly evolving technology has furthered the design of elegant transgenic mouse models that facilitate the study of neurogenesis by allowing the investigator to conditionally label, ablate, or genetically alter neural progenitor cells (NPCs) (Dranovsky et al., 2011; Dupret et al., 2008; Sahay, Scobie, et al., 2011; Schloesser, Lehmann, Martinowich, Manji, & Herkenham, 2010). Unfortunately, relatively few methodologies have yet to be developed and successfully utilized to examine adult neurogenesis *in vivo* in humans (Ho, Hooker, Sahay, Holt, & Roffman, 2013; Manganas et al., 2007; Pereira et al., 2007). However, several postmortem studies have examined the amount of new neurons and the number of proliferating cells in the human central nervous system in normal controls and in patients with brain disorders (Boldrini et al., 2009; Lucassen, Stumpel, Wang, & Aronica, 2010; Reif et al., 2006). As technologies continue to develop, future research will be targeted at obtaining a better understanding of the contribution of adult neurogenesis in humans. The ability of neural stem cells to self-renew, proliferate, and differentiate into all cell types of the central nervous system underscores the importance of understanding the mechanisms that control adult neurogenesis as this is critical information for developing potential therapies for neurodegenerative diseases where neuronal loss or injury occurs.

IDENTIFICATION AND CHARACTERIZATION OF ADULT NEUROGENESIS

Neurogenesis occurs in adulthood to varying degrees in nearly all vertebrates (Barker, Boonstra, & Wojtowicz, 2011). In nonmammals, the process can be relatively prolific. For example, fish, reptiles, and amphibian species retain proliferative capacity in multiple neurogenic centers, which contribute new neurons throughout life (Kaslin, Ganz, & Brand, 2008; Lindsey & Tropepe, 2006). In birds, adult neurogenesis occurs in a single periventricular niche, but the new neurons migrate and are distributed throughout the brain (Nottebohm, 2004). In contrast, the adult mammalian brain has relatively little neurogenic capacity, and is comprised mainly of postmitotic neurons that are generated during embryogenesis. The mammalian brain does contain several cell types with proliferative capacity, but these cells generally do not differentiate into neurons.

The seminal discovery of postnatal neurogenesis by Altman and Das in the 1960s overturned a long-standing belief that the adult mammalian brain lacked any capacity to generate neurons (Altman, 1962; Altman & Das, 1965). Ongoing neurogenic potential has been conclusively identified in two discrete neurogenic niches of the mammalian brain, namely the subgranular zone of the hippocampal dentate gyrus and the SVZ of the lateral brain ventricles (Aimone et al., 2014; Christian et al., 2014; Lledo et al., 2006; Vivar & van Praag, 2013). Within these niches, neural stem cells are capable of undergoing proliferation, migration, and differentiation into neurons.

Although neural stem cells have been identified in other regions under pathological conditions or following trauma (Rossi, Mahairaki, Zhou, Song, & Koliatsos, 2014), it remains controversial whether and to what extent active neurogenesis occurs outside of these two regions under physiologically normal conditions (Ming & Song, 2011). Some areas of the neocortex as well as the hypothalamus have been proposed to contain neurogenic niches (Cameron & Dayer, 2008; Gould, 2007; Gould, Reeves, Graziano, & Gross, 1999; Kokoeva, Yin, & Flier, 2005), but the extent of neurogenesis in these areas remains understudied and controversial (Rakic, 2002). The fact that generation of neurons after embryogenesis was lost from other brain regions early in mammalian evolution suggests that unique features of the dentate gyrus and the olfactory bulb circuitries may allow these regions to accommodate, benefit, and possibly depend on the addition of new neurons for their normal function (Barker et al., 2011; Kempermann, 2012). In this review, we focus specifically on adult neurogenesis in the subgranular zone of the dentate gyrus and its potential roles in hippocampal circuitry and cognitive functions. While many of the neurobiological processes are similar, the reader is directed to several excellent reviews on regulation and function specific to SVZ neurogenesis (Brann & Firestein, 2014; Curtis, Low, & Faull, 2012; Lim & Alvarez-Buylla, 2014).

DEVELOPMENT AND MATURATION OF ADULT-BORN NEURONS

In general, the process of adult neurogenesis follows a similar trajectory to newborn neurons in the embryo (Esposito et al., 2005). Neural stem cells give rise to intermediate progenitor cells, which divide to generate proliferating neuroblasts, then postmitotic immature neurons, and finally, a mature granule cell (Figure 3.1). It takes about 2 months for proliferating progenitors to develop into new granule cells, and several months for these new granule cells to acquire the full morphological and physiological profile of a mature granule cell (Zhao, Deng, & Gage, 2008). Strikingly, the vast majority of these adult-born cells (~80%) die at various stages before integrating into the existing circuitry (Sierra et al., 2010; Tashiro, Sandler, Toni, Zhao, & Gage, 2006). At each stage of development and differentiation, the cells are vulnerable to apoptosis. Around half of newborn cells die within 4 days of mitosis, and are then lost at a lower rate (Sierra et al., 2010; Tashiro et al., 2006). Substantial research efforts have been dedicated to understanding the mechanisms by which cells are influenced to promote their survival and eventual integration into the circuitry.

As discussed above, new neurons initially arise from a population of neural stem cells in the subgranular zone of the hippocampal dentate

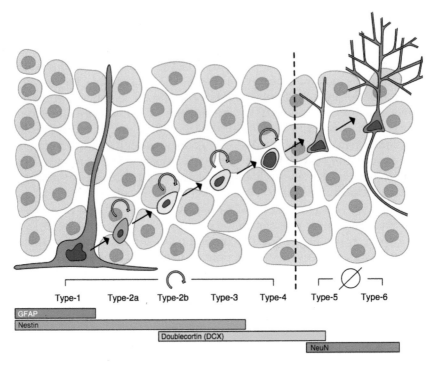

FIGURE 3.1   Developmental stages in adult neurogenesis. The schematic summarizes the developmental timeline from birth of new neurons through to their maturation. The schematic shows morphological characteristics as well as biochemical expression profiles over the maturation process.

gyrus (Bonaguidi et al., 2011). This population of cells is relatively quiescent, and displays characteristics of radial glial cells. Morphologically, radial-glial-like neural stem cells have a long process that extends and branches within the molecular layer of the dentate. Biochemically, these cells express glial-fibrillary acidic protein (GFAP), the intermediate filament protein Nestin, and the Sry-related HMG-box transcription factor Sox2 (Aimone et al., 2014; Garcia, Doan, Imura, Bush, & Sofroniew, 2004; Song, Zhong, et al., 2012; Suh et al., 2007). These radial-glial-like neural stem cells are often classified as "Type I" cells in the literature, and evidence now suggests that these cells are the primary neural stem cell population in the adult brain (Garcia, Doan, et al., 2004).

These "Type I" neural stem cells proliferate and generate "Type II" intermediate NPCs, which in turn give rise to "Type III" neuroblasts (Aimone et al., 2014; Christian et al., 2014). Discrete populations of "Type II" intermediate progenitors have been identified based on their unique morphological characteristics and electrophysiological properties as well as the expression of specific biochemical markers (Seri, Garcia-Verdugo, Collado-Morente, McEwen, & Alvarez-Buylla, 2004; Seri, Garcia-Verdugo, McEwen, & Alvarez-Buylla, 2001). The "Type III" cells are classified as migrating neuroblasts. These cells transiently express early neural lineage markers including doublecortin, PSA-NCAM, NeuroD, Prox1, and calretinin (Encinas, Vaahtokari, & Enikolopov, 2006; Lugert et al., 2010; Suh et al., 2007). Morphologically, these cells display processes of varying lengths and complexities, properties that are correlated with their migration status and maturity. This is a transitional cell type that ranges from migrating neuroblast to immature neuron, and as such these cells have limited proliferative activity (Song, Christian, Ming, & Song, 2012).

As migration progresses, these cells differentiate further and then begin to functionally integrate into the existing microcircuit. Activation of neural progenitors is initiated by specific factors inherent to the neurogenic niche and nonsynaptic activation by the neurotransmitter, gamma-aminobutyric acid (GABA) (Lugert et al., 2010; Mira et al., 2010; Song, Zhong, et al., 2012). As tonic activation increases, these cells begin to receive morphologically mature, functional synapses; GABAergic inputs are established within 8 days after birth, followed by glutamatergic inputs, which emerge at about 2 weeks of age (Esposito et al., 2005; Ge et al., 2006; Overstreet-Wadiche, Bromberg, Bensen, & Westbrook, 2006). The glutamatergic synapses gradually mature over the next several weeks, which is accompanied by an increase in the density of dendritic spines (see below) (Chancey, Poulsen, Wadiche, & Overstreet-Wadiche, 2014; Esposito et al., 2005; Ge et al., 2006). Similar to the developing brain, adult-born immature neurons in the dentate gyrus have elevated chloride currents and increased expression of the sodium/potassium transporter NKCC1. This phenomenon renders GABA to be initially depolarizing to newborn neurons, gradually shifting to hyperpolarizing responses within 2–3 weeks after birth (Ge et al., 2006; Overstreet Wadiche, Bromberg, Bensen, & Westbrook, 2005).

Cell signaling downstream of depolarizing GABA is critical for continued maturation of adult-born neurons as it promotes their survival as well as continued maturation and synapse formation (Chancey et al., 2013; Ge et al., 2006; Song et al., 2013; Tozuka, Fukuda, Namba, Seki, & Hisatsune, 2005). The developmental increase in chloride currents contributes to the switch of GABA from being tonic to phasic, and in activation of glutamatergic synapses. Fast, perisomatic GABAergic inhibitory currents can be recorded after new neurons begin receiving glutamatergic inputs and they continue to increase as the new neurons mature (Li, Aimone, Xu, Callaway, & Gage, 2012). The gradual increase in inhibition after initiation of glutamatergic innervation contributes to the increased excitability and potential for synaptic plasticity that is characteristic of immature newborn neurons (Dieni, Nietz, Panichi, Wadiche, & Overstreet-Wadiche, 2013;

Esposito et al., 2005; Ge, Yang, Hsu, Ming, & Song, 2007; Marin-Burgin, Mongiat, Pardi, & Schinder, 2012; Mongiat, Esposito, Lombardi, & Schinder, 2009; Schmidt-Hieber, Jonas, & Bischofberger, 2004). These unique physiological properties are specialized features that reflect the critical period for enhanced plasticity that is inherent to newborn neurons. As discussed further below, there is now evidence that these unique cellular properties may be critical for the specialized roles played by newborn neurons in certain forms of memory encoding and retrieval (Aimone, Wiles, & Gage, 2009).

Functional integration of newborn granule cells depends on the synaptic inputs they receive and the connections they make with their synaptic targets. Dendrites of newborn granule cells migrate into the molecular layer while their axons project to the hilus and continue on to CA3, which they reach within 7 days (Sun et al., 2013). Within a month after birth, newborn neurons acquire the morphological and physiological characteristics of mature granule cells (Esposito et al., 2005; Vivar et al., 2012; Zhao, Teng, Summers, Ming, & Gage, 2006). However, full cellular maturation and final synaptic incorporation into the circuitry is prolonged, taking up to 3 months. Dendritic spine density increases significantly between 3 and 4 weeks after birth, hitting its peak by 2 months of age. During these 2 months, the spines are highly plastic and very sensitive to activity-dependent signaling (Zhao et al., 2006). It is established that newborn granule cells make functional synapses with their postsynaptic targets. Specifically, they can make synaptic contact with hilar interneurons, mossy cells, and CA3 pyramidal cells, onto which they release the neurotransmitter glutamate (Toni et al., 2008).

ADULT MAMMALIAN HIPPOCAMPAL CIRCUITRY

A brief review of the adult hippocampal circuitry will be informative for discussing the potential roles of adult-born neurons in hippocampal function (Figure 3.2). Hippocampal principal cells are located in three primary subregions: granule cells in the dentate gyrus, and pyramidal cells in two principal cornu ammonis (CA) regions. Each of these regions displays specific cells types and forms of plasticity that contribute to hippocampal function, including learning and memory (Amaral & Witter, 1989). Within this system, neurons are synaptically connected to form a "trisynaptic circuit." Within this circuit, information flows from the entorhinal cortex, the major afferent input to the dentate gyrus via both the medial and lateral perforant pathways (Kohler, 1986). The dentate gyrus projects to CA3 via the axons of granule cells (the mossy fibers), and CA3 principal cells then project to CA1 (the Schaffer collateral projections) (Amaral & Witter, 1989; Claiborne, Amaral, & Cowan, 1986). CA1 principal cells project to the subiculum and from the subiculum, activity

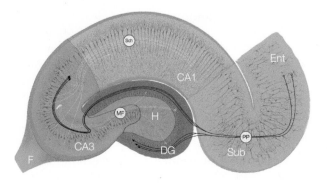

FIGURE 3.2 Trisynaptic circuits of the hippocampal network. A diagram of a transverse section of the rodent hippocampus shows different subregions. Both mature and newborn granule cells located in the dentate gyrus (DG) receive their major synaptic input from the perforant path (PP), which originates from the entorhinal cortex (Ent). Granule cells send their axons, mossy fibers (MF), through the hilus (H) into CA3, where they synapse onto CA3 pyramidal cells or CA3 GABAergic interneurons. In turn, the CA3 pyramidal cells send their axons, Schaffer collaterals (Sch), into CA1, which innervate proximal dendrites of CA1 pyramidal cells.

is transmitted to other cortical and subcortical regions or back to the entorhinal cortex. This projection pattern forms a closed circuit where incoming information from cortical areas converges at the entorhinal cortex and is processed through the hippocampal circuit, and loops back to return to the entorhinal cortex (Li, Mu, & Gage, 2009). In addition to the main excitatory projections pathways that comprise the trisynaptic circuit, inhibitory influence is provided by GABAergic interneurons, which play a major modulatory role over circuit dynamics (Coulter & Carlson, 2007). The hippocampus is further influenced by other neuromodulatory systems; for example, cholinergic projections from the medial septum, serotonergic projections from the dorsal and medial raphe, and dopaminergic inputs from the ventral tegmental area (Gasbarri, Sulli, & Packard, 1997; Leranth & Hajszan, 2007; Veena, Rao, & Srikumar, 2011). The hippocampal formation is noted for its exquisite sensitivity to neural activity and capacity for activity-dependent plasticity as well as by the continual structural and functional plasticity that is afforded by the ongoing addition of new granule cells throughout postnatal life.

FACTORS AFFECTING ADULT NEUROGENESIS

The process of adult neurogenesis can be influenced in different ways by a variety of factors, several of which also have known impacts on cognitive processes. In this section we first review some of the molecular and biochemical characteristics of the microenvironment that define the dentate gyrus neurogenic niche, which is endowed with the unique ability to support the process of adult neurogenesis. We then further examine some of the best-documented intrinsic and extrinsic factors, including stress, aging, and environmental influences, which impact various phases of adult newborn neuron birth and survival. Finally, we review evidence that various therapies used in the treatment of mood and anxiety disorders impact adult neurogenesis.

Neurogenic Niche

In the niche, signaling downstream from secreted factors as well as cell–cell contact impacts various phases of adult neurogenesis. Microglia are macrophages that represent the primary immune cells of the brain. Microglia make a significant contribution to the niche by providing a source of secreted cytokines and chemokines. These secretions balance levels of pro- versus anti-inflammatory conditions, which can facilitate or suppress neurogenesis (Battista, Ferrari, Gage, & Pitossi, 2006; Butovsky et al., 2006; Carpentier & Palmer, 2009). Astrocytes are another key contributor to the neurogenic niche. Astrocytes secrete a variety of factors that are important components of the niche and play a critical role in balancing levels of growth factors and other modulators (Barkho et al., 2006; Oh et al., 2010; Song, Stevens, & Gage, 2002). In addition to secreting factors into the niche, astrocytes are tightly coupled to the vasculature system as their endfeet make physical contact with endothelial cells that are coupled to the vasculatory system. The vasculature is a major source of extrinsic factors that can impact neurogenesis. NPCs are clustered near blood vessels, suggesting that they may receive critical signals from endothelial cells, which can instruct progenitor cells to proliferate or differentiate (Palmer, Willhoite, & Gage, 2000; Shen et al., 2008).

In addition to factors arising from the local circuitry, several distal neuromodulatory systems have effects on the hippocampus, and are known to affect the proliferation of neural stem and progenitor cells as well as influence the differentiation and maturation of immature neurons (Gasbarri et al., 1997; Leranth & Hajszan, 2007; Veena et al., 2011). In particular, the effects of serotonin have been especially well studied. Serotonergic fibers project to the dentate gyrus from the median and the dorsal raphe nuclei (Leranth & Hajszan, 2007). Attention to the role of serotonergic innervation was prompted by observations that selective serotonin reuptake inhibitors (SSRIs), commonly prescribed as antidepressants, could directly increase levels of proliferation (see detailed discussion below) (Malberg, Eisch, Nestler, & Duman, 2000). Less detailed information is available regarding the effects of other major neuromodulator systems (e.g., norepinephrine, dopamine, and acetylcholine), but nevertheless studies have implicated each of these systems in some regulatory role over neurogenesis (Aimone et al., 2014; Veena et al., 2011).

Stress

Stress is defined as the body's nonspecific response to any demand (Selye, 1956). Biological regulation of the stress response is largely mediated by the hypothalamic–pituitary–adrenal (HPA) axis. Regulation of the HPA axis begins with secretion of corticotropin releasing hormone (CRH) from the paraventricular nucleus of the hypothalamus in response to stress. CRH release triggers adrenocorticotropic hormone (ACTH) secretion from the pituitary gland, which in turn targets the adrenal cortex. This signals the adrenal cortex to increase secretion of glucocorticoid "stress" hormones. Both CRH and ACTH can be inhibited by glucocorticoids, constituting a dual-level inhibitory feedback mechanism (de Kloet, Joels, & Holsboer, 2005; Vale, Spiess, Rivier, & Rivier, 1981). Importantly, the central nervous system also has mechanisms for top-down influence over the stress response. In particular, the hippocampus, which contains the highest density of glucocorticoid receptors (GR) in the brain, exerts powerful inhibitory control over the HPA axis (Sapolsky, Krey, & McEwen, 1984).

A principal function of this endocrine system is the maintenance of homeostasis between competing bodily functions including metabolism, immune function, and cognition. In general, glucocorticoid secretion increases during exposure to stress. The purpose of this rise in circulating glucocorticoids is to facilitate gluconeogenesis, which provides the body with more energy, and simultaneously to inhibit immune system activity. It is important to appreciate that the duration and magnitude of stress and the associated changes in the HPA axis are important determinants of the final biological outcomes. Acute and moderate rises in glucocorticoids are necessary to ensure that the organism's biological functions remain within beneficial physiological boundaries. These rises are nonpathological, and are often referred to as "eustress" or good stress, which can facilitate temporary improvements of functions and increases in performance during times of high cognitive demand (de Kloet et al., 2005; McEwen, 1999).

Conversely, a variety of maladaptive molecular, cellular, and behavioral effects on the central nervous system have been documented in a phenomenon termed allostasis (McEwen, 1998), which occurs following chronic exposure to elevated glucocorticoids. For example, prolonged administration of glucocorticoids leads to structural remodeling of the dendritic arbor of principal cells in the hippocampal CA regions and poorer cognitive performance (Magarinos & McEwen, 1995a, 1995b; McEwen, 2005). In humans the sequelae of chronically elevated glucocorticoids is observed in Cushing's syndrome, which is caused either by aberrant secretion of glucocorticoids or by iatrogenically-induced exposure to high-dose corticosteroids. As expected, this leads to suppression of the immune system and metabolic changes that can result in diabetes mellitus. The well-established cognitive and behavioral effects include decreased memory capacity and increased susceptibility to depression (Newell-Price, Bertagna, Grossman, & Nieman, 2006). Importantly, exposure to stress and resulting changes in stress axis hormones impact several key aspects in the process of adult neurogenesis. In rodents, administration of glucocorticoids leads to a decrease in the proliferation, differentiation, as well as survival of adult-born cells (Cameron & Gould, 1994; Gould, Cameron, Daniels, Woolley, & McEwen, 1992). Furthermore, stress hormones can facilitate a fate-change in NPCs from a neuronal lineage toward an oligodendroglial lineage (Chetty et al., 2014).

Many chronic stress paradigms cause depressive- and anxiety-like behavior and impaired adult neurogenesis in rodents. Specifically, evidence for impairment of adult neurogenesis has been identified in models of repeated restraint stress, chronic unpredictable mild stress, social defeat stress, social isolation, and chronic corticosteroid administration (Cameron & Gould, 1994; Dranovsky et al., 2011; Lee et al., 2006; Pham, Nacher, Hof, & McEwen, 2003; Van Bokhoven et al., 2011). The mechanisms by which increased levels of glucocorticoids negatively impact neural progenitor proliferation and survival are not completely understood. Early reports argued for an indirect mechanism that occurred downstream of glutamate excitotoxicity because initial observations detected minimal expression of the mineralocorticoid receptors (MR) and glucocorticoid receptors (GR) in proliferating progenitor cells (Cameron, McEwen, & Gould, 1995; Cameron, Woolley, & Gould, 1993). However, studies by Kempermann et al. have indeed shown that neuronal progenitor cells in the dentate gyrus express GR and MR and hence, a direct effect of glucocorticoids on neural progenitors is plausible (Garcia, Steiner, Kronenberg, Bick-Sander, & Kempermann, 2004). Interestingly, under some circumstances, including running, living in an enriched environment, or mating, increases in adult neurogenesis have been identified despite robust increases in glucocorticoids. In these situations it has been argued that the rewarding experience of the stimulus may buffer the negative effects of stress hormones on neurogenesis (Lehmann, Brachman, Martinowich, Schloesser, & Herkenham, 2013; Schoenfeld & Gould, 2013).

Aging

Arguably the most dramatic, and one of the most well-studied negative regulators of neurogenesis is aging. Aging leads to declines in neurogenesis in both neurogenic niches of the mammalian brain, and is correlated with significant reductions in proliferation, differentiation, and survival of newborn cells (Aimone et al., 2014; Kuhn, Dickinson-Anson, & Gage, 1996; Lazarov, Mattson, Peterson, Pimplikar, & van Praag, 2010). Age-related decreases in the SVZ are functionally associated with decreased olfactory performance and decreases in hippocampal neurogenesis are associated with a decline in cognitive function (Bernal & Peterson, 2004; Bizon, Lee, & Gallagher, 2004; Dupret et al., 2008; Enwere et al., 2004). Furthermore, several models of age-related neurodegenerative disorders, including models of Alzheimer's, exhibit decreased levels of neurogenesis (Chen et al., 2008; Lazarov et al., 2012; Verret, Jankowsky, Xu, Borchelt, & Rampon, 2007; Zhang, McNeil, Dressler, & Siman, 2007). Such findings have led to increased interest in the hypothesis that a reduction of neurogenesis is responsible for the cognitive deficits observed in age-related disorders, and in understanding whether stimulation of neurogenesis could be utilized as a novel therapeutic strategy for preserving or repairing cognitive function. However, realization of such strategies would require a better understanding of the mechanisms initiated during aging that impact neurogenesis. In order to develop novel stimulation strategies, more information is needed to understand the basic biological mechanisms that allow specific manipulations to regulate neurogenesis.

Interestingly, there is a robust interaction between the effects of aging and exposure to glucocorticoids with adult neurogenesis. Specifically, it was shown that reducing corticosteroid levels in aged rodents is able to restore the declining rates of proliferation. This result contributes to the notion that the neuronal precursor population may remain intact into old age, but is actively suppressed by increased levels of circulating glucocorticoids (Cameron & McKay, 1999). However, other studies have found no correlation between the aging-induced decline in neurogenesis and HPA axis activation (Heine, Maslam, Joels, & Lucassen, 2004). Another study provided direct evidence that the number of neural stem cells does not decline with aging, suggesting that the main mechanism driving a decrease in neurogenesis is an enhancement in the quiescence of progenitor cells, as opposed to a reduction of the stem cell pool (Hattiangady & Shetty, 2008; Lugert et al., 2010). Despite a reduction in the number of newly generated neurons and the fact that new neurons require a longer time period for maturation in aged animals, once neurons are fully differentiated and integrated, they are functionally indistinguishable from cells that are generated in younger animals (Morgenstern, Lombardi, & Schinder, 2008; Toni et al., 2008).

The current evidence suggests that a combination of both intrinsic and extrinsic factors associated with increased age are likely to contribute to the observed decrease in hippocampal neurogenesis in aged animals. However, the identity of specific cell-intrinsic factors, changes in the microenvironment (i.e., the neurogenic niche), or how combinations of such factors act to impact neurogenesis during aging is not known. Several studies have begun to shed light on how the decrease in neurogenesis during aging may be mediated. Studies have pointed to alterations in cell-cycle activation, decrements in the vascular niche, as well as changes in the neurogenic milieu, particularly decreases in levels of growth and neurotrophic factors, which have been established in aging (Janzen et al., 2006; Lee, Clemenson, & Gage, 2012; Molofsky et al., 2006; Nishino, Kim, Chada, & Morrison, 2008; Shetty, Hattiangady, & Shetty, 2005).

Factors that have been shown to reduce the age-dependent decline in neurogenesis include environmental enrichment and physical activity (Kempermann, Gast, & Gage, 2002; Kronenberg et al., 2006; van Praag, Shubert, Zhao, & Gage, 2005). Importantly, stimulating neurogenesis in aged animals is associated with improvements in cognitive performance, suggesting that age-related declines that are correlated with deficits in neurogenesis may be reversible (van Praag, Christie, Sejnowski, & Gage, 1999; van Praag et al., 2005). In addition, even at relatively late stages of aging when the decreases in neurogenesis and cognitive decline are maximal, voluntary physical activity can ameliorate many of these impairments, providing an important incentive for continued study in this field (van Praag et al., 2005).

Environmental Enrichment, Physical Activity, and Diet

Some of the most robust positive regulators of adult neurogenesis are environmental enrichment, voluntary exercise, and diet. Enrichment is defined as an environment that provides elevated sensory, social, or motor stimulation. Most experimental setups for rodents consist of an enlarged living area that incorporates tunnels, enclosures, toys, running wheels, or an increased number of social partners. Living in environmental enrichment is also associated with an increase in levels of growth and neurotrophic factors, as well as an increase in synaptic plasticity, including long-term potentiation, which may be mechanistically linked with the positive effects on neurogenesis (Artola et al., 2006; Buschler, Goh, & Manahan-Vaughan, 2012; Cao et al., 2004; Kuzumaki et al., 2011; Novkovic, Mittmann, & Manahan-Vaughan, 2014). Most of the initial environmental enrichment studies incorporated running wheels into the experimental setups. It was subsequently determined that voluntary physical activity in the form of running on the wheel was one of the most important components of the positive effects on neurogenesis. Runners display a dramatic increase in the number of proliferating cells, which leads to an increase in the overall number of newborn neurons that survive and functionally integrate in the circuit (Kobilo et al., 2011; Mustroph et al., 2012; van Praag, Christie, et al., 1999; van Praag, Kempermann, & Gage, 1999).

Besides lifestyle choices that incorporate exercise and increased mental stimulation, diet may influence levels of neurogenesis (Stangl & Thuret, 2009). Rodents on diets that lack essential vitamins and minerals display decreased hippocampal neurogenesis, which is accompanied by deficits in learning and memory. Conversely, rodents that are fed a diet supplemented with polyphenols or omega three fatty acids show increased hippocampal neurogenesis, which is accompanied by enhanced cognitive performance (Stangl & Thuret, 2009). There have been many studies on the role of diet and cognition, but results of the studies can be hard to interpret because of the numerous individual factors that may be contributing to observed effects. Further research in this area is needed to determine which of the many components in an improved diet are causally related to improvement in cognition, and whether these improvements depend on the action of newborn neurons. Finally, additional work is clearly needed to understand the mechanisms by which components in an improved diet can exert these changes.

Antidepressants and Mood-Stabilizing Treatments

Early studies showed that adult neurogenesis is impaired in models of depression and anxiety, but accelerated after antidepressant therapies. These findings led to a proposal of the "neurogenic theory of depression." This theory hypothesized that depression results from impairments in the process of neurogenesis, and that restoration of adult neurogenesis is required for recovery (Jacobs, van Praag, & Gage, 2000). While significant data has been accumulated that support the theory, a complex set of often inconsistent results has led to the conclusion that this theory is likely overly simplistic. The general consensus is in agreement that links exist between adult neurogenesis, behavioral components of stress-induced depression or anxiety, and efficacy of therapeutic treatments. However, ongoing research has made it clear that this is a very complicated problem and additional work is required to elucidate specific roles and requirements for new neurons in anxiety and depression (Miller & Hen, 2014).

Early work demonstrated that some of the most effective treatments of affective disorders in humans, including electroconvulsive shock and the prototypical SSRI, fluoxetine, boost levels of adult neurogenesis in rodents (Madsen et al., 2000; Malberg et al., 2000; Scott, Wojtowicz, & Burnham, 2000). An array of additional pharmacological agents that have mood and anxiety ameliorating properties were subsequently shown to increase levels of neurogenesis (Benninghoff et al., 2013; Chen, Rajkowska, Du, Seraji-Bozorgzad, & Manji, 2000; Miller & Hen, 2014). Novel, rapid-acting antidepressants, including subanesthetic doses of ketamine, also elevate levels of adult neurogenesis (Keilhoff, Bernstein, Becker, Grecksch, & Wolf, 2004). These findings have led to increased efforts and interest in using adult neurogenesis as a screening tool for identifying novel antidepressant therapies. Interestingly, more recent experiments have demonstrated that adult neurogenesis is most consistently and robustly elevated by antidepressants in animal models where animals have been stressed, and subsequently display stress-induced depressive- and anxiety-like behavior (David et al., 2009; Surget et al., 2011).

With the finding of an interaction between adult neurogenesis and regulation of the HPA axis (Schloesser, Manji, & Martinowich, 2009; Snyder, Soumier, Brewer, Pickel, & Cameron, 2011; Surget et al., 2011), interest has risen in whether the ultimate outcomes of these agents on stress-induced behavior is a potential ability to normalize HPA axis deficits by boosting levels of neurogenesis (Dranovsky & Leonardo, 2012). Earlier studies were able to show that the most commonly prescribed mood-stabilizing agents, valproic acid and lithium, can robustly increase adult neurogenesis (Chen et al., 2000; Hao et al., 2004), and more recent studies have shown that both mood-stabilizers rescue the negative effects of glucocorticoids on proliferation of neural progenitors in the dentate gyrus (Boku et al., 2009, 2014). It was recently reported that the anti-GR drug mifepristone blocks the negative effects of glucocorticoids on adult neurogenesis (Anacker et al., 2013; Hu et al., 2012). Finally, an important link between the antidepressant effects of fluoxetine, adult neurogenesis, and regulation of the HPA axis was recently demonstrated. Specifically, it was shown that the effects of fluoxetine are dependent on new neurons, which are required to normalize chronic stress-induced disruptions in inhibitory feedback over the HPA axis (Khemissi, Farooq, Le Guisquet, Sakly, & Belzung, 2014; Surget et al., 2011).

While most studies interrogate the effects of particular compounds on proliferation, differentiation, and survival to make correlations with levels of adult neurogenesis, other studies have attempted to associate the observed cellular effects with related behavioral outcomes, including effects on mood and cognition. Most of these studies have not queried causal links between changes in adult neurogenesis and behavioral outcomes. However, by using transgenic models or irradiation models that allow for ablation of neurogenesis, some studies have demonstrated causality (Santarelli et al., 2003; Schloesser et al., 2010; Snyder et al., 2011). The most recent studies confirm that adult neurogenesis is indeed required for some of the behavioral effects of antidepressant treatments in stressed mice (David et al., 2009; Surget et al., 2011). As discussed further below, ablation of adult neurogenesis impairs various forms of cognition in rodents. This is important for studies of depression and anxiety and the neurogenic hypothesis because impaired cognition is an important component in both anxiety and depression. For example, in humans, overgeneralization is a cognitive impairment that is frequently observed in patients with anxiety and depressive disorders (Kheirbek, Klemenhagen, Sahay, & Hen, 2012). Recent experiments have utilized pattern separation to model the phenomenon of overgeneralization in mice (discussed below). The outcomes of these experiments have led to speculation that this cognitive impairment may be mechanistically related to reduced levels of adult neurogenesis.

PROPOSED FUNCTIONS OF ADULT NEUROGENESIS

Adult-born neurons could directly impact hippocampal function and its contribution to information processing by exerting specific effects on neuroplasticity in hippocampal circuits via their unique physiological properties. A defining feature of the dentate gyrus is its relative quiescence and sparse activation patterns (Chawla et al., 2005; Neunuebel & Knierim, 2012, 2014; Ramirez-Amaya et al., 2005). Although newborn neuron density is significantly lower than that of mature granule cells, immature neurons may be preferentially recruited in response to certain stimuli or activation patterns due to their lower threshold for excitability (Esposito et al., 2005; Mongiat et al., 2009; Mongiat & Schinder, 2011). Coupled with their enhanced capacity for experience-dependent plasticity (Ge et al., 2007; Schmidt-Hieber et al., 2004), these cellular and physiological properties may confer this population of cells with a critical role in information processing (Aimone et al., 2009).

In addition, newborn neurons could indirectly influence hippocampal function by exerting a modulatory role over local circuits. The unique cellular and physiological properties of new neurons coupled with environmental influences that can differentially impact rates of proliferation and survival could confer new neurons with the ability to control network function with high precision. Specifically, it has been suggested that newborn neurons may play a critical role in mediating the excitation/inhibition balance in the dentate gyrus, thus affecting subsequent mossy fiber innervation of CA3 pyramidal cells and inhibitory interneurons (Ikrar et al., 2013). This influence could strongly modulate function in existing circuitry by coordinating neuron firing and synchronization as well as oscillatory activity in the dentate gyrus and its downstream targets (Schloesser et al., 2014). How could this occur mechanistically? Observations from computational models of hippocampal circuitry may offer some clues. Hilar interneurons receive significant excitatory innervation from granule cells, and hence play a significant role in influencing overall levels of inhibition in the dentate gyrus (Buckmaster & Dudek, 1997; Sloviter, 1987). Thus, the synaptic connections between new neurons and hilar interneurons may be particularly important because these cells provide robust inhibitory control over granule cell hyperactivity. This is important because this activity dictates levels of excitation at the mossy fiber-CA3 synapse, which must be tightly controlled to prevent excitotoxicity.

It has been proposed that adult-born neurons may decrease excitation of mature granule cells via feedback inhibition (Ikrar et al., 2013; Kheirbek, Klemenhagen, et al., 2012; Sahay, Wilson, & Hen, 2011), and several recent studies have supported the hypothesis that young neurons may significantly affect levels of inhibition in the dentate gyrus. First, ablation of adult neurogenesis increases gamma frequency bursting and the synchronization of existing granule cell firing to this bursting (Lacefield, Itskov, Reardon, Hen, & Gordon, 2012). Second, ablation of neurogenesis decreases inhibitory drive onto mature granule cells (Singer et al., 2011), while increasing neurogenesis increases excitatory drive onto hilar interneurons (Ikrar et al., 2013). Third, ablation of neurogenesis leads to reduced neural activity in the dentate as well as in its downstream target region, CA3 (Burghardt, Park, Hen, & Fenton, 2012; Schloesser et al., 2014). Finally, chronic suppression of neurogenesis increases levels of stress-induced glutamate release in CA3 as well as dendritic retraction of CA3 pyramidal cells (Schloesser et al., 2014). Such data are in line with the previous findings and support the emerging notion that new neurons contribute to controlling network activity, specifically the balance between excitation/inhibition that determines dentate gyrus output.

Pattern Separation

The ability of the hippocampus to form distinct memories based on similar experiences is thought to require encoding that enables separation of overlapping inputs such that they can be stored as distinct patterns (Marr, 1971). Such a coding mechanism would enable recall of distinct events that are encoded by similar pieces of contextual information as discrete memories. A central role for the dentate gyrus in mediating this form of information processing, known as pattern separation, is relatively well accepted and supported by empirical evidence (Hunsaker & Kesner, 2013; Kesner, 2007, 2013). Computational models of hippocampal function and circuitry propose that the dentate gyrus participates in pattern separation, the ability to distinguish between similar contexts (Treves & Rolls, 1994). Conversely, its downstream target CA3 mediates pattern completion, the reinstatement of neural activity patterns that encode complete contexts by utilizing only disparate or degraded pieces of information (Rolls, 2013). Based on computational algorithms and experimentally determined structure and connectivity maps within the hippocampus, models emerged suggesting that the cell-rich dentate gyrus is poised to process and discriminate simultaneous lines of incoming information (i.e., pattern separation) (Marr, 1971; Treves & Rolls, 1992, 1994). The CA3 region, with its extensive recurrent connections, could support reactivation of stored patterns using only partial or degraded bits of information (i.e., pattern completion) (Marr, 1971; Treves & Rolls, 1992, 1994). Empirical support for the role of the dentate gyrus as a pattern separator came from experimental evidence showing that granule cells are sparsely activated and display low firing rates (Leutgeb, Leutgeb, Moser, & Moser, 2007; Neunuebel & Knierim, 2012, 2014). Many groups have now designed and completed experiments that support the view that the dentate gyrus supports pattern separation, while the CA3 region participates in pattern completion (Hunsaker & Kesner, 2013).

Behavioral paradigms have been designed to probe the concept of pattern separation in animal models (Hunsaker & Kesner, 2013). These behavioral tasks assess the ability of the animal to form distinct memories when contextual or spatial inputs are similar. Specifically, the tests quantify an animal's ability to acquire and recall memories for spatial cues or contexts that are similar versus dissimilar to each other. Such behavioral paradigms have been used to query a role in the mnemonic discrimination process for the dentate gyrus overall, and more recently, the specific role of newborn neurons (Aimone, Deng, & Gage, 2011; Sahay, Wilson, et al., 2011). Empirical evidence from behavioral experiments lends good support for the dentate gyrus as critical in mediating pattern separation (Hunsaker & Kesner, 2013; Kesner, 2007, 2013). Notably, recent evidence suggests that newborn neurons in the dentate gyrus may play a critical role in the ability of the dentate gyrus to function in its capacity as a pattern separator (Aimone et al., 2011; Sahay, Wilson, et al., 2011). Mice lacking neurogenesis, mice with reduced neurogenesis, and mice with impaired glutamatergic function in newborn neurons all show impaired behavior in context-discrimination tasks designed to probe deficits in pattern separation (Kheirbek, Tannenholz, & Hen, 2012; Sahay, Scobie, et al., 2011; Tronel et al., 2012). Conversely, better context discrimination is observed in mice in which adult neurogenesis is enhanced (Sahay, Scobie, et al., 2011).

Additional evidence for a role of neurogenesis in context discrimination comes from studies showing that ablation of adult neurogenesis leads to impairments in memory tasks that assess the ability to discriminate between similar spatial contexts. In these experiments, mice lacking hippocampal neurogenesis performed poorly when required to discriminate between stimuli that were spatially well separated. However, no difference was found between mice without neurogenesis and controls when the discriminatory choices were widely separated in space (Clelland et al., 2009). In another example, mice were engineered such that the synaptic release mechanism was selectively blocked in mature granule cells, but not in immature granule cells. This model renders mature granule cells unable to transfer synaptic information, rendering them effectively silenced. Surprisingly, these mice exhibit improved context discrimination while mice in which neurogenesis was ablated performed more poorly in this task. Together, these data suggest that newborn neurons may selectively mediate pattern separation, while mature granule cells may actually interfere with discriminator ability (Nakashiba et al., 2012). In combination with related studies (Creer, Romberg, Saksida, van Praag, & Bussey, 2010; Denny et al., 2014; Niibori et al., 2012), these findings strongly suggest that adult-born neurons are crucial for the ability to generate distinct memories from similar inputs, and are consistent with the view that these cells provide a marked contribution to the overall ability of the dentate gyrus to mediate the task of pattern separation.

Other Aspects of Spatial and Contextual Memory

Adult neurogenesis has been implicated in additional aspects of spatial and contextual memory besides pattern separation, consistent with the proposed role of the dentate gyrus in these functions (Kesner, 2007; Xavier & Costa, 2009). In particular, newborn neurons appear to play a critical role in distinguishing temporal information in time-dependent tasks (Denny, Burghardt, Schachter, Hen, & Drew, 2012; Drew, Denny, & Hen, 2010). New neurons also seem to be preferentially important for especially difficult tasks that rely on the ability to discern changing contingencies (Burghardt et al., 2012; Dupret et al., 2008). Varying the time between ablation of adult neurogenesis and behavioral tests has revealed specific roles for adult-born neurons at different stages of maturation. For example, studies have recently revealed that 4- to 6-week-old newborn neurons are crucial when the window for contextual coding is limited, suggesting that these immature neurons may be especially important when rapid acquisition and processing of contextual information is required (Denny et al., 2012). These results may reflect cellular functions carried out by newborn neurons at specific periods of maturation when they are bestowed with unique physiological properties that alter their excitability and capacity for plasticity. Another notable example in which new neurons may be preferentially recruited is in tasks with cognitive demands that require the ability to rapidly change temporal or contextual contingency. An example is requirement for intact neurogenesis in some forms of behavioral flexibility. It was recently shown that compared to control animals, animals lacking neurogenesis performed similarly in a task where the location of a negative reinforcer needed to be learned and avoided. However, animals lacking neurogenesis were significantly impaired when the contingency was changed and the animals needed to learn to inhibit the old information, and avoid the negative reinforcer in a new location (Burghardt et al., 2012).

Despite the recent progress, much remains to be learned about the specific roles of adult neurons in hippocampal functions in memory. The above studies suffer from the caveat that neurogenesis is ablated throughout the entire learning paradigm or behavioral task and thus cannot distinguish the role of new neurons in discrete phases (e.g., acquisition, consolidation, recall) (Kim, Christian, Ming, & Song, 2012). In addition, chronic ablation of neurogenesis may lead to compensatory network and circuit level changes that could mask potential effects of new neurons on behavior (Singer et al., 2011). New, powerful technologies that are rapidly coming online have significant potential to enhance our understanding of adult neurogenesis. In particular, optogenetic techniques allow investigators to rapidly and reversibly silence or activate newborn neurons. A recent study using such optogenetic techniques showed that suppression of 4-week-old, but not 2- or 8-week-old newborn neurons, impairs a specific form of contextual fear memory and spatial memory retrieval (Gu et al., 2012). A significant impediment to our understanding of the roles and mechanisms used by new neurons has been the inability to directly interrogate physiological properties of new neurons in awake and behaving animals due to technical limitations of recording from granule cells in the dentate gyrus (Leutgeb et al., 2007; Neunuebel & Knierim, 2012). While some *in vivo* recordings have been carried out in the dentate (Leutgeb et al., 2007; Neunuebel & Knierim, 2012), efforts to generate activity maps from significant numbers of granule cells, let alone newborn granule cells, during behavior have not been fully realized. Future studies, especially those incorporating optical sensors that allow visualization of newborn neuron neural activity in awake behaving animals will be particularly useful (Ziv et al., 2013). These techniques will combine calcium imaging with miniature microscopes and deep brain endoscopy to provide significant advances in our understanding of the roles of newborn neurons in specific behavior and their impact on hippocampal circuit activity. In conclusion, the available evidence points to several potential roles for adult neurogenesis. In particular, new neurons may be especially relevant when the task is difficult and requires changing the contingency of provided spatial and contextual information, requires rapid information processing, or demands highly efficient discrimination between similar contexts.

Regulation of Mood and the HPA-Axis

Early investigations showed a striking positive correlation between antidepressant therapies (including SSRIs, electroconvulsive therapy, and monoamine oxidase inhibitors) and adult neurogenesis (Madsen et al., 2000; Malberg et al., 2000; Scott et al., 2000). These findings, coupled with observation of a negative impact of stress on the proliferation and survival of adult-born neurons, generated significant interest in a potential functional role of adult neurogenesis in regulation of mood-related behavior (Cameron & Gould, 1994; Dranovsky et al., 2011; Lee et al., 2006; Pham et al., 2003; Van Bokhoven et al., 2011). Despite intensive investigation, the contribution of neurogenesis to baseline mood-related behavior is highly controversial, and a substantial amount of data now suggest that ablation of neurogenesis does not per se alter affective or mood-related behaviors. Conversely, the ability of newborn neurons to influence the response to antidepressants as well as to the response to stress and resultant maladaptive mood-related behavior is more convincing (Miller & Hen, 2014).

Emerging evidence suggests that adult neurogenesis may significantly impact the response to stress by modulating the function of the HPA axis (Schloesser et al., 2009; Snyder et al., 2011). These data have led to a renewed interest in understanding the mechanisms by which adult-born neurons impact the response to stress and the behaviors that are adopted in response to aversive experiences (Dranovsky & Leonardo, 2012). The first direct evidence that adult neurogenesis may play a regulatory role in the stress response showed that ablation of new neurons led to an increased release of circulating glucocorticoids following exposure to mild stress, suggesting that new neurons may participate in inhibiting the stress axis during stress exposure (Schloesser et al., 2009). Additional experiments suggested that this increase in release of stress hormones could be relevant for stress-related behaviors. In particular, mice without neurogenesis demonstrated increased anxiety-like behavior following exposure to mild stress (Snyder et al., 2011).

The roles of new neurons in mood-related behavior and pattern separation are not necessarily mutually exclusive, but rather, may be intimately related. Recent experiments show that the ability to discriminate between different environmental living conditions may affect susceptibility to maladaptive stress-induced coping mechanisms as well as the ability to capitalize on positive changes in the environment that promote resiliency (Schloesser et al., 2010). Specifically, environmental enrichment is effective in extinguishing submissive behavior that is induced in response to chronic exposure to psychosocial stress in mice. However, mice lacking neurogenesis are unable to benefit from the restorative effects of environmental enrichment on stress-induced depressive- and anxiety-like behaviors (Schloesser et al., 2010). These results support the hypothesis that a lack of new neurons renders the animal unable to distinguish between changes in contextual and temporal information that discriminate stressful versus positive or safe environments. Newborn neurons may be necessary for the animal to take advantage of the new contextual experiences provided by environmental enrichment to positively impact resiliency to maladaptive stress-induced depressive- and anxiety-like behavior.

Negative effects of an overactive stress response on hippocampal structure and function as well as on various measures of cognitive function are well-documented (de Kloet et al., 2005; McEwen, 1999). Hence, it is possible that the effects of adult neurogenesis on the stress axis are linked with the proposed roles of new neurons on cognition. Importantly, it was shown that chronic suppression of neurogenesis led to increased release of glutamate as well as dendritic retraction of pyramidal cells in the CA3 region (Schloesser et al., 2014). Preclinical studies have observed increases in glutamate release and transmission in response to stress exposure (Moghaddam, 1993;

Moghaddam, Bolinao, Stein-Behrens, & Sapolsky, 1994), and hence it has been proposed that maladaptive increases in glutamatergic transmission may precipitate the morphological changes in the hippocampus that are observed in chronic depressive disorders (McEwen & Magarinos, 1997; Sapolsky, 2000). HPA axis disruptions are frequently identified in patients with major depression and a role in stress-induced dendrite remodeling has been attributed to actions of glucocorticoid hormones, which are increased in response to stress (de Kloet et al., 2005; McEwen & Magarinos, 1997; Sapolsky, 2000). The elevated levels of glucocorticoids observed in aging and after exposure to chronic stress are linked to decreased levels of neurogenesis (Cameron & Gould, 1994; Cameron & McKay, 1999). Conversely, the stress-induced effects on neurogenesis and CA3 dendritic remodeling have been attributed to interactions between glucocorticoids and stress-induced glutamate release (McEwen & Magarinos, 1997; Schloesser et al., 2014). The human hippocampus is particularly sensitive to atrophy and shows greater impairments compared to other regions in response to a variety of insults, in numerous neurological and psychiatric disorders as well as during aging (Bogerts et al., 1993; Bremner et al., 1995; Convit et al., 1995; Golomb et al., 1994; Sheline, Wang, Gado, Csernansky, & Vannier, 1996; Starkman, Gebarski, Berent, & Schteingart, 1992). This is particularly important because an interaction between hippocampal atrophy and cognition is well established. Owing to its potential importance, this area will significantly benefit from ongoing research.

NEUROGENESIS, HIPPOCAMPAL DYSFUNCTION, AND ALZHEIMER'S DISEASE

Age-dependent decline in hippocampal neurogenesis may play a role in cognitive deterioration and enhance vulnerability to Alzheimer's disease (AD) by compromising hippocampal plasticity and function. Aging is the greatest risk factor for sporadic AD, the late onset form of the disease that encompasses 95% of AD cases. Transgenic mice harboring familial Alzheimer's disease (FAD)-linked mutant forms, such as APPswe/PS1ΔE9, 3XTg-AD, PS1M146V, PS1P117L, APP23, PS2APP, (PDGF)-APPSw, Ind exhibit impairments in neurogenesis (Demars, Hu, Gadadhar, & Lazarov, 2010; Ermini et al., 2008; Gan et al., 2008; Ghosal, Stathopoulos, & Pimplikar, 2010; Hamilton et al., 2010; Hartl et al., 2008; Haughey, Liu, Nath, Borchard, & Mattson, 2002; Haughey, Nath, et al., 2002; Krezymon et al., 2013; Manganas et al., 2007; Poirier, Veltman, Pflimlin, Knoflach, & Metzger, 2010; Rodriguez et al., 2008; Rodriguez, Jones, & Verkhratsky, 2009; Sun et al., 2009; Taniuchi et al., 2007; Veeraraghavalu, Choi, Zhang, & Sisodia, 2010; Verret et al., 2007; Wang, Dineley, Sweatt, & Zheng, 2004; Wen et al., 2004). Some of these studies show that impairments take place early in life, as early as 2 months of age, preceding AD hallmarks and cognitive deficits, suggesting a potential causative role for neurogenesis in the disease. In addition, deficits in neurogenesis were detected following neurodegeneration in conditional PS1/PS2KO (Chen et al., 2008), later in life in APP/PS1KI (Zhang et al., 2007), APPswe/PS1ΔE9 (Taniuchi et al., 2007; Verret et al., 2007), 3XTg-AD (Blanchard et al., 2010), APPswe/PS1L166P, APP23 (Ermini et al., 2008), APP/PS1, and at multiple time points in Tg2576 (APPswe) mice (Dong et al., 2004; Krezymon et al., 2013; Verret et al., 2007). Impairments in environmental enrichment-induced neurogenesis were also found in conditional PS1KO (Feng et al., 2001), PS1ΔE9, and PS1M146L mutant mice (Choi et al., 2008). Recent studies suggest that downregulating the expression of PS1 in NPCs in the hippocampus induces cognitive deficits, suggesting that compromising neurogenesis is sufficient for the induction of learning and memory impairments (Bonds et al., 2015). In contrast, some studies find increased numbers of newly-proliferating cells in the SGL and SVZ (Chevallier et al., 2005; Donovan et al., 2006; Jin et al., 2004; Kolecki et al., 2008; Lopez-Toledano & Shelanski, 2004, 2007). As described in the following paragraphs, the different roles that APP metabolites, PS1, and other critical molecules play in neurogenesis, may account, at least in part, for some of the observed variations in the fate of neurogenesis in the different animal models used in these studies. Importantly, the high responsiveness and sensitivity of neurogenesis to internal and external stimuli may require careful examination of the data as a function of age, neurodegeneration, extent and onset of amyloidosis, and other experimental conditions, such as regiment of neurogenic markers under study. Differences may be attributed to the time of examination of neurogenesis relative to stage of disease and pathology progression. Interestingly, FAD-linked impaired neurogenesis is rescued by experience of mice in environmental enrichment and learning regimens (for more information see Chapter 7 (Hu et al., 2010)).

Proteins Linked to FAD Regulate Neurogenesis

Recent studies show that beyond a role in the amyloidogenic pathway, proteins linked to FAD regulate adult and embryonic neurogenesis. Most of the evidence provided so far is in relation to sAPP and PS1. Nevertheless, other molecules linked to FAD have been implicated in neurogenesis (for review see Lazarov & Marr, 2010).

PS1: Experiments in mice with genomic deletions of *PSEN1* reveal severely abnormal somitogenic and neurogenic processes in the brains of these mice (Shen et al., 1997; Wong et al., 1997). Regulated intramembrane proteolysis of neurogenic molecules, such as the receptor tyrosine kinases ErbB4 (Ni, Murphy, Golde, & Carpenter, 2001; Sardi, Murtie, Koirala, Patten, & Corfas, 2006), IGF-1R (McElroy, Powell, & McCarthy, 2007), and insulin receptor (Kasuga, Kaneko, Nishizawa, Onodera, & Ikeuchi, 2007), Notch1 (De Strooper et al., 1999), L1 (Maretzky, Schulte, et al., 2005), and E-cadherin (Marambaud et al., 2002) is catalyzed by PS1/γ-secretase. Following ErbB4 activation and PS1/γ-secretase-dependent cleavage in neural stem cells, the C'-terminus of ErbB4 forms a complex that undergoes nuclear translocation, binds the promoters of GFAP and S100β, leading to inhibition of their differentiation into astrocytes (Sardi et al., 2006). Previous studies suggest that PS1/γ-secretase regulates EGFR signaling, a major regulator of neural stem proliferation, migration, and survival (for review see Ayuso-Sacido, Graham, Greenfield, & Boockvar, 2006). PS1 is further implicated in the Wnt/β-catenin signaling pathway, one of the principal regulators of hippocampal neurogenesis (Lie et al., 2005; Maretzky, Reiss, et al., 2005; Tesco, Kim, Diehlmann, Beyreuther, & Tanzi, 1998). In the adult brain, PS1 is capable of regulating NPC differentiation (Gadadhar, Marr, & Lazarov, 2011).

sAPPα: A crystal structure of the amino terminal of sAPPα reveals a domain that is similar to cysteine-rich growth factors, suggesting that sAPPα may act as a potential ligand for growth factor receptors (Rossjohn et al., 1999). In addition, APP, like a number of growth factors such as midkine19, hepatocyte growth factor, and vascular endothelial growth factor, possess disulfide-bonded, β-hairpin loops implicated in proteoglycan binding (Rossjohn et al., 1999). Deficiency of the sortilin-related receptor with type-A repeats results in enhancement of sAPP production, extracellular signal regulated kinase stimulation, and increased proliferation and survival of NPCs in both the SVZ and SGL of the dentate gyrus (Rohe et al., 2008). Recent studies show that the production of sAPPα by α-secretase processing is essential for the proliferation of NPCs derived from the adult mouse brain (Caille et al., 2004; Demars, Bartholomew, Strakova, & Lazarov, 2011; Demars, Hollands, Zhao, & Lazarov, 2013). Thus, a shift in the amyloidogenic APP processing may compromise the proliferation and survival of NPCs, leading to reduced neurogenesis.

METHODOLOGIES FOR STUDYING ADULT NEUROGENESIS

Several methodologies have been developed to study adult neurogenesis. Methods to label and quantify adult neurogenesis include radiological and immunohistological detection of progenitors cells and immature neurons. Irradiation and transgenic technologies have been adopted to both label and manipulate newborn cells in animals. Finally, radiological and other methods have been developed to quantify adult neurogenesis in the human being.

Labeling Dividing and Maturing Cells

The most common method for labeling dividing cells involves the incorporation of a traceable molecule into DNA. DNA synthesis usually occurs only when cells are actively undergoing mitosis, and hence this has been utilized as a marker of neurogenesis. Initial neurogenesis studies used tritiated (^{3}H) thymidine, which permitted the radiographic labeling of cells that had been born at the time of injection (Altman, 1962). The thymidine analog, BrdU, was developed in the 1990s to facilitate the ability to use immunohistochemistry, rather than radiolabeling, for detection (Kuhn & Cooper-Kuhn, 2007). This was a crucial advance since birth-dating could be coupled with immunodetection for other markers (e.g., GFAP, PSA-NCAM, Nestin, and NeuN), which assisted researchers in determining the identities and maturity levels of the newly born cells over time. Other histological markers, most commonly Ki67, are used to identify proliferating cells (Namba et al., 2005). Other markers can be used to identify cells at specific time points in their maturation. While these markers are specific, it is important to point out that they are transient in nature and therefore cannot be used for long-term labeling and quantification. For these types of tracing studies, investigators frequently utilize genetic labeling with systems such as Cre-lox (Dranovsky et al., 2011). In these studies, a transiently activated promoter is coupled to a reporter that can permanently label a cell with an indelible marker (e.g., green fluorescence protein). Another commonly used approach is retroviral labeling of dividing cells with a permanent marker (van Praag et al., 2002; Zhao et al., 2006). Drawbacks of retroviral labeling are the relative sparseness of labeling. However, it is highly accurate for birth-dating cells and has the advantage of allowing for excellent imaging of newborn cells, including their processes. Since the label is incorporated into the living animal, it can be utilized to identify cells for electrophysiological analysis in *ex vivo* experiments.

Manipulating Neurogenesis in Rodents

Early strategies for experimentally manipulating neurogenesis suppressed generation of new neurons by utilizing systemic administration of antimitotic agents (Shors et al., 2001). However, these treatments are confounded by nonspecific effects on proliferating cells in other regions of the brain and body. Subsequent developments included X-ray irradiation techniques, which kill dividing cells. The first attempts used whole body irradiation and subsequent focal refinements have been made to target more limited brain regions. However, despite advancements, X-ray irradiation still effects all dividing cells even if localized to a certain region, and can cause DNA damage and inflammation (Ford et al., 2011; Monje,

Mizumatsu, Fike, & Palmer, 2002; Parent, Tada, Fike, & Lowenstein, 1999; Peissner, Kocher, Treuer, & Gillardon, 1999). Genetic techniques were developed to complement these studies and generate models with fewer confounds. Popular models have incorporated the use of herpes simplex virus thymidine kinase, an enzyme that can catalyze the conversion of the prodrug ganciclovir to a toxic metabolite that can kill actively dividing cells. Expression of herpes simplex virus thymidine kinase can be directed only to specific populations of cells (e.g., NPCs), by driving its expression with promoters that are selectively expressed in relevant populations of NPCs (e.g., GFAP or Nestin) or in migrating neuroblasts (e.g., doublecortin) (Garcia, Doan, et al., 2004; Schloesser et al., 2009, 2010; Singer et al., 2009; Sun, Wang, Mao, Xie, & Jin, 2012). Conversely, genetic enhancement of neurogenesis can be achieved by deleting the pro-apoptotic gene, *Bax*, selectively in NPCs (Sahay, Scobie, et al., 2011).

The ability to tightly and reversibly control the activity of newborn neurons will be necessary in order to further understand the function of adult neurogenesis (Kim et al., 2012). New optogenetic strategies are currently being developed and tested to reversibly activate and silence specific populations of newborn neurons in awake and behaving animals. The ability to observe the activity of immature neurons within intact dentate gyrus circuits will significantly contribute to understanding of adult neurogenesis (Gu et al., 2012). A technique that is rapidly developing to facilitate such experiments is the use of large-scale calcium imaging of neural circuits. *In vivo* optical microscopy is combined with selective expression of florescent calcium indicators to record neural activity in specific populations of cells (e.g., immature neurons) (Ziv et al., 2013). Action potentials trigger the influx of calcium, and hence changes in calcium concentration can be used as a quantifiable readout of neural activity. Such technologies have great potential for increasing our understanding of the functional role of new neurons.

Human Technologies

Measuring neurogenesis in humans is considerably more difficult, which has limited our understanding of the extent and importance of neurogenesis in the human brain. However, this is of the utmost importance for understanding whether the process of neurogenesis is important for human health and disease. Human neurogenesis was originally confirmed using BrdU, but the sample sizes were too low for accurate quantification (Eriksson et al., 1998). Later studies developed a novel technique where levels of neurogenesis were estimated by taking advantage of increased radioactive carbon in the atmosphere as a result of nuclear testing activity (Spalding et al., 2013; Spalding, Bhardwaj, Buchholz, Druid, & Frisen, 2005). For approximately 15 years, ^{14}C levels steadily increased from this activity before decaying due to international treatises to limit nuclear testing. Upon entering the food supply, ^{14}C incorporates into DNA similarly to ^{3}H thymidine, mimicking previous rodent studies. Thus, this technique enabled the estimate of overall levels of neurogenesis, which suggested a turn-over rate of granule cells at approximately 1.75% per year (Spalding et al., 2013). Postmortem studies are capable of making correlations in levels of adult neurogenesis in various populations, for instance across different age groups, and in patients with diagnoses of neurological and psychiatric disorders. Novel techniques to directly examine adult neurogenesis in living humans are not readily available, but interesting approaches using SPECT and MRI are currently in development (Ho et al., 2013; Manganas et al., 2007; Pereira et al., 2007).

CONCLUSION

In summary, there is now strong evidence that adult neurogenesis contributes to a variety of hippocampal functions, particularly those related to stress-related emotional disruptions and cognitive functions that require fine discriminatory function. Continued research will undoubtedly increase our understanding of the neurobiological processes that regulate adult neurogenesis as well as its role in both normal and abnormal brain functions.

References

Aimone, J. B., Deng, W., & Gage, F. H. (2011). Resolving new memories: A critical look at the dentate gyrus, adult neurogenesis, and pattern separation. *Neuron, 70*(4), 589–596. http://dx.doi.org/10.1016/j.neuron.2011.05.010.

Aimone, J. B., Li, Y., Lee, S. W., Clemenson, G. D., Deng, W., & Gage, F. H. (2014). Regulation and function of adult neurogenesis: From genes to cognition. *Physiological Reviews, 94*(4), 991–1026. http://dx.doi.org/10.1152/physrev.00004.2014.

Aimone, J. B., Wiles, J., & Gage, F. H. (2009). Computational influence of adult neurogenesis on memory encoding. *Neuron, 61*(2), 187–202. http://dx.doi.org/10.1016/j.neuron.2008.11.026.

Altman, J. (1962). Are new neurons formed in the brains of adult mammals? *Science, 135*(3509), 1127–1128.

Altman, J., & Das, G. D. (1965). Autoradiographic and histological evidence of postnatal hippocampal neurogenesis in rats. *The Journal of Comparative Neurology, 124*(3), 319–335.

Amaral, D. G., & Witter, M. P. (1989). The three-dimensional organization of the hippocampal formation: A review of anatomical data. *Neuroscience, 31*(3), 571–591.

Anacker, C., Cattaneo, A., Luoni, A., Musaelyan, K., Zunszain, P. A., Milanesi, E., … Pariante, C. M. (2013). Glucocorticoid-related molecular signaling pathways regulating hippocampal neurogenesis. *Neuropsychopharmacology, 38*(5), 872–883. http://dx.doi.org/10.1038/npp.2012.253.

Artola, A., von Frijtag, J. C., Fermont, P. C., Gispen, W. H., Schrama, L. H., Kamal, A., et al. (2006). Long-lasting modulation of the induction of LTD and LTP in rat hippocampal

CA1 by behavioural stress and environmental enrichment. *The European Journal of Neuroscience*, 23(1), 261–272. http://dx.doi.org/10.1111/j.1460-9568.2005.04552.x.

Ayuso-Sacido, A., Graham, C., Greenfield, J. P., & Boockvar, J. A. (2006). The duality of epidermal growth factor receptor (EGFR) signaling and neural stem cell phenotype: Cell enhancer or cell transformer? *Current Stem Cell Research & Therapy*, 1(3), 387–394.

Barker, J. M., Boonstra, R., & Wojtowicz, J. M. (2011). From pattern to purpose: How comparative studies contribute to understanding the function of adult neurogenesis. *The European Journal of Neuroscience*, 34(6), 963–977. http://dx.doi.org/10.1111/j.1460-9568.2011.07823.x.

Barkho, B. Z., Song, H., Aimone, J. B., Smrt, R. D., Kuwabara, T., Nakashima, K., … Zhao, X. (2006). Identification of astrocyte-expressed factors that modulate neural stem/progenitor cell differentiation. *Stem Cells and Development*, 15((3), 407–421. http://dx.doi.org/10.1089/scd.2006.15.407.

Battista, D., Ferrari, C. C., Gage, F. H., & Pitossi, F. J. (2006). Neurogenic niche modulation by activated microglia: Transforming growth factor beta increases neurogenesis in the adult dentate gyrus. *The European Journal of Neuroscience*, 23(1), 83–93. http://dx.doi.org/10.1111/j.1460-9568.2005.04539.x.

Benninghoff, J., Grunze, H., Schindler, C., Genius, J., Schloesser, R. J., van der Ven, A., … Rujescu, D. (2013). Ziprasidone—not haloperidol—induces more *de-novo* neurogenesis of adult neural stem cells derived from murine hippocampus. *Pharmacopsychiatry*, 46(1), 10–15. http://dx.doi.org/10.1055/s-0032-1311607.

Bernal, G. M., & Peterson, D. A. (2004). Neural stem cells as therapeutic agents for age-related brain repair. *Aging Cell*, 3(6), 345–351. http://dx.doi.org/10.1111/j.1474-9728.2004.00132.x.

Bizon, J. L., Lee, H. J., & Gallagher, M. (2004). Neurogenesis in a rat model of age-related cognitive decline. *Aging Cell*, 3(4), 227–234. http://dx.doi.org/10.1111/j.1474-9728.2004.00099.x.

Blanchard, J., Wanka, L., Tung, Y. C., Cardenas-Aguayo Mdel, C., LaFerla, F. M., Iqbal, K., et al. (2010). Pharmacologic reversal of neurogenic and neuroplastic abnormalities and cognitive impairments without affecting Abeta and tau pathologies in 3xTg-AD mice. *Acta Neuropathologica*, 120(5), 605–621. http://dx.doi.org/10.1007/s00401-010-0734-6.

Bogerts, B., Lieberman, J. A., Ashtari, M., Bilder, R. M., Degreef, G., Lerner, G., … Masiar, S. (1993). Hippocampus-amygdala volumes and psychopathology in chronic schizophrenia. *Biological Psychiatry*, 33(4), 236–246.

Boku, S., Nakagawa, S., Masuda, T., Nishikawa, H., Kato, A., Kitaichi, Y., … Koyama, T. (2009). Glucocorticoids and lithium reciprocally regulate the proliferation of adult dentate gyrus-derived neural precursor cells through GSK-3beta and beta-catenin/TCF pathway. *Neuropsychopharmacology*, 34(3), 805–815. http://dx.doi.org/10.1038/npp.2008.198.

Boku, S., Nakagawa, S., Masuda, T., Nishikawa, H., Kato, A., Takamura, N., … Kusumi, I. (2014). Valproate recovers the inhibitory effect of dexamethasone on the proliferation of the adult dentate gyrus-derived neural precursor cells via GSK-3beta and beta-catenin pathway. *European Journal of Pharmacology*, 723, 425–430. http://dx.doi.org/10.1016/j.ejphar.2013.10.060.

Boldrini, M., Underwood, M. D., Hen, R., Rosoklija, G. B., Dwork, A. J., John Mann, J., et al. (2009). Antidepressants increase neural progenitor cells in the human hippocampus. *Neuropsychopharmacology*, 34(11), 2376–2389. http://dx.doi.org/10.1038/npp.2009.75.

Bonaguidi, M. A., Wheeler, M. A., Shapiro, J. S., Stadel, R. P., Sun, G. J., Ming, G. L., et al. (2011). *In vivo* clonal analysis reveals self-renewing and multipotent adult neural stem cell characteristics. *Cell*, 145(7), 1142–1155. http://dx.doi.org/10.1016/j.cell.2011.05.024.

Bonds, J. A., Kuttner-Hirshler, Y., Bartolotti, N., Tobin, M. K., Pizzi, M., Marr, R., et al. (2015). Presenilin-1 dependent neurogenesis regulates hippocampal learning and memory. *PLoS One*, 10(6), e0131266. http://dx.doi.org/10.1371/journal.pone.0131266.

Brann, J. H., & Firestein, S. J. (2014). A lifetime of neurogenesis in the olfactory system. *Frontiers in Neuroscience*, 8, 182. http://dx.doi.org/10.3389/fnins.2014.00182.

Bremner, J. D., Randall, P., Scott, T. M., Bronen, R. A., Seibyl, J. P., Southwick, S. M., ... Innis, R. B. (1995). MRI-based measurement of hippocampal volume in patients with combat-related posttraumatic stress disorder. *The American Journal of Psychiatry, 152*(7), 973–981.

Buckmaster, P. S., & Dudek, F. E. (1997). Neuron loss, granule cell axon reorganization, and functional changes in the dentate gyrus of epileptic kainate-treated rats. *The Journal of Comparative Neurology, 385*(3), 385–404.

Burghardt, N. S., Park, E. H., Hen, R., & Fenton, A. A. (2012). Adult-born hippocampal neurons promote cognitive flexibility in mice. *Hippocampus, 22*(9), 1795–1808. http://dx.doi.org/10.1002/hipo.22013.

Buschler, A., Goh, J. J., & Manahan-Vaughan, D. (2012). Frequency dependency of NMDA receptor-dependent synaptic plasticity in the hippocampal CA1 region of freely behaving mice. *Hippocampus, 22*(12), 2238–2248. http://dx.doi.org/10.1002/hipo.22041.

Butovsky, O., Ziv, Y., Schwartz, A., Landa, G., Talpalar, A. E., Pluchino, S., ... Schwartz, M. (2006). Microglia activated by IL-4 or IFN-gamma differentially induce neurogenesis and oligodendrogenesis from adult stem/progenitor cells. *Molecular and Cellular Neurosciences, 31*(1), 149–160. http://dx.doi.org/10.1016/j.mcn.2005.10.006.

Caille, I., Allinquant, B., Dupont, E., Bouillot, C., Langer, A., Muller, U., et al. (2004). Soluble form of amyloid precursor protein regulates proliferation of progenitors in the adult subventricular zone. *Development, 131*(9), 2173–2181.

Cameron, H. A., & Dayer, A. G. (2008). New interneurons in the adult neocortex: Small, sparse, but significant? *Biological Psychiatry, 63*(7), 650–655. http://dx.doi.org/10.1016/j.biopsych.2007.09.023.

Cameron, H. A., & Gould, E. (1994). Adult neurogenesis is regulated by adrenal steroids in the dentate gyrus. *Neuroscience, 61*(2), 203–209.

Cameron, H. A., McEwen, B. S., & Gould, E. (1995). Regulation of adult neurogenesis by excitatory input and NMDA receptor activation in the dentate gyrus. *The Journal of Neuroscience, 15*(6), 4687–4692.

Cameron, H. A., & McKay, R. D. (1999). Restoring production of hippocampal neurons in old age. *Nature Neuroscience, 2*(10), 894–897. http://dx.doi.org/10.1038/13197.

Cameron, H. A., & McKay, R. D. (2001). Adult neurogenesis produces a large pool of new granule cells in the dentate gyrus. *The Journal of Comparative Neurology, 435*(4), 406–417.

Cameron, H. A., Woolley, C. S., & Gould, E. (1993). Adrenal steroid receptor immunoreactivity in cells born in the adult rat dentate gyrus. *Brain Research, 611*(2), 342–346.

Cao, L., Jiao, X., Zuzga, D. S., Liu, Y., Fong, D. M., Young, D., et al. (2004). VEGF links hippocampal activity with neurogenesis, learning and memory. *Nature Genetics, 36*(8), 827–835. http://dx.doi.org/10.1038/ng1395.

Carpentier, P. A., & Palmer, T. D. (2009). Immune influence on adult neural stem cell regulation and function. *Neuron, 64*(1), 79–92. http://dx.doi.org/10.1016/j.neuron.2009.08.038.

Chancey, J. H., Adlaf, E. W., Sapp, M. C., Pugh, P. C., Wadiche, J. I., & Overstreet-Wadiche, L. S. (2013). GABA depolarization is required for experience-dependent synapse unsilencing in adult-born neurons. *The Journal of Neuroscience, 33*(15), 6614–6622. http://dx.doi.org/10.1523/JNEUROSCI.0781-13.2013.

Chancey, J. H., Poulsen, D. J., Wadiche, J. I., & Overstreet-Wadiche, L. (2014). Hilar mossy cells provide the first glutamatergic synapses to adult-born dentate granule cells. *The Journal of Neuroscience, 34*(6), 2349–2354. http://dx.doi.org/10.1523/JNEUROSCI.3620-13.2014.

Chawla, M. K., Guzowski, J. F., Ramirez-Amaya, V., Lipa, P., Hoffman, K. L., Marriott, L. K., ... Barnes, C. A. (2005). Sparse, environmentally selective expression of Arc RNA in the upper blade of the rodent fascia dentata by brief spatial experience. *Hippocampus, 15*(5), 579–586. http://dx.doi.org/10.1002/hipo.20091.

Chen, G., Rajkowska, G., Du, F., Seraji-Bozorgzad, N., & Manji, H. K. (2000). Enhancement of hippocampal neurogenesis by lithium. *Journal of Neurochemistry, 75*(4), 1729–1734.

Chen, Q., Nakajima, A., Choi, S. H., Xiong, X., Sisodia, S. S., & Tang, Y. P. (2008). Adult neurogenesis is functionally associated with AD-like neurodegeneration. *Neurobiology of Disease, 29*(2), 316–326.

Chetty, S., Friedman, A. R., Taravosh-Lahn, K., Kirby, E. D., Mirescu, C., Guo, F., … Kaufer, D. (2014). Stress and glucocorticoids promote oligodendrogenesis in the adult hippocampus. *Molecular Psychiatry, 19*(12), 1275–1283. http://dx.doi.org/10.1038/mp.2013.190.

Chevallier, N. L., Soriano, S., Kang, D. E., Masliah, E., Hu, G., & Koo, E. H. (2005). Perturbed neurogenesis in the adult hippocampus associated with presenilin-1 A246E mutation. *The American Journal of Pathology, 167*(1), 151–159.

Choi, S. H., Veeraraghavalu, K., Lazarov, O., Marler, S., Ransohoff, R. M., Ramirez, J. M., et al. (2008). Non-cell-autonomous effects of presenilin 1 variants on enrichment-mediated hippocampal progenitor cell proliferation and differentiation. *Neuron, 59*(4), 568–580.

Christian, K. M., Song, H., & Ming, G. L. (2014). Functions and dysfunctions of adult hippocampal neurogenesis. *Annual Review of Neuroscience, 37*, 243–262. http://dx.doi.org/10.1146/annurev-neuro-071013-014134.

Claiborne, B. J., Amaral, D. G., & Cowan, W. M. (1986). A light and electron microscopic analysis of the mossy fibers of the rat dentate gyrus. *The Journal of Comparative Neurology, 246*(4), 435–458. http://dx.doi.org/10.1002/cne.902460403.

Clelland, C. D., Choi, M., Romberg, C., Clemenson, G. D., Jr., Fragniere, A., Tyers, P., … Bussey, T. J. (2009). A functional role for adult hippocampal neurogenesis in spatial pattern separation. *Science, 325*(5937), 210–213. http://dx.doi.org/10.1126/science.1173215.

Convit, A., de Leon, M. J., Tarshish, C., De Santi, S., Kluger, A., Rusinek, H., et al. (1995). Hippocampal volume losses in minimally impaired elderly. *Lancet, 345*(8944), 266.

Coulter, D. A., & Carlson, G. C. (2007). Functional regulation of the dentate gyrus by GABA-mediated inhibition. *Progress in Brain Research, 163*, 235–243. http://dx.doi.org/10.1016/S0079-6123(07)63014-3.

Creer, D. J., Romberg, C., Saksida, L. M., van Praag, H., & Bussey, T. J. (2010). Running enhances spatial pattern separation in mice. *Proceedings of the National Academy of Sciences of the United States of America, 107*(5), 2367–2372. http://dx.doi.org/10.1073/pnas.0911725107.

Curtis, M. A., Low, V. F., & Faull, R. L. (2012). Neurogenesis and progenitor cells in the adult human brain: A comparison between hippocampal and subventricular progenitor proliferation. *Developmental Neurobiology, 72*(7), 990–1005. http://dx.doi.org/10.1002/dneu.22028.

David, D. J., Samuels, B. A., Rainer, Q., Wang, J. W., Marsteller, D., Mendez, I., … Hen, R. (2009). Neurogenesis-dependent and -independent effects of fluoxetine in an animal model of anxiety/depression. *Neuron, 62*(4), 479–493. http://dx.doi.org/10.1016/j.neuron.2009.04.017.

de Kloet, E. R., Joels, M., & Holsboer, F. (2005). Stress and the brain: From adaptation to disease. *Nature Reviews. Neuroscience, 6*(6), 463–475. http://dx.doi.org/10.1038/nrn1683.

Demars, M., Hu, Y. S., Gadadhar, A., & Lazarov, O. (2010). Impaired neurogenesis is an early event in the etiology of familial Alzheimer's disease in transgenic mice. *Journal of Neuroscience Research, 88*(10), 2103–2117. http://dx.doi.org/10.1002/jnr.22387.

Demars, M. P., Bartholomew, A., Strakova, Z., & Lazarov, O. (2011). Soluble amyloid precursor protein: A novel proliferation factor of adult progenitor cells of ectodermal and mesodermal origin. *Stem Cell Research & Therapy, 2*(4), 36. doi:scrt77 [pii] 10.1186/scrt77.

Demars, M. P., Hollands, C., Zhao, K. D., & Lazarov, O. (2013). Soluble amyloid precursor protein-alpha rescues age-linked decline in neural progenitor cell proliferation. *Neurobiology of Aging*. http://dx.doi.org/10.1016/j.neurobiolaging.2013.04.016.

Denny, C. A., Burghardt, N. S., Schachter, D. M., Hen, R., & Drew, M. R. (2012). 4- to 6-week-old adult-born hippocampal neurons influence novelty-evoked exploration and contextual fear conditioning. *Hippocampus, 22*(5), 1188–1201. http://dx.doi.org/10.1002/hipo.20964.

Denny, C. A., Kheirbek, M. A., Alba, E. L., Tanaka, K. F., Brachman, R. A., Laughman, K. B., ... Hen, R. (2014). Hippocampal memory traces are differentially modulated by experience, time, and adult neurogenesis. *Neuron, 83*(1), 189–201. http://dx.doi.org/10.1016/j.neuron.2014.05.018.

De Strooper, B., Annaert, W., Cupers, P., Saftig, P., Craessaerts, K., Mumm, J. S., ... Kopan, R. (1999). A presenilin-1-dependent gamma-secretase-like protease mediates release of Notch intracellular domain [see comments]. *Nature, 398*(6727), 518–522.

Dieni, C. V., Nietz, A. K., Panichi, R., Wadiche, J. I., & Overstreet-Wadiche, L. (2013). Distinct determinants of sparse activation during granule cell maturation. *The Journal of Neuroscience, 33*(49), 19131–19142. http://dx.doi.org/10.1523/JNEUROSCI.2289-13.2013.

Dong, H., Goico, B., Martin, M., Csernansky, C. A., Bertchume, A., & Csernansky, J. G. (2004). Modulation of hippocampal cell proliferation, memory, and amyloid plaque deposition in APPsw (Tg2576) mutant mice by isolation stress. *Neuroscience, 127*(3), 601–609. http://dx.doi.org/10.1016/j.neuroscience.2004.05.040.

Donovan, M. H., Yazdani, U., Norris, R. D., Games, D., German, D. C., & Eisch, A. J. (2006). Decreased adult hippocampal neurogenesis in the PDAPP mouse model of Alzheimer's disease. *The Journal of Comparative Neurology, 495*(1), 70–83.

Dranovsky, A., & Leonardo, E. D. (2012). Is there a role for young hippocampal neurons in adaptation to stress? *Behavioural Brain Research, 227*(2), 371–375. http://dx.doi.org/10.1016/j.bbr.2011.05.007.

Dranovsky, A., Picchini, A. M., Moadel, T., Sisti, A. C., Yamada, A., Kimura, S., ... Hen, R. (2011). Experience dictates stem cell fate in the adult hippocampus. *Neuron, 70*(5), 908–923. http://dx.doi.org/10.1016/j.neuron.2011.05.022.

Drew, L. J., Fusi, S., & Hen, R. (2013). Adult neurogenesis in the mammalian hippocampus: Why the dentate gyrus? *Learning & Memory, 20*(12), 710–729. http://dx.doi.org/10.1101/lm.026542.112.

Drew, M. R., Denny, C. A., & Hen, R. (2010). Arrest of adult hippocampal neurogenesis in mice impairs single- but not multiple-trial contextual fear conditioning. *Behavioral Neuroscience, 124*(4), 446–454. http://dx.doi.org/10.1037/a0020081.

Dupret, D., Revest, J. M., Koehl, M., Ichas, F., De Giorgi, F., Costet, P., ... Piazza, P. V. (2008). Spatial relational memory requires hippocampal adult neurogenesis. *PLoS One, 3*(4), e1959. http://dx.doi.org/10.1371/journal.pone.0001959.

Encinas, J. M., Vaahtokari, A., & Enikolopov, G. (2006). Fluoxetine targets early progenitor cells in the adult brain. *Proceedings of the National Academy of Sciences of the United States of America, 103*(21), 8233–8238. http://dx.doi.org/10.1073/pnas.0601992103.

Enwere, E., Shingo, T., Gregg, C., Fujikawa, H., Ohta, S., & Weiss, S. (2004). Aging results in reduced epidermal growth factor receptor signaling, diminished olfactory neurogenesis, and deficits in fine olfactory discrimination. *The Journal of Neuroscience, 24*(38), 8354–8365. http://dx.doi.org/10.1523/JNEUROSCI.2751-04.2004.

Eriksson, P. S., Perfilieva, E., Bjork-Eriksson, T., Alborn, A. M., Nordborg, C., Peterson, D. A., et al. (1998). Neurogenesis in the adult human hippocampus. *Nature Medicine, 4*(11), 1313–1317. http://dx.doi.org/10.1038/3305.

Ermini, F. V., Grathwohl, S., Radde, R., Yamaguchi, M., Staufenbiel, M., Palmer, T. D., et al. (2008). Neurogenesis and alterations of neural stem cells in mouse models of cerebral amyloidosis. *The American Journal of Pathology, 172*(6), 1520–1528.

Ernst, A., Alkass, K., Bernard, S., Salehpour, M., Perl, S., Tisdale, J., ... Frisen, J. (2014). Neurogenesis in the striatum of the adult human brain. *Cell, 156*(5), 1072–1083. http://dx.doi.org/10.1016/j.cell.2014.01.044.

Esposito, M. S., Piatti, V. C., Laplagne, D. A., Morgenstern, N. A., Ferrari, C. C., Pitossi, F. J., et al. (2005). Neuronal differentiation in the adult hippocampus recapitulates embryonic development. *The Journal of Neuroscience, 25*(44), 10074–10086. http://dx.doi.org/10.1523/JNEUROSCI.3114-05.2005.

Feng, R., Rampon, C., Tang, Y. P., Shrom, D., Jin, J., Kyin, M., ... Tsien, J. Z. (2001). Deficient neurogenesis in forebrain-specific presenilin-1 knockout mice is associated with reduced clearance of hippocampal memory traces. *Neuron, 32*(5), 911–926.

Ford, E. C., Achanta, P., Purger, D., Armour, M., Reyes, J., Fong, J., ... Quinones-Hinojosa, A. (2011). Localized CT-guided irradiation inhibits neurogenesis in specific regions of the adult mouse brain. *Radiation Research, 175*(6), 774–783. http://dx.doi.org/10.1667/RR2214.1.

Gadadhar, A., Marr, R. A., & Lazarov, O. (2011). Presenilin-1 regulates neural progenitor cell differentiation in the adult brain. *The Journal of Neuroscience, 31*(7), 2615–2623.

Gan, L., Qiao, S., Lan, X., Chi, L., Luo, C., Lien, L., ... Liu, R. (2008). Neurogenic responses to amyloid-beta plaques in the brain of Alzheimer's disease-like transgenic (pPDGF-APPSw,Ind) mice. *Neurobiology of Disease, 29*(1), 71–80.

Garcia, A., Steiner, B., Kronenberg, G., Bick-Sander, A., & Kempermann, G. (2004). Age-dependent expression of glucocorticoid- and mineralocorticoid receptors on neural precursor cell populations in the adult murine hippocampus. *Aging Cell, 3*(6), 363–371. http://dx.doi.org/10.1111/j.1474-9728.2004.00130.x.

Garcia, A. D., Doan, N. B., Imura, T., Bush, T. G., & Sofroniew, M. V. (2004). GFAP-expressing progenitors are the principal source of constitutive neurogenesis in adult mouse forebrain. *Nature Neuroscience, 7*(11), 1233–1241. http://dx.doi.org/10.1038/nn1340.

Gasbarri, A., Sulli, A., & Packard, M. G. (1997). The dopaminergic mesencephalic projections to the hippocampal formation in the rat. *Progress in Neuro-psychopharmacology & Biological Psychiatry, 21*(1), 1–22.

Ge, S., Goh, E. L., Sailor, K. A., Kitabatake, Y., Ming, G. L., & Song, H. (2006). GABA regulates synaptic integration of newly generated neurons in the adult brain. *Nature, 439*(7076), 589–593. http://dx.doi.org/10.1038/nature04404.

Ge, S., Yang, C. H., Hsu, K. S., Ming, G. L., & Song, H. (2007). A critical period for enhanced synaptic plasticity in newly generated neurons of the adult brain. *Neuron, 54*(4), 559–566. http://dx.doi.org/10.1016/j.neuron.2007.05.002.

Ghosal, K., Stathopoulos, A., & Pimplikar, S. W. (2010). APP intracellular domain impairs adult neurogenesis in transgenic mice by inducing neuroinflammation. *PLoS One, 5*(7), e11866. http://dx.doi.org/10.1371/journal.pone.0011866.

Golomb, J., Kluger, A., de Leon, M. J., Ferris, S. H., Convit, A., Mittelman, M. S., ... George, A. E. (1994). Hippocampal formation size in normal human aging: A correlate of delayed secondary memory performance. *Learning & Memory, 1*(1), 45–54.

Gould, E. (2007). How widespread is adult neurogenesis in mammals? *Nature Reviews. Neuroscience, 8*(6), 481–488. http://dx.doi.org/10.1038/nrn2147.

Gould, E., Cameron, H. A., Daniels, D. C., Woolley, C. S., & McEwen, B. S. (1992). Adrenal hormones suppress cell division in the adult rat dentate gyrus. *The Journal of Neuroscience, 12*(9), 3642–3650.

Gould, E., Reeves, A. J., Graziano, M. S., & Gross, C. G. (1999). Neurogenesis in the neocortex of adult primates. *Science, 286*(5439), 548–552.

Gu, Y., Arruda-Carvalho, M., Wang, J., Janoschka, S. R., Josselyn, S. A., Frankland, P. W., et al. (2012). Optical controlling reveals time-dependent roles for adult-born dentate granule cells. *Nature Neuroscience, 15*(12), 1700–1706. http://dx.doi.org/10.1038/nn.3260.

Hamilton, L. K., Aumont, A., Julien, C., Vadnais, A., Calon, F., & Fernandes, K. J. (2010). Widespread deficits in adult neurogenesis precede plaque and tangle formation in the 3xTg mouse model of Alzheimer's disease. *The European Journal of Neuroscience, 32*(6), 905–920. http://dx.doi.org/10.1111/j.1460-9568.2010.07379.x.

Hao, Y., Creson, T., Zhang, L., Li, P., Du, F., Yuan, P., ... Chen, G. (2004). Mood stabilizer valproate promotes ERK pathway-dependent cortical neuronal growth and neurogenesis. *The Journal of Neuroscience, 24*(29), 6590–6599. http://dx.doi.org/10.1523/JNEUROSCI.5747-03.2004.

Hartl, D., Rohe, M., Mao, L., Staufenbiel, M., Zabel, C., & Klose, J. (2008). Impairment of adolescent hippocampal plasticity in a mouse model for Alzheimer's disease precedes disease phenotype. *PLoS One, 3*(7), e2759. http://dx.doi.org/10.1371/journal.pone.0002759.

Hattiangady, B., & Shetty, A. K. (2008). Aging does not alter the number or phenotype of putative stem/progenitor cells in the neurogenic region of the hippocampus. *Neurobiology of Aging, 29*(1), 129–147. http://dx.doi.org/10.1016/j.neurobiolaging.2006.09.015.

Haughey, N. J., Liu, D., Nath, A., Borchard, A. C., & Mattson, M. P. (2002). Disruption of neurogenesis in the subventricular zone of adult mice, and in human cortical neuronal precursor cells in culture, by amyloid beta-peptide: Implications for the pathogenesis of Alzheimer's disease. *Neuromolecular Medicine, 1*(2), 125–135.

Haughey, N. J., Nath, A., Chan, S. L., Borchard, A. C., Rao, M. S., & Mattson, M. P. (2002). Disruption of neurogenesis by amyloid beta-peptide, and perturbed neural progenitor cell homeostasis, in models of Alzheimer's disease. *Journal of Neurochemistry, 83*(6), 1509–1524.

Heine, V. M., Maslam, S., Joels, M., & Lucassen, P. J. (2004). Prominent decline of newborn cell proliferation, differentiation, and apoptosis in the aging dentate gyrus, in absence of an age-related hypothalamus-pituitary-adrenal axis activation. *Neurobiology of Aging, 25*(3), 361–375. http://dx.doi.org/10.1016/S0197-4580(03)00090-3.

Ho, N. F., Hooker, J. M., Sahay, A., Holt, D. J., & Roffman, J. L. (2013). *In vivo* imaging of adult human hippocampal neurogenesis: Progress, pitfalls and promise. *Molecular Psychiatry, 18*(4), 404–416. http://dx.doi.org/10.1038/mp.2013.8.

Hu, P., Oomen, C., van Dam, A. M., Wester, J., Zhou, J. N., Joels, M., et al. (2012). A single-day treatment with mifepristone is sufficient to normalize chronic glucocorticoid induced suppression of hippocampal cell proliferation. *PLoS One, 7*(9), e46224. http://dx.doi.org/10.1371/journal.pone.0046224.

Hu, Y. S., Xu, P., Pigino, G., Brady, S. T., Larson, J., & Lazarov, O. (2010). Complex environment experience rescues impaired neurogenesis, enhances synaptic plasticity, and attenuates neuropathology in familial Alzheimer's disease-linked APPswe/PS1DeltaE9 mice. *The FASEB Journal, 24*(6), 1667–1681. http://dx.doi.org/10.1096/fj.09-136945.

Hunsaker, M. R., & Kesner, R. P. (2013). The operation of pattern separation and pattern completion processes associated with different attributes or domains of memory. *Neuroscience and Biobehavioral Reviews, 37*(1), 36–58. http://dx.doi.org/10.1016/j.neubiorev.2012.09.014.

Ikrar, T., Guo, N., He, K., Besnard, A., Levinson, S., Hill, A., … Sahay, A. (2013). Adult neurogenesis modifies excitability of the dentate gyrus. *Front Neural Circuits, 7*, 204. http://dx.doi.org/10.3389/fncir.2013.00204.

Jacobs, B. L., van Praag, H., & Gage, F. H. (2000). Adult brain neurogenesis and psychiatry: A novel theory of depression. *Molecular Psychiatry, 5*(3), 262–269.

Janzen, V., Forkert, R., Fleming, H. E., Saito, Y., Waring, M. T., Dombkowski, D. M., … Scadden, D. T. (2006). Stem-cell ageing modified by the cyclin-dependent kinase inhibitor p16INK4a. *Nature, 443*(7110), 421–426. http://dx.doi.org/10.1038/nature05159.

Jin, K., Galvan, V., Xie, L., Mao, X. O., Gorostiza, O. F., Bredesen, D. E., et al. (2004). Enhanced neurogenesis in Alzheimer's disease transgenic (PDGF-APPSw,Ind) mice. *Proceedings of the National Academy of Sciences of the United States of America, 101*(36), 13363–13367.

Kaslin, J., Ganz, J., & Brand, M. (2008). Proliferation, neurogenesis and regeneration in the non-mammalian vertebrate brain. *Philosophical Transactions of the Royal Society of London Series B, Biological Sciences, 363*(1489), 101–122. http://dx.doi.org/10.1098/rstb.2006.2015.

Kasuga, K., Kaneko, H., Nishizawa, M., Onodera, O., & Ikeuchi, T. (2007). Generation of intracellular domain of insulin receptor tyrosine kinase by gamma-secretase. *Biochemical and Biophysical Research Communications, 360*(1), 90–96.

Keilhoff, G., Bernstein, H. G., Becker, A., Grecksch, G., & Wolf, G. (2004). Increased neurogenesis in a rat ketamine model of schizophrenia. *Biological Psychiatry, 56*(5), 317–322. http://dx.doi.org/10.1016/j.biopsych.2004.06.010.

Kempermann, G. (2012). New neurons for 'survival of the fittest'. *Nature Reviews. Neuroscience*, *13*(10), 727–736. http://dx.doi.org/10.1038/nrn3319.

Kempermann, G., Gast, D., & Gage, F. H. (2002). Neuroplasticity in old age: Sustained five-fold induction of hippocampal neurogenesis by long-term environmental enrichment. *Annals of Neurology*, *52*(2), 135–143. http://dx.doi.org/10.1002/ana.10262.

Kesner, R. P. (2007). A behavioral analysis of dentate gyrus function. *Progress in Brain Research*, *163*, 567–576. http://dx.doi.org/10.1016/S0079-6123(07)63030-1.

Kesner, R. P. (2013). An analysis of the dentate gyrus function. *Behavioural Brain Research*, *254*(7), 1. http://dx.doi.org/10.1016/j.bbr.2013.01.012.

Kheirbek, M. A., Klemenhagen, K. C., Sahay, A., & Hen, R. (2012). Neurogenesis and generalization: A new approach to stratify and treat anxiety disorders. *Nature Neuroscience*, *15*(12), 1613–1620. http://dx.doi.org/10.1038/nn.3262.

Kheirbek, M. A., Tannenholz, L., & Hen, R. (2012). NR2B-dependent plasticity of adult-born granule cells is necessary for context discrimination. *The Journal of Neuroscience*, *32*(25), 8696–8702. http://dx.doi.org/10.1523/JNEUROSCI.1692-12.2012.

Khemissi, W., Farooq, R. K., Le Guisquet, A. M., Sakly, M., & Belzung, C. (2014). Dysregulation of the hypothalamus-pituitary-adrenal axis predicts some aspects of the behavioral response to chronic fluoxetine: Association with hippocampal cell proliferation. *Frontiers in Behavioral Neuroscience*, *8*, 340. http://dx.doi.org/10.3389/fnbeh.2014.00340.

Kim, W. R., Christian, K., Ming, G. L., & Song, H. (2012). Time-dependent involvement of adult-born dentate granule cells in behavior. *Behavioural Brain Research*, *227*(2), 470–479. http://dx.doi.org/10.1016/j.bbr.2011.07.012.

Kobilo, T., Liu, Q. R., Gandhi, K., Mughal, M., Shaham, Y., & van Praag, H. (2011). Running is the neurogenic and neurotrophic stimulus in environmental enrichment. *Learning & Memory*, *18*(9), 605–609. http://dx.doi.org/10.1101/lm.2283011.

Kohler, C. (1986). Intrinsic connections of the retrohippocampal region in the rat brain. II. The medial entorhinal area. *The Journal of Comparative Neurology*, *246*(2), 149–169. http://dx.doi.org/10.1002/cne.902460202.

Kokoeva, M. V., Yin, H., & Flier, J. S. (2005). Neurogenesis in the hypothalamus of adult mice: Potential role in energy balance. *Science*, *310*(5748), 679–683. http://dx.doi.org/10.1126/science.1115360.

Kolecki, R., Lafauci, G., Rubenstein, R., Mazur-Kolecka, B., Kaczmarski, W., & Frackowiak, J. (2008). The effect of amyloidosis-beta and ageing on proliferation of neuronal progenitor cells in APP-transgenic mouse hippocampus and in culture. *Acta Neuropathologica*, *116*(4), 419–424.

Krezymon, A., Richetin, K., Halley, H., Roybon, L., Lassalle, J. M., Frances, B., … Rampon, C. (2013). Modifications of hippocampal circuits and early disruption of adult neurogenesis in the tg2576 mouse model of Alzheimer's disease. *PLoS One*, *8*(9), e76497. http://dx.doi.org/10.1371/journal.pone.0076497.

Kronenberg, G., Bick-Sander, A., Bunk, E., Wolf, C., Ehninger, D., & Kempermann, G. (2006). Physical exercise prevents age-related decline in precursor cell activity in the mouse dentate gyrus. *Neurobiology of Aging*, *27*(10), 1505–1513. http://dx.doi.org/10.1016/j.neurobiolaging.2005.09.016.

Kuhn, H. G., & Cooper-Kuhn, C. M. (2007). Bromodeoxyuridine and the detection of neurogenesis. *Current Pharmaceutical Biotechnology*, *8*(3), 127–131.

Kuhn, H. G., Dickinson-Anson, H., & Gage, F. H. (1996). Neurogenesis in the dentate gyrus of the adult rat: Age-related decrease of neuronal progenitor proliferation. *The Journal of Neuroscience*, *16*(6), 2027–2033.

Kuzumaki, N., Ikegami, D., Tamura, R., Hareyama, N., Imai, S., Narita, M., … Narita, M. (2011). Hippocampal epigenetic modification at the brain-derived neurotrophic factor gene induced by an enriched environment. *Hippocampus*, *21*(2), 127–132. http://dx.doi.org/10.1002/hipo.20775.

Lacefield, C. O., Itskov, V., Reardon, T., Hen, R., & Gordon, J. A. (2012). Effects of adult-generated granule cells on coordinated network activity in the dentate gyrus. *Hippocampus*, *22*(1), 106–116. http://dx.doi.org/10.1002/hipo.20860.

Lazarov, O., Demars, M. P., Zhao Kda, T., Ali, H. M., Grauzas, V., Kney, A., et al. (2012). Impaired survival of neural progenitor cells in dentate gyrus of adult mice lacking fMRP. *Hippocampus*, *22*(6), 1220–1224. http://dx.doi.org/10.1002/hipo.20989.

Lazarov, O., & Marr, R. A. (2010). Neurogenesis and Alzheimer's disease: At the crossroads. *Experimental Neurology*, *223*(2), 267–281.

Lazarov, O., Mattson, M. P., Peterson, D. A., Pimplikar, S. W., & van Praag, H. (2010). When neurogenesis encounters aging and disease. *Trends in Neurosciences*, *33*(12), 569–579. http://dx.doi.org/10.1016/j.tins.2010.09.003.

Lee, K. J., Kim, S. J., Kim, S. W., Choi, S. H., Shin, Y. C., Park, S. H., … Shin, K. H. (2006). Chronic mild stress decreases survival, but not proliferation, of new-born cells in adult rat hippocampus. *Experimental & Molecular Medicine*, *38*(1), 44–54. http://dx.doi.org/10.1038/emm.2006.6.

Lee, S. W., Clemenson, G. D., & Gage, F. H. (2012). New neurons in an aged brain. *Behavioural Brain Research*, *227*(2), 497–507. http://dx.doi.org/10.1016/j.bbr.2011.10.009.

Lehmann, M. L., Brachman, R. A., Martinowich, K., Schloesser, R. J., & Herkenham, M. (2013). Glucocorticoids orchestrate divergent effects on mood through adult neurogenesis. *The Journal of Neuroscience*, *33*(7), 2961–2972. http://dx.doi.org/10.1523/JNEUROSCI.3878-12.2013.

Leranth, C., & Hajszan, T. (2007). Extrinsic afferent systems to the dentate gyrus. *Progress in Brain Research*, *163*, 63–84. http://dx.doi.org/10.1016/S0079-6123(07)63004-0.

Leutgeb, J. K., Leutgeb, S., Moser, M. B., & Moser, E. I. (2007). Pattern separation in the dentate gyrus and CA3 of the hippocampus. *Science*, *315*(5814), 961–966. http://dx.doi.org/10.1126/science.1135801.

Li, Y., Aimone, J. B., Xu, X., Callaway, E. M., & Gage, F. H. (2012). Development of GABAergic inputs controls the contribution of maturing neurons to the adult hippocampal network. *Proceedings of the National Academy of Sciences of the United States of America*, *109*(11), 4290–4295. http://dx.doi.org/10.1073/pnas.1120754109.

Li, Y., Mu, Y., & Gage, F. H. (2009). Development of neural circuits in the adult hippocampus. *Current Topics in Developmental Biology*, *87*, 149–174. http://dx.doi.org/10.1016/S0070-2153(09)01205-8.

Lie, D. C., Colamarino, S. A., Song, H. J., Desire, L., Mira, H., Consiglio, A., … Gage, F. H. (2005). Wnt signalling regulates adult hippocampal neurogenesis. *Nature*, *437*(7063), 1370–1375.

Lim, D. A., & Alvarez-Buylla, A. (2014). Adult neural stem cells stake their ground. *Trends in Neurosciences*, *37*(10), 563–571. http://dx.doi.org/10.1016/j.tins.2014.08.006.

Lindsey, B. W., & Tropepe, V. (2006). A comparative framework for understanding the biological principles of adult neurogenesis. *Progress in Neurobiology*, *80*(6), 281–307. http://dx.doi.org/10.1016/j.pneurobio.2006.11.007.

Lledo, P. M., Alonso, M., & Grubb, M. S. (2006). Adult neurogenesis and functional plasticity in neuronal circuits. *Nature Reviews. Neuroscience*, *7*(3), 179–193. http://dx.doi.org/10.1038/nrn1867.

Lopez-Toledano, M. A., & Shelanski, M. L. (2004). Neurogenic effect of beta-amyloid peptide in the development of neural stem cells. *The Journal of Neuroscience*, *24*(23), 5439–5444.

Lopez-Toledano, M. A., & Shelanski, M. L. (2007). Increased neurogenesis in young transgenic mice overexpressing human APP(Sw, Ind). *Journal of Alzheimer's Disease: JAD*, *12*(3), 229–240.

Lucassen, P. J., Stumpel, M. W., Wang, Q., & Aronica, E. (2010). Decreased numbers of progenitor cells but no response to antidepressant drugs in the hippocampus of elderly depressed patients. *Neuropharmacology*, *58*(6), 940–949. http://dx.doi.org/10.1016/j.neuropharm.2010.01.012.

Lugert, S., Basak, O., Knuckles, P., Haussler, U., Fabel, K., Gotz, M., ... Giachino, C. (2010). Quiescent and active hippocampal neural stem cells with distinct morphologies respond selectively to physiological and pathological stimuli and aging. *Cell Stem Cell, 6*(5), 445–456. http://dx.doi.org/10.1016/j.stem.2010.03.017.

Madsen, T. M., Treschow, A., Bengzon, J., Bolwig, T. G., Lindvall, O., & Tingstrom, A. (2000). Increased neurogenesis in a model of electroconvulsive therapy. *Biological Psychiatry, 47*(12), 1043–1049.

Magarinos, A. M., & McEwen, B. S. (1995a). Stress-induced atrophy of apical dendrites of hippocampal CA3c neurons: Comparison of stressors. *Neuroscience, 69*(1), 83–88.

Magarinos, A. M., & McEwen, B. S. (1995b). Stress-induced atrophy of apical dendrites of hippocampal CA3c neurons: Involvement of glucocorticoid secretion and excitatory amino acid receptors. *Neuroscience, 69*(1), 89–98.

Malberg, J. E., Eisch, A. J., Nestler, E. J., & Duman, R. S. (2000). Chronic antidepressant treatment increases neurogenesis in adult rat hippocampus. *The Journal of Neuroscience, 20*(24), 9104–9110.

Manganas, L. N., Zhang, X., Li, Y., Hazel, R. D., Smith, S. D., Wagshul, M. E., ... Maletic-Savatic, M. (2007). Magnetic resonance spectroscopy identifies neural progenitor cells in the live human brain. *Science, 318*(5852), 980–985. http://dx.doi.org/10.1126/science.1147851.

Marambaud, P., Shioi, J., Serban, G., Georgakopoulos, A., Sarner, S., Nagy, V., ... Robakis, N. K. (2002). A presenilin-1/gamma-secretase cleavage releases the E-cadherin intracellular domain and regulates disassembly of adherens junctions. *The EMBO Journal, 21*(8), 1948–1956.

Maretzky, T., Reiss, K., Ludwig, A., Buchholz, J., Scholz, F., Proksch, E., ... Saftig, P. (2005). ADAM10 mediates E-cadherin shedding and regulates epithelial cell-cell adhesion, migration, and beta-catenin translocation. *Proceedings of the National Academy of Sciences of the United States of America, 102*(26), 9182–9187.

Maretzky, T., Schulte, M., Ludwig, A., Rose-John, S., Blobel, C., Hartmann, D., ... Reiss, K. (2005). L1 is sequentially processed by two differently activated metalloproteases and presenilin/gamma-secretase and regulates neural cell adhesion, cell migration, and neurite outgrowth. *Molecular and Cellular Biology, 25*(20), 9040–9053.

Marin-Burgin, A., Mongiat, L. A., Pardi, M. B., & Schinder, A. F. (2012). Unique processing during a period of high excitation/inhibition balance in adult-born neurons. *Science, 335*(6073), 1238–1242. http://dx.doi.org/10.1126/science.1214956.

Marr, D. (1971). Simple memory: A theory for archicortex. *Philosophical Transactions of the Royal Society of London Series B, Biological Sciences, 262*(841), 23–81.

McElroy, B., Powell, J. C., & McCarthy, J. V. (2007). The insulin-like growth factor 1 (IGF-1) receptor is a substrate for gamma-secretase-mediated intramembrane proteolysis. *Biochemical and Biophysical Research Communications, 358*(4), 1136–1141.

McEwen, B. S. (1998). Stress, adaptation, and disease. Allostasis and allostatic load. *Annals of the New York Academy of Sciences, 840*, 33–44.

McEwen, B. S. (1999). Stress and hippocampal plasticity. *Annual Review of Neuroscience, 22*, 105–122. http://dx.doi.org/10.1146/annurev.neuro.22.1.105.

McEwen, B. S. (2005). Glucocorticoids, depression, and mood disorders: Structural remodeling in the brain. *Metabolism, 54*(5 Suppl. 1), 20–23. http://dx.doi.org/10.1016/j.metabol.2005.01.008.

McEwen, B. S., & Magarinos, A. M. (1997). Stress effects on morphology and function of the hippocampus. *Annals of the New York Academy of Sciences, 821*, 271–284.

Miller, B. R., & Hen, R. (2014). The current state of the neurogenic theory of depression and anxiety. *Current Opinion in Neurobiology, 30C*, 51–58. http://dx.doi.org/10.1016/j.conb.2014.08.012.

Ming, G. L., & Song, H. (2011). Adult neurogenesis in the mammalian brain: Significant answers and significant questions. *Neuron, 70*(4), 687–702. http://dx.doi.org/10.1016/j.neuron.2011.05.001.

Mira, H., Andreu, Z., Suh, H., Lie, D. C., Jessberger, S., Consiglio, A., … Gage, F. H. (2010). Signaling through BMPR-IA regulates quiescence and long-term activity of neural stem cells in the adult hippocampus. *Cell Stem Cell, 7*(1), 78–89. http://dx.doi.org/10.1016/j.stem.2010.04.016.

Moghaddam, B. (1993). Stress preferentially increases extraneuronal levels of excitatory amino acids in the prefrontal cortex: Comparison to hippocampus and basal ganglia. *Journal of Neurochemistry, 60*(5), 1650–1657.

Moghaddam, B., Bolinao, M. L., Stein-Behrens, B., & Sapolsky, R. (1994). Glucocorticoids mediate the stress-induced extracellular accumulation of glutamate. *Brain Research, 655*(1–2), 251–254.

Molofsky, A. V., Slutsky, S. G., Joseph, N. M., He, S., Pardal, R., Krishnamurthy, J., … Morrison, S. J. (2006). Increasing p16INK4a expression decreases forebrain progenitors and neurogenesis during ageing. *Nature, 443*(7110), 448–452. http://dx.doi.org/10.1038/nature05091.

Mongiat, L. A., Esposito, M. S., Lombardi, G., & Schinder, A. F. (2009). Reliable activation of immature neurons in the adult hippocampus. *PLoS One, 4*(4), e5320. http://dx.doi.org/10.1371/journal.pone.0005320.

Mongiat, L. A., & Schinder, A. F. (2011). Adult neurogenesis and the plasticity of the dentate gyrus network. *The European Journal of Neuroscience, 33*(6), 1055–1061. http://dx.doi.org/10.1111/j.1460-9568.2011.07603.x.

Monje, M. L., Mizumatsu, S., Fike, J. R., & Palmer, T. D. (2002). Irradiation induces neural precursor-cell dysfunction. *Nature Medicine, 8*(9), 955–962. http://dx.doi.org/10.1038/nm749.

Morgenstern, N. A., Lombardi, G., & Schinder, A. F. (2008). Newborn granule cells in the ageing dentate gyrus. *The Journal of Physiology, 586*(16), 3751–3757. http://dx.doi.org/10.1113/jphysiol.2008.154807.

Mustroph, M. L., Chen, S., Desai, S. C., Cay, E. B., DeYoung, E. K., & Rhodes, J. S. (2012). Aerobic exercise is the critical variable in an enriched environment that increases hippocampal neurogenesis and water maze learning in male C57BL/6J mice. *Neuroscience, 219*, 62–71. http://dx.doi.org/10.1016/j.neuroscience.2012.06.007.

Nakashiba, T., Cushman, J. D., Pelkey, K. A., Renaudineau, S., Buhl, D. L., McHugh, T. J., … Tonegawa, S. (2012). Young dentate granule cells mediate pattern separation, whereas old granule cells facilitate pattern completion. *Cell, 149*(1), 188–201. http://dx.doi.org/10.1016/j.cell.2012.01.046.

Namba, T., Mochizuki, H., Onodera, M., Mizuno, Y., Namiki, H., & Seki, T. (2005). The fate of neural progenitor cells expressing astrocytic and radial glial markers in the postnatal rat dentate gyrus. *The European Journal of Neuroscience, 22*(8), 1928–1941. http://dx.doi.org/10.1111/j.1460-9568.2005.04396.x.

Neunuebel, J. P., & Knierim, J. J. (2012). Spatial firing correlates of physiologically distinct cell types of the rat dentate gyrus. *The Journal of Neuroscience, 32*(11), 3848–3858. http://dx.doi.org/10.1523/JNEUROSCI.6038-11.2012.

Neunuebel, J. P., & Knierim, J. J. (2014). CA3 retrieves coherent representations from degraded input: Direct evidence for CA3 pattern completion and dentate gyrus pattern separation. *Neuron, 81*(2), 416–427. http://dx.doi.org/10.1016/j.neuron.2013.11.017.

Newell-Price, J., Bertagna, X., Grossman, A. B., & Nieman, L. K. (2006). Cushing's syndrome. *Lancet, 367*(9522), 1605–1617. doi:10.1016/S0140-6736(06)68699-6.

Ni, C. Y., Murphy, M. P., Golde, T. E., & Carpenter, G. (2001). gamma-Secretase cleavage and nuclear localization of ErbB-4 receptor tyrosine kinase. *Science, 294*(5549), 2179–2181.

Niibori, Y., Yu, T. S., Epp, J. R., Akers, K. G., Josselyn, S. A., & Frankland, P. W. (2012). Suppression of adult neurogenesis impairs population coding of similar contexts in hippocampal CA3 region. *Nature Communications, 3*, 1253. http://dx.doi.org/10.1038/ncomms2261.

Nishino, J., Kim, I., Chada, K., & Morrison, S. J. (2008). Hmga2 promotes neural stem cell self-renewal in young but not old mice by reducing p16Ink4a and p19Arf Expression. *Cell, 135*(2), 227–239. http://dx.doi.org/10.1016/j.cell.2008.09.017.

Nottebohm, F. (2004). The road we travelled: Discovery, choreography, and significance of brain replaceable neurons. *Annals of the New York Academy of Sciences, 1016*, 628–658. http://dx.doi.org/10.1196/annals.1298.027.

Novkovic, T., Mittmann, T., & Manahan-Vaughan, D. (2014). BDNF contributes to the facilitation of hippocampal synaptic plasticity and learning enabled by environmental enrichment. *Hippocampus.* http://dx.doi.org/10.1002/hipo.22342.

Oh, J., McCloskey, M. A., Blong, C. C., Bendickson, L., Nilsen-Hamilton, M., & Sakaguchi, D. S. (2010). Astrocyte-derived interleukin-6 promotes specific neuronal differentiation of neural progenitor cells from adult hippocampus. *Journal of Neuroscience Research, 88*(13), 2798–2809. http://dx.doi.org/10.1002/jnr.22447.

Overstreet Wadiche, L., Bromberg, D. A., Bensen, A. L., & Westbrook, G. L. (2005). GABAergic signaling to newborn neurons in dentate gyrus. *Journal of Neurophysiology, 94*(6), 4528–4532. http://dx.doi.org/10.1152/jn.00633.2005.

Overstreet-Wadiche, L. S., Bromberg, D. A., Bensen, A. L., & Westbrook, G. L. (2006). Seizures accelerate functional integration of adult-generated granule cells. *The Journal of Neuroscience, 26*(15), 4095–4103. http://dx.doi.org/10.1523/JNEUROSCI.5508-05.2006.

Palmer, T. D., Willhoite, A. R., & Gage, F. H. (2000). Vascular niche for adult hippocampal neurogenesis. *The Journal of Comparative Neurology, 425*(4), 479–494.

Parent, J. M., Tada, E., Fike, J. R., & Lowenstein, D. H. (1999). Inhibition of dentate granule cell neurogenesis with brain irradiation does not prevent seizure-induced mossy fiber synaptic reorganization in the rat. *The Journal of Neuroscience, 19*(11), 4508–4519.

Peissner, W., Kocher, M., Treuer, H., & Gillardon, F. (1999). Ionizing radiation-induced apoptosis of proliferating stem cells in the dentate gyrus of the adult rat hippocampus. *Brain Research. Molecular Brain Research, 71*(1), 61–68.

Pereira, A. C., Huddleston, D. E., Brickman, A. M., Sosunov, A. A., Hen, R., McKhann, G. M., … Small, S. A. (2007). An *in vivo* correlate of exercise-induced neurogenesis in the adult dentate gyrus. *Proceedings of the National Academy of Sciences of the United States of America, 104*(13), 5638–5643. http://dx.doi.org/10.1073/pnas.0611721104.

Pham, K., Nacher, J., Hof, P. R., & McEwen, B. S. (2003). Repeated restraint stress suppresses neurogenesis and induces biphasic PSA-NCAM expression in the adult rat dentate gyrus. *The European Journal of Neuroscience, 17*(4), 879–886.

Poirier, R., Veltman, I., Pflimlin, M. C., Knoflach, F., & Metzger, F. (2010). Enhanced dentate gyrus synaptic plasticity but reduced neurogenesis in a mouse model of amyloidosis. *Neurobiology of Disease, 40*(2), 386–393. doi:S0969-9961(10)00210-X [pii].

Rakic, P. (2002). Neurogenesis in adult primate neocortex: An evaluation of the evidence. *Nature Reviews. Neuroscience, 3*(1), 65–71. http://dx.doi.org/10.1038/nrn700.

Ramirez-Amaya, V., Vazdarjanova, A., Mikhael, D., Rosi, S., Worley, P. F., & Barnes, C. A. (2005). Spatial exploration-induced Arc mRNA and protein expression: Evidence for selective, network-specific reactivation. *The Journal of Neuroscience, 25*(7), 1761–1768. http://dx.doi.org/10.1523/JNEUROSCI.4342-04.2005.

Reif, A., Fritzen, S., Finger, M., Strobel, A., Lauer, M., Schmitt, A., et al. (2006). Neural stem cell proliferation is decreased in schizophrenia, but not in depression. *Molecular Psychiatry, 11*(5), 514–522. http://dx.doi.org/10.1038/sj.mp.4001791.

Rodriguez, J. J., Jones, V. C., Tabuchi, M., Allan, S. M., Knight, E. M., LaFerla, F. M., … Verkhratsky, A. (2008). Impaired adult neurogenesis in the dentate gyrus of a triple transgenic mouse model of Alzheimer's disease. *PLoS One, 3*(8), e2935.

Rodriguez, J. J., Jones, V. C., & Verkhratsky, A. (2009). Impaired cell proliferation in the subventricular zone in an Alzheimer's disease model. *Neuroreport, 20*(10), 907–912. http://dx.doi.org/10.1097/WNR.0b013e32832be77d.

Rohe, M., Carlo, A. S., Breyhan, H., Sporbert, A., Militz, D., Schmidt, V., … Andersen, O. M. (2008). Sortilin-related receptor with A-type repeats (SORLA) affects the amyloid precursor protein-dependent stimulation of ERK signaling and adult neurogenesis. *The Journal of Biological Chemistry, 283*(21), 14826–14834.

Rolls, E. T. (2013). The mechanisms for pattern completion and pattern separation in the hippocampus. *Frontiers in Systems Neuroscience, 7,* 74. http://dx.doi.org/10.3389/fnsys.2013.00074.

Rossi, S. L., Mahairaki, V., Zhou, L., Song, Y., & Koliatsos, V. E. (2014). Remodeling of the piriform cortex after lesion in adult rodents. *Neuroreport, 25*(13), 1006–1012. http://dx.doi.org/10.1097/WNR.0000000000000203.

Rossjohn, J., Cappai, R., Feil, S. C., Henry, A., McKinstry, W. J., Galatis, D., … Parker, M. W. (1999). Crystal structure of the N-terminal, growth factor-like domain of Alzheimer amyloid precursor protein. *Nature Structural Biology, 6*(4), 327–331. http://dx.doi.org/10.1038/7562.

Sahay, A., Scobie, K. N., Hill, A. S., O'Carroll, C. M., Kheirbek, M. A., Burghardt, N. S., … Hen, R. (2011). Increasing adult hippocampal neurogenesis is sufficient to improve pattern separation. *Nature, 472*(7344), 466–470. http://dx.doi.org/10.1038/nature09817.

Sahay, A., Wilson, D. A., & Hen, R. (2011). Pattern separation: A common function for new neurons in hippocampus and olfactory bulb. *Neuron, 70*(4), 582–588. http://dx.doi.org/10.1016/j.neuron.2011.05.012.

Santarelli, L., Saxe, M., Gross, C., Surget, A., Battaglia, F., Dulawa, S., … Hen, R. (2003). Requirement of hippocampal neurogenesis for the behavioral effects of antidepressants. *Science, 301*(5634), 805–809. http://dx.doi.org/10.1126/science.1083328.

Sapolsky, R. M. (2000). Glucocorticoids and hippocampal atrophy in neuropsychiatric disorders. *Archives of General Psychiatry, 57*(10), 925–935.

Sapolsky, R. M., Krey, L. C., & McEwen, B. S. (1984). Glucocorticoid-sensitive hippocampal neurons are involved in terminating the adrenocortical stress response. *Proceedings of the National Academy of Sciences of the United States of America, 81*(19), 6174–6177.

Sardi, S. P., Murtie, J., Koirala, S., Patten, B. A., & Corfas, G. (2006). Presenilin-dependent ErbB4 nuclear signaling regulates the timing of astrogenesis in the developing brain. *Cell, 127*(1), 185–197.

Schloesser, R. J., Jimenez, D. V., Hardy, N. F., Paredes, D., Catlow, B. J., Manji, H. K., … Martinowich, K. (2014). Atrophy of pyramidal neurons and increased stress-induced glutamate levels in CA3 following chronic suppression of adult neurogenesis. *Brain Structure & Function, 219*(3), 1139–1148. http://dx.doi.org/10.1007/s00429-013-0532-8.

Schloesser, R. J., Lehmann, M., Martinowich, K., Manji, H. K., & Herkenham, M. (2010). Environmental enrichment requires adult neurogenesis to facilitate the recovery from psychosocial stress. *Molecular Psychiatry, 15*(12), 1152–1163. http://dx.doi.org/10.1038/mp.2010.34.

Schloesser, R. J., Manji, H. K., & Martinowich, K. (2009). Suppression of adult neurogenesis leads to an increased hypothalamo-pituitary-adrenal axis response. *Neuroreport, 20*(6), 553–557. http://dx.doi.org/10.1097/WNR.0b013e3283293e59.

Schmidt-Hieber, C., Jonas, P., & Bischofberger, J. (2004). Enhanced synaptic plasticity in newly generated granule cells of the adult hippocampus. *Nature, 429*(6988), 184–187. http://dx.doi.org/10.1038/nature02553.

Schoenfeld, T. J., & Gould, E. (2013). Differential effects of stress and glucocorticoids on adult neurogenesis. *Current Topics in Behavioral Neurosciences, 15,* 139–164. http://dx.doi.org/10.1007/7854_2012_233.

Scott, B. W., Wojtowicz, J. M., & Burnham, W. M. (2000). Neurogenesis in the dentate gyrus of the rat following electroconvulsive shock seizures. *Experimental Neurology, 165*(2), 231–236. http://dx.doi.org/10.1006/exnr.2000.7458.

Selye, H. (1956). What is stress? *Metabolism, 5*(5), 525–530.

Seri, B., Garcia-Verdugo, J. M., Collado-Morente, L., McEwen, B. S., & Alvarez-Buylla, A. (2004). Cell types, lineage, and architecture of the germinal zone in the adult dentate gyrus. *The Journal of Comparative Neurology, 478*(4), 359–378. http://dx.doi.org/10.1002/cne.20288.

Seri, B., Garcia-Verdugo, J. M., McEwen, B. S., & Alvarez-Buylla, A. (2001). Astrocytes give rise to new neurons in the adult mammalian hippocampus. *The Journal of Neuroscience, 21*(18), 7153–7160.

Sheline, Y. I., Wang, P. W., Gado, M. H., Csernansky, J. G., & Vannier, M. W. (1996). Hippocampal atrophy in recurrent major depression. *Proceedings of the National Academy of Sciences of the United States of America, 93*(9), 3908–3913.

Shen, J., Bronson, R. T., Chen, D. F., Xia, W., Selkoe, D. J., & Tonegawa, S. (1997). Skeletal and CNS defects in Presenilin-1-deficient mice. *Cell, 89*(4), 629–639.

Shen, Q., Wang, Y., Kokovay, E., Lin, G., Chuang, S. M., Goderie, S. K., ... Temple, S. (2008). Adult SVZ stem cells lie in a vascular niche: A quantitative analysis of niche cell-cell interactions. *Cell Stem Cell, 3*(3), 289–300. http://dx.doi.org/10.1016/j.stem.2008.07.026.

Shetty, A. K., Hattiangady, B., & Shetty, G. A. (2005). Stem/progenitor cell proliferation factors FGF-2, IGF-1, and VEGF exhibit early decline during the course of aging in the hippocampus: Role of astrocytes. *Glia, 51*(3), 173–186. http://dx.doi.org/10.1002/glia.20187.

Shors, T. J., Miesegaes, G., Beylin, A., Zhao, M., Rydel, T., & Gould, E. (2001). Neurogenesis in the adult is involved in the formation of trace memories. *Nature, 410*(6826), 372–376. http://dx.doi.org/10.1038/35066584.

Sierra, A., Encinas, J. M., Deudero, J. J., Chancey, J. H., Enikolopov, G., Overstreet-Wadiche, L. S., ... Maletic-Savatic, M. (2010). Microglia shape adult hippocampal neurogenesis through apoptosis-coupled phagocytosis. *Cell Stem Cell, 7*(4), 483–495. http://dx.doi.org/10.1016/j.stem.2010.08.014.

Singer, B. H., Gamelli, A. E., Fuller, C. L., Temme, S. J., Parent, J. M., & Murphy, G. G. (2011). Compensatory network changes in the dentate gyrus restore long-term potentiation following ablation of neurogenesis in young-adult mice. *Proceedings of the National Academy of Sciences of the United States of America, 108*(13), 5437–5442. http://dx.doi.org/10.1073/pnas.1015425108.

Singer, B. H., Jutkiewicz, E. M., Fuller, C. L., Lichtenwalner, R. J., Zhang, H., Velander, A. J., ... Parent, J. M. (2009). Conditional ablation and recovery of forebrain neurogenesis in the mouse. *The Journal of Comparative Neurology, 514*(6), 567–582. http://dx.doi.org/10.1002/cne.22052.

Sloviter, R. S. (1987). Decreased hippocampal inhibition and a selective loss of interneurons in experimental epilepsy. *Science, 235*(4784), 73–76.

Snyder, J. S., Soumier, A., Brewer, M., Pickel, J., & Cameron, H. A. (2011). Adult hippocampal neurogenesis buffers stress responses and depressive behaviour. *Nature, 476*(7361), 458–461. http://dx.doi.org/10.1038/nature10287.

Song, H., Stevens, C. F., & Gage, F. H. (2002). Astroglia induce neurogenesis from adult neural stem cells. *Nature, 417*(6884), 39–44. http://dx.doi.org/10.1038/417039a.

Song, J., Christian, K. M., Ming, G. L., & Song, H. (2012). Modification of hippocampal circuitry by adult neurogenesis. *Developmental Neurobiology, 72*(7), 1032–1043. http://dx.doi.org/10.1002/dneu.22014.

Song, J., Sun, J., Moss, J., Wen, Z., Sun, G. J., Hsu, D., ... Song, H. (2013). Parvalbumin interneurons mediate neuronal circuitry-neurogenesis coupling in the adult hippocampus. *Nature Neuroscience, 16*(12), 1728–1730. http://dx.doi.org/10.1038/nn.3572.

Song, J., Zhong, C., Bonaguidi, M. A., Sun, G. J., Hsu, D., Gu, Y., ... Song, H. (2012). Neuronal circuitry mechanism regulating adult quiescent neural stem-cell fate decision. *Nature, 489*(7414), 150–154. http://dx.doi.org/10.1038/nature11306.

Spalding, K. L., Bergmann, O., Alkass, K., Bernard, S., Salehpour, M., Huttner, H. B., ... Frisen, J. (2013). Dynamics of hippocampal neurogenesis in adult humans. *Cell, 153*(6), 1219–1227. http://dx.doi.org/10.1016/j.cell.2013.05.002.

Spalding, K. L., Bhardwaj, R. D., Buchholz, B. A., Druid, H., & Frisen, J. (2005). Retrospective birth dating of cells in humans. *Cell, 122*(1), 133–143. http://dx.doi.org/10.1016/j.cell.2005.04.028.

Squire, L. R., Stark, C. E., & Clark, R. E. (2004). The medial temporal lobe. *Annual Review of Neuroscience, 27*, 279–306. http://dx.doi.org/10.1146/annurev.neuro.27.070203.144130.

Stangl, D., & Thuret, S. (2009). Impact of diet on adult hippocampal neurogenesis. *Genes & Nutrition, 4*(4), 271–282. http://dx.doi.org/10.1007/s12263-009-0134-5.

Starkman, M. N., Gebarski, S. S., Berent, S., & Schteingart, D. E. (1992). Hippocampal formation volume, memory dysfunction, and cortisol levels in patients with Cushing's syndrome. *Biological Psychiatry, 32*(9), 756–765.

Suh, H., Consiglio, A., Ray, J., Sawai, T., D'Amour, K. A., & Gage, F. H. (2007). *In vivo* fate analysis reveals the multipotent and self-renewal capacities of Sox2+ neural stem cells in the adult hippocampus. *Cell Stem Cell, 1*(5), 515–528. http://dx.doi.org/10.1016/j.stem.2007.09.002.

Sun, B., Halabisky, B., Zhou, Y., Palop, J. J., Yu, G., Mucke, L., et al. (2009). Imbalance between GABAergic and Glutamatergic transmission impairs adult neurogenesis in an animal model of Alzheimer's disease. *Cell Stem Cell, 5*(6), 624–633. http://dx.doi.org/10.1016/j.stem.2009.10.003.

Sun, F., Wang, X., Mao, X., Xie, L., & Jin, K. (2012). Ablation of neurogenesis attenuates recovery of motor function after focal cerebral ischemia in middle-aged mice. *PLoS One, 7*(10), e46326. http://dx.doi.org/10.1371/journal.pone.0046326.

Sun, G. J., Sailor, K. A., Mahmood, Q. A., Chavali, N., Christian, K. M., Song, H., et al. (2013). Seamless reconstruction of intact adult-born neurons by serial end-block imaging reveals complex axonal guidance and development in the adult hippocampus. *The Journal of Neuroscience, 33*(28), 11400–11411. http://dx.doi.org/10.1523/JNEUROSCI.1374-13.2013.

Surget, A., Tanti, A., Leonardo, E. D., Laugeray, A., Rainer, Q., Touma, C., ... Belzung, C. (2011). Antidepressants recruit new neurons to improve stress response regulation. *Molecular Psychiatry, 16*(12), 1177–1188. http://dx.doi.org/10.1038/mp.2011.48.

Taniuchi, N., Niidome, T., Goto, Y., Akaike, A., Kihara, T., & Sugimoto, H. (2007). Decreased proliferation of hippocampal progenitor cells in APPswe/PS1dE9 transgenic mice. *Neuroreport, 18*(17), 1801–1805.

Tashiro, A., Sandler, V. M., Toni, N., Zhao, C., & Gage, F. H. (2006). NMDA-receptor-mediated, cell-specific integration of new neurons in adult dentate gyrus. *Nature, 442*(7105), 929–933. http://dx.doi.org/10.1038/nature05028.

Tesco, G., Kim, T. W., Diehlmann, A., Beyreuther, K., & Tanzi, R. E. (1998). Abrogation of the presenilin 1/beta-catenin interaction and preservation of the heterodimeric presenilin 1 complex following caspase activation. *The Journal of Biological Chemistry, 273*(51), 33909–33914.

Toni, N., Laplagne, D. A., Zhao, C., Lombardi, G., Ribak, C. E., Gage, F. H., et al. (2008). Neurons born in the adult dentate gyrus form functional synapses with target cells. *Nature Neuroscience, 11*(8), 901–907. http://dx.doi.org/10.1038/nn.2156.

Tozuka, Y., Fukuda, S., Namba, T., Seki, T., & Hisatsune, T. (2005). GABAergic excitation promotes neuronal differentiation in adult hippocampal progenitor cells. *Neuron, 47*(6), 803–815. http://dx.doi.org/10.1016/j.neuron.2005.08.023.

Treves, A., & Rolls, E. T. (1992). Computational constraints suggest the need for two distinct input systems to the hippocampal CA3 network. *Hippocampus, 2*(2), 189–199. http://dx.doi.org/10.1002/hipo.450020209.

Treves, A., & Rolls, E. T. (1994). Computational analysis of the role of the hippocampus in memory. *Hippocampus, 4*(3), 374–391. http://dx.doi.org/10.1002/hipo.450040319.

Tronel, S., Belnoue, L., Grosjean, N., Revest, J. M., Piazza, P. V., Koehl, M., et al. (2012). Adult-born neurons are necessary for extended contextual discrimination. *Hippocampus, 22*(2), 292–298. http://dx.doi.org/10.1002/hipo.20895.

Vale, W., Spiess, J., Rivier, C., & Rivier, J. (1981). Characterization of a 41-residue ovine hypothalamic peptide that stimulates secretion of corticotropin and beta-endorphin. *Science, 213*(4514), 1394–1397.

Van Bokhoven, P., Oomen, C. A., Hoogendijk, W. J., Smit, A. B., Lucassen, P. J., & Spijker, S. (2011). Reduction in hippocampal neurogenesis after social defeat is long-lasting and responsive to late antidepressant treatment. *The European Journal of Neuroscience, 33*(10), 1833–1840. http://dx.doi.org/10.1111/j.1460-9568.2011.07668.x.

van Praag, H., Christie, B. R., Sejnowski, T. J., & Gage, F. H. (1999). Running enhances neurogenesis, learning, and long-term potentiation in mice. *Proceedings of the National Academy of Sciences of the United States of America, 96*(23), 13427–13431.

van Praag, H., Kempermann, G., & Gage, F. H. (1999). Running increases cell proliferation and neurogenesis in the adult mouse dentate gyrus. *Nature Neuroscience, 2*(3), 266–270. http://dx.doi.org/10.1038/6368.

van Praag, H., Schinder, A. F., Christie, B. R., Toni, N., Palmer, T. D., & Gage, F. H. (2002). Functional neurogenesis in the adult hippocampus. *Nature, 415*(6875), 1030–1034. http://dx.doi.org/10.1038/4151030a.

van Praag, H., Shubert, T., Zhao, C., & Gage, F. H. (2005). Exercise enhances learning and hippocampal neurogenesis in aged mice. *The Journal of Neuroscience, 25*(38), 8680–8685. http://dx.doi.org/10.1523/JNEUROSCI.1731-05.2005.

Veena, J., Rao, B. S., & Srikumar, B. N. (2011). Regulation of adult neurogenesis in the hippocampus by stress, acetylcholine and dopamine. *Journal of Natural Science, Biology and Medicine, 2*(1), 26–37. http://dx.doi.org/10.4103/0976-9668.82312.

Veeraraghavalu, K., Choi, S. H., Zhang, X., & Sisodia, S. S. (2010). Presenilin 1 mutants impair the self-renewal and differentiation of adult murine subventricular zone-neuronal progenitors via cell-autonomous mechanisms involving notch signaling. *The Journal of Neuroscience, 30*(20), 6903–6915. http://dx.doi.org/10.1523/JNEUROSCI.0527-10.2010.

Verret, L., Jankowsky, J. L., Xu, G. M., Borchelt, D. R., & Rampon, C. (2007). Alzheimer's-type amyloidosis in transgenic mice impairs survival of newborn neurons derived from adult hippocampal neurogenesis. *The Journal of Neuroscience, 27*(25), 6771–6780. http://dx.doi.org/10.1523/JNEUROSCI.5564-06.2007.

Vivar, C., Potter, M. C., Choi, J., Lee, J. Y., Stringer, T. P., Callaway, E. M., ... van Praag, H. (2012). Monosynaptic inputs to new neurons in the dentate gyrus. *Nature Communications, 3*, 1107. http://dx.doi.org/10.1038/ncomms2101.

Vivar, C., & van Praag, H. (2013). Functional circuits of new neurons in the dentate gyrus. *Front Neural Circuits, 7*, 15. http://dx.doi.org/10.3389/fncir.2013.00015.

Wang, R., Dineley, K. T., Sweatt, J. D., & Zheng, H. (2004). Presenilin 1 familial Alzheimer's disease mutation leads to defective associative learning and impaired adult neurogenesis. *Neuroscience, 126*(2), 305–312.

Wen, P. H., Hof, P. R., Chen, X., Gluck, K., Austin, G., Younkin, S. G., ... Elder, G. A. (2004). The presenilin-1 familial Alzheimer disease mutant P117L impairs neurogenesis in the hippocampus of adult mice. *Experimental Neurology, 188*(2), 224–237.

Wong, P. C., Zheng, H., Chen, H., Becher, M. W., Sirinathsinghji, D. J., Trumbauer, M. E., ... Sisodia, S. S. (1997). Presenilin 1 is required for Notch1 and Dll1 expression in the paraxial mesoderm. *Nature, 387*(6630), 288–292.

Xavier, G. F., & Costa, V. C. (2009). Dentate gyrus and spatial behaviour. *Progress in Neuro-psychopharmacology & Biological Psychiatry, 33*(5), 762–773. http://dx.doi.org/10.1016/j.pnpbp.2009.03.036.

Zhang, C., McNeil, E., Dressler, L., & Siman, R. (2007). Long-lasting impairment in hippocampal neurogenesis associated with amyloid deposition in a knock-in mouse model of familial Alzheimer's disease. *Experimental Neurology, 204*(1), 77–87. http://dx.doi.org/10.1016/j.expneurol.2006.09.018.

Zhao, C., Deng, W., & Gage, F. H. (2008). Mechanisms and functional implications of adult neurogenesis. *Cell, 132*(4), 645–660. http://dx.doi.org/10.1016/j.cell.2008.01.033.

Zhao, C., Teng, E. M., Summers, R. G., Jr., Ming, G. L., & Gage, F. H. (2006). Distinct morphological stages of dentate granule neuron maturation in the adult mouse hippocampus. *The Journal of Neuroscience, 26*(1), 3–11. http://dx.doi.org/10.1523/JNEUROSCI.3648-05.2006.

Ziv, Y., Burns, L. D., Cocker, E. D., Hamel, E. O., Ghosh, K. K., Kitch, L. J., … Schnitzer, M. J. (2013). Long-term dynamics of CA1 hippocampal place codes. *Nature Neuroscience, 16*(3), 264–266. http://dx.doi.org/10.1038/nn.3329.

HALLMARKS AND GENETIC FORMS OF AD

The Amyloid β Precursor Protein and Cognitive Function in Alzheimer's Disease

Robert A. Marr

Department of Neuroscience, Rosalind Franklin University of Medicine and Science, North Chicago, IL, United States

OUTLINE

Introduction	98
The APP Gene and Its Homologs	99
The APP Protein, Its Domains, and Its Proteolytic Fragments	100
Intracellular Trafficking and Processing of APP	102
Cholesterol and APP Processing	103
APP-Interacting Proteins: Signal Transduction and Processing	104
Interactors with the YENPTY Motif	*104*
Low-Density Lipoprotein Receptors	*104*
Functions of APP	106
Neurogenesis: Embryonic and Adult	*106*
Neurite Outgrowth	*107*
Axonal Transport	*108*
Synaptic Formation and Function	*109*
Transcription Factor	*109*
Metal Binding/Redox	*111*
Does Aβ Have a Function?	111
APP Knockout Mice	113

Genes, Environment and Alzheimer's Disease.
DOI: http://dx.doi.org/10.1016/B978-0-12-802851-3.00004-8

97

APP and AD   113
  Disrupted Axonal Transport   *113*
  Disrupted Neurogenesis   *114*
  Altered APP Processing/Translocation   *115*
  Altered Signal Transduction   *116*

References   117

INTRODUCTION

The identification of the importance of the amyloid precursor protein (APP) to Alzheimer's disease (AD) came from multiple disciplines of science. Near the beginning of the previous century, emerging histopathological techniques were being used to identify aberrant microscopic structures associated with neurologic disorders using specific dyes that stained these structures in brain sections. In November of 1906, Dr Alois Alzheimer first described the two pathological hallmarks that together are associated with AD. These pathologic structures were termed extracellular plaques and intracellular tangles. Decades later, the emerging field of electron microscopy characterized the fine structure of these plaques and tangles, showing that they were composed of organized amyloid fibrils (Kidd, 1963; Terry, Gonatas, & Weiss, 1964). The next breakthrough came from the discipline of biochemistry, which was used to identify the nature and compositions of these forms of amyloid. In 1984, the major component of the amyloid plaque was determined to be the amyloid-beta (Aβ) peptide (Glenner & Wong, 1984a, 1984b; Masters et al., 1985). This was followed by the identification of tau as being the primary component of neurofibrillary tangles (NFTs) (Grundke-Iqbal et al., 1986; Kosik, Joachim, & Selkoe, 1986). The field of genetics was the last major contributor to this story. The discovery of the protein components of plaques and tangles led to the identification of their parent genes (APP and tau). Since AD had been linked to mutations on chromosome 21, which carried APP, it was predicted that these mutations would be associated with APP (Goate et al., 1989; St George-Hyslop et al., 1990). Soon after, autosomal dominant heritable (familial) mutations that cause early onset AD were identified in APP (Chartier-Harlin et al., 1991; Goate et al., 1991; Murrell, Farlow, Ghetti, & Benson, 1991). Since then, other familial mutations or duplications have been identified in APP that cause AD as well as mutations in the APP processing proteins presenilin-1 (PS1) and presenilin-2 (PS2) (Goedert & Spillantini, 2006) (discussed below in the section 'APP and AD'). These discoveries established the central role APP plays in the progression of AD.

THE APP GENE AND ITS HOMOLOGS

The APP gene is expressed ubiquitously throughout the body, including peripheral tissues and the central nervous system (CNS) (reviewed in van der Kant & Goldstein, 2015). There are several splice forms of APP's 18 exons including the major 770 and 695 amino acid (a.a.) forms (Figure 4.1), with 770 being expressed throughout the body and 695 primarily restricted

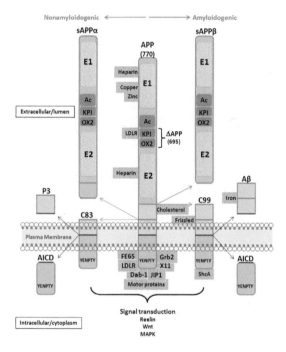

FIGURE 4.1 The amyloid beta (A4) precursor protein (APP). The major 770 amino acid isoform of APP is shown (center) including the region that is deleted in APP695 (amino-terminus on top, carboxy-terminus on bottom). The major protein domains and motifs are labeled (E2, heparin binding; E1, Cu/Zn, binding plus heparin binding; OX2, homology domain; KPI, Kunitz protease inhibitor region; Ac, acidic domain). The YENPTY phosphorylation/signaling motif is also shown. Alongside APP some major binding proteins/molecules/atoms are shown in purple boxes (LDLR, low-density lipoprotein receptor family members including LRP1, LRP1b, LRP2, ApoER2, and SorLA). Secretase cut sites are represented by colored lines (red, β-secretase; green, α-secretase; blue, γ-secretase). Both nonamyloidogenic and amyloidogenic cleavage products are shown on the left and right, respectively. First, APP is processed by α (primarily at the extracellular surface) or β (primarily in early endosomes) secretases and the products are shown (sAPPα + C83 and sAPPβ + C99, respectively). Second, the membrane associated C83 and C99 are processed by γ-secretase (within late endosomes/lysosomes and/or the trans-Golgi network) to produce P3 and Aβ, respectively, plus AICD. Some signal transduction pathways regulated through APP are also shown.

to the CNS. There are two paralogs of APP termed APP-like protein-1 and APP-like protein-2 (APLP1, APLP2). These related genes possess similar domains to APP with APLP1 being more closely related to APP695 (and restricted to the CNS as well) while APLP2 appears more similar to APP770 (including whole body expression). However, these APP-like proteins neither possess nor produce a homolog of the Aβ peptide. The APP gene family is well conserved evolutionarily with orthologs also present in invertebrates including *Caenorhabditis elegans* and *Drosophila melanogaster*. It has been suggested that the emergence of the APP family of genes coincides with the evolution of a nervous system and lipoproteins (the latter being involved in cholesterol transport and signal transduction among other functions) (van der Kant & Goldstein, 2015). There is functional redundancy between the APP gene family members. Knockout (KO) of any one family member is not embryonic lethal and produces subtle but measurable phenotypes (discussed below in the section 'APP Knockout Mice'). Studies on double KO mice suggest the APLP2 provides an essential function for viability that only APP and APLP1 combined can compensate for, whereas APLP2 can compensate for even the lack of both APP and APLP1.

THE APP PROTEIN, ITS DOMAINS, AND ITS PROTEOLYTIC FRAGMENTS

The APP protein is a type I transmembrane protein with a single membrane spanning region (Figure 4.1). Biochemical/structural analysis suggests that the extreme C terminal end curls back to also associate with the lipid bilayer (not shown) (Barrett et al., 2012). The large extracellular N-terminal region contains many domains including heparin binding domains (within E1/E2), metal binding domains (within E1), an acidic domain, a Kunitz protease inhibitor (KPI) region, and an OX2 sequence, with the latter two domains missing from APP695 (Figure 4.1). The YENPTY motif is located in the intracellular C-terminal domain and is a key region for phosphorylation, protein interactions, and signal transduction (discussed below in the section 'Interactors with the YENPTY Motif'). APP can also be separated into other domains based on the complex proteolytic processing that this protein can undergo. These proteolytic domains include the secreted extracellular sAPPα and sAPPβ regions as well as C83, C99, P3, Aβ, and the APP intracellular domain (AICD) (Figure 4.1). These fragments of APP are the product of three secretase activities termed alpha, beta, and gamma.

There are many proteases that contribute to alpha secretase activity comprising the "a disintegrin and metalloprotease domain" (ADAM) family. These include ADAM10 and ADAM17 (also known as TNFα converting enzyme or TACE) and ADAM21 (Lazarov & Marr, 2009). Alpha secretase cleaves APP near the extracellular plasma membrane within the

Aβ region (between Lysine16 and Leucine17 in the Aβ sequence) releasing the soluble/secreted sAPPα domain. This cleavage also leaves behind the C-terminal C83 peptide within the plasma membrane. After this the C83 fragment can undergo further processing by γ-secretase. The γ-secretase activity is facilitated by a relatively large protein complex consisting of either PS1 or PS2 in association with Pen-2, APH-1, and nicastrin. PS1 and PS2 are aspartyl proteases that contribute the catalytic component of the complexes (Lazarov & Marr, 2009). This processing of APP occurs within the hydrophobic region of the plasma membrane (intramembranous cleavage) releasing the P3 peptide (secreted) and AICD (intracellular). This process is described as the nonamyloidogenic pathway as the initial cleavage is within the Aβ peptide sequence of APP preventing its generation.

APP can also be processed by β-secretase activity, which is facilitated by the beta-site amyloid precursor protein cleaving enzyme 1 (BACE1) protein (Hussain et al., 1999). BACE1 cuts at the extracellular surface further from the plasma membrane, compared to α-secretase, producing the secreted sAPPβ fragment and the membrane associated C99 peptide (Figure 4.1). After this, the C99 fragment can be further processed by γ-secretase to liberate the secreted Aβ peptide and intracellular AICD. It should be noted that γ-secretase can cleave at multiple locations producing Aβ isoforms of 36, 38, 40, 42, and a.a. and more (Figure 4.2). The major focus of Aβ research has been on the 40 and 42 a.a. forms of Aβ. Under normal conditions, about 90% of secreted Aβ peptides are $Aβ_{40}$, which is a more soluble form that can slowly convert to a less soluble β-sheet conformation. In contrast, approximately 10% of secreted Aβ peptides

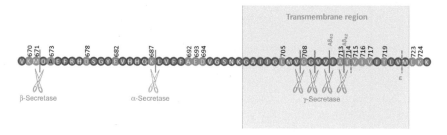

FIGURE 4.2   Location of APP mutants associated with familial AD. Amino acids 669–725 are shown (as numbered within APP770). Red amino acids indicate where mutations occur that cause familial AD. Green indicates the location where a protective mutation has been located that reduces β-secretase processing (Jonsson et al., 2012; Maloney et al., 2014). The secretase cut sites are shown. Due to promiscuous specificity in γ-secretase, multiple cut sites are shown with the ε activity cut site (major cut sites shown with solid lines, minor cut sites shown with dotted lines). The major $Aβ_{40}$ and $Aβ_{42}$ producing γ-secretase cut sites are labeled in red text.

are $A\beta_{42}$, which is highly prone to taking on a fibrillogenic β-sheet conformation and is more toxic to neurons compared to $A\beta_{40}$ (Clippingdale, Wade, & Barrow, 2001; Selkoe, 2001). The APP holoprotein as well as all the proteolytic products have been implicated in biological functions/dysfunctions and will be discussed below (in the section 'APP Knockout Mice').

INTRACELLULAR TRAFFICKING AND PROCESSING OF APP

The trafficking of APP in relation to its processing has been comprehensively reviewed but remains an area of continued investigation (van der Kant & Goldstein, 2015). The fragmentation of APP is segregated within distinct subcellular compartments (Haass, Kaether, Thinakaran, & Sisodia, 2012; Rajendran & Annaert, 2012). APP reaches the plasma membrane like most membrane proteins by transit through the endoplasmic reticulum and trans-Golgi network (TGN). It is at the plasma membrane that APP is cleaved by α-secretase (Sisodia, 1992). Cleavage by β-secretase occurs primarily in early endosomes (Rajendran et al., 2006; Sannerud et al., 2011), and this activity is dependent on the acidification that occurs within these early endosomes. APP internalization is thought to occur as a result of its association with complexes of membrane proteins known as lipid rafts and this process could separate APP from nonlipid raft associated α-secretases (Ehehalt, Keller, Haass, Thiele, & Simons, 2003). APP and BACE1 are found primarily in separate vesicle pools and synaptic activity can induce the merger of these vesicles in the somatodendritic compartment of neurons (Das et al., 2013). APP and BACE1 are also reported to be endocytosed separately from the plasma membrane (APP entering by a clathrin-dependent mechanism and BACE1 through an ADP ribosylation factor 6 (ARF6)-dependent mechanism), and they primarily interact when both are shuttled into common early endosomes (Sannerud et al., 2011). We might think that the production of sAPPβ in the early endosomes would doom this fragment to degradation; however, it can be secreted by neurons indicating that some fraction of sAPPβ is either generated at the plasma membrane, sorted into recycling endosomes from early endosomes, or sorted into the TGN from early endosomes (Sannerud et al., 2011). Following α/β-processing, the membrane-bound C-terminal fragments are then further processed by γ-secretase. The cellular compartment where γ-secretase cleave APP fragments is a contentious issue that is under continuing investigation. Some studies point to γ-secretase cleaving the APP C-terminal fragments in the late endosomal/lysosomal system (Vieira et al., 2010), while other studies point to γ-cleavage in the TGN (Choy, Cheng, & Schekman, 2012). Also, within neurons, APP can be localized to both pre- and postsynaptic regions (DeBoer, Dolios, Wang, & Sisodia,

2014; Yamazaki, Selkoe, & Koo, 1995). There is evidence for both pathways leading to γ-secretase cleavage of APP, and more research will be needed to establish whether Aβ is primarily generated in endosomes, the TGN, or perhaps in both (van der Kant & Goldstein, 2015). For AD progression, it is important to consider that γ-secretase cutting of the C-terminal fragments of APP in late endosomes as opposed to the TGN could alter the location of γ-secretase cleavage and change the ratio of $A\beta_{42}$ to $A\beta_{40}$. Additionally, this can be regulated by APP-interacting proteins like sortilin-related receptor (SorLA; discussed below in the section 'Altered APP Processing/Translocation').

CHOLESTEROL AND APP PROCESSING

Cholesterol is proposed to affect APP processing (reviewed in Maulik, Westaway, Jhamandas, & Kar, 2013). As discussed above, APP and the secretases are associated with different lipid raft domains within the plasma membrane (Bouillot, Prochiantz, Rougon, & Allinquant, 1996; Lee et al., 1998) and it is believed that levels of membrane cholesterol can modulate the processing of APP. Amyloidogenic processing enzymes are proposed to associate with high cholesterol containing lipid rafts. In addition, there is evidence that differing cholesterol levels within subcellular compartments can alter the trafficking of APP or its processing enzymes, leading to changes in amyloidogenic versus nonamyloidogenic processing of APP (Maulik et al., 2013). Cholesterol binds to a central motif located on residues 17–23 (LVFFAED) of the Aβ region of APP (Barrett et al., 2012; Beel, Sakakura, Barrett, & Sanders, 2010; Tang et al., 2014; van der Kant & Goldstein, 2015), and it has been reported that high cholesterol levels can result in elevated binding to APP, which is very near to the α-secretase cleavage site. This binding possibly leads to the inhibition of α-secretase cleavage and, as a result, increases amyloidogenic processing (Beel et al., 2010; Yao & Papadopoulos, 2002). Multiple studies using primary neurons or cell lines reported that an increase in cellular cholesterol can reduce α-secretase processing of APP and subsequently induce amyloidogenic processing (Bodovitz & Klein, 1996; Frears, Stephens, Walters, Davies, & Austen, 1999). Conversely, lowering cholesterol levels using chemicals or drugs can stimulate α-secretase cleavage and inhibit amyloidogenic processing of APP (Abad-Rodriguez et al., 2004; Fassbender et al., 2001; Frears et al., 1999; Kojro, Gimpl, Lammich, Marz, & Fahrenholz, 2001; Simons et al., 1998). However, *in vivo* studies have produced mixed results related to the effect of cholesterol on the amyloidogenic processing of APP. This may be a result of species and strain differences, peripheral versus central cholesterol manipulation, different subcellular alterations in cholesterol, different genetic models, and the potential effects of marginal versus more substantial alterations in cholesterol levels. Ultimately what can be said is that these *in vivo* results suggest that alterations in cholesterol levels and/or subcellular distribution can play an active role in APP processing and as a result affect AD progression (Maulik et al., 2013).

APP-INTERACTING PROTEINS: SIGNAL TRANSDUCTION AND PROCESSING

Interactors with the YENPTY Motif

The wide array of proteins that interact with APP (Figure 4.1) has been termed as the APP interactome. Many of the key APP-interacting proteins are phosphotyrosine binding proteins and some contain Src homology-2 (SH2) domains that bind to APP in a phosphorylated state (reviewed in van der Kant & Goldstein, 2015). These phosphorylations of APP primarily occur on the YENPTY motif located in the intracellular C-terminal domain. These interacting proteins include FE65, Disabled-1 (Dab-1), X11, JIP1, Shc, and GRB2. The phosphorylation of different tyrosine residues (Y) within the YENPTY sequence results in different phosphotyrosine binding proteins associating with the motif. For example, the Dab-1 protein is involved in signaling through the Reelin pathway, which mediates cellular migration and survival (Herz & Chen, 2006). FE65 binding alters APP processing and associates with AICD to potentially act as a modulator of gene expression (Cao & Sudhof, 2001; Gao & Pimplikar, 2001). X11 binding to APP alters its processing and inhibits amyloidogenic processing and JIP1 is associated with the control of anterograde and retrograde transport of proteins including APP. In addition, ShcA, which preferentially associates with C99, has been linked to the MAPK signaling pathway further showing how APP processing can affect signal transduction (van der Kant & Goldstein, 2015). The prolyl isomerase, Pin1, which also binds to the C-terminal region of APP can function to alter protein conformation and as a result modify the affinity of binding partners of APP like FE65 (Stukenberg & Kirschner, 2001). The C99 fragment of APP can associate with members of the Wnt-planar cell polarity complex including frizzled receptors and Vangl2. As the name would suggest, this complex controls cellular polarity and axonal growth (discussed below in the section 'Neurite Outgrowth') (Soldano et al., 2013). Therefore, this diverse and complex set of binding partners can affect several signal transduction pathways including Reelin, Wnt, and MAPK, and suggests a role for APP in multiple cellular functions.

Low-Density Lipoprotein Receptors

APP can also associate with a wide array of other proteins that are involved in the modulation of its properties. A major group of these interacting proteins are part of the low-density lipoprotein (LDL) receptor family (reviewed in Wagner & Pietrzik, 2012). One hallmark feature of the LDL receptor family is that they all bind to apolipoprotein E (apoE). ApoE is found in blood serum and cerebrospinal fluid and functions in cholesterol/triglyceride transport as well as signal transduction (Huang, Weisgraber, Mucke, & Mahley, 2004). Interestingly, common gene polymorphisms in apoE are the strongest known risk factors for the development of the more common sporadic late-onset form of AD. Also, the LDL receptors contain similar phosphorylation/signaling/protein interaction domains with APP (NPxY) and can be processed by the APP cleaving secretases. Many of the LDL receptors can associate with APP to regulate its functions and processing. For example LDL-receptor related protein-1 (LRP1) is the largest family member and directly interacts with APP, resulting in alterations in processing (Kounnas et al., 1995; Pietrzik, Busse, Merriam, Weggen, & Koo, 2002; Trommsdorff, Borg, Margolis, & Herz, 1998). LPR1 can bind the KPI domain and intracellular domain of APP directly or in association with APP through binding to FE65 and is reported to facilitate the internalization of APP and result in enhancement of amyloidogenic processing (Knauer, Orlando, & Glabe, 1996; Pietrzik et al., 2004; Ulery et al., 2000). LRP1b is very similar to LRP1 and is believed to interact with APP by similar mechanisms (Cam et al., 2004); however, LRP1b has the opposite effect on Aβ production, which is suggested to be a result of its much slower kinetics on internalization (Liu, Li, Obermoeller-McCormick, Schwartz, & Bu, 2001). LRP2/Megalin is also highly homologous to LRP1 and has also been reported to associate with APP through similar mechanisms (Alvira-Botero & Carro, 2010). Another LDL receptor, apoE-receptor-2 (apoER2), also associates with APP and modifies its processing. ApoER2 is a Reelin/F-spondin receptor and interacts with APP through both intracellular and extracellular domains including FE65 and Dab1 (Hoe, Magill, et al., 2006; Hoe & Rebeck, 2005; Hoe, Tran, Matsuoka, Howell, & Rebeck, 2006). ApoER2 and APP can associate through binding to F-spondin and this promotes retention at the cell surface and thus nonamyloidogenic processing of APP (Hoe et al., 2005).

Other APP-interacting proteins have been found that also regulate its processing (discussed further in the section 'APP and AD' below). Using transgenic mice that express an epitope tagged form of APP, new binding partners for APP have been identified using proteomics (Norstrom, Zhang, Tanzi, & Sisodia, 2010). Neuron-enriched endosomal protein of 21 kDa (NEEP21) was identified and found to enhance nonamyloidogenic processing of APP and it functions in regulating receptor endocytosis, recycling, and degradation (Muthusamy, Chen, Yin, Mei, & Bergson, 2015; Norstrom et al., 2010). In neurons, NEEP21 is localized to vesicles located in somatodendritic compartments (Steiner et al., 2002; Utvik et al., 2009).

FUNCTIONS OF APP

Neurogenesis: Embryonic and Adult

Multiple functions can be attributed to APP and some of these are summarized in Table 4.1, which is not an all-inclusive summary but lists some of the major proposed functions (many of which are interrelated). APP clearly functions as a neural developmental gene (reviewed in Nicolas & Hassan, 2014). Genetic ablation studies in zebrafish reveal that during embryogenesis, APP plays an essential role in the process by which embryonic tissue is remodeled by cellular migration that functions to converge along a single perpendicular axis (convergent–extension) (Joshi, Liang, DiMonte, Sullivan, & Pimplikar, 2009). The expression of APP is detected quite early during embryogenesis (during gastrulation in mice) (Ott & Bullock, 2001; Sarasa et al., 2000), and is highly expressed in the neural tube at E9.5 (Trapp & Hauer, 1994). The sAPPα fragment has been shown to act as a growth factor for embryonic neural stem cells (Hayashi et al., 1994; Ohsawa, Takamura, Morimoto, Ishiguro, & Kohsaka, 1999). In human embryonic stem cells, rapid neuronal specification can be induced by upregulation of APP, sAPPα, and sAPPβ expression (Chen, Boiteau, Lai, Barger, & Cataldo, 2006). APP has also been implicated in neuronal migration in the developing cortex. The inhibition of APP during embryogenesis was shown to prevent cortical ventricular zone neural progenitor cells from migrating to the cortical plate; overexpression facilitated this process (Young-Pearse et al.,

TABLE 4.1   Summary of the Proposed Functions of APP

Function	Role of APP
Neurogenesis: embryonic	APP/sAPPα(β) ↑ neural stem cell proliferation
	↑ neuronal differentiation
	↑ migration
Neurogenesis: adult	sAPPα—↑ neural stem cell/progenitor proliferation
	AICD—↓ neural stem cell/progenitor proliferation
Neurite outgrowth	sAPPα—↑ neurite outgrowth/dendritic spine density
Axonal transport	APP, APP C—terminal fragments: axon transport regulation
Synaptic formation/function	APP and sAPPα—↑ synaptic boutons
Transcription factor	AICD—↑ target gene expression?
Metal binding/redox	APP—bind/reduce copper
	Aβ—bind/reduce iron

2007) and this migration can be regulated through association with another group of APP-interacting proteins called pancortins (Rice et al., 2012).

During embryonic development the engagement of APP by the TAG1 protein results in elevated levels of AICD, leading to modulation of neurogenesis (Ma et al., 2008). Studies also suggest that γ-secretase processing of APP regulates the epidermal growth factor receptor (EGFR) (Li et al., 2007; Repetto, Yoon, Zheng, & Kang, 2007; Zhang et al., 2007). Therefore, familial AD induced alterations in APP and PS1 processing and/or function may affect EGFR expression and function, which might then affect neurogenesis. As APP is involved in neurogenesis during embryogenesis, it is not surprising that APP also plays a role in adult neurogenesis, which takes place throughout life in discrete locations of the brain (subgranular layer of the dentate gyrus and the subventricular zone (SVZ)) (Lazarov & Marr, 2009). The sAPPα fragment has been implicated in the regulation of cell proliferation (Meng, Kataoka, Itoh, & Koono, 2001; Pietrzik et al., 1998; Saitoh et al., 1989; Schmitz, Tikkanen, Kirfel, & Herzog, 2002; Slack, Breu, Muchnicki, & Wurtman, 1997). It has also been found that sAPPα is a proliferation factor for adult neural progenitor cells and this factor was also able to rescue age-associated deficits in neural progenitor cell proliferation *in vivo* (Demars, Bartholomew, Strakova, & Lazarov, 2011; Demars, Hollands, Zhao Kda, & Lazarov, 2013; Gakhar-Koppole et al., 2008; Rohe et al., 2008). This activity was not exhibited by the sAPPβ fragment. Also, the suppression of APP *in vivo* was reported to reduce the proliferation of neural progenitor cells in the SVZ (Caille et al., 2004). These data suggest APP and sAPPα, specifically, may act as a growth factor for neural progenitor cells.

Contrary to sAPPα, it is believed that AICD has the opposite effect on neural progenitors. AICD suppresses the expression of the epidermal growth factor receptor, which also drives the proliferation of neural progenitors (Zhang et al., 2007). In APP KO mice, the introduction of a recombinant AICD gene, alone, results in suppressed neural progenitor proliferation (Ghosal, Stathopoulos, & Pimplikar, 2010). It should also be mentioned that the APP processing enzyme PS1 has also been implicated in adult neurogenesis. The suppression of PS1 was shown to reduce proliferation while promoting the differentiation of neural progenitors into neurons and glia (Gadadhar, Marr, & Lazarov, 2011). This suppression is associated with learning and memory deficits in mice (Bonds et al., 2015). However, it should be noted that this effect could be mediated through reduced processing of multiple substrates of PS1/γ-secretase including NOTCH (discussed below in the section 'Transcription Factor').

Neurite Outgrowth

APP has also been implicated in the promotion of neurite outgrowth. The large extracellular fragments of APP were reported to interact with the p75 neurotrophin receptor to promote neurite outgrowth, and sAPPα has also been reported to increase dendritic spine density in cultured neurons from APP knockout mice (Hasebe et al., 2013; Tyan et al., 2012). APP is implicated in increasing neurite length/branching directly or through its interaction with other proteins such as Dab-1 (Chasseigneaux et al., 2011; Hoareau et al., 2008; Milward et al., 1992; Zhou et al., 2012). The overexpression of human APP in drosophila was reported to enhance axonal extensions and this activity was dependent on the YENPTY domain (Leyssen et al., 2005). Consistent with this notion, human APP delivery could compensate for axonal outgrowth deficits induced by knockdown of APP homologs. These studies have also implicated APP in binding to cytoskeletal elements, which are important for migration and neurite outgrowth (Nicolas & Hassan, 2014).

It should be noted that the α and β forms of soluble APP only differ by 17 a.a. on the C-terminal end yet they have distinct activities and possibly altered processing. For example, another intriguing possibility for APP signaling has been reported, which suggests that sAPPβ, but not sAPPα, can undergo additional cleavage to produce a N-terminal APP product (N-APP) of 35 kDa (Nikolaev, McLaughlin, O'Leary, & Tessier-Lavigne, 2009). N-APP was reported to bind to the death receptor DR6 and mediate a process of degeneration known as axon pruning under conditions where trophic factors were withdrawn. It is tempting to link this pathway to both normal processes of axonal pruning during development as well as axonal degeneration, which is an early process in AD as the sAPPβ isoform is specific to the pathogenic/amyloidogenic pathway of APP processing.

Axonal Transport

Binding to cytoskeletal elements also implicates APP in axonal transport. In the mammalian brain, APP was observed to be rapidly transported to synaptic sites (Koo et al., 1990) and was found associated with vesicular elements of axons and dendrites (Schubert et al., 1991). The overexpression of APP has been associated with the disruption of axonal transport that is independent of Aβ. APP CTFs have been implicated in interactions with motor proteins including myosin, dynein, and kinesin (Cottrell et al., 2005; Goldstein, 2012; Rodrigues, Weissmiller, & Goldstein, 2012; Szodorai et al., 2009). Kinesin-1 colocalizes with APP-positive vesicles within the axon (Szpankowski, Encalada, & Goldstein, 2012), and may directly interact with APP or indirectly interact with APP via JIP1 (Chiba et al., 2014; Fu & Holzbaur, 2013; Goldstein, 2012; Kamal, Almenar-Queralt, LeBlanc, Roberts, & Goldstein, 2001; Kamal, Stokin, Yang, Xia, & Goldstein, 2000). However, it should be noted that the direct involvement of kinesin in APP transportation has been disputed (Lazarov et al., 2005). Overall, accumulating evidence shows that these and other associations result in the formation of protein complexes that regulate motor transport within cells (Kamal et al., 2000; Vagnoni et al., 2012). It should be noted that axonal transport deficits and axon pathology are early pathological developments during the progression of AD. This role for APP in axonal transport implies that it may also function in synaptic formation and function.

Synaptic Formation and Function

APP is found at the synapse and, paradoxically, neuronal activity can promote amyloidogenic processing (Kamenetz et al., 2003). As mentioned above, APP KO mice have reduced dendritic spine density that is correlated with reduced synaptic function and sAPPα can reverse this phenotype (Tyan et al., 2012). This is supported by a study in wild-type mice in which the infusion of sAPPα or overexpression of ADAM10 (α-secretase) resulted in increased numbers of presynaptic boutons (Bell, Zheng, Fahrenholz, & Cuello, 2008). APP overproduction in transgenic mice results in elevated dendritic spine density (Lee et al., 2010). APP is localized to both the pre- and postsynaptic membrane where it can form heteromeric complexes with APLP1 to functions in adhesion (Soba et al., 2005). Other cellular adhesion molecules including NgCAM (Osterfield, Egelund, Young, & Flanagan, 2008) and NCAM (Ashley, Packard, Ataman, & Budnik, 2005) have been shown to associate with APP. Additionally, APP is important for the formation and maintenance of neuromuscular junctions and APP/APLP2 double KO neurons show reduced synaptic vesicles and deficits in neurotransmitter release (Wang et al., 2005; Weyer et al., 2011). As we might expect, APP KO mice also display reduced forelimb strength and locomotor activity (Zheng et al., 1995).

Transcription Factor

APP is also thought to function in the regulation of gene expression. Circumstantial evidence on this function of APP comes from similarities with the processing of NOTCH, which is a quintessential development factor mediating cell fate decisions (Collu, Hidalgo-Sastre, & Brennan, 2014). NOTCH is also processed by the secretases that process APP and its cleavage is reminiscent of APP processing. NOTCH is first cleaved by α-secretase releasing an extracellular portion and a membrane-bound portion. After this first cleavage, the membrane portion undergoes intramembranous γ-secretase cleavage, releasing a relatively short intracellular domain termed NOTCH intracellular domain (NICD), which is reminiscent of the AICD fragment of APP (Selkoe & Kopan, 2003). NICD can translocate into the nucleus and form protein complexes that induce gene expression related to cell differentiation. Evidence also suggests a role for AICD as a transcription factor (or repressor) that can be linked to AD. AICD has been shown to bind the neprilysin promoter and induce expression of this Aβ degrading enzyme (Belyaev, Nalivaeva, Makova, & Turner, 2009; Pardossi-Piquard et al., 2005); thus, this could represent a negative feedback loop where the processing of APP (which could produce elevated Aβ) stimulates the expression of an enzyme that can clear this peptide from the brain. This is supported by evidence that only the amyloidogenic processing of APP generates AICD capable of nuclear signaling, and this is due to the subcellular compartmentalization of its processing (Flammang et al., 2012; Konietzko, 2012). Transthyretin binds to $A\beta_{42}$ and is reported to facilitate its clearance from the brain as well as inhibit aggregation of this amyloidogenic peptide (Liu & Murphy, 2006; Philibert, Marr, Norstrom, & Glucksman, 2014; Tsuzuki et al., 2000) and was found to be protective in APP transgenic models of AD (Stein et al., 2004; Stein & Johnson, 2002).

Interestingly, transthyretin appears to be induced in AD and the AICD fragment of APP was shown to bind to the transthyretin promoter and facilitate expression of this Aβ clearing factor. This process is reminiscent of AICD's regulation of neprilysin (Kerridge, Belyaev, Nalivaeva, & Turner, 2014; Li, Masliah, Reixach, & Buxbaum, 2011). A transcriptional activating role of the AICD complex for many other targets has been reported, with many studies using artificial overexpression systems and recombinant gene reporter constructs (reviewed in Muller & Zheng, 2012). Other suggested transcriptional targets include APP itself (von Rotz et al., 2004), glycogen-synthase kinase-3β (GSK-3β) (Kim et al., 2003; Ryan & Pimplikar, 2005), KAI (Baek et al., 2002), p53 (Checler et al., 2007), LRP1 (Liu et al., 2007), cytoskeletal genes (Muller et al., 2007), and genes that regulate calcium (Leissring et al., 2002). It is important to note, however, that the evidence for the function of AICD as part of a transcription factor complex targeting specific genes remains highly controversial, unlike the case for NICD. The validity of these findings (including the targeting of neprilysin) have been questioned in many studies as well (Aydin et al., 2011; Chen & Selkoe, 2007; Giliberto et al., 2008; Hebert et al., 2006; Repetto et al., 2007; Tamboli et al., 2008; Waldron et al., 2008; Yang, Cool, Martin, & Hu, 2006). Therefore, the analogy between the functions of NICD and AICD remains ambiguous. Alternatively, outside of its activity in transcriptional modulation, AICD may also act as a direct binding inhibitor of other signaling pathways. For example, AICD is reported to bind to GSK3β and potentiate its activity, thus destabilizing β-catenin and suppressing canonical Wnt signaling, resulting in the stimulation of neurite outgrowth (Zhou et al., 2012).

Metal Binding/Redox

As described above, APP possesses metal binding sites in its extracellular domain (Hesse, Beher, Masters, & Multhaup, 1994) and we would presume that that there is a functional aspect to these domains. The copper binding site is able to reduce Cu^{2+} to Cu^+ and the APP copper binding site was able to protect against Cu^{2+}-induced neurotoxicity, including impairment of spatial memory, elevated neuronal cell death, and increased astrogliosis (Cerpa et al., 2004). Copper may also alter APP localization within the cell. Copper was found to promote APP trafficking in cultured epithelial cells and neuronal cells (Acevedo et al., 2011). This altered trafficking resulted in a widened redistribution of APP from perinuclear localizations to many other locations, including neurites. Copper was also reported to enhance APP dimerization and promotes amyloidogenic processing (Noda et al., 2013). Evidence also suggests that the copper binding by APP has an essential synaptic function. APP supports synaptogenic function through transdirected dimerization of the extracellular E1 domain including the copper-binding region, and copper binding promoted this process (Baumkotter et al., 2014). Also, this study found, *in vitro*, that APP potently promotes synaptogenesis depending on copper binding. This aspect of APP is relevant to AD as excess copper has been suggested to be a causative factor in the disease.

DOES Aβ HAVE A FUNCTION?

Finally, it should be mentioned that Aβ itself may have constructive physiological functions and not only function as a mediator of AD. While no clear function for Aβ has been firmly established, there are several possibilities. For example, it has been found that low (picomolar) levels of Aβ increase long-term potentiation (LTP)/synaptic plasticity resulting in improved memory in rodents (Morley et al., 2010; Puzzo et al., 2008). Furthermore, there is a possibility that Aβ may even play a role in the normal response to traumatic brain injury (TBI). While it is becoming clear that Aβ is associated with neurodegeneration in conjunction with chronic exposure to TBI events (i.e., linked to the development of chronic traumatic encephalopathy or AD), it may not always be deleterious in relation to the acute response post-TBI. In animals, it has been shown that APP and its processing enzymes BACE1 and PS1 are elevated in the days after TBI (Chen et al., 2004; Loane et al., 2009). Furthermore, 1–7 days following TBI in mice, elevated intraaxonal Aβ can be found (Tran, LaFerla, Holtzman, & Brody, 2011). Thus APP and all its processing machinery are found to be elevated and then fall back to normal in a coordinated fashion postinjury.

The conventional wisdom is that Aβ contributes to pathology acutely after TBI. This is supported by a study that showed BACE1 KO and PS1 inhibition (both lowering Aβ) protect against acute pathology after open-skull controlled cortical impact (CCI) (Loane et al., 2009). However, this is contradicted in another study where they showed BACE1 KO mice were more vulnerable to this type of pathology after TBI (Mannix, Zhang, Park, Lee, & Whalen, 2011). While this latter study used relatively younger mice, it still suggests that Aβ might have a protective role acutely after TBI. This study was followed up by a second study, which showed that the injection of $A\beta_{40}$ ameliorated some of the acute deficits after TBI in BACE1 KO mice, further supporting a protective role for Aβ (Mannix, Zhang, Berglass, Qui, & Whalen, 2013). In addition to this, a different study showed that APP deficiency also exacerbated pathology following open-skull "diffuse" CCI (Corrigan et al., 2012b), again supporting a possible beneficial effect of Aβ. However, this group produced follow-up studies suggesting that the lack of soluble sAPPα, not Aβ, was the reason for the exacerbated pathology (Corrigan et al., 2014, 2012a). Interestingly, a clinical study in humans with severe TBI showed that Aβ levels positively correlated with improved neurological status in the days following injury (Brody et al., 2008). Finally, a recent study on spinal cord injury in mice reported that reduced Aβ also exacerbated acute pathology (Pajoohesh-Ganji et al., 2014). Taken together these studies suggest that more investigation needs to be done into the potential beneficial functions of Aβ.

It has been reported that Aβ is like a metalloprotein that binds transition metal ions via three histidines and a tyrosine residue located in the N-terminal region of the peptide. Also, the presence of an iron response element in the 5′ untranslated region of the APP RNA transcript indicates that APP expression is responsive to intracellular iron (Rogers et al., 1999). The three histidine residues in Aβ may also control the redox activity of iron, and it was shown that iron-ion catalyzed hydroxyl radical generation was inhibited in the presence of both $A\beta_{40}$ and $A\beta_{42}$ (Nakamura et al., 2007). Therefore it is possible that the induction of Aβ by iron may serve to reduce free metal ions and so reduce oxidative stress. Also, primary neuronal cultures were protected from iron neurotoxicity by specific doses of $A\beta_{40}$ (Bishop & Robinson, 2004; Zou, Gong, Yanagisawa, & Michikawa, 2002). This suggested protective role is controversial as Aβ is also reported to be a promoter of oxidative stress and neurotoxicity. The same can be said for the above proposed potential beneficial role for Aβ in synaptic plasticity and in the response to brain injury since Aβ is believed to be toxic. However, the toxic properties of Aβ are integrally linked to its aberrant accumulation and aggregation. At lower levels Aβ could be beneficial at the synapse and could be produced as a compensatory/protective response to brain damage or metal-induced oxidative stress (Avramovich-Tirosh, Amit, Bar-Am, Weinreb, & Youdim, 2008).

APP KNOCKOUT MICE

Considering the functions of APP and its proteolytic fragments in neurogenesis, neuronal migration, neurite outgrowth, axonal transport, synaptic function, gene expression, and more, it is predicted that APP deficits would disrupt neurologic and cognitive function. However, the paradoxical fact that APP KO phenotypes show relatively modest dysfunction is presumably the result of functional redundancy with its paralogs APLP1 and APLP2 (discussed above in the section 'The APP Gene and Its Homologs'). That being said, APP and its family members show a wide spectrum of marginal deficits. APP expression is important for normal spatial learning ability and knockouts show deficits in synaptic plasticity that correlate with poor performance in spatial navigation, as tested in the Morris Water Maze, which can be ameliorated by sAPPα (Ring et al., 2007). Other studies show that the ablation of APP in mice induces impairments in both spatial learning and memory, which go hand-in-hand with defective LTP (Dawson et al., 1999; Phinney et al., 1999; Seabrook et al., 1999). APP KO mice were also reported to have elevated levels of copper in the brain (White et al., 1999), as well as cholesterol and sphingolipid (Grimm et al., 2005). Interestingly, APP null mice were found to be susceptible to kainate-induced seizure activity (Steinbach et al., 1998), implying a role for APP or its proteolytic products in feedback control of excitatory neuronal activity. Additionally, APP KO mice show a plethora of other deficits including but not limited to elevated gliosis, reduced synaptic markers, axonal growth/white matter defects, reduced brain mass, as well as axonal transport defects, and these alterations have been reproduced by multiple laboratories (Dawson et al., 1999; Goldstein, 2012; Kamal et al., 2001; Magara et al., 1999; Seabrook et al., 1999; Smith, Kallhoff, Zheng, & Pautler, 2007; Smith, Peethumnongsin, Lin, Zheng, & Pautler, 2010; Zheng et al., 1995). Altogether, these knockout studies exemplify the wide range of functions that APP has a role in mediating.

APP AND AD

Disrupted Axonal Transport

Much of the interest in APP has arisen from its connection with AD (discussed above in the 'Introduction'). There are 18 amino acid positions, comprising more than 30 mutations within the APP gene that are associated with autosomal dominant early-onset familial AD (van der Kant & Goldstein, 2015) (Figure 4.2). Because all the mutations in APP that cause familial AD (including all such mutations in PS1 and PS2) induce elevated Aβ, elevation of the $A\beta_{42}/A\beta_{40}$ ratio, or enhanced Aβ aggregation, much of the research on APP has focused on its relationship to Aβ production (Marr & Hafez, 2014), which is believed to be neuro/synaptotoxic. However, the vast majority of familial AD-causing mutations also result in alterations in the production of other APP proteolytic products and other properties related to this multifunctional factor. Since neurons have to transport material much further than other cell types, it is thought they are particularly vulnerable to disruptions in trafficking and endosomal function, which is linked to APP and its metabolites. As an example, the Swedish familial mutant of APP interacts poorly with FE65 and results in disruptions in axonal transport (Rodrigues et al., 2012; Zambrano et al., 1997).

Increases in APP levels resulting from gene duplications result in elevated Aβ and are associated with familial AD (Rovelet-Lecrux et al., 2006). However, such moderate elevations in APP have also been reported to interfere with processes regulated by other APP fragments such as axonal transport and long-distance signaling (Cataldo et al., 2004; Israel et al., 2012; Salehi et al., 2006). Using mutants of APP with enhanced or reduced β-secretase processing, it was found that elevated beta processing reduced anterograde axonal transport of APP, whereas inhibited β-cleavage stimulated APP anterograde axonal transport (Rodrigues et al., 2012). In the same study, the use of secretase inhibitors suggested that the amount of C99 produced these changes in transport. Consistent with this, C99 showed reduced anterograde axonal transport compared with full-length APP. Also, APP transgenic mice expressing the Swedish mutations of APP (that enhance β-cleavage) displayed axonal dystrophy in the absence of amyloid plaque deposition and synaptic degradation suggesting that the C99 fragment of APP can produce axonal pathology independent of Aβ.

Disrupted Neurogenesis

Mutations in APP may also influence the progression of AD through alterations in neurogenesis related to developmental alterations or deficits in adult neurogenesis (which is most relevant in the hippocampus that is affected early during the progression of this disease) (Lazarov & Marr, 2009). Consistent with this, adult neurogenesis is believed to contribute to cognitive function (Lazarov & Marr, 2013). Stem cell ablation in neurogenic areas using radiation or cytostatic/cytotoxic drugs produced deficits in learning and memory (Kim et al., 2008; Shors et al., 2001; Winocur, Wojtowicz, Sekeres, Snyder, & Wang, 2006). Transgenic approaches have provided more targeted assessments of the role of neurogenesis in cognition (Dupret et al., 2008; Imayoshi et al., 2008). This approach has shown that suppressed neurogenesis produced deficits in hippocampal-dependent learning while not affecting nonhippocampal-dependent processes (Zhang, Zou, He, Gage, & Evans, 2008). The use of both irradiation and genetic ablation of neural stem cells found that neurogenesis plays a significant role in the ability to perform pattern separation and distinguish between multiple similar memories (Burghardt, Park, Hen, & Fenton, 2012).

Altered APP Processing/Translocation

There are mutations in other genes that can affect AD through altering the processing of APP. These mutations are purported genetic risk factors for developing AD, which are distinct from the familial mutation in APP, PS1, and PS2 that are virtually 100% penetrant. These genetic risk factors have only marginal effects on the risk of developing AD and do not result in a definite sentence of getting the disease. Mutations in the apoE gene are the best known since they are the most common and have the greatest effects on odds of developing the disease (~three- to fivefold increased risk per inherited ε4 allele (Liu, Kanekiyo, Xu, & Bu, 2013)).

SorLA is a member of the LDL receptor family (perhaps the most distant member) (Yamazaki et al., 1996) that is also categorized as part of the vacuolar protein sorting 10 protein (Vps10p) domain receptors (Jacobsen et al., 1996). SorLA is expressed in neural tissue and functions to regulate the sorting of transport proteins (Wagner & Pietrzik, 2012) and is found associated with APP in endosomal and Golgi compartments (Andersen et al., 2005). This interaction is associated with reduced cell surface APP; however, *in vivo* this slowly internalized receptor may not affect APP through relocalization but is reported to compete with the access of APP to BACE1 and so inhibit amyloidogenic processing (Spoelgen et al., 2006). Supporting this, APP-transgenic mice also deficient for SorLA showed elevated amyloidogenic processing (Dodson et al., 2008), and AD patient cerebrospinal fluid was reported to contain lower levels of SorLA compared to controls (Ma et al., 2009). Also supporting this, polymorphisms in SorLA increase the risk of AD and the mutations were shown to directly affect amyloidogenic APP processing (Vardarajan et al., 2015). Therefore, factors affecting the localization of APP, and thus its processing, have relevance to AD.

Mutations in the APP-interacting proteins FE65 and Dab1 are also implicated in affecting risk for AD (Cousin et al., 2003; Gao, Tao, He, Song, & Saffen, 2015; Hu et al., 1998; Lambert et al., 2000). These and many other gene polymorphisms may affect the risk for developing AD to varying degrees but the vast majority affect risk by less than twofold.

However, a variant (V232M) in the phospholipase D3 (PLD3) gene was reported to double the risk for AD in a European cohort of 11,000 people (Cruchaga et al., 2014). PLD3 is elevated in brain regions that are particularly vulnerable to the pathology of AD, including the hippocampus and cerebral cortex, and PLD3 is significantly suppressed in neurons from AD brains compared to controls. Interestingly, this study also found that PLD3 regulates APP localization and processing. *In vitro* studies showed that the overexpression of PLD3 leads to significantly reduced intracellular APP as well as amyloidogenic processing, while knockdown of PLD3 produces an increase in amyloidogenic processing.

Similar to the case with PLD3, a mutation in the phosphatidylinositol clathrin assembly lymphoid-myeloid leukemia (PICALM) gene was found to significantly affect the odds of developing AD (odds ratio = 0.85) (Harold et al., 2009). It was reported that, after endocytosis is induced, PICALM colocalizes with APP in neuronal-like cells *in vitro*. Specific knockdown of PICALM resulted in reduced APP internalization and amyloidogenic processing (Xiao et al., 2012). Conversely, PICALM overproduction upregulated APP internalization and amyloidogenic processing. In APP transgenic mice, PICALM was shown to be expressed from neurons and colocalized with APP throughout the cortex and hippocampus. Using gene transfer vectors expressing PICALM shRNA or PICALM cDNA injected into the hippocampus of APP transgenic mice, alterations in PICALM expression were tested for the effects on amyloidogenic processing. PICALM knockdown decreased Aβ levels and deposited amyloid plaque load in the hippocampus, while PICALM overexpression elevated Aβ levels and amyloid plaque load.

Altered Signal Transduction

Many of the signal transduction pathways that interact with APP have been associated with AD. Reelin is a good example of this as it has been reported to reduce Aβ toxicity (Durakoglugil, Chen, White, Kavalali, & Herz, 2009), and its expression has been shown to be downregulated in AD (Chin et al., 2007). In addition, polymorphisms in the Reelin gene have been associated with the risk of developing AD (Kramer et al., 2011; Seripa et al., 2008). Lastly, it was found that F-spondin overexpression via viral vectors reduced Aβ levels and improved memory in mice (Hafez et al., 2012). Another good example of the role of APP modulated signaling in AD is Wnt signaling (Small & Duff, 2008). During the process of stimulation of the Wnt pathway, GSK3β is inhibited. This is important as GSK3β can phosphorylate tau protein (the principal component of NFTs), thus promoting its aggregation.

However, the effects of Wnt on AD pathology are not likely limited to modulation of tau phosphorylation alone (Wang et al., 2011). Activation of Wnt signaling through the use of lithium was shown to considerably repress levels of Aβ *in vitro* and in APP transgenic mice (Phiel, Wilson, Lee, & Klein, 2003; Su et al., 2004; Toledo & Inestrosa, 2010). In addition, it has also been reported that lithium could ameliorate neurodegeneration and improve memory performance in rodents (Caccamo, Oddo, Tran, & LaFerla, 2007; De Ferrari et al., 2003; Fiorentini, Rosi, Grossi, Luccarini, and Casamenti, 2010). Similar to Reelin, Wnt has been reported to interfere with Aβ toxicity (Cerpa et al., 2010). As we might predict, the Wnt inhibitory protein Dickkopf-1 was reported to promote neurodegeneration (Caricasole et al., 2004; Zhang, Wang, Khan, Mahesh, & Brann, 2008). The Wnt pathway may also be inhibited by the AD-promoting apoE-ε4 isoform through binding to the LDL-receptor family members, LRP5 and LRP6 (Caruso et al., 2006). Finally, genetic linkage analysis has identified an association between mutations in LRP6 and risk of developing AD (De Ferrari et al., 2007). Thus it appears that APP's effects on multiple signal transduction pathways can plausibly contribute to its role in AD progression.

In conclusion, it is not difficult to imagine how deficits related to the many proposed functions of APP related to neuronal/stem cell function, axonal transport, signal transduction, and more, could manifest in the cognitive deficits that are observed in AD. APP is a diverse multifunctional protein that can be linked to multiple processes that are critical for neuronal function. Therefore, its relevance to neurobiology and dementia go beyond Aβ and merit further study going forward.

References

Abad-Rodriguez, J., Ledesma, M. D., Craessaerts, K., Perga, S., Medina, M., Delacourte, A., … Dotti, C. G. (2004). Neuronal membrane cholesterol loss enhances amyloid peptide generation. *The Journal of Cell Biology*, 167(5), 953–960. http://dx.doi.org/10.1083/jcb.200404149.

Acevedo, K. M., Hung, Y. H., Dalziel, A. H., Li, Q. X., Laughton, K., Wikhe, K., … Camakaris, J. (2011). Copper promotes the trafficking of the amyloid precursor protein. *The Journal of Biological Chemistry*, 286(10), 8252–8262. http://dx.doi.org/10.1074/jbc.M110.128512.

Alvira-Botero, X., & Carro, E. M. (2010). Clearance of amyloid-beta peptide across the choroid plexus in Alzheimer's disease. *Current Aging Science*, 3(3), 219–229.

Andersen, O. M., Reiche, J., Schmidt, V., Gotthardt, M., Spoelgen, R., Behlke, J., … Willnow, T. E. (2005). Neuronal sorting protein-related receptor sorLA/LR11 regulates processing of the amyloid precursor protein. *Proceedings of the National Academy of Sciences of the United States of America*, 102(38), 13461–13466. http://dx.doi.org/10.1073/pnas.0503689102.

Ashley, J., Packard, M., Ataman, B., & Budnik, V. (2005). Fasciclin II signals new synapse formation through amyloid precursor protein and the scaffolding protein dX11/Mint. *The Journal of Neuroscience*, 25(25), 5943–5955. http://dx.doi.org/10.1523/JNEUROSCI.1144-05.2005.

Avramovich-Tirosh, Y., Amit, T., Bar-Am, O., Weinreb, O., & Youdim, M. B. (2008). Physiological and pathological aspects of Abeta in iron homeostasis via 5′UTR in the APP mRNA and the therapeutic use of iron-chelators. *BMC Neuroscience, 9*(Suppl. 2), S2.http://dx.doi.org/10.1186/1471-2202-9-S2-S2.

Aydin, D., Filippov, M. A., Tschape, J. A., Gretz, N., Prinz, M., Eils, R., … Müller, U. C. (2011). Comparative transcriptome profiling of amyloid precursor protein family members in the adult cortex. *BMC Genomics, 12,* 160.http://dx.doi.org/10.1186/1471-2164-12-160.

Baek, S. H., Ohgi, K. A., Rose, D. W., Koo, E. H., Glass, C. K., & Rosenfeld, M. G. (2002). Exchange of N-CoR corepressor and Tip60 coactivator complexes links gene expression by NF-kappaB and beta-amyloid precursor protein. *Cell, 110*(1), 55–67.

Barrett, P. J., Song, Y., Van Horn, W. D., Hustedt, E. J., Schafer, J. M., Hadziselimovic, A., … Sanders, C. R. (2012). The amyloid precursor protein has a flexible transmembrane domain and binds cholesterol. *Science, 336*(6085), 1168–1171. http://dx.doi.org/10.1126/science.1219988.

Baumkotter, F., Schmidt, N., Vargas, C., Schilling, S., Weber, R., Wagner, K., … Kins, S. (2014). Amyloid precursor protein dimerization and synaptogenic function depend on copper binding to the growth factor-like domain. *The Journal of Neuroscience, 34*(33), 11159–11172. http://dx.doi.org/10.1523/JNEUROSCI.0180-14.2014.

Beel, A. J., Sakakura, M., Barrett, P. J., & Sanders, C. R. (2010). Direct binding of cholesterol to the amyloid precursor protein: An important interaction in lipid-Alzheimer's disease relationships? *Biochimica et Biophysica Acta, 1801*(8), 975–982. http://dx.doi.org/10.1016/j.bbalip.2010.03.008.

Bell, K. F., Zheng, L., Fahrenholz, F., & Cuello, A. C. (2008). ADAM-10 over-expression increases cortical synaptogenesis. *Neurobiology of Aging, 29*(4), 554–565. http://dx.doi.org/10.1016/j.neurobiolaging.2006.11.004.

Belyaev, N. D., Nalivaeva, N. N., Makova, N. Z., & Turner, A. J. (2009). Neprilysin gene expression requires binding of the amyloid precursor protein intracellular domain to its promoter: Implications for Alzheimer disease. *EMBO Reports, 10*(1), 94–100. http://dx.doi.org/10.1038/embor.2008.222.

Bishop, G. M., & Robinson, S. R. (2004). The amyloid paradox: Amyloid-beta-metal complexes can be neurotoxic and neuroprotective. *Brain Pathology, 14*(4), 448–452.

Bodovitz, S., & Klein, W. L. (1996). Cholesterol modulates alpha-secretase cleavage of amyloid precursor protein. *The Journal of Biological Chemistry, 271*(8), 4436–4440.

Bonds, J. A., Kuttner-Hirshler, Y., Bartolotti, N., Tobin, M. K., Pizzi, M., Marr, R., et al. (2015). Presenilin-1 dependent neurogenesis regulates hippocampal learning and memory. *PLoS One, 10*(6), e0131266. http://dx.doi.org/10.1371/journal.pone.0131266.

Bouillot, C., Prochiantz, A., Rougon, G., & Allinquant, B. (1996). Axonal amyloid precursor protein expressed by neurons *in vitro* is present in a membrane fraction with caveolae-like properties. *The Journal of Biological Chemistry, 271*(13), 7640–7644.

Brody, D. L., Magnoni, S., Schwetye, K. E., Spinner, M. L., Esparza, T. J., Stocchetti, N., … Holtzman, D. M. (2008). Amyloid-beta dynamics correlate with neurological status in the injured human brain. *Science, 321*(5893), 1221–1224. http://dx.doi.org/10.1126/science.1161591.

Burghardt, N. S., Park, E. H., Hen, R., & Fenton, A. A. (2012). Adult-born hippocampal neurons promote cognitive flexibility in mice. *Hippocampus, 22*(9), 1795–1808. http://dx.doi.org/10.1002/hipo.22013.

Caccamo, A., Oddo, S., Tran, L. X., & LaFerla, F. M. (2007). Lithium reduces tau phosphorylation but not A beta or working memory deficits in a transgenic model with both plaques and tangles. *The American Journal of Pathology, 170*(5), 1669–1675.

Caille, I., Allinquant, B., Dupont, E., Bouillot, C., Langer, A., Muller, U., et al. (2004). Soluble form of amyloid precursor protein regulates proliferation of progenitors in the adult subventricular zone. *Development, 131*(9), 2173–2181. http://dx.doi.org/10.1242/dev.01103.

Cam, J. A., Zerbinatti, C. V., Knisely, J. M., Hecimovic, S., Li, Y., & Bu, G. (2004). The low density lipoprotein receptor-related protein 1B retains beta-amyloid precursor protein at the cell surface and reduces amyloid-beta peptide production. *The Journal of Biological Chemistry, 279*(28), 29639–29646. http://dx.doi.org/10.1074/jbc.M313893200.

Cao, X., & Sudhof, T. C. (2001). A transcriptionally [correction of transcriptively] active complex of APP with Fe65 and histone acetyltransferase Tip60. *Science, 293*(5527), 115–120. http://dx.doi.org/10.1126/science.1058783.

Caricasole, A., Copani, A., Caraci, F., Aronica, E., Rozemuller, A. J., Caruso, A., … Nicoletti, F. (2004). Induction of Dickkopf-1, a negative modulator of the Wnt pathway, is associated with neuronal degeneration in Alzheimer's brain. *The Journal of Neuroscience, 24*(26), 6021–6027.

Caruso, A., Motolese, M., Iacovelli, L., Caraci, F., Copani, A., Nicoletti, F., … Caricasole, A. (2006). Inhibition of the canonical Wnt signaling pathway by apolipoprotein E4 in PC12 cells. *Journal of Neurochemistry, 98*(2), 364–371.

Cataldo, A. M., Petanceska, S., Terio, N. B., Peterhoff, C. M., Durham, R., Mercken, M., … Nixon, R. A. (2004). Abeta localization in abnormal endosomes: Association with earliest Abeta elevations in AD and down syndrome. *Neurobiology of Aging, 25*(10), 1263–1272. http://dx.doi.org/10.1016/j.neurobiolaging.2004.02.027.

Cerpa, W., Farias, G. G., Godoy, J. A., Fuenzalida, M., Bonansco, C., & Inestrosa, N. C. (2010). Wnt-5a occludes Abeta oligomer-induced depression of glutamatergic transmission in hippocampal neurons. *Molecular Neurodegeneration, 5*, 3.

Cerpa, W. F., Barria, M. I., Chacon, M. A., Suazo, M., Gonzalez, M., Opazo, C., … Inestrosa, N. C. (2004). The N-terminal copper-binding domain of the amyloid precursor protein protects against Cu^{2+} neurotoxicity *in vivo*. *The FASEB Journal, 18*(14), 1701–1703. http://dx.doi.org/10.1096/fj.03-1349fje.

Chartier-Harlin, M. C., Crawford, F., Houlden, H., Warren, A., Hughes, D., Fidani, L., et al. (1991). Early-onset alzheimer's disease caused by mutations at codon 717 of the beta-amyloid precursor protein gene. *Nature, 353*(6347), 844–846. http://dx.doi.org/10.1038/353844a0.

Chasseigneaux, S., Dinc, L., Rose, C., Chabret, C., Coulpier, F., Topilko, P., … Allinquant, B. (2011). Secreted amyloid precursor protein beta and secreted amyloid precursor protein alpha induce axon outgrowth *in vitro* through Egr1 signaling pathway. *PLoS One, 6*(1), e16301. http://dx.doi.org/10.1371/journal.pone.0016301.

Checler, F., Sunyach, C., Pardossi-Piquard, R., Sevalle, J., Vincent, B., Kawarai, T., … da Costa, C. A. (2007). The gamma/epsilon-secretase-derived APP intracellular domain fragments regulate p53. *Current Alzheimer Research, 4*(4), 423–426.

Chen, A. C., & Selkoe, D. J. (2007). Response to: Pardossi-Piquard et al., "Presenilin-Dependent Transcriptional Control of the Abeta-Degrading Enzyme Neprilysin by Intracellular Domains of betaAPP and APLP". *Neuron, 46*, 541–554. *Neuron, 53*(4), 479–483. http://dx.doi.org/10.1016/j.neuron.2007.01.023.

Chen, C. W., Boiteau, R. M., Lai, W. F., Barger, S. W., & Cataldo, A. M. (2006). sAPPalpha enhances the transdifferentiation of adult bone marrow progenitor cells to neuronal phenotypes. *Current Alzheimer Research, 3*(1), 63–70.

Chen, X. H., Siman, R., Iwata, A., Meaney, D. F., Trojanowski, J. Q., & Smith, D. H. (2004). Long-term accumulation of amyloid-beta, beta-secretase, presenilin-1, and caspase-3 in damaged axons following brain trauma. *The American Journal of Pathology, 165*(2), 357–371.

Chiba, K., Araseki, M., Nozawa, K., Furukori, K., Araki, Y., Matsushima, T., … Suzuki, T. (2014). Quantitative analysis of APP axonal transport in neurons: Role of JIP1 in enhanced APP anterograde transport. *Molecular Biology of the Cell*, 25(22), 3569–3580. http://dx.doi.org/10.1091/mbc.E14-06-1111.

Chin, J., Massaro, C. M., Palop, J. J., Thwin, M. T., Yu, G. Q., Bien-Ly, N., … Mucke, L. (2007). Reelin depletion in the entorhinal cortex of human amyloid precursor protein transgenic mice and humans with Alzheimer's disease. *The Journal of Neuroscience*, 27(11), 2727–2733.

Choy, R. W., Cheng, Z., & Schekman, R. (2012). Amyloid precursor protein (APP) traffics from the cell surface via endosomes for amyloid beta (Abeta) production in the trans-Golgi network. *Proceedings of the National Academy of Sciences of the United States of America*, 109(30), E2077–E2082. http://dx.doi.org/10.1073/pnas.1208635109.

Clippingdale, A. B., Wade, J. D., & Barrow, C. J. (2001). The amyloid-beta peptide and its role in Alzheimer's disease. *Journal of Peptide Science*, 7(5), 227–249. http://dx.doi.org/10.1002/psc.324.

Collu, G. M., Hidalgo-Sastre, A., & Brennan, K. (2014). Wnt-NOTCH signalling crosstalk in development and disease. *Cellular and Molecular Life Sciences*, 71(18), 3553–3567. http://dx.doi.org/10.1007/s00018-014-1644-x.

Corrigan, F., Thornton, E., Roisman, L. C., Leonard, A. V., Vink, R., Blumbergs, P. C., … Cappai, R. (2014). The neuroprotective activity of the amyloid precursor protein against traumatic brain injury is mediated via the heparin binding site in residues 96-110. *Journal of Neurochemistry*, 128(1), 196–204. http://dx.doi.org/10.1111/jnc.12391.

Corrigan, F., Vink, R., Blumbergs, P. C., Masters, C. L., Cappai, R., & van den Heuvel, C. (2012a). Characterisation of the effect of knockout of the amyloid precursor protein on outcome following mild traumatic brain injury. *Brain Research*, 1451, 87–99. http://dx.doi.org/10.1016/j.brainres.2012.02.045.

Corrigan, F., Vink, R., Blumbergs, P. C., Masters, C. L., Cappai, R., & van den Heuvel, C. (2012b). sAPPalpha rescues deficits in amyloid precursor protein knockout mice following focal traumatic brain injury. *Journal of Neurochemistry*, 122(1), 208–220. http://dx.doi.org/10.1111/j.1471-4159.2012.07761.x.

Cottrell, B. A., Galvan, V., Banwait, S., Gorostiza, O., Lombardo, C. R., Williams, T., … Bredesen, D. E. (2005). A pilot proteomic study of amyloid precursor interactors in Alzheimer's disease. *Annals of Neurology*, 58(2), 277–289. http://dx.doi.org/10.1002/ana.20554.

Cousin, E., Hannequin, D., Ricard, S., Mace, S., Genin, E., Chansac, C., … Deleuze, J. F. (2003). A risk for early-onset alzheimer's disease associated with the APBB1 gene (FE65) intron 13 polymorphism. *Neuroscience Letters*, 342(1–2), 5–8.

Cruchaga, C., Karch, C. M., Jin, S. C., Benitez, B. A., Cai, Y., Guerreiro, R., … Goate, A. M. (2014). Rare coding variants in the phospholipase D3 gene confer risk for Alzheimer's disease. *Nature*, 505(7484), 550–554. http://dx.doi.org/10.1038/nature12825.

Das, U., Scott, D. A., Ganguly, A., Koo, E. H., Tang, Y., & Roy, S. (2013). Activity-induced convergence of APP and BACE-1 in acidic microdomains via an endocytosis-dependent pathway. *Neuron*, 79(3), 447–460. http://dx.doi.org/10.1016/j.neuron.2013.05.035.

Dawson, G. R., Seabrook, G. R., Zheng, H., Smith, D. W., Graham, S., O'Dowd, G., … Sirinathsinghji, D. J. (1999). Age-related cognitive deficits, impaired long-term potentiation and reduction in synaptic marker density in mice lacking the beta-amyloid precursor protein. *Neuroscience*, 90(1), 1–13.

DeBoer, S. R., Dolios, G., Wang, R., & Sisodia, S. S. (2014). Differential release of beta-amyloid from dendrite- versus axon-targeted APP. *The Journal of Neuroscience*, 34(37), 12313–12327. http://dx.doi.org/10.1523/JNEUROSCI.2255-14.2014.

De Ferrari, G. V., Chacon, M. A., Barria, M. I., Garrido, J. L., Godoy, J. A., Olivares, G., … Inestrosa, N. C. (2003). Activation of Wnt signaling rescues neurodegeneration and behavioral impairments induced by beta-amyloid fibrils. *Molecular Psychiatry, 8*(2), 195–208.

De Ferrari, G. V., Papassotiropoulos, A., Biechele, T., Wavrant De-Vrieze, F., Avila, M. E., Major, M. B., ... Moon, R. T. (2007). Common genetic variation within the low-density lipoprotein receptor-related protein 6 and late-onset alzheimer's disease. *Proceedings of the National Academy of Sciences of the United States of America, 104*(22), 9434–9439.

Demars, M. P., Bartholomew, A., Strakova, Z., & Lazarov, O. (2011). Soluble amyloid precursor protein: A novel proliferation factor of adult progenitor cells of ectodermal and mesodermal origin. *Stem Cell Research & Therapy, 2*(4), 36.http://dx.doi.org/10.1186/scrt77.

Demars, M. P., Hollands, C., Zhao Kda, T., & Lazarov, O. (2013). Soluble amyloid precursor protein-alpha rescues age-linked decline in neural progenitor cell proliferation. *Neurobiology of Aging, 34*(10), 2431–2440. http://dx.doi.org/10.1016/j.neurobiolaging.2013.04.016.

Dodson, S. E., Andersen, O. M., Karmali, V., Fritz, J. J., Cheng, D., Peng, J., ... Lah, J. J. (2008). Loss of LR11/SORLA enhances early pathology in a mouse model of amyloidosis: Evidence for a proximal role in Alzheimer's disease. *The Journal of Neuroscience, 28*(48), 12877–12886. http://dx.doi.org/10.1523/JNEUROSCI.4582-08.2008.

Dupret, D., Revest, J. M., Koehl, M., Ichas, F., De Giorgi, F., Costet, P., ... Piazza, P. V. (2008). Spatial relational memory requires hippocampal adult neurogenesis. *PLoS One, 3*(4), e1959.

Durakoglugil, M. S., Chen, Y., White, C. L., Kavalali, E. T., & Herz, J. (2009). Reelin signaling antagonizes beta-amyloid at the synapse. *Proceedings of the National Academy of Sciences of the United States of America, 106*(37), 15938–15943.

Ehehalt, R., Keller, P., Haass, C., Thiele, C., & Simons, K. (2003). Amyloidogenic processing of the Alzheimer beta-amyloid precursor protein depends on lipid rafts. *The Journal of Cell Biology, 160*(1), 113–123. http://dx.doi.org/10.1083/jcb.200207113.

Fassbender, K., Simons, M., Bergmann, C., Stroick, M., Lutjohann, D., Keller, P., ... Hartmann, T. (2001). Simvastatin strongly reduces levels of Alzheimer's disease beta -amyloid peptides Abeta 42 and Abeta 40 *in vitro* and *in vivo*. *Proceedings of the National Academy of Sciences of the United States of America, 98*(10), 5856–5861. http://dx.doi.org/10.1073/pnas.081620098.

Fiorentini, A., Rosi, M. C., Grossi, C., Luccarini, I., & Casamenti, F. (2010). Lithium improves hippocampal neurogenesis, neuropathology and cognitive functions in APP mutant mice. *PLoS One, 5*(12), e14382.

Flammang, B., Pardossi-Piquard, R., Sevalle, J., Debayle, D., Dabert-Gay, A. S., Thevenet, A., ... Checler, F. (2012). Evidence that the amyloid-beta protein precursor intracellular domain, AICD, derives from beta-secretase-generated C-terminal fragment. *Journal of Alzheimer's Disease, 30*(1), 145–153. http://dx.doi.org/10.3233/JAD-2012-112186.

Frears, E. R., Stephens, D. J., Walters, C. E., Davies, H., & Austen, B. M. (1999). The role of cholesterol in the biosynthesis of beta-amyloid. *Neuroreport, 10*(8), 1699–1705.

Fu, M. M., & Holzbaur, E. L. (2013). JIP1 regulates the directionality of APP axonal transport by coordinating kinesin and dynein motors. *The Journal of Cell Biology, 202*(3), 495–508. http://dx.doi.org/10.1083/jcb.201302078.

Gadadhar, A., Marr, R., & Lazarov, O. (2011). Presenilin-1 regulates neural progenitor cell differentiation in the adult brain. *The Journal of Neuroscience, 31*(7), 2615–2623.

Gakhar-Koppole, N., Hundeshagen, P., Mandl, C., Weyer, S. W., Allinquant, B., Muller, U., et al. (2008). Activity requires soluble amyloid precursor protein alpha to promote neurite outgrowth in neural stem cell-derived neurons via activation of the MAPK pathway. *The European Journal of Neuroscience, 28*(5), 871–882. http://dx.doi.org/10.1111/j.1460-9568.2008.06398.x.

Gao, H., Tao, Y., He, Q., Song, F., & Saffen, D. (2015). Functional enrichment analysis of three Alzheimer's disease genome-wide association studies identities DAB1 as a novel candidate liability/protective gene. *Biochemical and Biophysical Research Communications*, *463*(4), 490–495. http://dx.doi.org/10.1016/j.bbrc.2015.05.044.

Gao, Y., & Pimplikar, S. W. (2001). The gamma -secretase-cleaved C-terminal fragment of amyloid precursor protein mediates signaling to the nucleus. *Proceedings of the National Academy of Sciences of the United States of America*, *98*(26), 14979–14984. http://dx.doi.org/10.1073/pnas.261463298.

Ghosal, K., Stathopoulos, A., & Pimplikar, S. W. (2010). APP intracellular domain impairs adult neurogenesis in transgenic mice by inducing neuroinflammation. *PLoS One*, *5*(7), e11866. http://dx.doi.org/10.1371/journal.pone.0011866.

Giliberto, L., Zhou, D., Weldon, R., Tamagno, E., De Luca, P., Tabaton, M., et al. (2008). Evidence that the Amyloid beta precursor protein-intracellular domain lowers the stress threshold of neurons and has a "regulated" transcriptional role. *Molecular Neurodegeneration*, *3*, 12.http://dx.doi.org/10.1186/1750-1326-3-12.

Glenner, G. G., & Wong, C. W. (1984a). Alzheimer's disease and down's syndrome: Sharing of a unique cerebrovascular amyloid fibril protein. *Biochemical and Biophysical Research Communications*, *122*(3), 1131–1135.

Glenner, G. G., & Wong, C. W. (1984b). Alzheimer's disease: Initial report of the purification and characterization of a novel cerebrovascular amyloid protein. *Biochemical and Biophysical Research Communications*, *120*(3), 885–890.

Goate, A., Chartier-Harlin, M. C., Mullan, M., Brown, J., Crawford, F., Fidani, L., et al. (1991). Segregation of a missense mutation in the amyloid precursor protein gene with familial Alzheimer's disease. *Nature*, *349*(6311), 704–706. http://dx.doi.org/10.1038/349704a0.

Goate, A. M., Haynes, A. R., Owen, M. J., Farrall, M., James, L. A., Lai, L. Y., et al. (1989). Predisposing locus for Alzheimer's disease on chromosome 21. *Lancet*, *1*(8634), 352–355.

Goedert, M., & Spillantini, M. G. (2006). A century of Alzheimer's disease. *Science*, *314*(5800), 777–781. http://dx.doi.org/10.1126/science.1132814.

Goldstein, L. S. (2012). Axonal transport and neurodegenerative disease: Can we see the elephant? *Progress in Neurobiology*, *99*(3), 186–190. http://dx.doi.org/10.1016/j.pneurobio.2012.03.006.

Grimm, M. O., Grimm, H. S., Patzold, A. J., Zinser, E. G., Halonen, R., Duering, M., … Hartmann, T. (2005). Regulation of cholesterol and sphingomyelin metabolism by amyloid-beta and presenilin. *Nature Cell Biology*, *7*(11), 1118–1123. http://dx.doi.org/10.1038/ncb1313.

Grundke-Iqbal, I., Iqbal, K., Quinlan, M., Tung, Y. C., Zaidi, M. S., & Wisniewski, H. M. (1986). Microtubule-associated protein tau. A component of Alzheimer paired helical filaments. *The Journal of Biological Chemistry*, *261*(13), 6084–6089.

Haass, C., Kaether, C., Thinakaran, G., & Sisodia, S. (2012). Trafficking and proteolytic processing of APP. *Cold Spring Harbor Perspectives in Medicine*, *2*(5), a006270. http://dx.doi.org/10.1101/cshperspect.a006270.

Hafez, D. M., Huang, J. Y., Richardson, J. C., Masliah, E., Peterson, D. A., & Marr, R. A. (2012). F-spondin gene transfer improves memory performance and reduces amyloid-beta levels in mice. *Neuroscience*, *223*, 465–472. http://dx.doi.org/10.1016/j.neuroscience.2012.07.038.

Harold, D., Abraham, R., Hollingworth, P., Sims, R., Gerrish, A., Hamshere, M. L., … Williams, J. (2009). Genome-wide association study identifies variants at CLU and PICALM associated with Alzheimer's disease. *Nature Genetics*, *41*(10), 1088–1093. http://dx.doi.org/10.1038/ng.440.

Hasebe, N., Fujita, Y., Ueno, M., Yoshimura, K., Fujino, Y., & Yamashita, T. (2013). Soluble beta-amyloid precursor protein alpha binds to p75 neurotrophin receptor to promote neurite outgrowth. *PLoS One*, *8*(12), e82321. http://dx.doi.org/10.1371/journal.pone.0082321.

Hayashi, Y., Kashiwagi, K., Ohta, J., Nakajima, M., Kawashima, T., & Yoshikawa, K. (1994). Alzheimer amyloid protein precursor enhances proliferation of neural stem cells from fetal rat brain. *Biochemical and Biophysical Research Communications, 205*(1), 936–943. http://dx.doi.org/10.1006/bbrc.1994.2755.

Hebert, S. S., Serneels, L., Tolia, A., Craessaerts, K., Derks, C., Filippov, M. A., … De Strooper, B. (2006). Regulated intramembrane proteolysis of amyloid precursor protein and regulation of expression of putative target genes. *EMBO Reports, 7*(7), 739–745. http://dx.doi.org/10.1038/sj.embor.7400704.

Herz, J., & Chen, Y. (2006). Reelin, lipoprotein receptors and synaptic plasticity. *Nature Reviews Neuroscience, 7*(11), 850–859.

Hesse, L., Beher, D., Masters, C. L., & Multhaup, G. (1994). The beta A4 amyloid precursor protein binding to copper. *FEBS Letters, 349*(1), 109–116.

Hoareau, C., Borrell, V., Soriano, E., Krebs, M. O., Prochiantz, A., & Allinquant, B. (2008). Amyloid precursor protein cytoplasmic domain antagonizes reelin neurite outgrowth inhibition of hippocampal neurons. *Neurobiology of Aging, 29*(4), 542–553. http://dx.doi.org/10.1016/j.neurobiolaging.2006.11.012.

Hoe, H. S., Magill, L. A., Guenette, S., Fu, Z., Vicini, S., & Rebeck, G. W. (2006). FE65 interaction with the ApoE receptor ApoEr2. *The Journal of Biological Chemistry, 281*(34), 24521–24530. http://dx.doi.org/10.1074/jbc.M600728200.

Hoe, H. S., & Rebeck, G. W. (2005). Regulation of ApoE receptor proteolysis by ligand binding. *Brain Research Molecular Brain Research, 137*(1-2), 31–39. http://dx.doi.org/10.1016/j.molbrainres.2005.02.013.

Hoe, H. S., Tran, T. S., Matsuoka, Y., Howell, B. W., & Rebeck, G. W. (2006). DAB1 and Reelin effects on amyloid precursor protein and ApoE receptor 2 trafficking and processing. *The Journal of Biological Chemistry, 281*(46), 35176–35185. http://dx.doi.org/10.1074/jbc.M602162200.

Hoe, H. S., Wessner, D., Beffert, U., Becker, A. G., Matsuoka, Y., & Rebeck, G. W. (2005). F-spondin interaction with the apolipoprotein E receptor ApoEr2 affects processing of amyloid precursor protein. *Molecular and Cellular Biology, 25*(21), 9259–9268.

Hu, Q., Kukull, W. A., Bressler, S. L., Gray, M. D., Cam, J. A., Larson, E. B., … Deeb, S. S. (1998). The human FE65 gene: Genomic structure and an intronic biallelic polymorphism associated with sporadic dementia of the Alzheimer type. *Human Genetics, 103*(3), 295–303.

Huang, Y., Weisgraber, K. H., Mucke, L., & Mahley, R. W. (2004). Apolipoprotein E: Diversity of cellular origins, structural and biophysical properties, and effects in Alzheimer's disease. *Journal of Molecular Neuroscience, 23*(3), 189–204.

Hussain, I., Powell, D., Howlett, D. R., Tew, D. G., Meek, T. D., Chapman, C., … Christie, G. (1999). Identification of a novel aspartic protease (Asp 2) as beta-secretase. *Molecular and Cellular Neurosciences, 14*(6), 419–427. http://dx.doi.org/10.1006/mcne.1999.0811.

Imayoshi, I., Sakamoto, M., Ohtsuka, T., Takao, K., Miyakawa, T., Yamaguchi, M., … Kageyama, R. (2008). Roles of continuous neurogenesis in the structural and functional integrity of the adult forebrain. *Nature Neuroscience, 11*(10), 1153–1161.

Israel, M. A., Yuan, S. H., Bardy, C., Reyna, S. M., Mu, Y., Herrera, C., … Goldstein, L. S. (2012). Probing sporadic and familial Alzheimer's disease using induced pluripotent stem cells. *Nature, 482*(7384), 216–220. http://dx.doi.org/10.1038/nature10821.

Jacobsen, L., Madsen, P., Moestrup, S. K., Lund, A. H., Tommerup, N., Nykjaer, A., … Petersen, C. M. (1996). Molecular characterization of a novel human hybrid-type receptor that binds the alpha2-macroglobulin receptor-associated protein. *The Journal of Biological Chemistry, 271*(49), 31379–31383.

Jonsson, T., Atwal, J. K., Steinberg, S., Snaedal, J., Jonsson, P. V., Bjornsson, S., … Stefansson, K. (2012). A mutation in APP protects against Alzheimer's disease and age-related cognitive decline. *Nature, 488*(7409), 96–99. http://dx.doi.org/10.1038/nature11283.

Joshi, P., Liang, J. O., DiMonte, K., Sullivan, J., & Pimplikar, S. W. (2009). Amyloid precursor protein is required for convergent-extension movements during Zebrafish development. *Developmental Biology,* *335*(1), 1–11. http://dx.doi.org/10.1016/j.ydbio.2009.07.041.

Kamal, A., Almenar-Queralt, A., LeBlanc, J. F., Roberts, E. A., & Goldstein, L. S. (2001). Kinesin-mediated axonal transport of a membrane compartment containing beta-secretase and presenilin-1 requires APP. *Nature,* *414*(6864), 643–648. http://dx.doi.org/10.1038/414643a.

Kamal, A., Stokin, G. B., Yang, Z., Xia, C. H., & Goldstein, L. S. (2000). Axonal transport of amyloid precursor protein is mediated by direct binding to the kinesin light chain subunit of kinesin-I. *Neuron,* *28*(2), 449–459.

Kamenetz, F., Tomita, T., Hsieh, H., Seabrook, G., Borchelt, D., Iwatsubo, T., ... Malinow, R. (2003). APP processing and synaptic function. *Neuron,* *37*(6), 925–937.

Kerridge, C., Belyaev, N. D., Nalivaeva, N. N., & Turner, A. J. (2014). The Abeta-clearance protein transthyretin, like neprilysin, is epigenetically regulated by the amyloid precursor protein intracellular domain. *Journal of Neurochemistry,* *130*(3), 419–431. http://dx.doi.org/10.1111/jnc.12680.

Kidd, M. (1963). Paired helical filaments in electron microscopy of Alzheimer's disease. *Nature,* *197*, 192–193.

Kim, H. S., Kim, E. M., Lee, J. P., Park, C. H., Kim, S., Seo, J. H., ... Suh, Y. H. (2003). C-terminal fragments of amyloid precursor protein exert neurotoxicity by inducing glycogen synthase kinase-3beta expression. *The FASEB Journal,* *17*(13), 1951–1953. http://dx.doi.org/10.1096/fj.03-0106fje.

Kim, J. S., Lee, H. J., Kim, J. C., Kang, S. S., Bae, C. S., Shin, T., ... Moon, C. (2008). Transient impairment of hippocampus-dependent learning and memory in relatively low-dose of acute radiation syndrome is associated with inhibition of hippocampal neurogenesis. *Journal of Radiation Research,* *49*(5), 517–526. doi:JST.JSTAGE/jrr/08020 [pii].

Knauer, M. F., Orlando, R. A., & Glabe, C. G. (1996). Cell surface APP751 forms complexes with protease nexin 2 ligands and is internalized via the low density lipoprotein receptor-related protein (LRP). *Brain Research,* *740*(1-2), 6–14.

Kojro, E., Gimpl, G., Lammich, S., Marz, W., & Fahrenholz, F. (2001). Low cholesterol stimulates the nonamyloidogenic pathway by its effect on the alpha-secretase ADAM 10. *Proceedings of the National Academy of Sciences of the United States of America,* *98*(10), 5815–5820. http://dx.doi.org/10.1073/pnas.081612998.

Konietzko, U. (2012). AICD nuclear signaling and its possible contribution to Alzheimer's disease. *Current Alzheimer Research,* *9*(2), 200–216.

Koo, E. H., Sisodia, S. S., Archer, D. R., Martin, L. J., Weidemann, A., Beyreuther, ... Price, D. L. (1990). Precursor of amyloid protein in Alzheimer disease undergoes fast anterograde axonal transport. *Proceedings of the National Academy of Sciences of the United States of America,* *87*(4), 1561–1565.

Kosik, K. S., Joachim, C. L., & Selkoe, D. J. (1986). Microtubule-associated protein tau (tau) is a major antigenic component of paired helical filaments in Alzheimer disease. *Proceedings of the National Academy of Sciences of the United States of America,* *83*(11), 4044–4048.

Kounnas, M. Z., Moir, R. D., Rebeck, G. W., Bush, A. I., Argraves, W. S., Tanzi, R. E., ... Strickland, D. K. (1995). LDL receptor-related protein, a multifunctional ApoE receptor, binds secreted beta-amyloid precursor protein and mediates its degradation. *Cell,* *82*(2), 331–340.

Kramer, P. L., Xu, H., Woltjer, R. L., Westaway, S. K., Clark, D., Erten-Lyons, D., ... Ott, J. (2011). Alzheimer disease pathology in cognitively healthy elderly: A genome-wide study. *Neurobiology of Aging,* *32*(12), 2113–2122. http://dx.doi.org/10.1016/j.neurobiolaging.2010.01.010.

Lambert, J. C., Mann, D., Goumidi, L., Harris, J., Pasquier, F., Frigard, B., ... Chartier-Harlin, M. C. (2000). A FE65 polymorphism associated with risk of developing sporadic late-onset alzheimer's disease but not with Abeta loading in brains. *Neuroscience Letters,* *293*(1), 29–32.

Lazarov, O., & Marr, R. A. (2009). Neurogenesis and Alzheimer's disease: At the crossroads. *Experimental Neurology, 223*(2), 267–281.

Lazarov, O., & Marr, R. A. (2013). Of mice and men: Neurogenesis, cognition and Alzheimer's disease. *Frontiers in Aging Neuroscience, 5,* 43.http://dx.doi.org/10.3389/fnagi.2013.00043.

Lazarov, O., Morfini, G. A., Lee, E. B., Farah, M. H., Szodorai, A., DeBoer, S. R., … Sisodia, S. S. (2005). Axonal transport, amyloid precursor protein, kinesin-1, and the processing apparatus: Revisited. *The Journal of Neuroscience, 25*(9), 2386–2395. http://dx.doi.org/10.1523/JNEUROSCI.3089-04.2005.

Lee, K. J., Moussa, C. E., Lee, Y., Sung, Y., Howell, B. W., Turner, R. S., … Hoe, H. S. (2010). Beta amyloid-independent role of amyloid precursor protein in generation and maintenance of dendritic spines. *Neuroscience, 169*(1), 344–356. http://dx.doi.org/10.1016/j.neuroscience.2010.04.078.

Lee, S. J., Liyanage, U., Bickel, P. E., Xia, W., Lansbury, P. T., Jr., & Kosik, K. S. (1998). A detergent-insoluble membrane compartment contains A beta *in vivo. Nature Medicine, 4*(6), 730–734.

Leissring, M. A., Murphy, M. P., Mead, T. R., Akbari, Y., Sugarman, M. C., Jannatipour, M., … LaFerla, F. M. (2002). A physiologic signaling role for the gamma-secretase-derived intracellular fragment of APP. *Proceedings of the National Academy of Sciences of the United States of America, 99*(7), 4697–4702. http://dx.doi.org/10.1073/pnas.072033799.

Leyssen, M., Ayaz, D., Hebert, S. S., Reeve, S., De Strooper, B., & Hassan, B. A. (2005). Amyloid precursor protein promotes post-developmental neurite arborization in the drosophila brain. *The EMBO Journal, 24*(16), 2944–2955. http://dx.doi.org/10.1038/sj.emboj.7600757.

Li, T., Wen, H., Brayton, C., Das, P., Smithson, L. A., Fauq, A., … Wong, P. C. (2007). Epidermal growth factor receptor and NOTCH pathways participate in the tumor suppressor function of gamma-secretase. *The Journal of Biological Chemistry, 282*(44), 32264–32273. http://dx.doi.org/10.1074/jbc.M703649200.

Li, X., Masliah, E., Reixach, N., & Buxbaum, J. N. (2011). Neuronal production of transthyretin in human and murine Alzheimer's disease: Is it protective? *The Journal of Neuroscience, 31*(35), 12483–12490. http://dx.doi.org/10.1523/JNEUROSCI.2417-11.2011.

Liu, C. C., Kanekiyo, T., Xu, H., & Bu, G. (2013). Apolipoprotein E and Alzheimer disease: Risk, mechanisms and therapy. *Nature Reviews Neurology, 9*(2), 106–118. http://dx.doi.org/10.1038/nrneurol.2012.263.

Liu, C. X., Li, Y., Obermoeller-McCormick, L. M., Schwartz, A. L., & Bu, G. (2001). The putative tumor suppressor LRP1B, a novel member of the low density lipoprotein (LDL) receptor family, exhibits both overlapping and distinct properties with the LDL receptor-related protein. *The Journal of Biological Chemistry, 276*(31), 28889–28896. http://dx.doi.org/10.1074/jbc.M102727200.

Liu, L., & Murphy, R. M. (2006). Kinetics of inhibition of beta-amyloid aggregation by transthyretin. *Biochemistry, 45*(51), 15702–15709. http://dx.doi.org/10.1021/bi0618520.

Liu, Q., Zerbinatti, C. V., Zhang, J., Hoe, H. S., Wang, B., Cole, S. L., et al. (2007). Amyloid precursor protein regulates brain apolipoprotein E and cholesterol metabolism through lipoprotein receptor LRP1. *Neuron, 56*(1), 66–78. http://dx.doi.org/10.1016/j.neuron.2007.08.008.

Loane, D. J., Pocivavsek, A., Moussa, C. E., Thompson, R., Matsuoka, Y., Faden, A. I., … Burns, M. P. (2009). Amyloid precursor protein secretases as therapeutic targets for traumatic brain injury. *Nature Medicine, 15*(4), 377–379. http://dx.doi.org/10.1038/nm.1940.

Ma, Q. H., Futagawa, T., Yang, W. L., Jiang, X. D., Zeng, L., Takeda, Y., ... Xiao, Z. C. (2008). A TAG1-APP signalling pathway through Fe65 negatively modulates neurogenesis. *Nature Cell Biology, 10*(3), 283–294. http://dx.doi.org/10.1038/ncb1690.

Ma, Q. L., Galasko, D. R., Ringman, J. M., Vinters, H. V., Edland, S. D., Pomakian, J., ... Cole, G. M. (2009). Reduction of SorLA/LR11, a sorting protein limiting beta-amyloid production, in Alzheimer disease cerebrospinal fluid. *Archives of Neurology, 66*(4), 448–457. http://dx.doi.org/10.1001/archneurol.2009.22.

Magara, F., Muller, U., Li, Z. W., Lipp, H. P., Weissmann, C., Stagljar, M., et al. (1999). Genetic background changes the pattern of forebrain commissure defects in transgenic mice underexpressing the beta-amyloid-precursor protein. *Proceedings of the National Academy of Sciences of the United States of America, 96*(8), 4656–4661.

Maloney, J. A., Bainbridge, T., Gustafson, A., Zhang, S., Kyauk, R., Steiner, P., ... Atwal, J. K. (2014). Molecular mechanisms of Alzheimer disease protection by the A673T allele of amyloid precursor protein. *The Journal of Biological Chemistry, 289*(45), 30990–31000. http://dx.doi.org/10.1074/jbc.M114.589069.

Mannix, R. C., Zhang, J., Berglass, J., Qui, J., & Whalen, M. J. (2013). Beneficial effect of amyloid beta after controlled cortical impact. *Brain Injury, 27*(6), 743–748. http://dx.doi.org/10.3109/02699052.2013.771797.

Mannix, R. C., Zhang, J., Park, J., Lee, C., & Whalen, M. J. (2011). Detrimental effect of genetic inhibition of B-site APP-cleaving enzyme 1 on functional outcome after controlled cortical impact in young adult mice. *Journal of Neurotrauma, 28*(9), 1855–1861. http://dx.doi.org/10.1089/neu.2011.1759.

Marr, R. A., & Hafez, D. M. (2014). Amyloid-beta and Alzheimer's disease: The role of neprilysin-2 in amyloid-beta clearance. *Frontiers in Aging Neuroscience, 6,* 187.http://dx.doi.org/10.3389/fnagi.2014.00187.

Masters, C. L., Simms, G., Weinman, N. A., Multhaup, G., McDonald, B. L., & Beyreuther, K. (1985). Amyloid plaque core protein in Alzheimer disease and Down syndrome. *Proceedings of the National Academy of Sciences of the United States of America, 82*(12), 4245–4249.

Maulik, M., Westaway, D., Jhamandas, J. H., & Kar, S. (2013). Role of cholesterol in APP metabolism and its significance in Alzheimer's disease pathogenesis. *Molecular Neurobiology, 47*(1), 37–63. http://dx.doi.org/10.1007/s12035-012-8337-y.

Meng, J. Y., Kataoka, H., Itoh, H., & Koono, M. (2001). Amyloid beta protein precursor is involved in the growth of human colon carcinoma cell *in vitro* and *in vivo. International Journal of Cancer, 92*(1), 31–39.

Milward, E. A., Papadopoulos, R., Fuller, S. J., Moir, R. D., Small, D., Beyreuther, K., et al. (1992). The amyloid protein precursor of Alzheimer's disease is a mediator of the effects of nerve growth factor on neurite outgrowth. *Neuron, 9*(1), 129–137.

Morley, J. E., Farr, S. A., Banks, W. A., Johnson, S. N., Yamada, K. A., & Xu, L. (2010). A physiological role for amyloid-beta protein: Enhancement of learning and memory. *Journal of Alzheimer's Disease, 19*(2), 441–449. http://dx.doi.org/10.3233/JAD-2009-1230.

Muller, T., Concannon, C. G., Ward, M. W., Walsh, C. M., Tirniceriu, A. L., Tribl, F., ... Egensperger, R. (2007). Modulation of gene expression and cytoskeletal dynamics by the amyloid precursor protein intracellular domain (AICD). *Molecular Biology of the Cell, 18*(1), 201–210. http://dx.doi.org/10.1091/mbc.E06-04-0283.

Muller, U. C., & Zheng, H. (2012). Physiological functions of APP family proteins. *Cold Spring Harbor Perspectives in Medicine, 2*(2), a006288. http://dx.doi.org/10.1101/cshperspect.a006288.

Murrell, J., Farlow, M., Ghetti, B., & Benson, M. D. (1991). A mutation in the amyloid precursor protein associated with hereditary Alzheimer's disease. *Science, 254*(5028), 97–99.

Muthusamy, N., Chen, Y. J., Yin, D. M., Mei, L., & Bergson, C. (2015). Complementary roles of the neuron-enriched endosomal proteins NEEP21 and calcyon in neuronal vesicle trafficking. *Journal of Neurochemistry, 132*(1), 20–31. http://dx.doi.org/10.1111/jnc.12989.

Nakamura, M., Shishido, N., Nunomura, A., Smith, M. A., Perry, G., Hayashi, Y., … Hayashi, T. (2007). Three histidine residues of amyloid-beta peptide control the redox activity of copper and iron. *Biochemistry, 46*(44), 12737–12743. http://dx.doi.org/10.1021/bi701079z.

Nicolas, M., & Hassan, B. A. (2014). Amyloid precursor protein and neural development. *Development, 141*(13), 2543–2548. http://dx.doi.org/10.1242/dev.108712.

Nikolaev, A., McLaughlin, T., O'Leary, D. D., & Tessier-Lavigne, M. (2009). APP binds DR6 to trigger axon pruning and neuron death via distinct caspases. *Nature, 457*(7232), 981–989. http://dx.doi.org/10.1038/nature07767.

Noda, Y., Asada, M., Kubota, M., Maesako, M., Watanabe, K., Uemura, M., … Uemura, K. (2013). Copper enhances APP dimerization and promotes Abeta production. *Neuroscience Letters, 547*, 10–15. http://dx.doi.org/10.1016/j.neulet.2013.04.057.

Norstrom, E. M., Zhang, C., Tanzi, R., & Sisodia, S. S. (2010). Identification of NEEP21 as a ss-amyloid precursor protein-interacting protein *in vivo* that modulates amyloidogenic processing *in vitro*. *The Journal of Neuroscience, 30*(46), 15677–15685. http://dx.doi.org/10.1523/JNEUROSCI.4464-10.2010.

Ohsawa, I., Takamura, C., Morimoto, T., Ishiguro, M., & Kohsaka, S. (1999). Amino-terminal region of secreted form of amyloid precursor protein stimulates proliferation of neural stem cells. *The European Journal of Neuroscience, 11*(6), 1907–1913.

Osterfield, M., Egelund, R., Young, L. M., & Flanagan, J. G. (2008). Interaction of amyloid precursor protein with contactins and NgCAM in the retinotectal system. *Development, 135*(6), 1189–1199. http://dx.doi.org/10.1242/dev.007401.

Ott, M. O., & Bullock, S. L. (2001). A gene trap insertion reveals that amyloid precursor protein expression is a very early event in murine embryogenesis. *Development Genes and Evolution, 211*(7), 355–357.

Pajoohesh-Ganji, A., Burns, M. P., Pal-Ghosh, S., Tadvalkar, G., Hokenbury, N. G., Stepp, M. A., et al. (2014). Inhibition of amyloid precursor protein secretases reduces recovery after spinal cord injury. *Brain Research, 1560*, 73–82. http://dx.doi.org/10.1016/j.brainres.2014.02.049.

Pardossi-Piquard, R., Petit, A., Kawarai, T., Sunyach, C., Alves da Costa, C., Vincent, B., … Checler, F. (2005). Presenilin-dependent transcriptional control of the Abeta-degrading enzyme neprilysin by intracellular domains of betaAPP and APLP. *Neuron, 46*(4), 541–554. http://dx.doi.org/10.1016/j.neuron.2005.04.008.

Phiel, C. J., Wilson, C. A., Lee, V. M., & Klein, P. S. (2003). GSK-3alpha regulates production of Alzheimer's disease amyloid-beta peptides. *Nature, 423*(6938), 435–439.

Philibert, K. D., Marr, R. A., Norstrom, E. M., & Glucksman, M. J. (2014). Identification and characterization of Abeta peptide interactors in Alzheimer's disease by structural approaches. *Frontiers in Aging Neuroscience, 6*, 265.http://dx.doi.org/10.3389/fnagi.2014.00265.

Phinney, A. L., Calhoun, M. E., Wolfer, D. P., Lipp, H. P., Zheng, H., & Jucker, M. (1999). No hippocampal neuron or synaptic bouton loss in learning-impaired aged beta-amyloid precursor protein-null mice. *Neuroscience, 90*(4), 1207–1216.

Pietrzik, C. U., Busse, T., Merriam, D. E., Weggen, S., & Koo, E. H. (2002). The cytoplasmic domain of the LDL receptor-related protein regulates multiple steps in APP processing. *The EMBO Journal, 21*(21), 5691–5700.

Pietrzik, C. U., Hoffmann, J., Stober, K., Chen, C. Y., Bauer, C., Otero, D. A., … , Herzog, V., et al. (1998). From differentiation to proliferation: The secretory amyloid precursor protein as a local mediator of growth in thyroid epithelial cells. *Proceedings of the National Academy of Sciences of the United States of America, 95*(4), 1770–1775.

Pietrzik, C. U., Yoon, I. S., Jaeger, S., Busse, T., Weggen, S., & Koo, E. H. (2004). FE65 constitutes the functional link between the low-density lipoprotein receptor-related protein and the amyloid precursor protein. *The Journal of Neuroscience*, 24(17), 4259–4265. http://dx.doi.org/10.1523/JNEUROSCI.5451-03.2004.

Puzzo, D., Privitera, L., Leznik, E., Fa, M., Staniszewski, A., Palmeri, A., et al. (2008). Picomolar amyloid-beta positively modulates synaptic plasticity and memory in hippocampus. *The Journal of Neuroscience*, 28(53), 14537–14545. http://dx.doi.org/10.1523/JNEUROSCI.2692-08.2008.

Rajendran, L., & Annaert, W. (2012). Membrane trafficking pathways in Alzheimer's disease. *Traffic*, 13(6), 759–770. http://dx.doi.org/10.1111/j.1600-0854.2012.01332.x.

Rajendran, L., Honsho, M., Zahn, T. R., Keller, P., Geiger, K. D., Verkade, P., et al. (2006). Alzheimer's disease beta-amyloid peptides are released in association with exosomes. *Proc Natl Acad Sci USA*, 103(30), 11172–11177. http://dx.doi.org/10.1073/pnas.0603838103.

Repetto, E., Yoon, I. S., Zheng, H., & Kang, D. E. (2007). Presenilin 1 regulates epidermal growth factor receptor turnover and signaling in the endosomal-lysosomal pathway. *The Journal of Biological Chemistry*, 282(43), 31504–31516. http://dx.doi.org/10.1074/jbc.M704273200.

Rice, H. C., Townsend, M., Bai, J., Suth, S., Cavanaugh, W., Selkoe, D. J., et al. (2012). Pancortins interact with amyloid precursor protein and modulate cortical cell migration. *Development*, 139(21), 3986–3996. http://dx.doi.org/10.1242/dev.082909.

Ring, S., Weyer, S. W., Kilian, S. B., Waldron, E., Pietrzik, C. U., Filippov, M. A., ... Muller, U. C. (2007). The secreted beta-amyloid precursor protein ectodomain APPs alpha is sufficient to rescue the anatomical, behavioral, and electrophysiological abnormalities of APP-deficient mice. *The Journal of Neuroscience*, 27(29), 7817–7826. http://dx.doi.org/10.1523/JNEUROSCI.1026-07.2007.

Rodrigues, E. M., Weissmiller, A. M., & Goldstein, L. S. (2012). Enhanced beta-secretase processing alters APP axonal transport and leads to axonal defects. *Human Molecular Genetics*, 21(21), 4587–4601. http://dx.doi.org/10.1093/hmg/dds297.

Rogers, J. T., Leiter, L. M., McPhee, J., Cahill, C. M., Zhan, S. S., Potter, H., et al. (1999). Translation of the alzheimer amyloid precursor protein mRNA is up-regulated by interleukin-1 through 5'-untranslated region sequences. *The Journal of Biological Chemistry*, 274(10), 6421–6431.

Rohe, M., Carlo, A. S., Breyhan, H., Sporbert, A., Militz, D., Schmidt, V., ... Andersen, O. M. (2008). Sortilin-related receptor with A-type repeats (SORLA) affects the amyloid precursor protein-dependent stimulation of ERK signaling and adult neurogenesis. *The Journal of Biological Chemistry*, 283(21), 14826–14834. http://dx.doi.org/10.1074/jbc.M710574200.

Rovelet-Lecrux, A., Hannequin, D., Raux, G., Le Meur, N., Laquerriere, A., Vital, A., ... Campion, D. (2006). APP locus duplication causes autosomal dominant early-onset alzheimer disease with cerebral amyloid angiopathy. *Nature Genetics*, 38(1), 24–26.

Ryan, K. A., & Pimplikar, S. W. (2005). Activation of GSK-3 and phosphorylation of CRMP2 in transgenic mice expressing APP intracellular domain. *The Journal of Cell Biology*, 171(2), 327–335. http://dx.doi.org/10.1083/jcb.200505078.

Saitoh, T., Sundsmo, M., Roch, J. M., Kimura, N., Cole, G., Schubert, D., ... Schenk, D. B. (1989). Secreted form of amyloid beta protein precursor is involved in the growth regulation of fibroblasts. *Cell*, 58(4), 615–622.

Salehi, A., Delcroix, J. D., Belichenko, P. V., Zhan, K., Wu, C., Valletta, J. S., ... Mobley, W. C. (2006). Increased App expression in a mouse model of down's syndrome disrupts NGF transport and causes cholinergic neuron degeneration. *Neuron*, 51(1), 29–42. http://dx.doi.org/10.1016/j.neuron.2006.05.022.

Sannerud, R., Declerck, I., Peric, A., Raemaekers, T., Menendez, G., Zhou, L., ... Annaert, W. (2011). ADP ribosylation factor 6 (ARF6) controls amyloid precursor protein (APP) processing by mediating the endosomal sorting of BACE1. *Proceedings of the National*

Academy of Sciences of the United States of America, 108(34), E559–E568. http://dx.doi. org/10.1073/pnas.1100745108.

Sarasa, M., Sorribas, V., Terradoa, J., Climent, S., Palacios, J. M., & Mengod, G. (2000). Alzheimer beta-amyloid precursor proteins display specific patterns of expression during embryogenesis. *Mechanisms of Development, 94*(1-2), 233–236.

Schmitz, A., Tikkanen, R., Kirfel, G., & Herzog, V. (2002). The biological role of the Alzheimer amyloid precursor protein in epithelial cells. *Histochemistry and Cell Biology, 117*(2), 171–180. http://dx.doi.org/10.1007/s00418-001-0351-5.

Schubert, W., Prior, R., Weidemann, A., Dircksen, H., Multhaup, G., Masters, C. L., et al. (1991). Localization of Alzheimer beta A4 amyloid precursor protein at central and peripheral synaptic sites. *Brain Research, 563*(1-2), 184–194.

Seabrook, G. R., Smith, D. W., Bowery, B. J., Easter, A., Reynolds, T., Fitzjohn, S. M., … Hill, R. G. (1999). Mechanisms contributing to the deficits in hippocampal synaptic plasticity in mice lacking amyloid precursor protein. *Neuropharmacology, 38*(3), 349–359.

Selkoe, D., & Kopan, R. (2003). NOTCH and presenilin: Regulated intramembrane proteolysis links development and degeneration. *Annual Review of Neuroscience, 26*, 565–597. http://dx.doi.org/10.1146/annurev.neuro.26.041002.131334.

Selkoe, D. J. (2001). Alzheimer's disease: Genes, proteins, and therapy. *Physiological Reviews, 81*(2), 741–766.

Seripa, D., Matera, M. G., Franceschi, M., Daniele, A., Bizzarro, A., Rinaldi, M., … Pilotto, A. (2008). The RELN locus in Alzheimer's disease. *Journal of Alzheimer's Disease, 14*(3), 335–344.

Shors, T. J., Miesegaes, G., Beylin, A., Zhao, M., Rydel, T., & Gould, E. (2001). Neurogenesis in the adult is involved in the formation of trace memories. *Nature, 410*(6826), 372–376.

Simons, M., Keller, P., De Strooper, B., Beyreuther, K., Dotti, C. G., & Simons, K. (1998). Cholesterol depletion inhibits the generation of beta-amyloid in hippocampal neurons. *Proceedings of the National Academy of Sciences of the United States of America, 95*(11), 6460–6464.

Sisodia, S. S. (1992). Beta-amyloid precursor protein cleavage by a membrane-bound protease. *Proceedings of the National Academy of Sciences of the United States of America, 89*(13), 6075–6079.

Slack, B. E., Breu, J., Muchnicki, L., & Wurtman, R. J. (1997). Rapid stimulation of amyloid precursor protein release by epidermal growth factor: Role of protein kinase C. *The Biochemical Journal, 327*(Pt 1), 245–249.

Small, S. A., & Duff, K. (2008). Linking Abeta and tau in late-onset alzheimer's disease: A dual pathway hypothesis. *Neuron, 60*(4), 534–542.

Smith, K. D., Kallhoff, V., Zheng, H., & Pautler, R. G. (2007). *In vivo* axonal transport rates decrease in a mouse model of Alzheimer's disease. *Neuroimage, 35*(4), 1401–1408. http:// dx.doi.org/10.1016/j.neuroimage.2007.01.046.

Smith, K. D., Peethumnongsin, E., Lin, H., Zheng, H., & Pautler, R. G. (2010). Increased human wildtype tau attenuates axonal transport deficits caused by loss of APP in mouse models. *Magnetic Resonance Insights, 4*, 11–18.

Soba, P., Eggert, S., Wagner, K., Zentgraf, H., Siehl, K., Kreger, S., … Beyreuther, K. (2005). Homo- and heterodimerization of APP family members promotes intercellular adhesion. *The EMBO Journal, 24*(20), 3624–3634. http://dx.doi.org/10.1038/sj.emboj.7600824.

Soldano, A., Okray, Z., Janovska, P., Tmejova, K., Reynaud, E., Claeys, A., … Hassan, B. A. (2013). The drosophila homologue of the amyloid precursor protein is a conserved modulator of Wnt PCP signaling. *PLoS Biology, 11*(5), e1001562. http://dx.doi.org/10.1371/journal.pbio.1001562.

Spoelgen, R., von Arnim, C. A., Thomas, A. V., Peltan, I. D., Koker, M., Deng, A., … Hyman, B. T. (2006). Interaction of the cytosolic domains of sorLA/LR11 with the amyloid precursor protein (APP) and beta-secretase beta-site APP-cleaving enzyme. *The Journal of Neuroscience*, 26(2), 418–428. http://dx.doi.org/10.1523/JNEUROSCI.3882-05.2006.

Stein, T. D., Anders, N. J., DeCarli, C., Chan, S. L., Mattson, M. P., & Johnson, J. A. (2004). Neutralization of transthyretin reverses the neuroprotective effects of secreted amyloid precursor protein (APP) in APPSW mice resulting in tau phosphorylation and loss of hippocampal neurons: Support for the amyloid hypothesis. *The Journal of Neuroscience*, 24(35), 7707–7717. http://dx.doi.org/10.1523/JNEUROSCI.2211-04.2004.

Stein, T. D., & Johnson, J. A. (2002). Lack of neurodegeneration in transgenic mice over-expressing mutant amyloid precursor protein is associated with increased levels of transthyretin and the activation of cell survival pathways. *The Journal of Neuroscience*, 22(17), 7380–7388.

Steinbach, J. P., Muller, U., Leist, M., Li, Z. W., Nicotera, P., & Aguzzi, A. (1998). Hypersensitivity to seizures in beta-amyloid precursor protein deficient mice. *Cell Death and Differentiation*, 5(10), 858–866. http://dx.doi.org/10.1038/sj.cdd.4400391.

Steiner, P., Sarria, J. C., Glauser, L., Magnin, S., Catsicas, S., & Hirling, H. (2002). Modulation of receptor cycling by neuron-enriched endosomal protein of 21 kD. *The Journal of Cell Biology*, 157(7), 1197–1209. http://dx.doi.org/10.1083/jcb.200202022.

St George-Hyslop, P. H., Haines, J. L., Farrer, L. A., Polinsky, R., Van Broeckhoven, C., Goate, A., et al. (1990). Genetic linkage studies suggest that Alzheimer's disease is not a single homogeneous disorder. *Nature*, 347(6289), 194–197. http://dx.doi.org/10.1038/347194a0.

Stukenberg, P. T., & Kirschner, M. W. (2001). Pin1 acts catalytically to promote a conformational change in Cdc25. *Molecular Cell*, 7(5), 1071–1083.

Su, Y., Ryder, J., Li, B., Wu, X., Fox, N., Solenberg, P., ... Ni, B. (2004). Lithium, a common drug for bipolar disorder treatment, regulates amyloid-beta precursor protein processing. *Biochemistry*, 43(22), 6899–6908.

Szodorai, A., Kuan, Y. H., Hunzelmann, S., Engel, U., Sakane, A., Sasaki, T., ... Kins, S. (2009). APP anterograde transport requires Rab3A GTPase activity for assembly of the transport vesicle. *The Journal of Neuroscience*, 29(46), 14534–14544. http://dx.doi.org/10.1523/JNEUROSCI.1546-09.2009.

Szpankowski, L., Encalada, S. E., & Goldstein, L. S. (2012). Subpixel colocalization reveals amyloid precursor protein-dependent kinesin-1 and dynein association with axonal vesicles. *Proceedings of the National Academy of Sciences of the United States of America*, 109(22), 8582–8587. http://dx.doi.org/10.1073/pnas.1120510109.

Tamboli, I. Y., Prager, K., Thal, D. R., Thelen, K. M., Dewachter, I., Pietrzik, C. U., ... Walter, J. (2008). Loss of gamma-secretase function impairs endocytosis of lipoprotein particles and membrane cholesterol homeostasis. *The Journal of Neuroscience*, 28(46), 12097–12106. http://dx.doi.org/10.1523/JNEUROSCI.2635-08.2008.

Tang, T. C., Hu, Y., Kienlen-Campard, P., El Haylani, L., Decock, M., Van Hees, J., ... Smith, S. O. (2014). Conformational changes induced by the A21G flemish mutation in the amyloid precursor protein lead to increased Abeta production. *Structure*, 22(3), 387–396. http://dx.doi.org/10.1016/j.str.2013.12.012.

Terry, R. D., Gonatas, N. K., & Weiss, M. (1964). Ultrastructural studies in Alzheimer's presenile dementia. *The American Journal of Pathology*, 44, 269–297.

Toledo, E. M., & Inestrosa, N. C. (2010). Activation of Wnt signaling by lithium and rosiglitazone reduced spatial memory impairment and neurodegeneration in brains of an APPswe/PSEN1DeltaE9 mouse model of Alzheimer's disease. *Molecular Psychiatry*, 15(3), 272–285. 228.

Tran, H. T., LaFerla, F. M., Holtzman, D. M., & Brody, D. L. (2011). Controlled cortical impact traumatic brain injury in 3xTg-AD mice causes acute intra-axonal amyloid-beta accumulation and independently accelerates the development of tau abnormalities. *The Journal of Neuroscience*, *31*(26), 9513–9525. http://dx.doi.org/10.1523/JNEUROSCI.0858-11.2011.

Trapp, B. D., & Hauer, P. E. (1994). Amyloid precursor protein is enriched in radial glia: Implications for neuronal development. *Journal of Neuroscience Research*, *37*(4), 538–550. http://dx.doi.org/10.1002/jnr.490370413.

Trommsdorff, M., Borg, J. P., Margolis, B., & Herz, J. (1998). Interaction of cytosolic adaptor proteins with neuronal apolipoprotein E receptors and the amyloid precursor protein. *The Journal of Biological Chemistry*, *273*(50), 33556–33560.

Tsuzuki, K., Fukatsu, R., Yamaguchi, H., Tateno, M., Imai, K., Fujii, N., et al. (2000). Transthyretin binds amyloid beta peptides, Abeta1-42 and Abeta1-40 to form complex in the autopsied human kidney—possible role of transthyretin for abeta sequestration. *Neuroscience Letters*, *281*(2-3), 171–174.

Tyan, S. H., Shih, A. Y., Walsh, J. J., Maruyama, H., Sarsoza, F., Ku, L., ... Dickstein, D. L. (2012). Amyloid precursor protein (APP) regulates synaptic structure and function. *Molecular and Cellular Neurosciences*, *51*(1-2), 43–52. http://dx.doi.org/10.1016/j.mcn.2012.07.009.

Ulery, P. G., Beers, J., Mikhailenko, I., Tanzi, R. E., Rebeck, G. W., Hyman, B. T., et al. (2000). Modulation of beta-amyloid precursor protein processing by the low density lipoprotein receptor-related protein (LRP). Evidence that LRP contributes to the pathogenesis of Alzheimer's disease. *The Journal of Biological Chemistry*, *275*(10), 7410–7415.

Utvik, J. K., Haglerod, C., Mylonakou, M. N., Holen, T., Kropf, M., Hirling, H., ... Davanger, S. (2009). Neuronal enriched endosomal protein of 21 kDa colocalizes with glutamate receptor subunit GLUR2/3 at the postsynaptic membrane. *Neuroscience*, *158*(1), 96–104. http://dx.doi.org/10.1016/j.neuroscience.2008.11.030.

Vagnoni, A., Perkinton, M. S., Gray, E. H., Francis, P. T., Noble, W., & Miller, C. C. (2012). Calsyntenin-1 mediates axonal transport of the amyloid precursor protein and regulates Abeta production. *Human Molecular Genetics*, *21*(13), 2845–2854. http://dx.doi.org/10.1093/hmg/dds109.

van der Kant, R., & Goldstein, L. S. (2015). Cellular functions of the amyloid precursor protein from development to dementia. *Developmental Cell*, *32*(4), 502–515. http://dx.doi.org/10.1016/j.devcel.2015.01.022.

Vardarajan, B. N., Zhang, Y., Lee, J. H., Cheng, R., Bohm, C., Ghani, M., ... Mayeux, R. (2015). Coding mutations in SORL1 and Alzheimer disease. *Annals of Neurology*, *77*(2), 215–227. http://dx.doi.org/10.1002/ana.24305.

Vieira, S. I., Rebelo, S., Esselmann, H., Wiltfang, J., Lah, J., Lane, R., ... da Cruz, E. S. O. A. (2010). Retrieval of the Alzheimer's amyloid precursor protein from the endosome to the TGN is S655 phosphorylation state-dependent and retromer-mediated. *Molecular Neurodegeneration*, *5*, 40.http://dx.doi.org/10.1186/1750-1326-5-40.

von Rotz, R. C., Kohli, B. M., Bosset, J., Meier, M., Suzuki, T., Nitsch, R. M., et al. (2004). The APP intracellular domain forms nuclear multiprotein complexes and regulates the transcription of its own precursor. *Journal of Cell Science*, *117*(Pt 19), 4435–4448. http://dx.doi.org/10.1242/jcs.01323.

Wagner, T., & Pietrzik, C. U. (2012). The role of lipoprotein receptors on the physiological function of APP. *Experimental Brain Research*, *217*(3–4), 377–387. http://dx.doi.org/10.1007/s00221-011-2876-8.

Waldron, E., Isbert, S., Kern, A., Jaeger, S., Martin, A. M., Hebert, S. S., ... Pietrzik, C. U. (2008). Increased AICD generation does not result in increased nuclear translocation or activation of target gene transcription. *Experimental Cell Research*, *314*(13), 2419–2433. http://dx.doi.org/10.1016/j.yexcr.2008.05.003.

Wang, C. Y., Zheng, W., Wang, T., Xie, J. W., Wang, S. L., Zhao, B. L., ... Wang, Z. Y. (2011). Huperzine A activates Wnt/beta-catenin signaling and enhances the nonamyloidogenic pathway in an Alzheimer transgenic mouse model. *Neuropsychopharmacology, 36*(5), 1073–1089. http://dx.doi.org/10.1038/npp.2010.245.

Wang, P., Yang, G., Mosier, D. R., Chang, P., Zaidi, T., Gong, Y. D., … Zheng, H. (2005). Defective neuromuscular synapses in mice lacking amyloid precursor protein (APP) and APP-Like protein 2. *The Journal of Neuroscience, 25*(5), 1219–1225. http://dx.doi.org/10.1523/JNEUROSCI.4660-04.2005.

Weyer, S. W., Klevanski, M., Delekate, A., Voikar, V., Aydin, D., Hick, M., … Muller, U. C. (2011). APP and APLP2 are essential at PNS and CNS synapses for transmission, spatial learning and LTP. *The EMBO Journal, 30*(11), 2266–2280. http://dx.doi.org/10.1038/emboj.2011.119.

White, A. R., Reyes, R., Mercer, J. F., Camakaris, J., Zheng, H., Bush, A. I., … Cappai, R. (1999). Copper levels are increased in the cerebral cortex and liver of APP and APLP2 knockout mice. *Brain Research, 842*(2), 439–444.

Winocur, G., Wojtowicz, J. M., Sekeres, M., Snyder, J. S., & Wang, S. (2006). Inhibition of neurogenesis interferes with hippocampus-dependent memory function. *Hippocampus, 16*(3), 296–304.

Xiao, Q., Gil, S. C., Yan, P., Wang, Y., Han, S., Gonzales, E., … Lee, J. M. (2012). Role of phosphatidylinositol clathrin assembly lymphoid-myeloid leukemia (PICALM) in intracellular amyloid precursor protein (APP) processing and amyloid plaque pathogenesis. *The Journal of Biological Chemistry, 287*(25), 21279–21289. http://dx.doi.org/10.1074/jbc.M111.338376.

Yamazaki, H., Bujo, H., Kusunoki, J., Seimiya, K., Kanaki, T., Morisaki, N., … Saito, Y. (1996). Elements of neural adhesion molecules and a yeast vacuolar protein sorting receptor are present in a novel mammalian low density lipoprotein receptor family member. *The Journal of Biological Chemistry, 271*(40), 24761–24768.

Yamazaki, T., Selkoe, D. J., & Koo, E. H. (1995). Trafficking of cell surface beta-amyloid precursor protein: Retrograde and transcytotic transport in cultured neurons. *The Journal of Cell Biology, 129*(2), 431–442.

Yang, Z., Cool, B. H., Martin, G. M., & Hu, Q. (2006). A dominant role for FE65 (APBB1) in nuclear signaling. *The Journal of Biological Chemistry, 281*(7), 4207–4214. http://dx.doi.org/10.1074/jbc.M508445200.

Yao, Z. X., & Papadopoulos, V. (2002). Function of beta-amyloid in cholesterol transport: A lead to neurotoxicity. *The FASEB Journal, 16*(12), 1677–1679. http://dx.doi.org/10.1096/fj.02-0285fje.

Young-Pearse, T. L., Bai, J., Chang, R., Zheng, J. B., LoTurco, J. J., & Selkoe, D. J. (2007). A critical function for beta-amyloid precursor protein in neuronal migration revealed by in utero RNA interference. *The Journal of Neuroscience, 27*(52), 14459–14469. http://dx.doi.org/10.1523/JNEUROSCI.4701-07.2007.

Zambrano, N., Buxbaum, J. D., Minopoli, G., Fiore, F., De Candia, P., De Renzis, S., … Russo, T. (1997). Interaction of the phosphotyrosine interaction/phosphotyrosine binding-related domains of Fe65 with wild-type and mutant Alzheimer's beta-amyloid precursor proteins. *The Journal of Biological Chemistry, 272*(10), 6399–6405.

Zhang, C. L., Zou, Y., He, W., Gage, F. H., & Evans, R. M. (2008). A role for adult TLX-positive neural stem cells in learning and behaviour. *Nature, 451*(7181), 1004–1007.

Zhang, Q. G., Wang, R., Khan, M., Mahesh, V., & Brann, D. W. (2008). Role of Dickkopf-1, an antagonist of the Wnt/beta-catenin signaling pathway, in estrogen-induced neuroprotection and attenuation of tau phosphorylation. *The Journal of Neuroscience, 28*(34), 8430–8441.

Zhang, Y. W., Wang, R., Liu, Q., Zhang, H., Liao, F. F., & Xu, H. (2007). Presenilin/gamma-secretase-dependent processing of beta-amyloid precursor protein regulates EGF receptor expression. *Proceedings of the National Academy of Sciences of the United States of America, 104*(25), 10613–10618. http://dx.doi.org/10.1073/pnas.0703903104.

Zheng, H., Jiang, M., Trumbauer, M. E., Sirinathsinghji, D. J., Hopkins, R., Smith, D. W., … Van der Ploeg, L. H. (1995). beta-Amyloid precursor protein-deficient mice show reactive gliosis and decreased locomotor activity. *Cell, 81*(4), 525–531.

Zhou, F., Gong, K., Song, B., Ma, T., van Laar, T., Gong, Y., et al. (2012). The APP intracellular domain (AICD) inhibits Wnt signalling and promotes neurite outgrowth. *Biochimica et Biophysica Acta, 1823*(8), 1233–1241. http://dx.doi.org/10.1016/j.bbamcr.2012.05.011.

Zou, K., Gong, J. S., Yanagisawa, K., & Michikawa, M. (2002). A novel function of monomeric amyloid beta-protein serving as an antioxidant molecule against metal-induced oxidative damage. *The Journal of Neuroscience, 22*(12), 4833–4841.

Molecular Pathways in Alzheimer's Disease and Cognitive Function: New Insights into Pathobiology of Tau

Xu Chen, Meredith C. Reichert and Li Gan

Department of Neurology, Gladstone Institute of Neurological Disease, University of California, San Francisco, CA, United States

OUTLINE

Introduction	136
Normal Production and Function of Tau	136
Pathogenic Tau	138
Tau-Mediated Neuronal Deficits	143
Tau Loss-of-Function Versus Gain-of-Function	*143*
Tau Aggregation	*143*
Regulation of Tau Homeostasis	*146*
Posttranslational Modifications of Tau	*148*
Tau Missorting	*151*
Targeting Tau in AD: Therapeutic Implications	154
References	156

INTRODUCTION

Neurofibrillary tangles (NFTs) and amyloid plaques are the hallmarks of Alzheimer's disease (AD). NFTs are composed of aggregated, hyper-phosphorylated tau protein and are found within neurons (Brion, 1992). NFTs first appear in the entorhinal cortex; as the disease progresses, they spread throughout the hippocampus, limbic system, and association cortices (Braak & Braak, 1991). Amyloid plaques are extracellular, consist primarily of amyloid-beta (Aβ) peptides, and are classified as either neuritic or diffuse (Glenner & Wong, 1984; Suenaga, Hirano, Llena, Yen, & Dickson, 1990). Neuritic plaques are fibrillar and have compact nuclei usually surrounded by dystrophic neurites, whereas diffuse plaques lack fibrillar structure and are not surrounded by dystrophic neurites and have been termed "preamyloid deposits" (Bugiani, Giaccone, Frangione, Ghetti, & Tagliavini, 1989; Masliah, Terry, Mallory, Alford, & Hansen, 1990). However, there is no relationship between the location or number of amyloid plaques and NFTs (Probst, Anderton, Brion, & Ulrich, 1989). The neurodegeneration specific to AD is characterized by the loss of cholinergic neurons in the basal forebrain and hippocampal pyramidal neurons (Schliebs & Arendt, 2006; West, Coleman, Flood, & Troncoso, 1994); these losses lead to severe memory deficits, mood swings usually including irritability and aggression, language break-down, and general withdrawal (Waldemar et al., 2007).

Over the past two decades, the predominant view about the cause of AD has been the amyloid cascade hypothesis, which holds that amyloid pathology is upstream from tau pathology and neuronal loss. A huge effort was undertaken to research and develop disease-modifying strategies targeting the synthesis, aggregation, and clearance of Aβ (Hardy & Selkoe, 2002; Querfurth & LaFerla, 2010). Despite the success of Aβ-based therapies in preclinical studies, the near-universal failure of these drugs in clinical trials has shown that cognitive loss cannot be ameliorated by targeting Aβ alone (reviewed in Jia, Deng, & Qing, 2014). For this reason, there has been a resurgence of interest in the role of tau in neurodegeneration and cognition. Here we will describe the complex mechanisms of tau biology and how gain or loss of these mechanisms can lead to neuronal dysfunction.

NORMAL PRODUCTION AND FUNCTION OF TAU

Tau protein is encoded by *MAPT*, a gene located on chromosome 17. Different splice variants of *MAPT* produce multiple isoforms of tau, six of which are common in the human central nervous system. These isoforms differ by the presence of exons 2 and 3 (0N, 1N, or 2N) and possess either

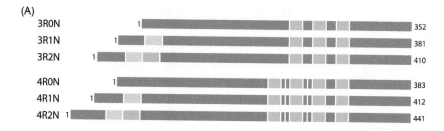

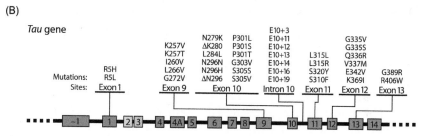

FIGURE 5.1 Tau isoforms and disease-causing mutations. (A) Six isoforms of tau protein with various exon 2 and 3 motifs (N) and microtubule-binding repeat motifs (R). (B) Genetic mutations of *Tau* gene associated with FTLD-17.

three or four microtubule-binding repeat motifs (3R or 4R) (Figure 5.1). The extra microtubule-binding motif in 4R tau allows it to bind more strongly to microtubules and enhances their polymerization dynamics (Trinczek, Biernat, Baumann, Mandelkow, & Mandelkow, 1995). In neurons, this interaction is critical for axon guidance during development and for axonal transport during adulthood (Dixit, Ross, Goldman, & Holzbaur, 2008; Ittner et al., 2008).

Interestingly, tau contains an amino-terminal projection domain, which makes it distinct from the many other microtubule-associated proteins (MAPs), although they are homologous in their microtubule-binding domains (Chapin & Bulinski, 1991; Lewis, Wang, & Cowan, 1988). This projection domain has suggested a dendritic function for tau, whereby tau binds Fyn kinase and transports it to the postsynaptic compartment (Ittner et al., 2010; Sharma, Litersky, Bhaskar, & Lee, 2007).

Tau has an unusually high number of possible phosphorylation sites: the longest isoform contains 45 serines, 35 threonines, and 4 tyrosines. Because its phosphorylation state can vary with temperature, metabolism, and stress, tau has been called a "thermometer of the cell" (Delacourte, 2008). The phosphorylation state affects the conformation of tau and its ability to bind to microtubules (Trinczek et al., 1995). An increase in posttranslational modifications, including phosphorylation and acetylation, can lead to aberrant accumulation of tau, as discussed in more detail later in this chapter (in the section 'Posttranslational Modifications of Tau').

PATHOGENIC TAU

Tau protein accumulates pathologically in several neurodegenerative disorders, collectively termed tauopathies (Lee, Goedert, & Trojanowski, 2001). In AD, tau accumulation was first thought to be caused solely by amyloid pathology, as no genetic mutations in tau cause AD. The amyloid cascade hypothesis originally predicted that Aβ plaques trigger downstream tau pathology and neurodegeneration. Aβ is generated by sequential cleavage of the amyloid precursor protein (APP). APP is a type I transmembrane protein with a large extracellular domain; the portion that generates Aβ begins in the ectodomain and continues into the transmembrane region. The mutations in familial AD enhance the amyloidogenic pathway in several ways: by increasing the amount of $A\beta_{42}$ produced, by increasing the $A\beta_{42}/A\beta_{40}$ ratio, and by promoting Aβ aggregation (Goate et al., 1991; Younkin, 1991, 1995). These mutations are found in presenilin (PS) 1 and PS2, in the catalytic components of γ-secretase and in APP itself. Because all familial AD mutations enhance Aβ pathogenesis, it was reasoned that Aβ is the central upstream trigger for AD pathology (Hardy & Higgins, 1992). Paradoxically, however, clinical and pathological evidence has shown that Aβ plaque burden does not correlate with cognition (Arriagada, Growdon, Hedley-Whyte, & Hyman, 1992; Berg, McKeel, Miller, Baty, & Morris, 1993; Delaère, Duyckaerts, He, Piette, & Hauw, 1991; Dickson et al., 1992; Terry et al., 1991), suggesting that other factors are more crucial for neurodegeneration and cognitive impairment. Many studies have shown that Aβ can facilitate tau pathology, but more recently tau has been shown to be the critical mediator of Aβ pathology (Ittner et al., 2010; Lewis et al., 2001; Oddo et al., 2003; Roberson et al., 2007). In diseases such as frontotemporal dementia (FTD) and chronic traumatic encephalopathy, the accumulation of tau is independent of the accumulation of Aβ, further suggesting that tau has a significant role in cognitive dysfunction.

Mutations in *MAPT* lead to many different tauopathies, all with slightly different patterns of tau aggregation. To aggregate, tau must be hyperphosphorylated and released from microtubules. Once enough free tau accumulates in the cell, it begins to form aggregates whose conformation becomes more structured than normal, soluble tau. The β-sheet structure allows the aggregated tau to form paired-helical filaments (PHFs) and eventually NFTs (Figure 5.2).

Although mutations in tau do not cause AD, many mutations found in other tauopathies have enhanced our understanding of tau biology (Figure 5.1 and Table 5.1). These mutations have largely been discovered through large family genetic studies of patients who have FTD with parkinsonism linked to chromosome 17 (FTDP-17). The mutations segregate

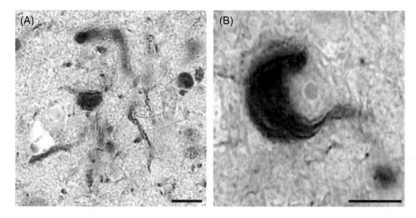

FIGURE 5.2   Neurofibrillary tangles are enriched with acetylated tau in Alzheimer's disease. Immunostaining with MAb 359 antibody showing acetylated-tau inclusions located in inferior temporal cortex of Alzheimer's disease brain. (A) Neuritic plaque; (B) Neurofibrillary tangle. Scale bars 10 μm. *Source: Reprinted from Grinberg et al. (2013).*

into two groups, which have different biological consequences. The first group consists of intronic mutations, which affect the splicing of exon 10 (Hutton et al., 1998). Exon 10 encodes one of the microtubule-binding repeats, the presence or absence of which leads to 3R or 4R tau. The normal ratio of 3R to 4R tau in the adult brain is 1:1; therefore, mutations that modify this ratio alter microtubule stability and tau aggregation. The second group of mutations is found in the coding region of the microtubule-binding domain and hence alter tau's affinity for microtubules. Other tau mutations cause progressive supranuclear palsy (PSP) and corticobasal degeneration, both of which are largely sporadic (Dickson, 1992). The changes in tau pathology are similar in these diseases, resulting in both a loss-of-function of tau by reducing microtubule assembly and a gain-of-function of tau by enhancing its aggregation into filamentous inclusions in neurons and glia. However, the nature and location of these tau aggregates are distinct in each tauopathy, resulting in a difference in clinical symptoms and disease onset (Dickson, Rademakers, & Hutton, 2007; Scaravilli, Tolosa, & Ferrer, 2005). Recently, a patient diagnosed with PSP was found to have a tau mutation in A152T, outside the microtubule region of tau (Coppola et al., 2012). Further genetic and pathological analysis in thousands of patients showed that this variant may be a risk factor for both AD and FTD. Functionally, the changes in tau pathology in these patients also include reduced binding of tau to microtubules and enhanced tau aggregation, suggesting that regions outside the microtubule-binding region are still critical for interactions between tau and microtubules.

TABLE 5.1 Known Tau Mutations Associated with Tauopathies

Mutation	Disease	Biology	Region	References
R5H	FTD	Reduced microtubule assembly; increased fibril formation	Exon 1	Hayashi et al. (2002)
R5L	PSP	Reduced microtubule assembly; no effect on ratio	Exon 1	Poorkaj et al. (2002)
G55R	FTD	4R form enhances microtubule assembly	Exon 2	Iyer et al. (2013)
A152T[a]	AD, FTD, CBD, PSP	Reduced microtubule assembly; decreased fibril formation; increased oligomer formation	Exon 7	Coppola et al. (2012)
K257T	Pick's disease	Reduced microtubule assembly	Exon 9	Pickering-Brown et al. (2000), Rizzini et al. (2000)
I260V	FTD	Increased aggregation of 4R tau	Exon 9	Grover et al. (2003)
L266V	FTD	Increased ratio of 4R tau; reduced microtubule assembly	Exon 9	Hogg et al. (2003), Kobayashi et al. (2003)
N279K	FTD	Increased ratio of 4R tau	Exon 10	Wszolek et al. (1992)
K280del	AD, FTD	Increased ratio of 3R tau; reduced microtubule assembly	Exon 10	Momeni et al. (2009), Rizzu et al. (1999)
L284L	FTD	Silent mutation; increased ratio of 4R tau	Exon 10	Rohrer et al. (2011)
N296del	PSP	Reduced microtubule assembly; increased fibril formation	Exon 10	Pastor et al. (2001)
N296H	FTD	Increased ratio of 4R tau; reduced microtubule assembly	Exon 10	Iseki et al. (2001)
N296N	FTD	Silent mutation; increased ratio of 4R tau	Exon 4a	Grover et al. (2002), Spillantini et al. (2000)
P301L	FTD	Promotes β-sheet formation; increased fibril formation	Exon 10	Clark et al. (1998), Dumanchin et al. (1998), Hutton et al. (1998), Spillantini et al. (1998)

Mutation	Disease	Effect	Exon	Reference
P301S	FTD	Reduced microtubule assembly	Exon 10	Bugiani et al. (1999), Lossos et al. (2003), Sperfeld et al. (1999)
S305I	AGD	Increased ratio of 4R tau	Exon 10	Kovacs et al. (2008)
S305S	FTD, PSP	Silent mutation; increased ratio of 4R tau	Exon 10	Skoglund et al. (2008), Stanford et al. (2000)
L315R	FTD	Reduced microtubule assembly	Exon 11	van Herpen et al. (2003)
S320F	FTD	Reduced microtubule assembly; removal of phosphorylation site	Exon 11	Rosso et al. (2002)
P332S	FTD	Reduced binding to microtubules	Exon 11	Deramecourt et al. (2012)
G335S	FTD	Reduced microtubule assembly	Exon 12	Spina et al. (2007)
G335V	FTD	Reduced microtubule assembly; increased fibril formation	Exon 12	Neumann et al. (2005)
Q336R	FTD	Increase microtubule assembly; increased fibril formation	Exon 12	Pickering-Brown et al. (2004)
V337M	FTD	Increased fibril formation; enhanced phosphorylation	Exon 12	Poorkaj et al. (1998), Sumi et al. (1992)
S352L	Tauopathy	Reduced microtubule binding; increased fibril formation	Exon 12	Nicholl et al. (2003)
P364S	FTD	Reduced microtubule assembly; increased fibril formation; increased chromosomal instability	Exon 12	Rossi et al. (2012)
G366R	FTD	Reduced microtubule assembly; increased chromosomal instability	Exon 12	Rossi et al. (2012)
K369I	FTD	Reduced microtubule assembly	Exon 12	Neumann et al. (2001)
G389R	FTD, Pick's disease	Reduced microtubule assembly	Exon 13	Pickering-Brown et al. (2000)
R406W	AD, FTD	Reduced microtubule assembly	Exon 13	Hutton et al. (1998)

(Continued)

TABLE 5.1   Continued

Mutation	Disease	Biology	Region	References
N410H	CBD	Increased ratio of 4R tau; reduced microtubule assembly; increased fibril formation; reduced microtubule polymerization	Exon 13	Kouri et al. (2014)
IVS9-10	FTD	Increased ratio of 4R tau	Intron 9	Malkani et al. (2006)
IVS10+3	FTD	Increased ratio of 4R tau	Intron 10	Spillantini et al. (1998)
IVS10+11	FTD	Increased ratio of 4R tau	Intron 10	Kowalska et al. (2002), Miyamoto et al. (2001)
IVS10+12	FTD	Increased ratio of 4R tau	Intron 10	Yasuda et al. (2000)
IVS10+13	FTD	Increased ratio of 4R tau	Intron 10	Yasuda et al. (2000)
IVS10+14	FTD	Increased ratio of 4R tau	Intron 10	Hutton et al. (1998)
IVS10+16	AD, FTD, PSP	Increased ratio of 4R tau	Intron 10	Hutton et al. (1998)
IVS10+19	FTD	Increased ratio of 3R tau	Intron 10	Stanford et al. (2003)

[a]Indicates risk factor.

TAU-MEDIATED NEURONAL DEFICITS

Tau Loss-of-Function Versus Gain-of-Function

Tau loss-of-function leads to destabilization of microtubules and deficits in axonal transport and has been implicated in AD pathogenesis (Lee & Cleveland, 1994). Microtubule-stabilizing drugs, such as paclitaxel, reverse axonal transport deficits in transgenic mice (Zhang et al., 2005). Interestingly, tau knockout mice are phenotypically normal, suggesting that tau loss-of-function is not a primary driver of cognitive decline. Notably, genetic reduction of endogenous tau prevents Aβ toxicity *in vivo*, suggesting that tau is necessary for Aβ-induced cognitive deficits and is therefore a key facilitator of neurodegeneration (Ittner et al., 2010; Roberson et al., 2007). Crossing a model of mutant human APP with a mouse tau knockout line rescued memory deficits, increased mortality, and prevented excitotoxic seizures (Roberson et al., 2007). Similarly, crossing a tau knockout mouse and a truncated tau (1–255) mouse with the APP23 model produced a comparable rescue of memory and mortality. The tau knockout and Δtau mice prevented postsynaptic targeting of the protein kinase Fyn, which was present in both the mutant tau transgenic mice and wildtype mice with endogenous murine tau (Ittner et al., 2010). These studies indicate that tau gain-of-function is responsible for cognitive decline, and more specifically, that the projection domain of tau is a primary driver of tau mislocalization and subsequent neuronal dysfunction.

Many animal models have been created to recapitulate salient aspects of AD pathology, allowing us to study the interactions between cognition and different tau mechanisms (Table 5.2). Using these models we can discern which forms of tau cause neuronal deficits and which forms do not. These models will be discussed in the context of the pathologies they are designed to emulate.

Tau Aggregation

NFTs are one of the most recognizable AD pathologies in the brain, but it was unknown whether these aggregates cause cognitive decline. A regulatable mutant tau model, rTg4510, was created that exhibits age-related synaptic loss and neurodegeneration (Ramsden et al., 2005). Experiments in this model demonstrated that tau aggregation correlates with loss of synapses and that turning off transgene expression reverses this loss, even though the NFTs persist (Santacruz et al., 2005). These findings suggest that a more soluble form of tau, and not the NFTs, cause synapse loss and were confirmed in a different mouse model of tau aggregation, Tau(RD) (Sydow et al., 2011). The Tau(RD) mice exhibited tau

TABLE 5.2 Overview of Transgenic Animal Models for Tauopathies

Species	Construct/Mutation	Promoter	Tau pathology	References
Mouse	2N4R	Thy1	Hyperphosphorylated PHFs; axonopathy without formation of neuronal NFTs; axonopathy containing neurofilament and tau-immunoreactive spheroids	Gotz et al. (1995), Probst et al. (2000), Spittaels et al. (1999)
	0N3R (fetal tau)	Prion protein	NFT in brain at 18 months of age	Ishihara et al. (1999, 2001)
	P301L	Prion protein	Tangle pathology detectable at 2.5 months of age	Lewis et al. (2001), Santacruz et al. (2005)
	P301L	Thy1.2	Tangle pathology detectable at 3 months of age	Gotz et al. (2001b)
	Inducible overexpression of P301L	CamKII	Tangle pathology detectable at 2.5 months of age	Takashima et al. (1998)
	Genomic tau	Endogenous	Tau-immunoreactive axonal swelling	Duff et al. (2000)
	P301S	Thy1.2	Tau pathology detectable at 5 months of age	Allen et al. (2002)
	P301S	Prion protein	Early indications of degeneration, synapse loss, and microglia activation	Yoshiyama et al. (2007)
	G272V	Prion protein	Oligodendroglial fibrillary lesions	Gotz et al. (2001a)
	R406W	CamKII	Hyperphosphorylated tau inclusions appear in the forebrain at 18 months of age	Tatebayashi et al. (2002)
	G272V P301L R406W	Thy1	Tau pathology detectable at 1.5 months of age	Lim et al. (2001)
	G272V P301S	Thy1.2	Hyperphosphorylated tau, tangles, and PHFs	Schindowski et al. (2006)
	hTau40/ΔK280, hTau40/ΔK280/PP	Inducible	Proaggregant full-length hTau40/ΔK280 leads to synapse loss, mislocalization of tau into the somatodendritic compartment, pretangle formation	Hochgrafe, Sydow, and Mandelkow (2013)

	Construct	Expression	Phenotype	Reference
	TauRD/ΔK280, TauRD/ΔK280/PP (tau repeat domain)	Inducible	The repeat domain TauRD/DK280 causes massive formation of neurofibrillary tangles and neuronal loss in the hippocampus	Hochgrafe et al. (2013)
Caenorhabditis elegans	WT, V337M or P301L	Pan-neuronal	Decreased motility (unc phenotype) and life span, presynaptic deficit in cholinergic transmission, increased insoluble tau accumulation	Kraemer et al. (2003)
	WT, P301L, R406W	Mechano-sensory neurons	Decrease in touch response, neurite abnormalities, microtubule loss	Miyasaka et al. (2005)
	F3ΔK280 (amyloidogenic tau)	Pan-neuronal	Age-dependent development of uncoordinated locomotion (unc phenotype) in the absence of neuronal degeneration	Brandt et al. (2009)
	WT, pseudohyperphosphorylated (PHP) tau	Pan-neuronal	Accelerated tau aggregation, severely impaired motility, perturbed axonal transport of mitochondria	Fatouros et al. (2012)
Drosophila	0N3R (WT)	Sensory neuron	Axon loss, abnormal bundling, and swelling	Williams, Tyrer, and Shepherd (2000)
	2N4R (WT)	Retinal	Neurodegeneration (rough eye), exacerbated by overexpression of GSK-3	Jackson et al. (2002)
	2N4R (P301L)	Retinal	Neurodegeneration (rough eye)	Chatterjee et al. (2009)
	0N4R (WT, V337M, R406W)	Pan-neuronal	Shortened life span, neurodegeneration (vacuoles) without tangle formation	Wittmann et al. (2001)

hyperphosphorylation and aggregation, synaptic and neuronal loss, as well as deficits in memory and long-term potentiation. When the aggregant tau transgene was switched on for 10 months and off for 4 months, the aggregates persisted, but the deficits in memory and long-term potentiation and synapse loss were rescued (Sydow et al., 2011). These studies demonstrate that an early, pretangle form of tau is the most detrimental to normal synaptic function.

Interestingly, tau fragments are toxic both *in vitro* and *in vivo*, and are found in brains from AD patients. Tau can be cleaved by calpain, cathepsins, and caspases (Gamblin et al., 2003; Rissman et al., 2004). Caspase-cleaved tau has enhanced polymerization kinetics, which leads to tangle formation (de Calignon et al., 2010; Gamblin et al., 2003). Work from the lab of Dr Bradley Hyman suggests that this tangle formation is "off pathway" to acute neuronal death. Using multiphoton imaging *in vivo* in different tau transgenic mouse models, they showed that caspase activation precedes tangle formation. Caspase activity appears to be suppressed after tangle formation, and the tangle-bearing neurons remain alive (de Calignon et al., 2010). However, fragmented tau has been identified in AD synapses, suggesting that a pretangle intermediate may be synaptotoxic.

Regulation of Tau Homeostasis

Tau is degraded via two main pathways: the ubiquitin-proteasome pathway and the autophagy-lysosome pathway. Preference for either pathway depends on several factors: the type of cell and its activation state and the conformation and posttranslational modifications of tau all affect whether and how tau is degraded.

The Ubiquitin-Proteasome Pathway

The ubiquitin-proteasome pathway acts as a major quality control center of the cell by regulating protein levels and degrading misfolded proteins. Proteins are targeted to this pathway through a ubiquitin-tagging system. E1, E2, and E3 enzymes coordinate to transfer a free ubiquitin molecule to the substrate. E1 enzymes activate ubiquitin, E2 enzymes conjugate the activated ubiquitin and transfer it to the E3 ligase, which attaches the ubiquitin to a lysine on the target protein. Humans have two E1 enzymes, fewer than 40 E2 enzymes, and hundreds of E3 ligases. The E3 ligases are substrate specific; for example, tau is ubiquitinated by the E3 ligase CHIP (C-terminus of heat shock cognate 70 interacting protein) (Petrucelli et al., 2004; Shimura, Schwartz, Gygi, & Kosik, 2004). CHIP facilitates degradation of full-length soluble tau through interactions with Hsp70; together, they ubiquitinate and transport tau to the proteasome. Inducing expression of Hsp70 in transgenic mice reduces levels of endogenous tau (Petrucelli et al., 2004). However, although ubiquitination targets tau for degradation, it may enhance tau aggregation, as mono- and polyubiquitinated tau is found in PHFs and NFTs (Morishima-Kawashima et al., 1993). This aggregation could also be caused by Aβ accumulation that blocks the proteasome, leaving tau to be sequestered instead of degraded. Aβ oligomers specifically block the proteasome in 3xTg-AD mice, leading to increased tau accumulation (Tseng, Green, Chan, Blurton-Jones, & LaFerla, 2008). In turn, PHFs inhibit proteasomal activity (Keck, Nitsch, Grune, & Ullrich, 2003). This detrimental feedback loop likely helps perpetuate AD pathology, as proteasomal activity is significantly decreased in human AD brains (Bence, Sampat, & Kopito, 2001; Keck et al., 2003; Keller, Hanni, & Markesbery, 2000).

Another chaperone that is critical for stabilizing oncoproteins in malignant cells, Hsp90, also stabilizes mutant tau in AD (Luo, Zhou, Yang, & Wang, 2007). Crystallization studies of the Hsp90/tau complex show that Hsp90 contains a long substrate-binding region that binds to aggregation-prone repeats in tau (Karagoz et al., 2014). This interaction stabilizes tau in a conformation that buries the Hsp70 binding sites, preventing tau degradation. Pharmacological inhibition of Hsp90 in transgenic mice enhances the Hsp70/Tau interaction and reduces tau levels (Dickey et al., 2007). Facilitating proteasomal degradation of tau, either by inhibiting Hsp90 or by enhancing Hsp70, is a promising therapeutic strategy for AD.

The Autophagy-Lysosome Pathway

Many pathological forms of tau cannot be degraded in the proteasome, and are therefore cleared via the autophagy-lysosome pathway. This pathway encompasses three main protein degradation mechanisms whereby long-lived or misfolded proteins are targeted to the lysosome for degradation. The three mechanisms are microautophagy, whereby the lysosomal membrane directly engulfs proteins in the cytosol; macroautophagy, whereby proteins are engulfed in autophagosomes and delivered to the lysosome; and chaperone-mediate autophagy. Chaperone-mediated autophagy is unique in that it is selective for proteins containing a specific pentapeptide targeting motif, KFERQ. Hsc70 recognizes these motifs and transfers them to the lysosome. Macroautophagy is the most common and well studied of these pathways, and is often referred to as autophagy, as we will do here.

Several lines of evidence suggests the importance of autophagy in degrading tau. First, tau colocalizes with lysosomes in AD (Ikeda, Akiyama, Arai, & Nishimura, 1998; Ikeda et al., 2000). Second, inhibition of autophagy increases tau accumulation (Hamano et al., 2008; Wang et al., 2010). These studies suggest that when the proteasome is overwhelmed or dysfunctional, autophagy is required to remove aberrant tau from the cytosol. Interestingly, a therapeutic that inhibits tau aggregation, methylene blue, induces autophagy, which lowers the levels of total and phosphorylated tau (Congdon et al., 2012). Additionally, tau contains two CMA-targeting motifs, indicating that chaperone-mediated autophagy may also help remove pathological tau (Wang et al., 2009).

Posttranslational Modifications of Tau

Hyperphosphorylation of Tau

Tau is a phosphoprotein with 85 sites that can be phosphorylated, including 80 serines/threonines and 5 tyrosines. In normal brain, the phosphor-tau concentration is about 2–3 moles per mole of the protein. Hyperphosphorylation of tau is about threefold higher in AD brains than in normal brains. More than 40 phosphorylation sites have been detected in AD brains by mass-spectrum analysis and antibodies specific for phosphor-tau.

Multiple kinases and phosphatases help regulate tau phosphorylation. Many *in vitro* and *in vivo* studies have been carried out to identify phosphorylation sites under both physiological and pathological conditions. Among the kinases that phosphorylate tau, glycogen synthase kinase-3β (GSK-3β), cyclin-dependent kinase 5 (Cdk5), casein kinase 1 (CK1), and cAMP-dependent protein kinase A (PKA) are the most well studied and are considered the most promising for targeting tau hyperphosphorylation (Ishiguro et al., 1993, 1994; Jicha et al., 1999). GSK-3β and Cdk5 can phosphorylate tau at most of the known sites found in AD (Gong, Liu, Grundke-Iqbal, & Iqbal, 2005). Among several known tau phosphatases, PP2A is the major one, accounting for more than 70% of dephosphorylated tau at most of its phosphorylation sites (Liu, Grundke-Iqbal, Iqbal, & Gong, 2005). In AD brains, PP2A activity is significantly compromised (Gong et al., 1995), and dephosphorylation of AD p-tau by PP2A inhibits PHF formation and restores microtubule stability; these beneficial changes can be reversed by sequential phosphorylation by PKA, GSK-3β, and Cdk5 (Wang, Grundke-Iqbal, & Iqbal, 2007).

Several tau kinases work in concert to regulate tau phosphorylation, as some phospho-epitopes are dependent on the sequential phosphorylation of other sites. For example, prephosphorylation of Ser400 is required for the phosphorylation of Ser396, whose phosphorylation is markedly increased in AD brains (Bramblett et al., 1993). In turn, phosphorylation of Ser400 is dependent on Cdk5-mediated phosphorylation of Ser404 (Li, Hawkes, Qureshi, Kar, & Paudel, 2006; Li & Paudel, 2006). Similarly, phosphorylation at Ser235, which is critical in regulating microtubule stability, is facilitated by primed phosphorylation of Thr231 (Cho & Johnson, 2004). In a rat model, GSK-3β-catalyzed phosphorylation of tau is facilitated by prephosphorylation by PKA, which is associated with spatial memory loss (Liu et al., 2004). The collaborative phosphorylation by several tau kinases suggests a therapeutic strategy if targeting multiple tau kinases, rather than inhibiting specific individual kinases, to reduce the overall level of tau phosphorylation. On the other hand, GSK-3β and PP2A can mutually inhibit each other (Wang et al., 2015), suggesting that tau kinases and phosphatases act synergistically to regulate hyperphosphorylation of tau.

Detrimental Roles of Tau Hyperphosphorylation in AD Brains and Animal Models

Phosphorylation of tau is necessary for its toxicity. In rodent tauopathy models, tau hyperphosphorylation correlates with the decline in memory retention. The PS19 mice model, which overexpresses the FTDP-17 tau mutation, has spatial memory deficits and hippocampal and entorhinal cortical atrophy at 9–12 months of age (Yoshiyama et al., 2007). In these mice, P301S tau becomes insoluble and hyperphosphorylated as early as 3–6 months. It is believed that the mutation makes tau a better substrate for kinases that hyperphosphorylate tau. In a mouse model in which GSK-3β is conditionally overexpressed, increased tau phosphorylation is associated with increased neurodegeneration and spatial learning deficits, which are independent of filament formation (Hernandez, Borrell, Guaza, Avila, & Lucas, 2002).

Importantly, hyperphosphorylated tau that is oligomeric and nonfilamentous precedes and promotes the formation of PHFs (Bancher et al., 1989), making it an important marker of disease progression. In patients with mild cognitive impairment, levels of phosphorylated tau in cerebrospinal fluid are strongly associated with subsequent development of AD (Hansson et al., 2006; Mattsson et al., 2009). Thus, hyperphosphorylated tau serves as a diagnostic predictor of AD.

Tau Hyperphosphorylation Impairs Synaptic Function

Hyperphosphorylated tau has long been known to mislocalize from axons to somatodendritic areas; only recently, however, was it shown that tau accumulates in dendritic spines and impairs synaptic function (Hoover et al., 2010). The necessity of hyperphosphorylation was demonstrated by using a pseudohyperphosphorylated construct of tau in which 14 disease-related serines and threonines were mutated to glutamates and a phosphorylation-deficient construct in which these same sites were mutated to alanines. Only the pseudohyperphosphorylated construct mislocalized to spines and caused a decrease in miniature excitatory postsynaptic currents in primary neurons (Hoover et al., 2010). This result is consistent with a proposed novel role of tau in transporting the protein kinase Fyn to dendritic spines, where it phosphorylates NR2B and stabilizes its interaction with PSD95 (Ittner et al., 2010). This anchoring of N-methyl-D-aspartate receptors in the postsynaptic density could enhance Aβ mediated excitotoxicity, a toxic gain-of-function for mislocalized tau.

Tau Hyperphosphorylation Leads to Cytoskeletal Deficits and Blocks Intraneuronal Trafficking

The hyperphosphorylation of tau compromises its ability to bind and stabilize microtubules in neurons. It is believed that phosphorylation at the microtubule-binding domain of tau (residues 244–386) is crucial in the regulation of microtubule stabilization. When microtubule dynamics (i.e., assembly and disassembly) is misregulated, normal neuronal functions are compromised and neuronal degeneration eventually ensues. In addition, hyperphosphorylated tau is prone to aggregate and to form PHFs, which may block axonal and dendritic transport in neurons. Furthermore, PHF tau disrupts intracellular compartments of neurons, leading to adverse effects on morphology, cell growth, and metabolism. One important phosphorylation site is the 12E8 epitope (pS262/356) in the KXGS motifs of the microtubule-binding domain of tau. In AD brains, phosphorylation of this motif occurs at a very early stage of NFT formation and is required to initiate the pathogenic cascade of hyperphosphorylation, which is ultimately associated with NFT formation in tauopathies (Augustinack, Schneider, Mandelkow, & Hyman, 2002). In Drosophila, phosphorylation of this motif by Par-1/MARK2 is required for proteotoxicity of tau (Nishimura, Yang, & Lu, 2004).

Another mechanism by which tau may target dendritic spines is through interactions with the actin cytoskeleton. In both Drosophila and mouse models, tau-induced neurodegeneration is associated with the accumulation of filamentous actin (f-actin) and the formation of actin-rich rods (Fulga et al., 2007) resembling the Hirano bodies found in AD brains (Galloway, Perry, & Gambetti, 1987). Coexpression of the tau transgene with the actin-stabilizing protein cofilin in the Drosophila model protected against neurodegeneration, while coexpression of tau with an actin transgene greatly enhanced neurodegeneration. Interestingly, coexpression of actin and a phosphorylation-deficient form of tau did not cause neurodegeneration, indicating that actin accumulates downstream from the initial tau pathology (Fulga et al., 2007). Clearly, hyperphosphorylation and mislocalization of tau are critical for neurodegeneration.

Hyperphosphorylation Inhibits the Proteasomal Degradation of Tau and Promotes Its Aggregation

The ubiquitin-proteasome and autophagy-lysosome pathways are the major molecular machineries for tau degradation. Proteasomal activity is stimulated by moderate phosphorylation of tau and inhibited by hyperphosphorylation of tau (Ren, Liao, Chen, Liu, & Wang, 2007). Hyperphosphorylated tau is found before the formation of PHF (Bancher et al., 1989). Phosphorylation at the KXGS motif inhibits tau degradation by rendering tau unrecognizable by the degradation machinery, including

CHIP and the Hsp70/90 chaperone complex, making these tau species prone to accumulate (Dickey et al., 2007). In a Drosophila model, when tau phosphorylation is stimulated by overexpression of GSK-3 or cdk5, neurofibrillary inclusions formation increases (Jackson et al., 2002; Nishimura et al., 2004). Conversely, inhibition of GSK-3 by LiCl in P301L mice reduces tau phosphorylation and tau aggregation (Perez, Hernandez, Lim, Diaz-Nido, & Avila, 2003).

Tau Acetylation

Besides phosphorylation, tau undergoes multiple posttranslational modifications, including acetylation, glycosylation, sumoylation, methylation, and nitration, allowing interplay between different modifications and making the regulation of posttranslational modification of tau extremely complex. Tau acetylation was first reported in 2010 by Min et al., who showed that acetylated tau appears early in AD brains and can be acetylated by the histone acetyltransferase p300 and deacetylated by sirtuin 1 (Min et al., 2010). Acetylation at certain lysines (detected by antibodies against acetylated tau) competes with the ubiquitination and subsequent degradation of tau (Min et al., 2010). Later, tau was found to be acetylated at K280 (Cohen et al., 2011; Irwin et al., 2012) and K274 (Grinberg et al., 2013). Both residues are associated with pathologically relevant tau. More recently, acetylation of tau was identified at the KXGS motif in the microtubule-binding domain; this acetylation appears to be regulated by HDAC6 and competes with phosphorylation of S262/356 (Cook et al., 2014). In AD brains and a mouse model of tauopathy (rTg4510), KXGS motifs are hypoacetylated and hyperphosphorylated (Cook et al., 2014). Overall, the acetylation profile of tau is regulated by multiple enzymes, and the functional consequence of tau acetylation appears to be residue/domain-specific. The importance of acetylation in regulating the phosphorylation and degradation of tau suggests novel therapeutic opportunities for targeting pathogenic tau species.

Tau Missorting

Neurons are highly polarized cells that can be divided into axons and the somatodendritic compartment, and their cellular contents are distributed asymmetrically. Unlike other MAPs, tau is sorted into axons under physiological conditions in adult brains. During embryonic development, the fetal 3R Tau isoform (0N3R) is evenly distributed throughout the cell body and neurites. As neurons mature, tau become restricted to axons, which are separated from the soma by the axon-initiation-segment, which acts as a diffusion barrier that allows tau to pass in a single direction from the cell body to the axon.

How is tau sorted and maintained in the axons? Several RNA-based models have been proposed, such as preferential transport and translation of tau mRNA at axons (Aronov, Aranda, Behar, & Ginzburg, 2001). Protein-based mechanisms include preferential degradation or stabilization of tau in axons (Hirokawa, Funakoshi, Sato-Harada, & Kanai, 1996) and the diffusion barrier that retains tau in axons (Kanai & Hirokawa, 1995). Studies of microinjected tagged tau suggest that compartment-specific degradation pathways (such as the proteasome and autophagy) or folding pathways are responsible for the axonal distribution of injected tau, which was initially distributed over all cell compartments (Hirokawa et al., 1996). On the other hand, tau has a higher affinity for axonal microtubules than dendritic microtubules, presumably because the phosphorylation status differs in axons and dendrites (Kanai & Hirokawa, 1995). In hibernating animals, the axonal distribution of tau is lost, and the loss is accompanied by changes in tau phosphorylation status, shorter dendrites, and decreased spine density (Arendt et al., 2003). Evidently, the phosphorylation status is important for tau sorting.

Tau Missorting Under Pathological Conditions and Disease Models

In AD brains, NFTs—which are composed of tau aggregates and hyper-phosphorylated tau—are found mainly in the somatodendritic compartment of neurons, where abnormally phosphorylated tau is mislocalized at an earlier stage of pathology. Thus, it is speculated that missorting of tau is important in tau pathology and contributes to cognitive deficits. Indeed, the formation of NFTs in the somatodendritic compartment and the phosphorylation of missorted tau have been used as diagnostic criteria and staging of disease progression in AD (Braak, Alafuzoff, Arzberger, Kretzschmar, & Del Tredici, 2006; Braak & Braak, 1991).

Tau missorting occurs late in life in patients with AD and FTD. Aβ and other stressors such as oxidative stress are thought to induce missorting of tau, presumably through impairments in Ca^{2+} signaling (Carroll et al., 2011; Zempel, Thies, Mandelkow, & Mandelkow, 2010). Therefore, chronic stress and trauma exacerbate tau pathology.

Several mouse models that express human tau have been generated to study the missorting of tau in relation to its phosphorylation, aggregation, and pathology. In a model with slight (<10%) overexpression of the longest isoform of human tau, tau is missorted into the somatodendritic compartment and hyperphosphorylated (Gotz et al., 1995). In contrast, in a tau replacement model (tau knockout-knockin mice) in which hTau expression is comparable to that of endogenous murine tau in the murine tau knockout background, tau was not missorted, other aspects of pathology were absent, and there were few motor deficits, even when the FTDP-17 mutation P301L was introduced (Terwel et al., 2005). However, double mutations associated with FTDP-17 (K257T/P301S) resulted in FTLD-Tau pathology, including missorting (Rosenmann et al., 2008). High levels of tau overexpression also lead to tau missorting, hyperphosphorylation, and neurodegeneration (Polydoro, Acker, Duff, Castillo, & Davies, 2009).

Consequence of Tau Missorting

Loss of Dendritic Microtubules and Spines

Mice overexpressing wildtype or mutant tau showed dendritic missorting of tau associated with microtubule breakdown (Lacroix et al., 2010; Zempel et al., 2013). In cultured neurons overexpressing tau, missorting of tau correlated with loss of dendritic microtubules and disappearance of spines (Thies & Mandelkow, 2007). These findings suggest that missorted tau in dendrites induces destabilization of microtubules and loss of spines. The mechanism of these missorting-induced defects is thought to be dendritic recruitment of the kinase Fyn (Ittner et al., 2010) and the microtubule-severing enzyme spastin (Lacroix et al., 2010; Zempel et al., 2013) and is dependent on phosphorylation of tau at the KXGS motif (Zempel et al., 2013). These changes lead to deficits of synaptic transmission that underlie cognitive impairment and motor deficits.

Inflammation

In multiple transgenic tau models, there is prominent neuroinflammation along with tau pathology. In tauopathy brains, filamentous tau inclusions are found in astrocytes and oligodendrocytes. These missorted pathological tau species could trigger inflammatory reactions that contribute to the neuronal deficits in AD. Indeed, glia-specific expression of tau leads to neuronal dysfunction and axonal degeneration (Forman et al., 2005; Higuchi et al., 2005). On the other hand, extracellular inflammatory signals such as tumor necrosis factor α and adenosine triphosphate can induce accumulation and missorting of tau in neuronal cultures (Gorlovoy, Larionov, Pham, & Neumann, 2009).

Altered Tau Phosphorylation

Axonal tau is highly phosphorylated at the PHF-1 epitope (S396/S404), whereas missorted tau in the somatodendritic compartment is highly phosphorylated at the 12E8 epitope (S262 in the KXGS motif in the repeat domain) (Zempel et al., 2010). This difference in phosphorylation status primarily reflects an asymmetrical distribution of tau phosphatases and tau kinases. PP2A, a major tau phosphatase that is predominantly responsible for the dephosphorylation of S262, is preferentially localized in axons (Zhu et al., 2010). On the other hand, kinases such as MARK and SAD, which are mainly responsible for the phosphorylation in the repeat domain, are localized in dendrites and spines (Hayashi et al., 2011; Kishi, Pan, Crump, & Sanes, 2005). The sorting of those kinases and phosphatases leads to differences in the phosphorylation status of the 12E8 epitope. Conversely, the preferential phosphorylation of the PHF-1 epitope is caused by the ubiquitous presence of GSK3β and the dendritic localization of PP2b, the kinase/phosphatase pair responsible for phosphorylation at S396/S404 (Martin et al., 2013). In normal brains, axonal tau is less phosphorylated at KXGS motifs, which gives tau a higher affinity for axonal microtubules. Under pathological conditions, phosphorylation at these sites is stronger in missorted dendritic tau, making tau more diffusible, which in turn contributes to its missorting (Schneider, Biernat, von Bergen, Mandelkow, & Mandelkow, 1999).

TARGETING TAU IN AD: THERAPEUTIC IMPLICATIONS

Targeting pathological tau has emerged as an important strategy for treating AD, especially after the recent failures of amyloid-β-targeted therapeutics in Phase III clinical trials. Preclinical studies identified tau as a major driver of neurodegeneration and showed that reducing endogenous tau levels ameliorates Aβ-induced deficits—underscoring the importance of targeting tau as a therapeutic strategy (Ittner & Gotz, 2011; Roberson et al., 2007).

The main therapeutic strategies to inhibit aberrant tau pathology are stabilizing microtubules, increasing tau degradation, and inhibiting posttranslational modifications of tau (Figure 5.3). In healthy neurons, tau binds and stabilizes microtubules; loss of this normal function reduces

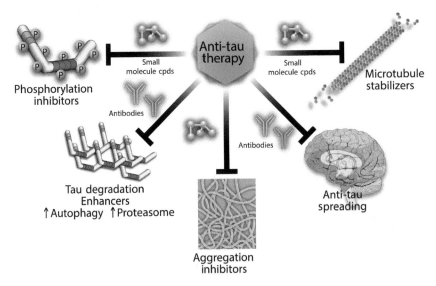

FIGURE 5.3 Therapeutic strategies targeting tau-mediated neurodegeneration in Alzheimer's disease. Small molecule compounds and antibodies are being developed to target multiple pathogenic mechanisms underlying tau-mediated neuronal deficits.

microtubule density and stabilization, resulting in deficits in axonal transport (Feinstein & Wilson, 2005; Lee, Daughenbaugh, & Trojanowski, 1994; Ma et al., 2014). These findings led to the hypothesis that loss of tau function contributes to neuronal dysfunction and ultimately cognitive decline. Attempts to stabilize microtubules by using small molecules such as paclitaxel in tau transgenic mice have improved both axonal transport and motor function (Zhang et al., 2005). However, the biggest challenge in using microtubule-stabilizing drugs is that most of them do not penetrate the blood–brain barrier well and cause significant side effects on mitosis in the periphery.

Recently, two BBB-permeable drugs, Epothilone D and Davunetide (also called NAP) have shown preclinical success. Davunetide is an octapeptide that promotes microtubule assembly; in transgenic mice, it reduced tau phosphorylation and $A\beta$ levels (Matsuoka et al., 2007; Shiryaev et al., 2009), suggesting that drug-induced stabilization of microtubules could be a strategy for treating tauopathies. Unfortunately, clinical development of davunetide was halted after a Phase 2/3 clinical trial as PSP patients demonstrated negative results for all end points (Boxer et al., 2014). Epothilone D also showed great success in preclinical studies. By stabilizing microtubules, Epothilone D increased microtubule density and improved cognition in two tau transgenic models (Barten et al., 2012; Brunden et al., 2010; Zhang et al., 2012). BMS-241027, a clinical drug developed from Epothilone D, was shown to be safe in a Phase I trial, but did not meet any clinical end points in AD patients (Malamut, 2013). Therefore, the field is moving away from microtubule stabilizers as possible therapeutics, and is instead pursuing drugs to enhance the degradation of tau or inhibit its aggregation.

As discussed above, Hsp90 is a molecular chaperone that stabilizes mutant tau, as well as different oncoproteins in cancer. In preclinical studies, inhibition of Hsp90 lowered tau levels in tau transgenic mice (Dickey et al., 2007). The development of small molecules to inhibit Hsp90 in humans has largely been pursued for cancer treatment, but these drugs are still under consideration as a treatment for tauopathy patients. One molecule that has shown preclinical success is PU-H71 (Taldone et al., 2011); a Phase 1 clinical trial is currently underway in patients with advanced malignancies.

Tau aggregation could be inhibited and tau degradation could be encouraged by inhibiting posttranslational modifications, which cause conformational changes in tau that increase aggregation and inhibit degradation. Kinases and phosphatases were first considered therapeutic targets, especially Gsk3b, Cdk5, and PP2A (Mazanetz & Fischer, 2007). However, it is extremely difficult to develop small molecules to selectively inhibit a specific kinase. Furthermore, these kinases have multiple targets, so even a molecule that selectively inhibits Gsk3b or Cdk5 might have unwanted side effects. One GSK-3 inhibitor, tideglusib, was tolerated in clinical trials, but did not meet any clinical end points in a cohort of PSP patients (Tolosa et al., 2014). However, chemists are still pursuing the crystal structure of these kinases with the hope of developing an effective small molecule therapeutic (Mazanetz, Laughton, & Fischer, 2014).

Biologics, specifically antibodies that target a specific posttranslational modification of tau, are also being developed to inhibit tau. The advantage of antibodies is specificity, as they would target the aberrant tau directly instead of an upstream enzyme with multiple substrates. However, they have much higher risks and costs than small molecules. One such monoclonal antibody, IPN007, is being developed by iPierian, now acquired by Bristol-Myers Squibb, and is expected to enter Phase I clinical trials in 2015, likely in PSP patients. Many labs are actively performing preclinical work with antibodies against different forms of phosphorylated and acetylated tau. Active immunotherapy against pathological tau is also being investigated. This strategy has a higher risk of immunogenicity than passive antibody treatment, but allows for fewer treatments with long-term beneficial effects. AADvac-1 was designed to inhibit a truncated form of tau believed to seed tangle formation. Preclinical studies in rats, rabbits, and dogs have proven successful, and a Phase I clinical trial in humans is currently underway. ACI-35, a vaccine targeting tau phosphorylated at S396/S404, was successful in preclinical studies and is being tested in human trials (Theunis et al., 2013). This burst of interest in targeting posttranslationally modified tau could lead to successful treatments for AD and other tauopathy patients.

References

Allen, B., et al. (2002). Abundant tau filaments and nonapoptotic neurodegeneration in transgenic mice expressing human P301S tau protein. *The Journal of Neuroscience*, 22(21), 9340–9351.

Arendt, T., Stieler, J., Strijkstra, A. M., Hut, R. A., Rudiger, J., Van der Zee, E. A., ... Hartig, W. (2003). Reversible paired helical filament-like phosphorylation of tau is an adaptive process associated with neuronal plasticity in hibernating animals. *The Journal of Neuroscience*, 23(18), 6972–6981.

Aronov, S., Aranda, G., Behar, L., & Ginzburg, I. (2001). Axonal tau mRNA localization coincides with tau protein in living neuronal cells and depends on axonal targeting signal. *The Journal of Neuroscience*, 21(17), 6577–6587.

Arriagada, P. V., Growdon, J. H., Hedley-Whyte, E. T., & Hyman, B. T. (1992). Neurofibrillary tangles but not senile plaques parallel duration and severity of Alzheimer's disease. *Neurology*, 42, 631–639.

Augustinack, J. C., Schneider, A., Mandelkow, E. M., & Hyman, B. T. (2002). Specific tau phosphorylation sites correlate with severity of neuronal cytopathology in Alzheimer's disease. *Acta Neuropathologica*, 103(1), 26–35.

Bancher, C., Brunner, C., Lassmann, H., Budka, H., Jellinger, K., Wiche, G., ... Wisniewski, H. M. (1989). Accumulation of abnormally phosphorylated tau precedes the formation of neurofibrillary tangles in Alzheimer's disease. *Brain Research*, 477(1–2), 90–99.

Barten, D. M., Fanara, P., Andorfer, C., Hoque, N., Wong, P. Y., Husted, K. H., ... Albright, C. F. (2012). Hyperdynamic microtubules, cognitive deficits, and pathology are improved in tau transgenic mice with low doses of the microtubule-stabilizing agent BMS-241027. *The Journal of Neuroscience, 32*(21), 7137–7145. http://dx.doi.org/10.1523/JNEUROSCI.0188-12.2012.

Bence, N. F., Sampat, R. M., & Kopito, R. R. (2001). Impairment of the ubiquitin-proteasome system by protein aggregation. *Science, 292*(5521), 1552–1555. http://dx.doi.org/10.1126/science.292.5521.1552.

Berg, L., McKeel, D. W., Jr., Miller, J. P., Baty, J., & Morris, J. C. (1993). Neuropathological indexes of Alzheimer's disease in demented and nondemented persons aged 80 years and older. *Archives of Neurology, 50*(4), 349–358.

Boxer, A. L., Lang, A. E., Grossman, M., Knopman, D. S., Miller, B. L., Schneider, L. S., ... Investigators, A. L. (2014). Davunetide in patients with progressive supranuclear palsy: A randomised, double-blind, placebo-controlled phase 2/3 trial. *Lancet Neurology, 13*(7), 676–685. http://dx.doi.org/10.1016/S1474-4422(14)70088-2.

Braak, H., Alafuzoff, I., Arzberger, T., Kretzschmar, H., & Del Tredici, K. (2006). Staging of Alzheimer disease-associated neurofibrillary pathology using paraffin sections and immunocytochemistry. *Acta Neuropathologica, 112*(4), 389–404. http://dx.doi.org/10.1007/s00401-006-0127-z.

Braak, H., & Braak, E. (1991). Neuropathological stageing of Alzheimer-related changes. *Acta Neuropathologica, 82*(4), 239–259.

Bramblett, G. T., Goedert, M., Jakes, R., Merrick, S. E., Trojanowski, J. Q., & Lee, V. M. (1993). Abnormal tau phosphorylation at Ser396 in Alzheimer's disease recapitulates development and contributes to reduced microtubule binding. *Neuron, 10*(6), 1089–1099.

Brandt, R., et al. (2009). A Caenorhabditis elegans model of tau hyperphosphorylation: Induction of developmental defects by transgenic overexpression of Alzheimer's disease-like modified tau. *Neurobiology of Aging, 30*(1), 22–33.

Brion, J. P. (1992). The pathology of the neuronal cytoskeleton in Alzheimer's disease. *Biochimica et Biophysica Acta, 1160*(1), 134–142.

Brookmeyer, R., Johnson, E., Ziegler-Graham, K., & Arrighi, H. M. (2007). Forecasting the global burden of Alzheimer's disease. *Alzheimer's & Dementia, 3*(3), 186–191. http://dx.doi.org/10.1016/j.jalz.2007.04.381.

Brunden, K. R., Zhang, B., Carroll, J., Yao, Y., Potuzak, J. S., Hogan, A. M., ... Trojanowski, J. Q. (2010). Epothilone D improves microtubule density, axonal integrity, and cognition in a transgenic mouse model of tauopathy. *The Journal of Neuroscience, 30*(41), 13861–13866. http://dx.doi.org/10.1523/JNEUROSCI.3059-10.2010.

Bugiani, O., Giaccone, G., Frangione, B., Ghetti, B., & Tagliavini, F. (1989). Alzheimer patients: Preamyloid deposits are more widely distributed than senile plaques throughout the central nervous system. *Neuroscience Letters, 103*(3), 263–268. doi:0304-3940(89)90110-9 [pii].

Bugiani, O., et al. (1999). Frontotemporal dementia and corticobasal degeneration in a family with a P301S mutation in tau. *Journal of Neuropathology and Experimental Neurology, 58*(6), 667–677.

Carroll, J. C., Iba, M., Bangasser, D. A., Valentino, R. J., James, M. J., Brunden, K. R., ... Trojanowski, J. Q. (2011). Chronic stress exacerbates tau pathology, neurodegeneration, and cognitive performance through a corticotropin-releasing factor receptor-dependent mechanism in a transgenic mouse model of tauopathy. *The Journal of Neuroscience, 31*(40), 14436–14449. http://dx.doi.org/10.1523/JNEUROSCI.3836-11.2011.

Chapin, S. J., & Bulinski, J. C. (1991). Non-neuronal 210 × 10(3) Mr microtubule-associated protein (MAP4) contains a domain homologous to the microtubule-binding domains of neuronal MAP2 and tau. *Journal of Cell Science, 98*(Pt 1), 27–36.

Chatterjee, S., et al. (2009). Dissociation of tau toxicity and phosphorylation: Role of GSK-3beta, MARK and Cdk5 in a Drosophila model. *Human Molecular Genetics, 18*(1), 164–177.

Cho, J. H., & Johnson, G. V. (2004). Primed phosphorylation of tau at Thr231 by glycogen synthase kinase 3beta (GSK3beta) plays a critical role in regulating tau's ability to bind and stabilize microtubules. *Journal of Neurochemistry, 88*(2), 349–358.

Clark, L. N., et al. (1998). Pathogenic implications of mutations in the tau gene in pallido-ponto-nigral degeneration and related neurodegenerative disorders linked to chromosome 17. *Proceedings of the National Academy of Sciences of the United States of America, 95*(22), 13103–13107.

Cohen, T. J., Guo, J. L., Hurtado, D. E., Kwong, L. K., Mills, I. P., Trojanowski, J. Q., et al. (2011). The acetylation of tau inhibits its function and promotes pathological tau aggregation. *Nature Communication, 2,* 252.http://dx.doi.org/10.1038/ncomms1255.

Congdon, E. E., Wu, J. W., Myeku, N., Figueroa, Y. H., Herman, M., Marinec, P. S., ... Duff, K. E. (2012). Methylthioninium chloride (methylene blue) induces autophagy and attenuates tauopathy *in vitro* and *in vivo. Autophagy, 8*(4), 609–622. http://dx.doi.org/10.4161/auto.19048.

Cook, C., Carlomagno, Y., Gendron, T. F., Dunmore, J., Scheffel, K., Stetler, C., ... Petrucelli, L. (2014). Acetylation of the KXGS motifs in tau is a critical determinant in modulation of tau aggregation and clearance. *Human Molecular Genetics, 23*(1), 104–116. http://dx.doi.org/10.1093/hmg/ddt402.

Coppola, G., Chinnathambi, S., Lee, J. J., Dombroski, B. A., Baker, M. C., Soto-Ortolaza, A. I., ... Geschwind, D. H. (2012). Evidence for a role of the rare p.A152T variant in MAPT in increasing the risk for FTD-spectrum and Alzheimer's diseases. *Human Molecular Genetics, 21*(15), 3500–3512. http://dx.doi.org/10.1093/hmg/dds161.

Coppola, G., et al. (2012). Evidence for a role of the rare p.A152T variant in MAPT in increasing the risk for FTD-spectrum and Alzheimer's diseases. *Human Molecular Genetics, 21*(15), 3500–3512.

de Calignon, A., Fox, L. M., Pitstick, R., Carlson, G. A., Bacskai, B. J., Spires-Jones, T. L., et al. (2010). Caspase activation precedes and leads to tangles. *Nature, 464*(7292), 1201–1204. http://dx.doi.org/10.1038/nature08890.

Delacourte, A. (2008). Tau pathology and neurodegeneration: An obvious but misunderstood link. *Journal of Alzheimer's Disease, 14*(4), 437–440.

Delaère, P., Duyckaerts, C., He, Y., Piette, F., & Hauw, J. J. (1991). Subtypes and differential laminar distributions of beta A4 deposits in Alzheimer's disease: Relationship with the intellectual status of 26 cases. *Acta Neuropathologica, 81,* 328–335.

Deramecourt, V., et al. (2012). Clinical, neuropathological, and biochemical characterization of the novel tau mutation P332S. *Journal of Alzheimer's Disease, 31*(4), 741–749.

Dickey, C. A., Kamal, A., Lundgren, K., Klosak, N., Bailey, R. M., Dunmore, J., ... Petrucelli, L. (2007). The high-affinity HSP90-CHIP complex recognizes and selectively degrades phosphorylated tau client proteins. *The Journal of Clinical Investigation, 117*(3), 648–658. http://dx.doi.org/10.1172/JCI29715.

Dickson, D. W. (1999). Neuropathologic differentiation of progressive supranuclear palsy and corticobasal degeneration. *Journal of Neurology, 246*(Suppl 2), II6–15.

Dickson, D. W., Crystal, H. A., Mattiace, L. A., Masur, D. M., Blau, A. D., Davies, P. Aronson, M. K. (1992). Identification of normal and pathological aging in prospectively studied nondemented elderly humans. *Neurobiology of Aging, 13,* 179–189.

Dickson, D. W., Rademakers, R., & Hutton, M. L. (2007). Progressive supranuclear palsy: Pathology and genetics. *Brain Pathology, 17*(1), 74–82. http://dx.doi.org/10.1111/j.1750-3639.2007.00054.x.

Dixit, R., Ross, J. L., Goldman, Y. E., & Holzbaur, E. L. (2008). Differential regulation of dynein and kinesin motor proteins by tau. *Science, 319*(5866), 1086–1089. http://dx.doi.org/10.1126/science.1152993.

Duff, K., et al. (2000). Characterization of pathology in transgenic mice over-expressing human genomic and cDNA tau transgenes. *Neurobiology of Disease, 7*(2), 87–98.

Dumanchin, C., et al. (1998). Segregation of a missense mutation in the microtubule-associated protein tau gene with familial frontotemporal dementia and parkinsonism. *Human Molecular Genetics, 7*(11), 1825–1829.

Fatouros, C., et al. (2012). Inhibition of tau aggregation in a novel *Caenorhabditis elegans* model of tauopathy mitigates proteotoxicity. *Human Molecular Genetics, 21*(16), 3587–3603.

Feinstein, S. C., & Wilson, L. (2005). Inability of tau to properly regulate neuronal microtubule dynamics: A loss-of-function mechanism by which tau might mediate neuronal cell death. *Biochimica et Biophysica Acta, 1739*(2–3), 268–279. http://dx.doi.org/10.1016/j.bbadis.2004.07.002.

Forman, M. S., Lal, D., Zhang, B., Dabir, D. V., Swanson, E., Lee, V. M., et al. (2005). Transgenic mouse model of tau pathology in astrocytes leading to nervous system degeneration. *The Journal of Neuroscience, 25*(14), 3539–3550. http://dx.doi.org/10.1523/JNEUROSCI.0081-05.2005.

Fulga, T. A., Elson-Schwab, I., Khurana, V., Steinhilb, M. L., Spires, T. L., Hyman, B. T., et al. (2007). Abnormal bundling and accumulation of F-actin mediates tau-induced neuronal degeneration *in vivo. Nature Cell Biology, 9*(2), 139–148. http://dx.doi.org/10.1038/ncb1528.

Galloway, P. G., Perry, G., & Gambetti, P. (1987). Hirano body filaments contain actin and actin-associated proteins. *Journal of Neuropathology and Experimental Neurology, 46*(2), 185–199.

Gamblin, T. C., Chen, F., Zambrano, A., Abraha, A., Lagalwar, S., Guillozet, A. L., ... Cryns, V. L. (2003). Caspase cleavage of tau: Linking amyloid and neurofibrillary tangles in Alzheimer's disease. *Proceedings of the National Academy of Sciences of the United States of America, 100*(17), 10032–10037. http://dx.doi.org/10.1073/pnas.16304281001630428100.

Glenner, G. G., & Wong, C. W. (1984). Alzheimer's disease: Initial report of the purification and characterization of a novel cerebrovascular amyloid protein. *Biochemical and Biophysical Research Communications, 120*(3), 885–890. doi:S0006-291X(84)80190-4 [pii].

Goate, A., Chartier-Harlin, M. C., Mullan, M., Brown, J., Crawford, F., Fidani, L., et al. (1991). Segregation of a missense mutation in the amyloid precursor protein gene with familial Alzheimer's disease. *Nature, 349*(6311), 704–706. http://dx.doi.org/10.1038/349704a0.

Gong, C. X., Liu, F., Grundke-Iqbal, I., & Iqbal, K. (2005). Post-translational modifications of tau protein in Alzheimer's disease. *Journal of Neural Transmission, 112*(6), 813–838. http://dx.doi.org/10.1007/s00702-004-0221-0.

Gong, C. X., Shaikh, S., Wang, J. Z., Zaidi, T., Grundke-Iqbal, I., & Iqbal, K. (1995). Phosphatase activity toward abnormally phosphorylated tau: Decrease in Alzheimer disease brain. *Journal of Neurochemistry, 65*(2), 732–738.

Gorlovoy, P., Larionov, S., Pham, T. T., & Neumann, H. (2009). Accumulation of tau induced in neurites by microglial proinflammatory mediators. *The FASEB Journal, 23*(8), 2502–2513. http://dx.doi.org/10.1096/fj.08-123877.

Gotz, J., Probst, A., Spillantini, M. G., Schafer, T., Jakes, R., Burki, K., et al. (1995). Somatodendritic localization and hyperphosphorylation of tau protein in transgenic mice expressing the longest human brain tau isoform. *The EMBO Journal, 14*(7), 1304–1313.

Gotz, J., et al. (1995). Somatodendritic localization and hyperphosphorylation of tau protein in transgenic mice expressing the longest human brain tau isoform. *The EMBO Journal, 14*(7), 1304–1313.

Gotz, J., et al. (2001a). Oligodendroglial tau filament formation in transgenic mice expressing G272V tau. *The European Journal of Neuroscience, 13*(11), 2131–2140.

Gotz, J., et al. (2001b). Tau filament formation in transgenic mice expressing P301L tau. *The Journal of Biological Chemistry, 276*(1), 529–534.

Grinberg, L. T., Wang, X., Wang, C., Sohn, P. D., Theofilas, P., Sidhu, M., ... Seeley, W. W. (2013). Argyrophilic grain disease differs from other tauopathies by lacking tau acetylation. *Acta Neuropathologica, 125*(4), 581–593. http://dx.doi.org/10.1007/s00401-013-1080-2.

Grover, A., et al. (2002). Effects on splicing and protein function of three mutations in codon N296 of tau *in vitro*. *Neuroscience Letters, 323*(1), 33–36.

Grover, A., et al. (2003). A novel tau mutation in exon 9 (1260V) causes a four-repeat tauopathy. *Experimental Neurology, 184*(1), 131–140.

Hamano, T., Gendron, T. F., Causevic, E., Yen, S. H., Lin, W. L., Isidoro, C., … Ko, L. W. (2008). Autophagic-lysosomal perturbation enhances tau aggregation in transfectants with induced wild-type tau expression. *The European Journal of Neuroscience, 27*(5), 1119–1130. http://dx.doi.org/10.1111/j.1460-9568.2008.06084.x.

Hansson, O., Zetterberg, H., Buchhave, P., Londos, E., Blennow, K., & Minthon, L. (2006). Association between CSF biomarkers and incipient Alzheimer's disease in patients with mild cognitive impairment: A follow-up study. *Lancet Neurology, 5*(3), 228–234. http://dx.doi.org/10.1016/S1474-4422(06)70355-6.

Hardy, J., & Selkoe, D. J. (2002). The amyloid hypothesis of Alzheimer's disease: Progress and problems on the road to therapeutics. *Science, 297*(5580), 353–356. http://dx.doi.org/10.1126/science.1072994297/5580/353.

Hardy, J. A., & Higgins, G. A. (1992). Alzheimer's disease: The amyloid cascade hypothesis. *Science, 256*(5054), 184–185.

Hayashi, K., Suzuki, A., Hirai, S., Kurihara, Y., Hoogenraad, C. C., & Ohno, S. (2011). Maintenance of dendritic spine morphology by partitioning-defective 1b through regulation of microtubule growth. *The Journal of Neuroscience, 31*(34), 12094–12103. http://dx.doi.org/10.1523/JNEUROSCI.0751-11.2011.

Hayashi, S., et al. (2002). Late-onset frontotemporal dementia with a novel exon 1 (Arg5His) tau gene mutation. *Annals of Neurology, 51*(4), 525–530.

Hebert, L. E., Scherr, P. A., Bienias, J. L., Bennett, D. A., & Evans, D. A. (2003). Alzheimer disease in the US population: Prevalence estimates using the 2000 census. *Archives of Neurology, 60*(8), 1119–1122. http://dx.doi.org/10.1001/archneur.60.8.1119. 60/8/1119.

Hernandez, F., Borrell, J., Guaza, C., Avila, J., & Lucas, J. J. (2002). Spatial learning deficit in transgenic mice that conditionally over-express GSK-3beta in the brain but do not form tau filaments. *Journal of Neurochemistry, 83*(6), 1529–1533.

Higuchi, M., Zhang, B., Forman, M. S., Yoshiyama, Y., Trojanowski, J. Q., & Lee, V. M. (2005). Axonal degeneration induced by targeted expression of mutant human tau in oligodendrocytes of transgenic mice that model glial tauopathies. *The Journal of Neuroscience, 25*(41), 9434–9443. http://dx.doi.org/10.1523/JNEUROSCI.2691-05.2005.

Hirokawa, N., Funakoshi, T., Sato-Harada, R., & Kanai, Y. (1996). Selective stabilization of tau in axons and microtubule-associated protein 2C in cell bodies and dendrites contributes to polarized localization of cytoskeletal proteins in mature neurons. *The Journal of Cell Biology, 132*(4), 667–679.

Hochgrafe, K., Sydow, A., & Mandelkow, E. M. (2013). Regulatable transgenic mouse models of Alzheimer disease: Onset, reversibility and spreading of Tau pathology. *The FEBS Journal, 280*(18), 4371–4381.

Hogg, M., et al. (2003). The L266V tau mutation is associated with frontotemporal dementia and Pick-like 3R and 4R tauopathy. *Acta Neuropathologica, 106*(4), 323–336.

Hoover, B. R., Reed, M. N., Su, J., Penrod, R. D., Kotilinek, L. A., Grant, M. K., … Liao, D. (2010). Tau mislocalization to dendritic spines mediates synaptic dysfunction independently of neurodegeneration. *Neuron, 68*(6), 1067–1081. http://dx.doi.org/10.1016/j.neuron.2010.11.030.

Hutton, M., et al. (1998). Association of missense and 5′-splice-site mutations in tau with the inherited dementia FTDP-17. *Nature, 393*(6686), 702–705. Reviewed in *Cold Spring Harbor Perspectives in Medicine,* 2011.

Ikeda, K., Akiyama, H., Arai, T., Kondo, H., Haga, C., Tsuchiya, K., … Hori, A. (2000). Neurons containing Alz-50-immunoreactive granules around the cerebral infarction: Evidence for the lysosomal degradation of altered tau in human brain? *Neuroscience Letters, 284*(3), 187–189.

Ikeda, K., Akiyama, H., Arai, T., & Nishimura, T. (1998). Glial tau pathology in neurode-generative diseases: Their nature and comparison with neuronal tangles. *Neurobiology of Aging, 19*(1 Suppl.), S85–S91.

Irwin, D. J., Cohen, T. J., Grossman, M., Arnold, S. E., Xie, S. X., Lee, V. M., et al. (2012). Acetylated tau, a novel pathological signature in Alzheimer's disease and other tauopathies. *Brain, 135*(Pt 3), 807–818. http://dx.doi.org/10.1093/brain/aws013.

Iseki, E., et al. (2001). Familial frontotemporal dementia and parkinsonism with a novel N296H mutation in exon 10 of the tau gene and a widespread tau accumulation in the glial cells. *Acta Neuropathologica, 102*(3), 285–292.

Ishiguro, K., Kobayashi, S., Omori, A., Takamatsu, M., Yonekura, S., Anzai, K., … Uchida, T. (1994). Identification of the 23 kDa subunit of tau protein kinase II as a putative activator of cdk5 in bovine brain. *FEBS Letters, 342*(2), 203–208.

Ishiguro, K., Shiratsuchi, A., Sato, S., Omori, A., Arioka, M., Kobayashi, S., … Imahori, K. (1993). Glycogen synthase kinase 3 beta is identical to tau protein kinase I generating several epitopes of paired helical filaments. *FEBS Letters, 325*(3), 167–172. http://dx.doi.org/0014-5793(93)81066-9.

Ishihara, T., et al. (1999). Age-dependent emergence and progression of a tauopathy in transgenic mice overexpressing the shortest human tau isoform. *Neuron, 24*(3), 751–762.

Ishihara, T., et al. (2001). Age-dependent induction of congophilic neurofibrillary tau inclusions in tau transgenic mice. *The American Journal of Pathology, 158*(2), 555–562.

Ittner, L. M., Fath, T., Ke, Y. D., Bi, M., van Eersel, J., Li, K. M., … Gotz, J. (2008). Parkinsonism and impaired axonal transport in a mouse model of frontotemporal dementia. *Proceedings of the National Academy of Sciences of the United States of America, 105*(41), 15997–16002. http://dx.doi.org/10.1073/pnas.0808084105.

Ittner, L. M., & Gotz, J. (2011). Amyloid-beta and tau—a toxic pas de deux in Alzheimer's disease. *Nature Reviews Neuroscience, 12*(2), 65–72. http://dx.doi.org/10.1038/nrn2967.

Ittner, L. M., Ke, Y. D., Delerue, F., Bi, M., Gladbach, A., van Eersel, J., … Gotz, J. (2010). Dendritic function of tau mediates amyloid-beta toxicity in Alzheimer's disease mouse models. *Cell, 142*(3), 387–397. http://dx.doi.org/10.1016/j.cell.2010.06.036.

Iyer, A., et al. (2013). A novel MAPT mutation, G55R, in a frontotemporal dementia patient leads to altered Tau function. *PLoS One, 8*(9), e76409.

Jackson, G. R., Wiedau-Pazos, M., Sang, T. K., Wagle, N., Brown, C. A., Massachi, S., et al. (2002). Human wild-type tau interacts with wingless pathway components and produces neurofibrillary pathology in Drosophila. *Neuron, 34*(4), 509–519.

Jackson, G. R., et al. (2002). Human wild-type tau interacts with wingless pathway components and produces neurofibrillary pathology in Drosophila. *Neuron, 34*(4), 509–519.

Jia, Q., Deng, Y., & Qing, H. (2014). Potential therapeutic strategies for Alzheimer's disease targeting or beyond beta-amyloid: Insights from clinical trials. *Biomedical Research International, 2014*, 837157.http://dx.doi.org/10.1155/2014/837157.

Jicha, G. A., Weaver, C., Lane, E., Vianna, C., Kress, Y., Rockwood, J., et al. (1999). cAMP-dependent protein kinase phosphorylations on tau in Alzheimer's disease. *The Journal of Neuroscience, 19*(17), 7486–7494.

Kanai, Y., & Hirokawa, N. (1995). Sorting mechanisms of tau and MAP2 in neurons: Suppressed axonal transit of MAP2 and locally regulated microtubule binding. *Neuron, 14*(2), 421–432.

Karagoz, G. E., Duarte, A. M., Akoury, E., Ippel, H., Biernat, J., Moran Luengo, T., … Rudiger, S. G. (2014). Hsp90-Tau complex reveals molecular basis for specificity in chaperone action. *Cell, 156*(5), 963–974. http://dx.doi.org/10.1016/j.cell.2014.01.037.

Keck, S., Nitsch, R., Grune, T., & Ullrich, O. (2003). Proteasome inhibition by paired helical filament-tau in brains of patients with Alzheimer's disease. *Journal of Neurochemistry, 85*(1), 115–122.

Keller, J. N., Hanni, K. B., & Markesbery, W. R. (2000). Impaired proteasome function in Alzheimer's disease. *Journal of Neurochemistry, 75*(1), 436–439.

Kishi, M., Pan, Y. A., Crump, J. G., & Sanes, J. R. (2005). Mammalian SAD kinases are required for neuronal polarization. *Science*, *307*(5711), 929–932. http://dx.doi.org/10.1126/science.1107403.

Kobayashi, T., et al. (2003). A novel L266V mutation of the tau gene causes frontotemporal dementia with a unique tau pathology. *Annals of Neurology*, *53*(1), 133–137.

Kouri, N., et al. (2014). Novel mutation in MAPT exon 13 (p.N410H) causes corticobasal degeneration. *Acta Neuropathologica*, *127*(2), 271–282.

Kovacs, G. G., et al. (2008). MAPT S305I mutation: Implications for argyrophilic grain disease. *Acta Neuropathologica*, *116*(1), 103–118.

Kowalska, A., et al. (2002). A novel mutation at position +11 in the intron following exon 10 of the tau gene in FTDP-17. *Journal of Applied Genetics*, *43*(4), 535–543.

Kraemer, B. C., et al. (2003). Neurodegeneration and defective neurotransmission in a *Caenorhabditis elegans* model of tauopathy. *Proceedings of the National Academy of Sciences of the United States of America*, *100*(17), 9980–9985.

Lacroix, B., van Dijk, J., Gold, N. D., Guizetti, J., Aldrian-Herrada, G., Rogowski, K., … Janke, C. (2010). Tubulin polyglutamylation stimulates spastin-mediated microtubule severing. *The Journal of Cell Biology*, *189*(6), 945–954. http://dx.doi.org/10.1083/jcb.201001024.

Lee, M. K., & Cleveland, D. W. (1994). Neurofilament function and dysfunction: Involvement in axonal growth and neuronal disease. *Current Opinion in Cell Biology*, *6*(1), 34–40.

Lee, V. M., Daughenbaugh, R., & Trojanowski, J. Q. (1994). Microtubule stabilizing drugs for the treatment of Alzheimer's disease. *Neurobiology of Aging*, *15*(Suppl. 2), S87–S89.

Lee, V. M., Goedert, M., & Trojanowski, J. Q. (2001). Neurodegenerative tauopathies. *Annual Review of Neuroscience*, *24*, 1121–1159. http://dx.doi.org/10.1146/annurev.neuro.24.1.112124/1/1121.

Lewis, J., Dickson, D. W., Lin, W. L., Chisholm, L., Corral, A., Jones, G., … McGowan, E. (2001). Enhanced neurofibrillary degeneration in transgenic mice expressing mutant tau and APP. *Science*, *293*, 1487–1491.

Lewis, S. A., Wang, D. H., & Cowan, N. J. (1988). Microtubule-associated protein MAP2 shares a microtubule binding motif with tau protein. *Science*, *242*(4880), 936–939.

Li, T., Hawkes, C., Qureshi, H. Y., Kar, S., & Paudel, H. K. (2006). Cyclin-dependent protein kinase 5 primes microtubule-associated protein tau site-specifically for glycogen synthase kinase 3beta. *Biochemistry*, *45*(10), 3134–3145. http://dx.doi.org/10.1021/bi051635j.

Li, T., & Paudel, H. K. (2006). Glycogen synthase kinase 3beta phosphorylates Alzheimer's disease-specific Ser396 of microtubule-associated protein tau by a sequential mechanism. *Biochemistry*, *45*(10), 3125–3133. http://dx.doi.org/10.1021/bi051634r.

Lim, F., et al. (2001). FTDP-17 mutations in tau transgenic mice provoke lysosomal abnormalities and Tau filaments in forebrain. *Molecular and Cellular Neurosciences*, *18*(6), 702–714.

Liu, F., Grundke-Iqbal, I., Iqbal, K., & Gong, C. X. (2005). Contributions of protein phosphatases PP1, PP2A, PP2B and PP5 to the regulation of tau phosphorylation. *The European Journal of Neuroscience*, *22*(8), 1942–1950. http://dx.doi.org/10.1111/j.1460-9568.2005.04391.x.

Liu, S. J., Zhang, J. Y., Li, H. L., Fang, Z. Y., Wang, Q., Deng, H. M., … Wang, J. Z. (2004). Tau becomes a more favorable substrate for GSK-3 when it is prephosphorylated by PKA in rat brain. *The Journal of Biological Chemistry*, *279*(48), 50078–50088. http://dx.doi.org/10.1074/jbc.M406109200.

Lossos, A., et al. (2003). Frontotemporal dementia and parkinsonism with the P301S tau gene mutation in a Jewish family. *Journal of Neurology*, *250*(6), 733–740.

Luo, Y., Zhou, X., Yang, X., & Wang, J. (2007). Homocysteine induces tau hyperphosphorylation in rats. *Neuroreport*, *18*(18), 2005–2008. http://dx.doi.org/10.1097/WNR.0b013e3282f29100.

Ma, Q. L., Zuo, X., Yang, F., Ubeda, O. J., Gant, D. J., Alaverdyan, M., … Cole, G. M. (2014). Loss of MAP function leads to hippocampal synapse loss and deficits in the morris water maze with aging. *The Journal of Neuroscience*, *34*(21), 7124–7136. http://dx.doi.org/10.1523/JNEUROSCI.3439-13.2014.

Malamut, R., Wang, J. -S., Savant, I., Xiao, H., Sverdlov O., Tendolkar, A. V., Keswani, S. C. (2013). A randomized, double-blind, placebo-controlled, multiple ascending dose study to evaluate the safety, tolerability and pharmacokinetics of a microtubule stabilizer (BMS-241027) in healthy females. In: *Paper Presented at the Alzheimer's Association International Conference 2013*.

Malkani, R., et al. (2006). A MAPT mutation in a regulatory element upstream of exon 10 causes frontotemporal dementia. *Neurobiology of Disease*, 22(2), 401–403.

Martin, L., Latypova, X., Wilson, C. M., Magnaudeix, A., Perrin, M. L., & Terro, F. (2013). Tau protein phosphatases in Alzheimer's disease: The leading role of PP2A. *Ageing Research Reviews*, 12(1), 39–49. http://dx.doi.org/10.1016/j.arr.2012.06.008.

Masliah, E., Terry, R. D., Mallory, M., Alford, M., & Hansen, L. A. (1990). Diffuse plaques do not accentuate synapse loss in Alzheimer's disease. *The American Journal of Pathology*, 137(6), 1293–1297.

Matsuoka, Y., Gray, A. J., Hirata-Fukae, C., Minami, S. S., Waterhouse, E. G., Mattson, M. P., … Aisen, P. S. (2007). Intranasal NAP administration reduces accumulation of amyloid peptide and tau hyperphosphorylation in a transgenic mouse model of Alzheimer's disease at early pathological stage. *Journal of Molecular Neuroscience*, 31(2), 165–170.

Mattsson, N., Zetterberg, H., Hansson, O., Andreasen, N., Parnetti, L., Jonsson, M., … Blennow, K. (2009). CSF biomarkers and incipient Alzheimer disease in patients with mild cognitive impairment. *JAMA*, 302(4), 385–393. http://dx.doi.org/10.1001/jama.2009.1064.

Mazanetz, M. P., & Fischer, P. M. (2007). Untangling tau hyperphosphorylation in drug design for neurodegenerative diseases. *Nature Reviews Drug Discovery*, 6(6), 464–479. http://dx.doi.org/10.1038/nrd2111.

Mazanetz, M. P., Laughton, C. A., & Fischer, P. M. (2014). Investigation of the flexibility of protein kinases implicated in the pathology of Alzheimer's disease. *Molecules*, 19(7), 9134–9159. http://dx.doi.org/10.3390/molecules19079134.

Min, S. W., Cho, S. H., Zhou, Y., Schroeder, S., Haroutunian, V., Seeley, W. W., … Gan, L. (2010). Acetylation of tau inhibits its degradation and contributes to tauopathy. *Neuron*, 67(6), 953–966. http://dx.doi.org/10.1016/j.neuron.2010.08.044.

Miyamoto, K., et al. (2001). Familial frontotemporal dementia and parkinsonism with a novel mutation at an intron 10+ 11-splice site in the tau gene. *Annals of Neurology*, 50(1), 117–120.

Miyasaka, T., et al. (2005). Progressive neurodegeneration in *C. elegans* model of tauopathy. *Neurobiology of Disease*, 20(2), 372–383.

Momeni, P., et al. (2009). Clinical and pathological features of an Alzheimer's disease patient with the MAPT Delta K280 mutation. *Neurobiology of Aging*, 30(3), 388–393.

Morishima-Kawashima, M., Hasegawa, M., Takio, K., Suzuki, M., Titani, K., & Ihara, Y. (1993). Ubiquitin is conjugated with amino-terminally processed tau in paired helical filaments. *Neuron*, 10(6), 1151–1160.

Neumann, M., et al. (2001). Pick's disease associated with the novel Tau gene mutation K369I. *Annals of Neurology*, 50(4), 503–513.

Neumann, M., et al. (2005). Novel G335V mutation in the tau gene associated with early onset familial frontotemporal dementia. *Neurogenetics*, 6(2), 91–95.

Nicholl, D. J., et al. (2003). An English kindred with a novel recessive tauopathy and respiratory failure. *Annals of Neurology*, 54(5), 682–686.

Nishimura, I., Yang, Y., & Lu, B. (2004). PAR-1 kinase plays an initiator role in a temporally ordered phosphorylation process that confers tau toxicity in Drosophila. *Cell*, 116(5), 671–682.

Oddo, S., Caccamo, A., Kitazawa, M., Tseng, B. P., & LaFerla, F. M. (2003). Amyloid deposition precedes tangle formation in a triple transgenic model of Alzheimer's disease. *Neurobiology of Aging*, 24, 1063–1070.

Pastor, P., et al. (2001). Familial atypical progressive supranuclear palsy associated with homozigosity for the delN296 mutation in the tau gene. *Annals of Neurology, 49*(2), 263–267.

Perez, M., Hernandez, F., Lim, F., Diaz-Nido, J., & Avila, J. (2003). Chronic lithium treatment decreases mutant tau protein aggregation in a transgenic mouse model. *Journal of Alzheimer's Disease, 5*(4), 301–308.

Petrucelli, L., Dickson, D., Kehoe, K., Taylor, J., Snyder, H., Grover, A., … Hutton, M. (2004). CHIP and Hsp70 regulate tau ubiquitination, degradation and aggregation. *Human Molecular Genetics, 13*(7), 703–714. http://dx.doi.org/10.1093/hmg/ddh083.

Pickering-Brown, S., et al. (2000). Pick's disease is associated with mutations in the tau gene. *Annals of Neurology, 48*(6), 859–867.

Pickering-Brown, S. M., et al. (2004). Frontotemporal dementia with Pick-type histology associated with Q336R mutation in the tau gene. *Brain, 127*(Pt 6), 1415–1426.

Polydoro, M., Acker, C. M., Duff, K., Castillo, P. E., & Davies, P. (2009). Age-dependent impairment of cognitive and synaptic function in the htau mouse model of tau pathology. *The Journal of Neuroscience, 29*(34), 10741–10749. http://dx.doi.org/10.1523/JNEUROSCI.1065-09.2009.

Poorkaj, P., et al. (1998). Tau is a candidate gene for chromosome 17 frontotemporal dementia. *Annals of Neurology, 43*(6), 815–825.

Poorkaj, P., et al. (2002). An R5L tau mutation in a subject with a progressive supranuclear palsy phenotype. *Annals of Neurology, 52*(4), 511–516.

Probst, A., Anderton, B. H., Brion, J. P., & Ulrich, J. (1989). Senile plaque neurites fail to demonstrate anti-paired helical filament and anti-microtubule-associated protein-tau immunoreactive proteins in the absence of neurofibrillary tangles in the neocortex. *Acta Neuropathologica, 77*(4), 430–436.

Probst, A., et al. (2000). Axonopathy and amyotrophy in mice transgenic for human four-repeat tau protein. *Acta Neuropathologica, 99*(5), 469–481.

Querfurth, H. W., & LaFerla, F. M. (2010). Alzheimer's disease. *The New England Journal of Medicine, 362*(4), 329–344. http://dx.doi.org/10.1056/NEJMra0909142.

Ramsden, M., Kotilinek, L., Forster, C., Paulson, J., McGowan, E., SantaCruz, K., … Ashe, K. H. (2005). Age-dependent neurofibrillary tangle formation, neuron loss, and memory impairment in a mouse model of human tauopathy (P301L). *The Journal of Neuroscience, 25*(46), 10637–10647. http://dx.doi.org/10.1523/JNEUROSCI.3279-05.2005.

Ren, Q. G., Liao, X. M., Chen, X. Q., Liu, G. P., & Wang, J. Z. (2007). Effects of tau phosphorylation on proteasome activity. *FEBS Letters, 581*(7), 1521–1528. http://dx.doi.org/10.1016/j.febslet.2007.02.065.

Rissman, R. A., Poon, W. W., Blurton-Jones, M., Oddo, S., Torp, R., Vitek, M. P., … Cotman, C. W. (2004). Caspase-cleavage of tau is an early event in Alzheimer disease tangle pathology. *The Journal of Clinical Investigation, 114*(1), 121–130. http://dx.doi.org/10.1172/JCI20640.

Rizzini, C., et al. (2000). Tau gene mutation K257T causes a tauopathy similar to Pick's disease. *Journal of Neuropathology and Experimental Neurology, 59*(11), 990–1001.

Rizzu, P., et al. (1999). High prevalence of mutations in the microtubule-associated protein tau in a population study of frontotemporal dementia in the Netherlands. *American Journal of Human Genetics, 64*(2), 414–421.

Roberson, E. D., Scearce-Levie, K., Palop, J. J., Yan, F., Cheng, I. H., Wu, T., … Mucke, L. (2007). Reducing endogenous tau ameliorates amyloid beta-induced deficits in an Alzheimer's disease mouse model. *Science, 316*(5825), 750–754. http://dx.doi.org/10.1126/science.1141736.

Rohrer, J. D., et al. (2011). Novel L284R MAPT mutation in a family with an autosomal dominant progressive supranuclear palsy syndrome. *Neuro-degenerative Diseases, 8*(3), 149–152.

Rosenmann, H., Grigoriadis, N., Eldar-Levy, H., Avital, A., Rozenstein, L., Touloumi, O., … Abramsky, O. (2008). A novel transgenic mouse expressing double mutant tau driven by its natural promoter exhibits tauopathy characteristics. *Experimental Neurology, 212*(1), 71–84. http://dx.doi.org/10.1016/j.expneurol.2008.03.007.

Rossi, G., et al. (2012). New mutations in MAPT gene causing frontotemporal lobar degeneration: Biochemical and structural characterization. *Neurobiology of Aging, 33*(4), 834.e1–834.e6.

Rosso, S. M., et al. (2002). A novel tau mutation, S320F, causes a tauopathy with inclusions similar to those in Pick's disease. *Annals of Neurology, 51*(3), 373–376.

Santacruz, K., Lewis, J., Spires, T., Paulson, J., Kotilinek, L., Ingelsson, M., … Ashe, K. H. (2005). Tau suppression in a neurodegenerative mouse model improves memory function. *Science, 309*(5733), 476–481. http://dx.doi.org/10.1126/science.1113694.

Santacruz, K., et al. (2005). Tau suppression in a neurodegenerative mouse model improves memory function. *Science, 309*(5733), 476–481.

Scaravilli, T., Tolosa, E., & Ferrer, I. (2005). Progressive supranuclear palsy and corticobasal degeneration: Lumping versus splitting. *Movement Disorders, 20*(Suppl. 12), S21–S28. http://dx.doi.org/10.1002/mds.20536.

Schindowski, K., et al. (2006). Alzheimer's disease-like tau neuropathology leads to memory deficits and loss of functional synapses in a novel mutated tau transgenic mouse without any motor deficits. *The American Journal of Pathology, 169*(2), 599–616.

Schliebs, R., & Arendt, T. (2006). The significance of the cholinergic system in the brain during aging and in Alzheimer's disease. *Journal of Neural Transmission, 113*(11), 1625–1644. http://dx.doi.org/10.1007/s00702-006-0579-2.

Schneider, A., Biernat, J., von Bergen, M., Mandelkow, E., & Mandelkow, E. M. (1999). Phosphorylation that detaches tau protein from microtubules (Ser262, Ser214) also protects it against aggregation into Alzheimer paired helical filaments. *Biochemistry, 38*(12), 3549–3558. http://dx.doi.org/10.1021/bi981874p.

Sharma, V. M., Litersky, J. M., Bhaskar, K., & Lee, G. (2007). Tau impacts on growth-factor-stimulated actin remodeling. *Journal of Cell Science, 120*(Pt 5), 748–757. http://dx.doi.org/10.1242/jcs.03378.

Shimura, H., Schwartz, D., Gygi, S. P., & Kosik, K. S. (2004). CHIP-Hsc70 complex ubiquitinates phosphorylated tau and enhances cell survival. *The Journal of Biological Chemistry, 279*(6), 4869–4876. http://dx.doi.org/10.1074/jbc.M305838200.

Shiryaev, N., Jouroukhin, Y., Giladi, E., Polyzoidou, E., Grigoriadis, N. C., Rosenmann, H., et al. (2009). NAP protects memory, increases soluble tau and reduces tau hyperphosphorylation in a tauopathy model. *Neurobiology of Disease, 34*(2), 381–388. http://dx.doi.org/10.1016/j.nbd.2009.02.011.

Skoglund, L., et al. (2008). The tau S305S mutation causes frontotemporal dementia with parkinsonism. *European Journal of Neurology, 15*(2), 156–161.

Sperfeld, A. D., et al. (1999). FTDP-17: An early-onset phenotype with parkinsonism and epileptic seizures caused by a novel mutation. *Annals of Neurology, 46*(5), 708–715.

Spillantini, M. G., et al. (1998). Tau pathology in two Dutch families with mutations in the microtubule-binding region of tau. *The American Journal of Pathology, 153*(5), 1359–1363.

Spillantini, M. G., et al. (1998). Mutation in the tau gene in familial multiple system tauopathy with presenile dementia. *Proceedings of the National Academy of Sciences of the United States of America, 95*(13), 7737–7741.

Spillantini, M. G., et al. (2000). A novel tau mutation (N296N) in familial dementia with swollen achromatic neurons and corticobasal inclusion bodies. *Annals of Neurology, 48*(6), 939–943.

Spina, S., et al. (2007). The novel Tau mutation G335S: Clinical, neuropathological and molecular characterization. *Acta Neuropathologica, 113*(4), 461–470.

Spittaels, K., et al. (1999). Prominent axonopathy in the brain and spinal cord of transgenic mice overexpressing four-repeat human tau protein. *The American Journal of Pathology, 155*(6), 2153–2165.

Stanford, P. M., et al. (2000). Progressive supranuclear palsy pathology caused by a novel silent mutation in exon 10 of the tau gene: Expansion of the disease phenotype caused by tau gene mutations. *Brain, 123*(Pt 5), 880–893.

Stanford, P. M., et al. (2003). Mutations in the tau gene that cause an increase in three repeat tau and frontotemporal dementia. *Brain, 126*(Pt 4), 814–826.

Suenaga, T., Hirano, A., Llena, J. F., Yen, S. H., & Dickson, D. W. (1990). Modified Bielschowsky stain and immunohistochemical studies on striatal plaques in Alzheimer's disease. *Acta Neuropathologica, 80*(3), 280–286.

Sumi, S. M., et al. (1992). Familial presenile dementia with psychosis associated with cortical neurofibrillary tangles and degeneration of the amygdala. *Neurology, 42*(1), 120–127.

Sydow, A., Van der Jeugd, A., Zheng, F., Ahmed, T., Balschun, D., Petrova, O., ... Mandelkow, E. M. (2011). Tau-induced defects in synaptic plasticity, learning, and memory are reversible in transgenic mice after switching off the toxic Tau mutant. *The Journal of Neuroscience, 31*(7), 2511–2525. http://dx.doi.org/10.1523/JNEUROSCI.5245-10.2011.

Takashima, A., et al. (1998). Presenilin 1 associates with glycogen synthase kinase-3beta and its substrate tau. *Proceedings of the National Academy of Sciences of the United States of America, 95*(16), 9637–9641.

Taldone, T., Zatorska, D., Patel, P. D., Zong, H., Rodina, A., Ahn, J. H., ... Chiosis, G. (2011). Design, synthesis, and evaluation of small molecule Hsp90 probes. *Bioorganic & Medicinal Chemistry, 19*(8), 2603–2614. http://dx.doi.org/10.1016/j.bmc.2011.03.013.

Tatebayashi, Y., et al. (2002). Tau filament formation and associative memory deficit in aged mice expressing mutant (R406W) human tau. *Proceedings of the National Academy of Sciences of the United States of America, 99*(21), 13896–13901.

Terry, R. D., Masliah, E., Salmon, D. P., Butters, N., DeTeresa, R., Hill, R., ... Katzman, R. (1991). Physical basis of cognitive alterations in Alzheimer's disease: Synapse loss is the major correlate of cognitive impairment. *Annals of Neurology, 30*(4), 572–580. http://dx.doi.org/10.1002/ana.410300410.

Terwel, D., Lasrado, R., Snauwaert, J., Vandeweert, E., Van Haesendonck, C., Borghgraef, P., et al. (2005). Changed conformation of mutant Tau-P301L underlies the moribund tauopathy, absent in progressive, nonlethal axonopathy of Tau-4R/2N transgenic mice. *The Journal of Biological Chemistry, 280*(5), 3963–3973. http://dx.doi.org/10.1074/jbc.M409876200.

Theunis, C., Crespo-Biel, N., Gafner, V., Pihlgren, M., Lopez-Deber, M. P., Reis, P., ... Muhs, A. (2013). Efficacy and safety of a liposome-based vaccine against protein Tau, assessed in tau.P301L mice that model tauopathy. *PLoS One, 8*(8), e72301.http://dx.doi.org/10.1371/journal.pone.0072301.

Thies, E., & Mandelkow, E. M. (2007). Missorting of tau in neurons causes degeneration of synapses that can be rescued by the kinase MARK2/Par-1. *The Journal of Neuroscience, 27*(11), 2896–2907. http://dx.doi.org/10.1523/JNEUROSCI.4674-06.2007.

Tolosa, E., Litvan, I., Hoglinger, G. U., Burn, D., Lees, A., Andres, M. V., ... Investigators, T. (2014). A phase 2 trial of the GSK-3 inhibitor tideglusib in progressive supranuclear palsy. *Movement Disorders, 29*(4), 470–478. http://dx.doi.org/10.1002/mds.25824.

Trinczek, B., Biernat, J., Baumann, K., Mandelkow, E. M., & Mandelkow, E. (1995). Domains of tau protein, differential phosphorylation, and dynamic instability of microtubules. *Molecular Biology of the Cell, 6*(12), 1887–1902.

Tseng, B. P., Green, K. N., Chan, J. L., Blurton-Jones, M., & LaFerla, F. M. (2008). Abeta inhibits the proteasome and enhances amyloid and tau accumulation. *Neurobiology of Aging, 29*(11), 1607–1618. http://dx.doi.org/10.1016/j.neurobiolaging.2007.04.014.

van Herpen, E., et al. (2003). Variable phenotypic expression and extensive tau pathology in two families with the novel tau mutation L315R. *Annals of Neurology, 54*(5), 573–581.

Waldemar, G., Dubois, B., Emre, M., Georges, J., McKeith, I. G., Rossor, M., ... Winblad, B. (2007). Recommendations for the diagnosis and management of Alzheimer's disease and other disorders associated with dementia: EFNS guideline. *European Journal of Neurology, 14*(1), e1–26. http://dx.doi.org/10.1111/j.1468-1331.2006.01605.x.

Walsh, J. S., Welch, H. G., & Larson, E. B. (1990). Survival of outpatients with Alzheimer-type dementia. *Annals of Internal Medicine, 113*(6), 429–434.

Wang, J. Z., Grundke-Iqbal, I., & Iqbal, K. (2007). Kinases and phosphatases and tau sites involved in Alzheimer neurofibrillary degeneration. *The European Journal of Neuroscience, 25*(1), 59–68. http://dx.doi.org/10.1111/j.1460-9568.2006.05226.x.

Wang, Y., Martinez-Vicente, M., Kruger, U., Kaushik, S., Wong, E., Mandelkow, E. M., … Mandelkow, E. (2009). Tau fragmentation, aggregation and clearance: The dual role of lysosomal processing. *Human Molecular Genetics, 18*(21), 4153–4170. http://dx.doi.org/10.1093/hmg/ddp367.

Wang, Y., Martinez-Vicente, M., Kruger, U., Kaushik, S., Wong, E., Mandelkow, E. M., … Mandelkow, E. (2010). Synergy and antagonism of macroautophagy and chaperone-mediated autophagy in a cell model of pathological tau aggregation. *Autophagy, 6*(1), 182–183.

Wang, Y., Yang, R., Gu, J., Yin, X., Jin, N., Xie, S., … Liu, F. (2015). Cross talk between PI3K-AKT-GSK-3beta and PP2A pathways determines tau hyperphosphorylation. *Neurobiology of Aging, 36*(1), 188–200. http://dx.doi.org/10.1016/j.neurobiolaging.2014.07.035.

West, M. J., Coleman, P. D., Flood, D. G., & Troncoso, J. C. (1994). Differences in the pattern of hippocampal neuronal loss in normal ageing and Alzheimer's disease. *Lancet, 344*(8925), 769–772. doi:S0140-6736(94)92338-8 [pii].

Williams, D. W., Tyrer, M., & Shepherd, D. (2000). Tau and tau reporters disrupt central projections of sensory neurons in Drosophila. *The Journal of Comparative Neurology, 428*(4), 630–640.

Wittmann, C. W., et al. (2001). Tauopathy in Drosophila: Neurodegeneration without neurofibrillary tangles. *Science, 293*(5530), 711–714.

Wszolek, Z. K., et al. (1992). Rapidly progressive autosomal dominant parkinsonism and dementia with pallido-ponto-nigral degeneration. *Annals of Neurology, 32*(3), 312–320.

Yasuda, M., et al. (2000). A novel mutation at position + 12 in the intron following exon 10 of the tau gene in familial frontotemporal dementia (FTD-Kumamoto). *Annals of Neurology, 47*(4), 422–429.

Yoshiyama, Y., Higuchi, M., Zhang, B., Huang, S. M., Iwata, N., Saido, T. C., … Lee, V. M. (2007). Synapse loss and microglial activation precede tangles in a P301S tauopathy mouse model. *Neuron, 53*(3), 337–351. http://dx.doi.org/10.1016/j.neuron.2007.01.010.

Younkin, S. G. (1991). Processing of the Alzheimer's disease beta A4 amyloid protein precursor (APP). *Brain Pathology, 1*(4), 253–262.

Younkin, S. G. (1995). Evidence that A beta 42 is the real culprit in Alzheimer's disease. *Annals of Neurology, 37*(3), 287–288. http://dx.doi.org/10.1002/ana.410370303.

Zempel, H., Luedtke, J., Kumar, Y., Biernat, J., Dawson, H., Mandelkow, E., et al. (2013). Amyloid-beta oligomers induce synaptic damage via Tau-dependent microtubule severing by TTLL6 and spastin. *The EMBO Journal, 32*(22), 2920–2937. http://dx.doi.org/10.1038/emboj.2013.207.

Zempel, H., Thies, E., Mandelkow, E., & Mandelkow, E. M. (2010). Abeta oligomers cause localized Ca(2+) elevation, missorting of endogenous Tau into dendrites, Tau phosphorylation, and destruction of microtubules and spines. *The Journal of Neuroscience, 30*(36), 11938–11950. http://dx.doi.org/10.1523/JNEUROSCI.2357-10.2010.

Zhang, B., Carroll, J., Trojanowski, J. Q., Yao, Y., Iba, M., Potuzak, J. S., … Brunden, K. R. (2012). The microtubule-stabilizing agent, epothilone D, reduces axonal dysfunction, neurotoxicity, cognitive deficits, and Alzheimer-like pathology in an interventional study with aged tau transgenic mice. *The Journal of Neuroscience, 32*(11), 3601–3611. http://dx.doi.org/10.1523/JNEUROSCI.4922-11.2012.

Zhang, B., Maiti, A., Shively, S., Lakhani, F., McDonald-Jones, G., Bruce, J., … , Trojanowski, J. Q., et al. (2005). Microtubule-binding drugs offset tau sequestration by stabilizing microtubules and reversing fast axonal transport deficits in a tauopathy model. *Proceedings of the National Academy of Sciences of the United States of America, 102*(1), 227–231. http://dx.doi.org/10.1073/pnas.0406361102.

Zhu, L. Q., Zheng, H. Y., Peng, C. X., Liu, D., Li, H. L., Wang, Q., et al. (2010). Protein phosphatase 2A facilitates axonogenesis by dephosphorylating CRMP2. *The Journal of Neuroscience, 30*(10), 3839–3848. http://dx.doi.org/10.1523/JNEUROSCI.5174-09.2010.

GENETIC AND ENVIRONMENTAL RISK FACTORS

6

Apolipoprotein E and Amyloid-β-Independent Mechanisms in Alzheimer's Disease

Takahisa Kanekiyo and Guojun Bu

Department of Neuroscience, Mayo Clinic Jacksonville,
Jacksonville, FL, United States

OUTLINE

Introduction	172
APOE Genotypes and Cognitive Functions in Nondemented Individuals	173
APOE and Cholesterol Metabolism in Synaptic Functions	174
APOE, Brain Glucose Metabolism, and Insulin Signaling	176
APOE, Mitochondria Dysfunction, and Tau Phosphorylation	177
APOE, ApoE Receptors, and Synaptic Functions	179
APOE and Cerebrovascular Functions	181
APOE and Inflammatory Response	183
Summary and Perspectives	185
Acknowledgments	187
References	188

Genes, Environment and Alzheimer's Disease.
DOI: http://dx.doi.org/10.1016/B978-0-12-802851-3.00006-1

171

INTRODUCTION

The accumulation and deposition of amyloid-β (Aβ) peptides in the brain are key events in the development of Alzheimer's disease (AD) (Hardy & Selkoe, 2002). Nonetheless, increasing evidence demonstrates the contribution of "Aβ-independent pathways" to the disease pathogenesis. Indeed, synaptic dysfunction (Arendt, 2009; Hardy & Selkoe, 2002), reduced cerebral glucose metabolism (Cohen & Klunk, 2014), and/or cerebral hypoperfusion (de la Torre, 2002) are observed prior to Aβ deposition. Since recent phase 3 clinical trial employing a humanized anti-Aβ monoclonal antibody Bapineuzumab did not improve clinical outcomes in AD patients despite its positive effects on decreasing brain Aβ deposition (Salloway et al., 2014), there may be a limitation to develop therapeutic methods by targeting only Aβ. Thus, it is essential to understand how these Aβ-independent factors contribute to the disease pathogenesis to design mechanism-based AD therapy.

While the pathogenesis of AD is multifactorial, the ε4 allele of the apolipoprotein E (*APOE*) gene is the strongest genetic risk factor among the three human gene alleles (ε2, ε3, ε4) (Bu, 2009; Liu, Kanekiyo, Xu, & Bu, 2013). A systematic meta-analysis has shown that the possession of *APOE* ε4 increases the risk of AD to 3.68-fold, when contrasted with referent variant *APOE* ε3/ε3. On the other hand, the ε2 allele of *APOE* is protective against AD, where carrying *APOE* ε2 reduces the risk for AD by nearly 50% (OR = 0.456–0.85) compared to *APOE* ε3/ε3 (Figure 6.1) (AlzGene; http://www.alzgene.org/geneoverview.asp?geneid=83). ApoE is composed of 299 amino acids, in which the isoforms are distinguished by two different amino acid substitutions (Arg or Cys) at positions 112 and 158 due to two corresponding SNPs for *APOE* localized on Chromosome 19 (rs429358 and rs7412); apoE2 (Cys112/Cys158), apoE3 (Cys112/Arg158), and apoE4 (Arg112/Arg158) (Figure 6.1) (Das, McPherson, Bruns, Karathanasis, & Breslow, 1985; Rall, Weisgraber, & Mahley, 1982; Weisgraber, Rall, & Mahley, 1981). Since a main function of apoE is to transport lipids by carrying them and binding to cell surface receptors, these *APOE* genotypes not only influence AD risk but also associate with blood lipid levels and the risk of coronary heart disease (Bennet et al., 2007). Furthermore, *APOE* genotypes are also likely to influence human longevity (Schachter et al., 1994). A proportional hazard mode showed that the *APOE* ε2 carriers survive 1 year longer, but *APOE* ε4 carriers have a decreased lifespan of 1.2 years compared to *APOE* ε3/ε3 genotype (Drenos & Kirkwood, 2010). Given that *APOE4* has been known to hasten the age of AD onset as well as increase the incidence (Corder et al., 1993), potential effects of *APOE* on cellular senescence or telomere length may also influence AD risk, although further studies are needed.

Overwhelming *in vitro* and *in vivo* studies have shown that apoE is a key molecule that regulates brain Aβ metabolism, where the presence of

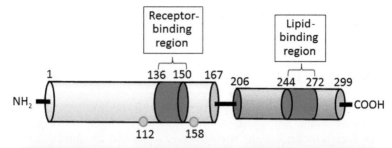

	Amino acid difference (dbSNP)		Allele frequency		AD risk
	112 (rs429358)	158 (rs7412)	CTL	AD	OR (95% CI)
ApoE2	Cys (TGC)	Cys (TGC)	0.07	0.04	0.621 (0.456–0.85)
ApoE3	Cys (TGC)	Arg (CGC)	0.79	0.58	—
ApoE4	Arg (CGC)	Arg (CGC)	0.14	0.38	3.68 (3.30–4.11)

FIGURE 6.1   Human apoE isoforms and AD risk. The 299-amino acid human apoE has two structural domains: an N-terminal domain containing the receptor-binding region (residues 136–150) and a C-terminal domain including the lipid-binding region (residues 244–272). The three human isoforms are distinguished by amino acid substitutions (Arg or Cys) at positions 112 and 158 in the N-terminal domain due to two corresponding SNPs (rs429358 and rs7412). A meta-analysis demonstrated significant associations between *APOE* genotypes and AD risk with *APOE4* representing a risky allele and *APOE2* a protective allele (AlzGene; http://www.alzgene.org/geneoverview.asp?geneid=83).

apoE4 correlates with less efficient Aβ elimination than apoE3 (Kanekiyo, Xu, & Bu, 2014). Moreover, there is considerable evidence indicating Aβ-independent mechanisms underlying the apoE isoform-specific effects in AD pathogenesis (Figure 6.1). ApoE isoforms contribute to the maintenance of brain homeostasis differently, by influencing neuronal functions, lipid/glucose metabolism, the cerebrovascular system, and neuroinflammation (Bu, 2009; Huang & Mucke, 2012). In this chapter, we will discuss how apoE isoforms differentially regulate AD pathogenic pathways with a particular focus on Aβ-independent mechanisms.

APOE GENOTYPES AND COGNITIVE FUNCTIONS IN NONDEMENTED INDIVIDUALS

Because of the strong association between *APOE4* and AD risk, apoE isoform-dependent effects on cognitive performance in nondemented individuals have also been actively studied. In a healthy aged population, *APOE4* carriers displayed reduced working memory functions (Reinvang, Winjevoll, Rootwelt, & Espeseth, 2010) and suppressed episodic memory functions (Zehnder et al., 2009) when compared to noncarriers. Further, *APOE4* carriers showed greater acceleration of longitudinal cognitive decline rate than the noncarriers in nondemented individuals (Praetorius, Thorvaldsson, Hassing, & Johansson, 2013; Schiepers et al., 2012). A meta-analysis demonstrated that differences between *APOE4* carriers and noncarriers on measures of episodic memory and global cognitive ability become significantly larger depending on age (Wisdom, Callahan, & Hawkins, 2011). Although the effects observed in these studies may be largely related to incident of dementia, another report by Caselli et al. have shown that the higher relative rate of cognitive decline in *APOE4* carriers is independent of developing mild cognitive impairment (MCI) and dementia (Caselli et al., 2009).

While *APOE2* is protective against AD, the genotype is also likely associated with changes in cognitive performance in healthy older individuals. *APOE2* carriers showed slower decline in episodic memory tests compared to *APOE* ε3/ε3 genotype in nondemented individuals (Wilson, Bienias, Berry-Kravis, Evans, & Bennett, 2002). A population-based study in nondemented elderly subjects has also shown that subjects with *APOE* ε2/ε2 and ε2/ε3 genotypes display better verbal learning ability than those with *APOE* ε3/ε3 (Helkala et al., 1995). Furthermore, *APOE2* carriers, as opposed to noncarriers, could maintain their verbal learning performance for a 3-year follow-up period (Helkala et al., 1996). Since 66% of these individuals likely have brain amyloid deposition detected through Pittsburgh Compound-B (PIB)-Positron Emission Tomography (PET) imaging by 89 years old without cognitive impairment (Jack et al., 2014), it is questionable whether the apoE-isoform effects are Aβ-independent in these clinical studies through nondemented subjects. Nonetheless, several animal experiments clearly demonstrated that expression of human apoE4 induces cognitive dysfunctions compared to apoE3 in the absence of Aβ (Huang, 2010). Therefore, the evidence clearly indicates that apoE isoforms have a strong correlation with the risk for cognitive decline in the elderly, although it should be argued whether *APOE* genotypes affect cognitive functions per se or influence the incidence of preclinical AD.

APOE AND CHOLESTEROL METABOLISM IN SYNAPTIC FUNCTIONS

Synaptic degeneration is a common feature of diverse neurodegenerative diseases. Whereas synaptic dysfunctions occur at a very early stage of AD prior to Aβ deposition and then likely progress during the disease

(Arendt, 2009; Overk & Masliah, 2014), synaptic activity also contributes to endocytosis of amyloid precursor protein (APP) and subsequent Aβ production in the pre- and postsynaptic regions (Bero et al., 2011; Cirrito et al., 2008; Verges, Restivo, Goebel, Holtzman, & Cirrito, 2011). Although soluble Aβ oligomers are one of the critical factors disturbing synaptic plasticity in AD (Hardy & Selkoe, 2002), it is also important to understand the mechanisms of synaptic degeneration through Aβ-independent pathways in AD pathogenesis.

Given that cholesterol is a critical component of synapses that supports their plasticity and functions, and that the main function of apoE produced by astrocytes is to transport cholesterol to neurons in the brain (Bu, 2009; Mauch et al., 2001), apoE-dependent synaptic phenotypes might be critically involved in cognitive performance. In the postmortem brains of cognitively normal individuals, *APOE4* carriers had reduced pre- and postsynaptic proteins, while *APOE2* was associated with significantly elevated postsynaptic marker PSD95 (Love et al., 2006). In human apoE-targeted replacement (TR) mice, apoE isoforms differently affect spine formation; apoE4-TR mice have reduced spine density compared to apoE2-TR and apoE3-TR mice at as young as 4 weeks of age (Dumanis et al., 2009). While apoE knockout mice had a lower density of dendritic spines, astrocyte-specific overexpression of human apoE3, but not apoE4, restored the phenotype (Ji et al., 2003). Consistently, apoE4-TR mice displayed suppressed excitatory synaptic transmission compared with apoE3-TR mice, which was independent of amyloid pathology (Wang et al., 2005). Long-term potentiation (LTP) in the hippocampus was also suppressed in young apoE4-TR mice (Trommer et al., 2004), although there are conflicting reports (Kitamura et al., 2004; Korwek, Trotter, Ladu, Sullivan, & Weeber, 2009). In spite of the neuroprotective functions of apoE2, LTP was reduced in apoE2-TR mice compared to apoE3-TR and wild-type mice (Trommer et al., 2004). In fact, the transcriptional profile in postmortem brains from normal individuals identified the reductions in expression of LTP-related molecules in *APOE2* carriers compared to *APOE* ε3/ε3 genotype (Conejero-Goldberg et al., 2014). Although further studies are necessary to assess apoE2 effects on synapses, it is possible that phenotype represented by suppressed LTP may be beneficial for AD by preventing neuronal damages due to hyperexcitation and reducing neuronal activity-dependent Aβ production.

For synaptogenesis, sufficient amounts of cholesterol associated with apoE-lipoprotein particles produced by astrocytes are essential (Mauch et al., 2001). However, it remains elucidated if increased neuronal cholesterol induces advantageous or deleterious effects to AD risk. Pharmacological approaches using agonists for liver X receptor (LXR) or retinoid X receptor (RXR) ameliorate amyloid pathology and cognitive function in amyloid AD model mice likely through increased brain cholesterol and/or apoE levels (Cramer et al., 2012; Riddell et al., 2007; Terwel et al., 2011; Vanmierlo et al., 2011). On the other hand, *APOE2* carriers displayed lower cholesterol levels in the synaptic region than those with *APOE* ε3/ε3 genotype in postmortem brains without amyloid pathology, although synaptic cholesterol levels were substantially decreased in AD patients when compared to control brains among *APOE* ε3/ε3 subjects (Oikawa, Hatsuta, Murayama, Suzuki, & Yanagisawa, 2014). In addition to providing cholesterol to neurons, apoE also likely has a function to inhibit lipid-particle-mediated cholesterol release from neurons in an isoform-dependent manner (apoE4 > apoE3) (Gong et al., 2007), which affects the exchange of cholesterol at synapses. Thus, the efficient turnover of synaptic cholesterol, rather than its amount per se, may contribute to cognitive functions under physiological conditions, while it is likely that supplying fresh cholesterol is essential to repair injured synapses during AD.

APOE, BRAIN GLUCOSE METABOLISM, AND INSULIN SIGNALING

Glucose is an essential source of energy for neurons (McEwen & Reagan, 2004). Therefore, impaired brain glucose metabolism would contribute to AD pathogenesis. Indeed, accumulating evidence indicates that diabetes and insulin resistance are strongly associated with risk for dementia including AD. (For more information about the association between diabetes and AD see Chapter 14; Biessels, Staekenborg, Brunner, Brayne, & Scheltens, 2006.) Studies using fluorodeoxyglucose (FDG)-PET imaging also revealed that cerebral glucose metabolism is suppressed in the brains of AD patients compared with that of cognitively healthy individuals. Importantly, these changes are observed several years prior to the first symptoms of AD and predict the disease progression (Cohen & Klunk, 2014; Mosconi et al., 2010, 2009).

While neuronal activities are associated with brain glucose metabolism, systemic glucose intolerance observed in type 2 diabetes also likely influences brain glucose metabolism. In fact, higher fasting serum glucose levels were shown to be significantly correlated with the lower regional cerebral metabolic rate for glucose measured by FDG-PET in cognitively normal nondiabetic adults (Burns et al., 2013). Thus, these findings suggest that altered brain glucose metabolism is not just a consequence of the suppressed neuronal activity during the progression of AD. Interestingly, the association between diabetes and AD is stronger in the presence of *APOE4* (Peila, Rodriguez, & Launer, 2002), although *APOE4* unlikely influences the risk for diabetes (Peila et al., 2002; Xu, Qiu, Wahlin, Winblad, & Fratiglioni, 2004). While aging lowers the glucose metabolic ratio broadly in cortical and subcortical regions in cognitively normal individuals, *APOE4* contributes to the reductions in glucose metabolic ratio specifically in the posterior cingulate and/or precuneus, and lateral parietal regions independent of age and amyloid burden (Knopman et al., 2014). Since these brain regions are preferentially affected in AD patients, aging and *APOE4*-related declines of glucose metabolism may be involved in the development of AD.

Although there are few studies regarding apoE2 and brain glucose metabolism, a meta-analysis showed that *APOE2* carriers are at an elevated risk for diabetes (Anthopoulos, Hamodrakas, & Bagos, 2010). When fed a Western-type diet, apoE2-TR mice displayed greater hyperinsulinemia and insulin intolerance compared to apoE3-TR mice, which corresponds to their robust hyperlipidemia (Kuhel et al., 2013). Because *APOE2* is protective for AD risk, the potential implication of these findings in glucose metabolism to AD requires further investigation.

In addition to abnormal glucose metabolism, brain insulin resistance has also been demonstrated to be a common feature of AD. In fact, impaired insulin signaling was documented in both postmortem analysis and in animal models of AD (Hoyer, 2004; Steen et al., 2005). These observations are further supported by the fact that intranasal administration of insulin improves cognitive functions and brain metabolism in patients with MCI and early AD (Craft et al., 2012; Reger, Watson, Green, Wilkinson et al., 2008). Interestingly, apoE isoforms influence the efficacy of intranasal insulin administration; among memory-impaired subjects, insulin significantly ameliorates verbal memory dysfunction in *APOE4* noncarriers, but not in *APOE4* carriers (Reger, Watson, Green, Baker et al., 2008). As reduced brain insulin signaling is observed in apoE4-TR mice (Ong, Chan, Lim, Cole, & Wong, 2014), apoE isoform-dependent responses to the nasal insulin treatment may be due to their differential insulin sensitivity. However, another study through an *ex vivo* insulin stimulation experiment using postmortem brains has shown that brain insulin resistance observed in MCI and AD cases are independent of diabetes or *APOE4* status (Talbot et al., 2012). Since insulin signaling modulates neuronal functions and synaptic integrity by controlling glucose metabolism, neurotransmitter channel activity, brain cholesterol synthesis, and/or mitochondrial function (Baker et al., 2011; Kleinridders, Ferris, Cai, & Kahn, 2014), further studies focusing on apoE isoform-dependent effects on these pathways are needed.

APOE, MITOCHONDRIA DYSFUNCTION, AND TAU PHOSPHORYLATION

Postmortem studies have shown the declines of cytochrome oxidase activity in the posterior cingulate in AD patients relative to controls

(Valla, Berndt, & Gonzalez-Lima, 2001). In fact, a genome-wide transcriptomic approach using autopsy brains of AD patients and normal controls revealed that AD cases had significantly lower expression of 70% of the nuclear genes encoding mitochondrial electron transport chain in posterior cingulate cortex compared with controls, 65% of those in the middle temporal gyrus, 61% of those in hippocampal CA1, 23% of those in entorhinal cortex, 16% of those in visual cortex, and 5% of those in the superior frontal gyrus (Liang et al., 2008). The activity of the α-ketoglutarate dehydrogenase complex (KGDHC), a mitochondrial enzyme, is also diminished in AD brains. Interestingly, cognitive performances in AD patients carrying *APOE4* are more strongly correlated with KGDHC activity than with the amyloid pathology, although the inverse results are observed in the case of *APOE4* noncarriers (Gibson et al., 2000). Consistently, *APOE4* carriers had lower mitochondrial cytochrome oxidase activity than the noncarriers in posterior cingulate cortex from young adults without any evidence of amyloid or tau pathology (Valla et al., 2010). Therefore, *APOE4* may induce mitochondrial/oxidative damage before Aβ deposition, resulting in cognitive dysfunction in AD patients.

Although neurons produce very limited amounts of apoE under physiological conditions, apoE expression can be induced in neurons under stressed conditions (Huang & Mucke, 2012). For instance, significant apoE expression was detected in neurons when kainic acid-induced excitotoxic injury was induced in mouse hippocampus (Xu et al., 2006). While neuronal apoE likely has protective functions to repair the damage and remodel components, apoE is fragmented in the intraneuronal space, which eventually causes neurotoxicity (Huang, 2010; Huang & Mahley, 2014). Indeed, more apoE fragments are observed in the brains of AD patients than controls, dependent on apoE isoform (apoE4 > apoE3) (Harris et al., 2003). To support these observations, *in vitro* experiments using apoE transfected neuronal cells and transgenic mouse models expressing apoE in neurons have demonstrated that neuronal apoE4 is more susceptible to proteolytic fragmentation than apoE3 (Huang, 2010; Huang & Mahley, 2014), although there is a conflicting report regarding apoE isoform-dependent proteolysis of apoE (Elliott et al., 2011). ApoE4 fragments containing 1–272 portions interact with mitochondria; overexpression of these fragments in neuronal cells cause mitochondrial dysfunction, leading to neurotoxicity (Chang et al., 2005; Nakamura, Watanabe, Fujino, Hosono, & Michikawa, 2009). When full-length apoE4 was expressed in neurons with neuron-specific promoter NSE, these mice displayed severe apoE fragmentation (Brecht et al., 2004), as well as impairments in learning by a water maze task and in vertical exploratory behavior (Raber et al., 1998). Furthermore, the generation of apo4 fragments likely cause tau phosphorylation in transgenic mice expressing apoE4 in neurons (Brecht et al., 2004). Consistently, transgenic mice expressing the apoE4 N-terminal fragment (Δ272–299) also displayed the accumulation of phosphorylated tau in the cortex and hippocampus (Harris et al., 2003).

Neuronal apoE4 stimulates tau phosphorylation not only through its fragmentation but also by activating the extracellular signal regulated kinase (ERK) pathway (Harris, Brecht, Xu, Mahley, & Huang, 2004). Interestingly, *in vitro* experiments have demonstrated that apoE interacts with nonphosphorylated tau at the microtubule binding site and interferes with tau hyperphosphorylation (Strittmatter et al., 1994); however, related apoE isoform-dependent effects are controversial (Flaherty, Lu, Soria, & Wood, 1999). Although the apoE effects on tau were not observed when apoE is expressed in astrocytes (Brecht et al., 2004; Tesseur et al., 2000), human apoE knock-in mice demonstrated that apoE4 facilitates tau phosphorylation, which impairs the GABAergic electrophysiological inputs in the hippocampus, resulting in the disturbances of hippocampal neurogenesis (Li et al., 2009) and learning and memory deficits (Andrews-Zwilling et al., 2010). When apoE4 is deleted in GABAergic interneurons from mouse brains using the cre-lox system, their age-related learning and memory deficits were rescued, whereas deletion of apoE4 in astrocytes did not lead to such effects (Knoferle et al., 2014). Although further studies are needed to define how the expression of apoE isoforms is regulated in GABAergic neurons under physiological and pathological conditions including AD, these studies imply that apoE4 may directly or indirectly contribute to tau phosphorylation, aggregation, and neurofibrillary tangle formation in AD.

APOE, APOE RECEPTORS, AND SYNAPTIC FUNCTIONS

The low-density lipoprotein receptor (LDLR) family members including the LDLR, low-density lipoprotein receptor-related protein 1 (LRP1), apolipoprotein E receptor 2 (ApoER2) and very low-density lipoprotein receptor (VLDLR) serve as apoE receptors in neurons (Figure 6.2) (Bu, 2009; Lane-Donovan, Philips, & Herz, 2014). Since apoE-mediated increase in synaptogenesis can be blocked by the receptor antagonist RAP (Mauch et al., 2001), apoE receptors likely play a critical role in synaptic plasticity by regulating lipid metabolism. In particular, LRP1 is abundantly expressed in neurons and efficiently mediates apoE endocytosis (Kanekiyo & Bu, 2014). Consistently, altered brain lipid metabolism has been demonstrated in LRP1 neuron-specific knockout mice (Liu et al., 2010). LRP1 is also involved in Aβ metabolism in neurons and the disturbances of the pathway result in brain Aβ accumulation (Kanekiyo et al., 2013, 2011). Furthermore, LRP1 can moderate glutamate receptor signaling pathways in neurons by coupling with N-Methyl-D-Aspartate

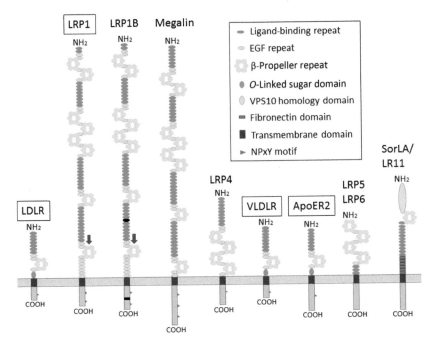

FIGURE 6.2   Members of the low-density lipoprotein receptor family. The major neuronal apoE receptors, LDLR, LRP1, VLDLR, and ApoER2 belong to the LDLR family. These receptors are type I transmembrane receptors composed of ligand-binding repeats, epidermal growth factor (EGF) repeats, and YWTD propeller (β-propeller) domains in the extracellular region, and a single transmembrane domain and a relatively short cytoplasmic tail. The majority of the cytoplasmic domains contain at least one NPxY motif, which serves as the dominant endocytosis signal. The LDLR, VLDLR, and ApoER2 contain an O-linked sugar domain. LRP1 and LRP1B have the furin cleavage sites as indicated with arrows. LRP1B is highly homologous to LRP1 with the two extra sequence motifs encoded by two exons. LRP5/6 and SorLA/LR11 are structurally distinct from other members; the SorLA/LR11 contains fibronectin domains and vacuolar protein sorting (VPS) domains, and LRP5/6 has primarily β-propeller domains (Bu, 2009; Lane-Donovan et al., 2014).

(NMDA) receptor NMDAR1 (Liu et al., 2010) and α-Amino-3-Hydroxy-5-Methyl-4-Isoxazole Propionic Acid (AMPA) receptor GluA1 (Gan, Jiang, McLean, Kanekiyo, & Bu, 2014) to regulate synaptic functions. Future studies will likely reveal how apoE isoforms contribute to the LRP1-mediated neuronal functions.

Among the LDLR family members, ApoER2 and VLDLR have been shown to interact with the glycoprotein Reelin, which has an important function enhancing LTP to maintain synaptic plasticity (Lane-Donovan et al., 2014). Since apoE also binds to ApoER2 and VLDLR influencing their recycling to the synapses, apoE likely competes with Reelin for its signaling pathway. Interestingly, apoE4 interferes with endosomal recycling of ApoER2 and VLDLR in an isoform-specific manner, resulting in reduced surface pool of the receptors (Chen, Durakoglugil, Xian, & Herz, 2010). Further studies to determine how apoE isoforms share the apoE receptors with other ligands such as Reelin on synapses in AD pathogenesis are desired.

Taken together, these studies indicate the importance of apoE receptors as well as apoE in regulating synaptic integrity and functions. Thus, to further define apoE isoform-dependent mechanisms in cognitive decline during aging and AD, it is critical to carry out additional studies regarding the functions of individual apoE receptors at the synapses.

APOE AND CEREBROVASCULAR FUNCTIONS

The cerebrovascular system plays an essential role in maintaining cognitive functions. Approximately 5% of aged population older than 65 years is estimated to have vascular cognitive impairment (Rockwood et al., 2000), which refers to a heterogeneous cognitive decline caused by cerebrovascular lesions (Moorhouse & Rockwood, 2008). The Rotterdam Study has also shown that greater cerebral blood flow (CBF) velocity measured by Transcranial Doppler echo was correlated with significantly slower cognitive decline over a 6.5-year follow-up period in nondemented subjects, suggesting that cerebral hypoperfusion might precede and contribute to the onset of dementia (Ruitenberg et al., 2005). A recent study using Dynamic Contrast-Enhanced (DCE)-Magnetic Resonance Imaging (MRI) also demonstrated that blood–brain barrier (BBB) integrity in the hippocampus is disturbed in cognitively normal older individuals and further in MCI patients compared to young controls (Montagne et al., 2015). Thus, cerebrovascular dysfunctions are likely early events during aging in human brains and might contribute to age-dependent cognitive impairment. Furthermore, the rate of cognitive decline was accelerated in those with amyloid or vascular pathologies in normal elderly subjects when monitored by PIB-PET and Fluid-Attenuated Inversion Recovery (FLAIR)-MRI. While amyloid deposition and vascular damages are likely independent processes contributing to cognitive decline in the elderly, the presence of both pathologies extensively disturbs cognitive functions in an additive manner (Vemuri et al., 2015).

Furthermore, epidemiological studies have shown that the prevalence of AD is strongly correlated with the presence of diverse vascular risk factors or indicators of vascular diseases including atherosclerosis, hypertension, diabetes, and hypercholesteremia (de la Torre, 2002; Ostergaard et al., 2013). Cerebrovascular lesions often coexist with AD, where 25–80% of demented subjects show both AD and cerebrovascular lesions (Jellinger, 2008). Neuroimaging using Single Photon Emission

Computed Tomography showed the longitudinal decline of regional CBF in MCI patients compared with healthy controls (Kogure et al., 2000). Furthermore, the regional decreases of CBF were significantly associated with the conversion of MCI to AD during the follow-up period (Hirao et al., 2005; Johnson et al., 1998). Thus, the regional cerebral hypoperfusion closely relates to the disease process of AD.

While the cerebrovascular system might be regulated by multiple factors, apoE is significantly involved in this mechanism. The cerebrovascular unit is composed of endothelial cells and vascular mural cells (vascular smooth muscle cells and pericytes) surrounded by endfeet of astrocytes (Zlokovic, 2011), where astrocytes and vascular mural cells are the major sources of apoE (Xu et al., 2006). Although the functions of apoE in the perivascular region are still unclear, increasing evidence has shown the substantial contribution of apoE isoforms to the homeostasis of the cerebrovascular system. Reduced CBF was found in the older *APOE4* carriers relative to noncarriers despite preserved gray matter volume using functional MRI (Filippini et al., 2011). When longitudinal changes in regional CBF were assessed in nondemented older adults using $H_2^{15}O$-PET, *APOE4* carriers showed greater CBF decline than noncarriers in brain regions vulnerable to pathological changes in AD (Thambisetty, Beason-Held, An, Kraut, & Resnick, 2010). Although *APOE4* effects on CBF is controversial in healthy young individuals (Filippini et al., 2011; Scarmeas, Habeck, Stern, & Anderson, 2003; Wierenga et al., 2013), *APOE4* likely exacerbates age-dependent CBF decline before the onset of measurable cognitive impairment.

Consistent with the human brain imaging studies, animal experiments using apoE-TR mice also showed apoE isoform-dependent modifications in the cerebrovascular system. ApoE4-TR mice exhibit compromised BBB integrity and reduced CBF compared to apoE2-TR and apoE3-TR mice (Bell et al., 2012). This study further showed that a pro-inflammatory cytokine cyclophilin A (CypA)-mediated pathway that includes activation of nuclear factor-kappa B (NFκB) and secretion of matrix metalloproteinase 9 (MMP9) is activated in capillary pericytes in apoE4-TR mice. In cognitively healthy individuals, *APOE4* carriers displayed an age-dependent increase of cerebrospinal fluid (CSF)/plasma albumin quotient, indicating BBB breakdown, with positive correlations with CSF levels of CypA and MMP9 (Halliday et al., 2013). Other groups have also demonstrated the diminished BBB integrity by apoE4 using *in vitro* (Nishitsuji, Hosono, Nakamura, Bu, & Michikawa, 2011) and *in vivo* models (Alata, Ye, St-Amour, Vandal, & Calon, 2015; Nishitsuji et al., 2011). Interestingly, apoE4-TR mice also showed the structurally reduced cerebral vascularization and thinner basement membranes in an age-dependent manner compared with apoE2-TR and apoE3-TR mice (Alata et al., 2015). Furthermore, apoE4 is associated with hyperlipidemia and hypercholesterolemia, leading to substantial vascular damages such as atherosclerosis, coronary heart disease, and stroke (Bennet et al., 2007). Therefore, apoE isoforms likely regulate several aspects of cerebrovascular functions. Although further studies to assess the molecular mechanisms are needed, cerebrovascular damages caused by apoE4 through multiple pathways likely contribute to AD onset.

APOE AND INFLAMMATORY RESPONSE

Inflammation is a major pathway contributing to cognitive decline and dementia. Systemic inflammation has been shown to correlate with cognitive impairment or as a risk factor for developing neurodegenerative diseases including AD (Ownby, 2010). Brain inflammatory response is characterized by the activation of astrocytes and microglia, and the production of inflammasomes and cytokines (Liu & Chan, 2014). In particular, the induction of major pro-inflammatory cytokines interleukin (IL)-1β, IL-6 or tumor necrosis factor-α (TNF-α) are known to be sufficient to cause cognitive decline (Gemma & Bickford, 2007; Jankowsky & Patterson, 1999). In addition to Aβ deposition and neurofibrillary tangle formation, activated glial cells surrounding the senile plaques in brains are also one of pathological hallmarks of AD. While glial cells play a critical role in Aβ clearance and debris-clearing in response to local neurodegeneration (Nagele, Wegiel, Venkataraman, Imaki, & Wang, 2004), the excess immune responses likely trigger the cascade of deleterious neuroinflammation during the disease progression.

Because of the involvement of inflammation in cognitive impairment, several clinical trials have been carried out to assess potential therapeutic effects of nonsteroidal anti-inflammatory drug (NSAID) treatment on age-dependent cognitive decline including AD (Gorelick, 2010). Chicago Health and Aging Project showed that long-term use of ibuprofen for over 5 years was associated with decreased rates of cognitive decline in older individuals (Grodstein et al., 2008). Long-term NSAID use was also likely to be protective against AD in which ibuprofen has a clear effect compared to arylpropionic acids and naproxen (Vlad, Miller, Kowall, & Felson, 2008). More importantly, several studies have shown positive effects of NSAID on cognition more so in *APOE4* carriers (Hayden et al., 2007; Szekely et al., 2008). In spite of negative effects of inflammation on cognition, there is a conflicting report showing that elevated serum level of a major systemic inflammatory marker C-reactive protein (CRP) is linked to a lower risk of cognitive decline in *APOE4* noncarriers over a 4-year follow-up period in older subjects (Lima et al., 2014). While increased CRP was associated with increased risk for coronary artery disease (CAD), the concentration of CRP was lower in individuals with *APOE4* compared to the noncarriers both in the presence or absence of CAD (Marz, Scharnagl, Hoffmann, Boehm, & Winkelmann, 2004). Although further studies are needed, these studies indicate that apoE is involved in the mechanism underlying inflammation-related cognitive decline, which might link to AD pathogenesis.

In fact, mounting evidence has shown differential effects of apoE isoforms on systemic inflammation. *APOE4* enhances the human innate immune response to multiple Toll-like receptor (TLR) ligands. After intravenous lipopolysaccharide (LPS) administration in healthy individuals, *APOE* ε3/ε4 had higher hyperthermia and plasma TNF-α levels and earlier plasma IL-6 than *APOE* ε3/ε3 subjects (Gale et al., 2014). Consistently, apoE4-TR mice displayed significantly greater elevations of the pro-inflammatory cytokines as compared with apoE3-TR mice (Gale et al., 2014; Lynch et al., 2003). Interestingly, intravenous administration of apoE-mimetic peptides composed of its receptor-binding region of apoE (133–149) also suppressed both systemic and brain inflammatory responses in mice after LPS administration (Lynch et al., 2003). An *in vitro* experiment has demonstrated that exogenous human apoE and the apoE-mimetic peptides inhibit macrophage inflammatory responses to LPS (TLR-4 agonist) and poly(I-C) (TLR-3 agonist) through the binding to LRP1/LDLR and heparan sulfate proteoglycan, respectively (Zhu, Kodvawala, & Hui, 2010).

Neuroinflammation is also significantly affected by apoE isoforms. ApoE4-TR mice displayed enhanced glial activation and increased pro-inflammatory cytokines in hippocampus in response to intracerebroventricular LPS stimulation compared to apoE2-TR and apoE3-TR mice, resulting in exacerbated synaptic loss (Zhu et al., 2012). However, another study demonstrated that the activation of primary astrocytes from apoE2-TR mice with LPS led to the greater secretion of pro-inflammatory cytokines than those from apoE3-TR or apoE4-TR mice (Maezawa, Maeda, Montine, & Montine, 2006). Meanwhile, exogenous apoE not only prevents inflammatory responses but also stimulates production of IL-1β in an isoform-dependent manner, with apoE4 inducing a significantly greater response than apoE3 in cultured rat glial cells (Guo, LaDu, & Van Eldik, 2004). These conflicting findings may be due to their experimental conditions, which seem to substantially affect glial activation status. Future studies should address how apoE4 regulates neuroinflammation through its gain of functions in the pro-inflammatory response and loss of functions in the anti-inflammatory abilities compared to other isoforms. Furthermore, the mobility of microglia is also apoE-isoform dependent. Primary cultures of microglia from apoE4-TR or apoE2-TR mice exhibited significantly reduced complement C5a- and adenosine triphosphate-stimulated migration compared with those from apoE3-TR mice (Cudaback, Li, Montine, Montine, & Keene, 2011). Together, these findings clearly demonstrate apoE isoform-dependent effects on inflammatory responses through diverse mechanisms, which could contribute to its association with AD and other neurodegenerative diseases.

SUMMARY AND PERSPECTIVES

Although we still do not fully understand how apoE isoforms regulate multiple AD-related pathways, it is evident that *APOE4* is associated with increased risk for cognitive decline in aged population in both Aβ-dependent and independent manners. In this chapter, we have discussed apoE isoform-dependent effects on cholesterol metabolism, glucose metabolism, mitochondria functions, cerebrovascular system, and inflammation, which are important pathways implicated in the cognitive functions and AD pathogenesis (Figure 6.3). Given that all the pathways are major components of metabolic processes essential for maintaining the homeostasis of the whole body as well as the central nervous system, these pathways likely interact with one another. Because of their inseparable link, it remains to be elucidated if apoE isoforms influence some specific pathways resulting in subsequent alternation of related metabolic pathways or that these events are independent during the disease process. Furthermore, it is interesting to note that *APOE4* carrier status increases the risk of cognitive decline in healthy older females to a greater extent than males (Altmann, Tian, Henderson, & Greicius, 2014). Since sex substantially influences the metabolic homeostasis, the sex-dependent differences in apoE4-related risk of developing AD might also be induced as a consequence of the altered metabolic pathways.

Most interestingly, these risk factors for Aβ-independent pathways for AD also sufficiently impact Aβ production and/or degradation (Figure 6.3). For example, increased neuronal cholesterol accelerates Aβ production by increasing β- and γ-secretase activities (Grimm, Zimmer, Lehmann, Grimm, & Hartmann, 2013). In type 2 diabetes, increased insulin levels prevent Aβ clearance by competing with Aβ for insulin-degrading enzyme, which is a major Aβ degrading enzyme in the brain (Saido & Leissring, 2012). Indeed, when Aβ is co-injected with insulin into rat brains, Aβ clearing rate was reduced, resulting in increased brain Aβ levels (Shiiki et al., 2004). Since the cerebrovascular system critically mediates brain Aβ clearance (Zlokovic, 2013), the disturbances of cerebrovascular cells, in particular vascular mural cell, result in exacerbated Aβ deposition as senile plaques in brain parenchyma as well as cerebral amyloid antipathy along the cerebrovasculature (Kanekiyo, Liu, Shinohara, Li, & Bu, 2012; Sagare et al., 2013). Furthermore, immune responses through inflammasomes and cytokines are also critically involved in Aβ deposition by affecting glial cell-mediated Aβ clearance and APP processing

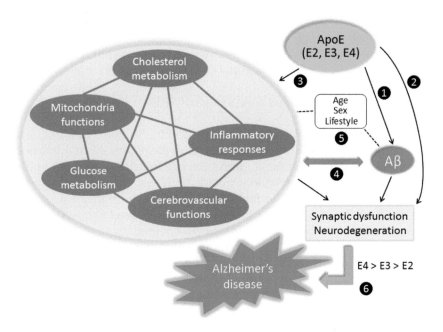

FIGURE 6.3   ApoE isoform-specific pathogenic pathways for AD through Aβ-dependent and independent mechanisms. ApoE isoforms are involved in AD pathogenesis through diverse pathways. (1) ApoE isoforms influence brain Aβ deposition by regulating Aβ aggregation and clearance; (2) ApoE4 causes synaptic dysfunction by affecting apoE receptors and/or induces neurodegeneration through its neurotoxic fragments; (3) ApoE isoforms contribute to the maintenance of the brain homeostasis by regulating cholesterol metabolism, glucose metabolism, mitochondria functions, cerebrovascular system, and inflammation. These apoE-regulated metabolic pathways are likely intertwined in AD pathogenesis. Importantly, (4) dysfunctions of these metabolic pathways also exacerbate Aβ accumulation, which may trigger the vicious circle in AD as accumulated Aβ likely in turn disturbs these pathways. Epidemiological studies have shown that (5) age, sex, and lifestyle (i.e., exercise, education, sleep) significantly affect apoE-related AD pathogenesis. (6) Through these complex pathogenic pathways, apoE likely contributes to AD risk in an isoform-dependent manner (E4 > E3 > E2).

(Heneka, Golenbock, & Latz, 2015; Zhang & Jiang, 2015). Therefore, these apoE-related indirect effects on Aβ metabolism should be further investigated in order to have a complete understanding of the contribution of apoE isoforms to AD pathogenesis.

Since apoE isoforms influence brain activity and several aspects of metabolic pathways, our lifestyle likely influences apoE isoform-dependent AD pathogenesis (Figure 6.3). Physical activity has been shown to be associated with lower risks of cognitive impairment, AD, and other dementia compared with no exercise in older subjects (Laurin, Verreault, Lindsay, MacPherson, & Rockwood, 2001). In fact, adequate exercise is significantly associated with reduced cerebral amyloid burden detected by PIB-PET in cognitively normal older adults, where a predominant effect of exercise engagement was observed in *APOE4* carriers but not in the noncarriers (Head et al., 2012). Since exercise likely invigorates brain and ameliorates symptoms of metabolic syndrome, exercise not only suppresses brain Aβ deposition, but might also reverse apoE4 effects on AD through Aβ-independent pathways, although further studies are required. Consistently, an animal experiment demonstrated that exercise improves cognition and synaptic plasticity in apoE4-TR mice (Nichol, Deeny, Seif, Camaclang, & Cotman, 2009). In addition to exercise, prospective longitudinal cohort studies showed that better sleep consolidation (Lim et al., 2013) and high lifetime intellectual enrichment (Vemuri et al., 2014) also attenuate the effect of *APOE4* on cognition in a nondemented aged population. Thus, to prevent AD, we might need to gain further knowledge on the molecular mechanisms underlying the link between these lifestyle factors and apoE isoforms.

Paradoxically, there is evidence showing that *APOE4* may possess cognitive advantages in young populations. It is interesting to note that *APOE4* displayed better episodic memory compared with *APOE2* and *APOE3* in young healthy persons with average age around 22 years (Mondadori et al., 2007). The IQs in *APOE4* carriers are also higher than that of the noncarriers in young females aged 19–21 years (Yu, Lin, Chen, Hong, & Tsai, 2000). While cholinergic stimulation through nicotine has been known to improve cognitive performance (Levin, McClernon, & Rezvani, 2006), *APOE4* shows a larger cognitive benefit from procholinergic nicotinic stimulation through nasal spray in healthy nonsmoking young adults (Marchant, King, Tabet, & Rusted, 2010). These conflicting phenotypes in *APOE4* carriers may be a consequence of the compensation; other factors might excessively rescue neuronal functions that are innately disturbed by apoE4. Another possibility is that apoE4 stimulates synapses, thus improving cognitive performance; however, the prolonged stimulation may irreversibly exhaust neurons during aging and/or promote synaptic activity-regulated Aβ production. These hypotheses should be considered in developing effective apoE-based therapies.

Taken together, apoE isoforms likely contribute to AD pathogenesis through diverse and complex pathways. Although decreased apoE expression seems to be profitable to reduce brain Aβ deposition, lower apoE levels may exacerbate Aβ-independent pathways in AD. Comprehensive studies are desired to determine whether increased brain apoE isoforms are deleterious or beneficial regarding cognitive functions independent of Aβ.

Acknowledgments

This work was supported by NIH grants R01AG035355, R01AG027924, R01AG046205, P01AG030128, and P01NS074969, and an IIRG from the Alzheimer's Association (to G.B.), and an NIRG from the Alzheimer's Association (to T.K.). We thank Mary D. Davis and Caroline T. Stetler for careful readings of this manuscript.

References

Alata, W., Ye, Y., St-Amour, I., Vandal, M., & Calon, F. (2015). Human apolipoprotein E varepsilon4 expression impairs cerebral vascularization and blood-brain barrier function in mice. *Journal of Cerebral Blood Flow and Metabolism, 35*(1), 86–94.

Altmann, A., Tian, L., Henderson, V. W., & Greicius, M. D. (2014). Sex modifies the APOE-related risk of developing Alzheimer disease. *Annals of Neurology, 75*(4), 563–573.

Andrews-Zwilling, Y., Bien-Ly, N., Xu, Q., Li, G., Bernardo, A., Yoon, S. Y., et al. (2010). Apolipoprotein E4 causes age- and Tau-dependent impairment of GABAergic interneurons, leading to learning and memory deficits in mice. *The Journal of Neuroscience, 30*(41), 13707–13717.

Anthopoulos, P. G., Hamodrakas, S. J., & Bagos, P. G. (2010). Apolipoprotein E polymorphisms and type 2 diabetes: A meta-analysis of 30 studies including 5423 cases and 8197 controls. *Molecular Genetics and Metabolism, 100*(3), 283–291.

Arendt, T. (2009). Synaptic degeneration in Alzheimer's disease. *Acta Neuropathologica, 118*(1), 167–179.

Baker, L. D., Cross, D. J., Minoshima, S., Belongia, D., Watson, G. S., & Craft, S. (2011). Insulin resistance and Alzheimer-like reductions in regional cerebral glucose metabolism for cognitively normal adults with prediabetes or early type 2 diabetes. *Archives of Neurology, 68*(1), 51–57.

Bell, R. D., Winkler, E. A., Singh, I., Sagare, A. P., Deane, R., Wu, Z., et al. (2012). Apolipoprotein E controls cerebrovascular integrity via cyclophilin A. *Nature, 485*(7399), 512–516.

Bennet, A. M., Di Angelantonio, E., Ye, Z., Wensley, F., Dahlin, A., Ahlbom, A., et al. (2007). Association of apolipoprotein E genotypes with lipid levels and coronary risk. *JAMA, 298*(11), 1300–1311.

Bero, A. W., Yan, P., Roh, J. H., Cirrito, J. R., Stewart, F. R., Raichle, M. E., et al. (2011). Neuronal activity regulates the regional vulnerability to amyloid-beta deposition. *Nature Neuroscience, 14*(6), 750–756.

Biessels, G. J., Staekenborg, S., Brunner, E., Brayne, C., & Scheltens, P. (2006). Risk of dementia in diabetes mellitus: A systematic review. *Lancet Neurology, 5*(1), 64–74.

Brecht, W. J., Harris, F. M., Chang, S., Tesseur, I., Yu, G. Q., Xu, Q., et al. (2004). Neuron-specific apolipoprotein e4 proteolysis is associated with increased tau phosphorylation in brains of transgenic mice. *The Journal of Neuroscience, 24*(10), 2527–2534.

Bu, G. (2009). Apolipoprotein E and its receptors in Alzheimer's disease: Pathways, pathogenesis and therapy. *Nature Reviews Neuroscience, 10*(5), 333–344.

Burns, C. M., Chen, K., Kaszniak, A. W., Lee, W., Alexander, G. E., Bandy, D., et al. (2013). Higher serum glucose levels are associated with cerebral hypometabolism in Alzheimer regions. *Neurology, 80*(17), 1557–1564.

Caselli, R. J., Dueck, A. C., Osborne, D., Sabbagh, M. N., Connor, D. J., Ahern, G. L., et al. (2009). Longitudinal modeling of age-related memory decline and the APOE epsilon4 effect. *The New England Journal of Medicine, 361*(3), 255–263.

Chang, S., ran Ma, T., Miranda, R. D., Balestra, M. E., Mahley, R. W., & Huang, Y. (2005). Lipid- and receptor-binding regions of apolipoprotein E4 fragments act in concert to cause mitochondrial dysfunction and neurotoxicity. *Proceedings of the National Academy of Sciences of the United States of America, 102*(51), 18694–18699.

Chen, Y., Durakoglugil, M. S., Xian, X., & Herz, J. (2010). ApoE4 reduces glutamate receptor function and synaptic plasticity by selectively impairing ApoE receptor recycling. *Proceedings of the National Academy of Sciences of the United States of America, 107*(26), 12011–12016.

Cirrito, J. R., Kang, J. E., Lee, J., Stewart, F. R., Verges, D. K., Silverio, L. M., et al. (2008). Endocytosis is required for synaptic activity-dependent release of amyloid-beta *in vivo*. *Neuron, 58*(1), 42–51.

Cohen, A. D., & Klunk, W. E. (2014). Early detection of Alzheimer's disease using PiB and FDG PET. *Neurobiology of Disease, 72 Pt A*, 117–122.

Conejero-Goldberg, C., Gomar, J. J., Bobes-Bascaran, T., Hyde, T. M., Kleinman, J. E., Herman, M. M., et al. (2014). APOE2 enhances neuroprotection against Alzheimer's disease through multiple molecular mechanisms. *Molecular Psychiatry, 19*(11), 1243–1250.

Corder, E. H., Saunders, A. M., Strittmatter, W. J., Schmechel, D. E., Gaskell, P. C., Small, G. W., et al. (1993). Gene dose of apolipoprotein E type 4 allele and the risk of Alzheimer's disease in late onset families. *Science, 261*(5123), 921–923.

Craft, S., Baker, L. D., Montine, T. J., Minoshima, S., Watson, G. S., Claxton, A., et al. (2012). Intranasal insulin therapy for Alzheimer disease and amnestic mild cognitive impairment: A pilot clinical trial. *Archives of Neurology, 69*(1), 29–38.

Cramer, P. E., Cirrito, J. R., Wesson, D. W., Lee, C. Y., Karlo, J. C., Zinn, A. E., et al. (2012). ApoE-directed therapeutics rapidly clear beta-amyloid and reverse deficits in AD mouse models. *Science, 335*(6075), 1503–1506.

Cudaback, E., Li, X., Montine, K. S., Montine, T. J., & Keene, C. D. (2011). Apolipoprotein E isoform-dependent microglia migration. *FASEB Journal, 25*(6), 2082–2091.

Das, H. K., McPherson, J., Bruns, G. A., Karathanasis, S. K., & Breslow, J. L. (1985). Isolation, characterization, and mapping to chromosome 19 of the human apolipoprotein E gene. *The Journal of Biological Chemistry, 260*(10), 6240–6247.

de la Torre, J. C. (2002). Alzheimer disease as a vascular disorder: Nosological evidence. *Stroke, 33*(4), 1152–1162.

Drenos, F., & Kirkwood, T. B. (2010). Selection on alleles affecting human longevity and late-life disease: The example of apolipoprotein E. *PLoS One, 5*(4), e10022.

Dumanis, S. B., Tesoriero, J. A., Babus, L. W., Nguyen, M. T., Trotter, J. H., Ladu, M. J., et al. (2009). ApoE4 decreases spine density and dendritic complexity in cortical neurons *in vivo*. *The Journal of Neuroscience, 29*(48), 15317–15322.

Elliott, D. A., Tsoi, K., Holinkova, S., Chan, S. L., Kim, W. S., Halliday, G. M., et al. (2011). Isoform-specific proteolysis of apolipoprotein-E in the brain. *Neurobiology of Aging, 32*(2), 257–271.

Filippini, N., Ebmeier, K. P., MacIntosh, B. J., Trachtenberg, A. J., Frisoni, G. B., Wilcock, G. K., et al. (2011). Differential effects of the APOE genotype on brain function across the lifespan. *Neuroimage, 54*(1), 602–610.

Flaherty, D., Lu, Q., Soria, J., & Wood, J. G. (1999). Regulation of tau phosphorylation in microtubule fractions by apolipoprotein E. *Journal of Neuroscience Research, 56*(3), 271–274.

Gale, S. C., Gao, L., Mikacenic, C., Coyle, S. M., Rafaels, N., Murray Dudenkov, T., et al. (2014). APOepsilon4 is associated with enhanced *in vivo* innate immune responses in human subjects. *The Journal of Allergy and Clinical Immunology, 134*(1), 127–134.

Gan, M., Jiang, P., McLean, P., Kanekiyo, T., & Bu, G. (2014). Low-density lipoprotein receptor-related protein 1 (LRP1) regulates the stability and function of GluA1 alpha-amino-3-hydroxy-5-methyl-4-isoxazole propionic acid (AMPA) receptor in neurons. *PLoS One, 9*(12), e113237.

Gemma, C., & Bickford, P. C. (2007). Interleukin-1beta and caspase-1: Players in the regulation of age-related cognitive dysfunction. *Reviews in the Neurosciences, 18*(2), 137–148.

Gibson, G. E., Haroutunian, V., Zhang, H., Park, L. C., Shi, Q., Lesser, M., et al. (2000). Mitochondrial damage in Alzheimer's disease varies with apolipoprotein E genotype. *Annals of Neurology, 48*(3), 297–303.

Gong, J. S., Morita, S. Y., Kobayashi, M., Handa, T., Fujita, S. C., Yanagisawa, K., et al. (2007). Novel action of apolipoprotein E (ApoE): ApoE isoform specifically inhibits lipid-particle-mediated cholesterol release from neurons. *Molecular Neurodegeneration, 2*, 9.

Gorelick, P. B. (2010). Role of inflammation in cognitive impairment: Results of observational epidemiological studies and clinical trials. *Annals of the New York Academy of Sciences, 1207*, 155–162.

II. GENETIC AND ENVIRONMENTAL RISK FACTORS

Grimm, M. O., Zimmer, V. C., Lehmann, J., Grimm, H. S., & Hartmann, T. (2013). The impact of cholesterol, DHA, and sphingolipids on Alzheimer's disease. *BioMed Research International, 2013*, 814390.

Grodstein, F., Skarupski, K. A., Bienias, J. L., Wilson, R. S., Bennett, D. A., & Evans, D. A. (2008). Anti-inflammatory agents and cognitive decline in a bi-racial population. *Neuroepidemiology, 30*(1), 45–50.

Guo, L., LaDu, M. J., & Van Eldik, L. J. (2004). A dual role for apolipoprotein e in neuro-inflammation: Anti- and pro-inflammatory activity. *Journal of Molecular Neuroscience, 23*(3), 205–212.

Halliday, M. R., Pomara, N., Sagare, A. P., Mack, W. J., Frangione, B., & Zlokovic, B. V. (2013). Relationship between cyclophilin a levels and matrix metalloproteinase 9 activity in cerebrospinal fluid of cognitively normal apolipoprotein e4 carriers and blood-brain barrier breakdown. *JAMA Neurology, 70*(9), 1198–1200.

Hardy, J., & Selkoe, D. J. (2002). The amyloid hypothesis of Alzheimer's disease: Progress and problems on the road to therapeutics. *Science, 297*(5580), 353–356.

Harris, F. M., Brecht, W. J., Xu, Q., Mahley, R. W., & Huang, Y. (2004). Increased tau phosphorylation in apolipoprotein E4 transgenic mice is associated with activation of extracellular signal-regulated kinase: Modulation by zinc. *The Journal of Biological Chemistry, 279*(43), 44795–44801.

Harris, F. M., Brecht, W. J., Xu, Q., Tesseur, I., Kekonius, L., Wyss-Coray, T., et al. (2003). Carboxyl-terminal-truncated apolipoprotein E4 causes Alzheimer's disease-like neurodegeneration and behavioral deficits in transgenic mice. *Proceedings of the National Academy of Sciences of the United States of America, 100*(19), 10966–10971.

Hayden, K. M., Zandi, P. P., Khachaturian, A. S., Szekely, C. A., Fotuhi, M., Norton, M. C., et al. (2007). Does NSAID use modify cognitive trajectories in the elderly? The Cache County study. *Neurology, 69*(3), 275–282.

Head, D., Bugg, J. M., Goate, A. M., Fagan, A. M., Mintun, M. A., Benzinger, T., et al. (2012). Exercise engagement as a moderator of the effects of APOE genotype on amyloid deposition. *Archives of Neurology, 69*(5), 636–643.

Helkala, E. L., Koivisto, K., Hanninen, T., Vanhanen, M., Kervinen, K., Kuusisto, J., et al. (1995). The association of apolipoprotein E polymorphism with memory: A population based study. *Neuroscience Letters, 191*(3), 141–144.

Helkala, E. L., Koivisto, K., Hanninen, T., Vanhanen, M., Kervinen, K., Kuusisto, J., et al. (1996). Memory functions in human subjects with different apolipoprotein E phenotypes during a 3-year population-based follow-up study. *Neuroscience Letters, 204*(3), 177–180.

Heneka, M. T., Golenbock, D. T., & Latz, E. (2015). Innate immunity in Alzheimer's disease. *Nature Immunology, 16*(3), 229–236.

Hirao, K., Ohnishi, T., Hirata, Y., Yamashita, F., Mori, T., Moriguchi, Y., et al. (2005). The prediction of rapid conversion to Alzheimer's disease in mild cognitive impairment using regional cerebral blood flow SPECT. *Neuroimage, 28*(4), 1014–1021.

Hoyer, S. (2004). Glucose metabolism and insulin receptor signal transduction in Alzheimer disease. *European Journal of Pharmacology, 490*(1-3), 115–125.

Huang, Y. (2010). Abeta-independent roles of apolipoprotein E4 in the pathogenesis of Alzheimer's disease. *Trends in Molecular Medicine, 16*(6), 287–294.

Huang, Y., & Mahley, R. W. (2014). Apolipoprotein E: Structure and function in lipid metabolism, neurobiology, and Alzheimer's diseases. *Neurobiology of Disease, 72 Pt A*, 3–12.

Huang, Y., & Mucke, L. (2012). Alzheimer mechanisms and therapeutic strategies. *Cell, 148*(6), 1204–1222.

Jack, C. R., Wiste, H. J., Weigand, S. D., Rocca, W. A., Knopman, D. S., Mielke, M. M., et al. (2014). Age-specific population frequencies of cerebral beta-amyloidosis and neurodegeneration among people with normal cognitive function aged 50-89 years: A cross-sectional study. *Lancet Neurology, 13*(10), 997–1005.

Jankowsky, J. L., & Patterson, P. H. (1999). Cytokine and growth factor involvement in long-term potentiation. *Molecular and Cellular Neurosciences, 14*(4-5), 273–286.

Jellinger, K. A. (2008). Morphologic diagnosis of "vascular dementia"—A critical update. *Journal of the Neurological Sciences, 270*(1–2), 1–12.

Ji, Y., Gong, Y., Gan, W., Beach, T., Holtzman, D. M., & Wisniewski, T. (2003). Apolipoprotein E isoform-specific regulation of dendritic spine morphology in apolipoprotein E transgenic mice and Alzheimer's disease patients. *Neuroscience, 122*(2), 305–315.

Johnson, K. A., Jones, K., Holman, B. L., Becker, J. A., Spiers, P. A., Satlin, A., et al. (1998). Preclinical prediction of Alzheimer's disease using SPECT. *Neurology, 50*(6), 1563–1571.

Kanekiyo, T., & Bu, G. (2014). The low-density lipoprotein receptor-related protein 1 and amyloid-beta clearance in Alzheimer's disease. *Frontiers in Aging Neuroscience, 6*, 93.

Kanekiyo, T., Cirrito, J. R., Liu, C. C., Shinohara, M., Li, J., Schuler, D. R., et al. (2013). Neuronal clearance of amyloid-beta by endocytic receptor LRP1. *The Journal of Neuroscience, 33*(49), 19276–19283.

Kanekiyo, T., Liu, C. C., Shinohara, M., Li, J., & Bu, G. (2012). LRP1 in brain vascular smooth muscle cells mediates local clearance of Alzheimer's amyloid-beta. *The Journal of Neuroscience, 32*(46), 16458–16465.

Kanekiyo, T., Xu, H., & Bu, G. (2014). ApoE and Abeta in Alzheimer's disease: Accidental encounters or partners? *Neuron, 81*(4), 740–754.

Kanekiyo, T., Zhang, J., Liu, Q., Liu, C. C., Zhang, L., & Bu, G. (2011). Heparan sulphate proteoglycan and the low-density lipoprotein receptor-related protein 1 constitute major pathways for neuronal amyloid-beta uptake. *The Journal of Neuroscience, 31*(5), 1644–1651.

Kitamura, H. W., Hamanaka, H., Watanabe, M., Wada, K., Yamazaki, C., Fujita, S. C., et al. (2004). Age-dependent enhancement of hippocampal long-term potentiation in knock-in mice expressing human apolipoprotein E4 instead of mouse apolipoprotein E. *Neuroscience Letters, 369*(3), 173–178.

Kleinridders, A., Ferris, H. A., Cai, W., & Kahn, C. R. (2014). Insulin action in brain regulates systemic metabolism and brain function. *Diabetes, 63*(7), 2232–2243.

Knoferle, J., Yoon, S. Y., Walker, D., Leung, L., Gillespie, A. K., Tong, L. M., et al. (2014). Apolipoprotein E4 produced in GABAergic interneurons causes learning and memory deficits in mice. *The Journal of Neuroscience, 34*(42), 14069–14078.

Knopman, D. S., Jack, C. R., Wiste, H. J., Lundt, E. S., Weigand, S. D., Vemuri, P., et al. (2014). 18F-fluorodeoxyglucose positron emission tomography, aging, and apolipoprotein E genotype in cognitively normal persons. *Neurobiology of Aging, 35*(9), 2096–2106.

Kogure, D., Matsuda, H., Ohnishi, T., Asada, T., Uno, M., Kunihiro, T., et al. (2000). Longitudinal evaluation of early Alzheimer's disease using brain perfusion SPECT. *Journal of Nuclear Medicine, 41*(7), 1155–1162.

Korwek, K. M., Trotter, J. H., Ladu, M. J., Sullivan, P. M., & Weeber, E. J. (2009). ApoE isoform-dependent changes in hippocampal synaptic function. *Molecular Neurodegeneration, 4*, 21.

Kuhel, D. G., Konaniah, E. S., Basford, J. E., McVey, C., Goodin, C. T., Chatterjee, T. K., et al. (2013). Apolipoprotein E2 accentuates postprandial inflammation and diet-induced obesity to promote hyperinsulinemia in mice. *Diabetes, 62*(2), 382–391.

Lane-Donovan, C., Philips, G. T., & Herz, J. (2014). More than cholesterol transporters: Lipoprotein receptors in CNS function and neurodegeneration. *Neuron, 83*(4), 771–787.

Laurin, D., Verreault, R., Lindsay, J., MacPherson, K., & Rockwood, K. (2001). Physical activity and risk of cognitive impairment and dementia in elderly persons. *Archives of Neurology, 58*(3), 498–504.

Levin, E. D., McClernon, F. J., & Rezvani, A. H. (2006). Nicotinic effects on cognitive function: Behavioral characterization, pharmacological specification, and anatomic localization. *Psychopharmacology, 184*(3–4), 523–539.

Li, G., Bien-Ly, N., Andrews-Zwilling, Y., Xu, Q., Bernardo, A., Ring, K., et al. (2009). GABAergic interneuron dysfunction impairs hippocampal neurogenesis in adult apolipoprotein E4 knockin mice. *Cell Stem Cell, 5*(6), 634–645.

Liang, W. S., Reiman, E. M., Valla, J., Dunckley, T., Beach, T. G., Grover, A., et al. (2008). Alzheimer's disease is associated with reduced expression of energy metabolism genes in posterior cingulate neurons. *Proceedings of the National Academy of Sciences of the United States of America, 105*(11), 4441–4446.

Lim, A. S., Yu, L., Kowgier, M., Schneider, J. A., Buchman, A. S., & Bennett, D. A. (2013). Modification of the relationship of the apolipoprotein E epsilon4 allele to the risk of Alzheimer disease and neurofibrillary tangle density by sleep. *JAMA Neurology, 70*(12), 1544–1551.

Lima, T. A., Adler, A. L., Minett, T., Matthews, F. E., Brayne, C., & Marioni, R. E. (2014). C-reactive protein, APOE genotype and longitudinal cognitive change in an older population. *Age and Ageing, 43*(2), 289–292.

Liu, C. C., Kanekiyo, T., Xu, H., & Bu, G. (2013). Apolipoprotein E and Alzheimer disease: Risk, mechanisms and therapy. *Nature Reviews Neurology, 9*(2), 106–118.

Liu, L., & Chan, C. (2014). The role of inflammasome in Alzheimer's disease. *Ageing Research Reviews, 15*, 6–15.

Liu, Q., Trotter, J., Zhang, J., Peters, M. M., Cheng, H., Bao, J., et al. (2010). Neuronal LRP1 knockout in adult mice leads to impaired brain lipid metabolism and progressive, age-dependent synapse loss and neurodegeneration. *The Journal of Neuroscience, 30*(50), 17068–17078.

Love, S., Siew, L. K., Dawbarn, D., Wilcock, G. K., Ben-Shlomo, Y., & Allen, S. J. (2006). Premorbid effects of APOE on synaptic proteins in human temporal neocortex. *Neurobiology of Aging, 27*(6), 797–803.

Lynch, J. R., Tang, W., Wang, H., Vitek, M. P., Bennett, E. R., Sullivan, P. M., et al. (2003). APOE genotype and an ApoE-mimetic peptide modify the systemic and central nervous system inflammatory response. *The Journal of Biological Chemistry, 278*(49), 48529–48533.

Maezawa, I., Maeda, N., Montine, T. J., & Montine, K. S. (2006). Apolipoprotein E-specific innate immune response in astrocytes from targeted replacement mice. *Journal of Neuroinflammation, 3*, 10.

Marchant, N. L., King, S. L., Tabet, N., & Rusted, J. M. (2010). Positive effects of cholinergic stimulation favor young APOE epsilon4 carriers. *Neuropsychopharmacology, 35*(5), 1090–1096.

Marz, W., Scharnagl, H., Hoffmann, M. M., Boehm, B. O., & Winkelmann, B. R. (2004). The apolipoprotein E polymorphism is associated with circulating C-reactive protein (the Ludwigshafen risk and cardiovascular health study). *European Heart Journal, 25*(23), 2109–2119.

Mauch, D. H., Nagler, K., Schumacher, S., Goritz, C., Muller, E. C., Otto, A., et al. (2001). CNS synaptogenesis promoted by glia-derived cholesterol. *Science, 294*(5545), 1354–1357.

McEwen, B. S., & Reagan, L. P. (2004). Glucose transporter expression in the central nervous system: Relationship to synaptic function. *European Journal of Pharmacology, 490*(1-3), 13–24.

Mondadori, C. R., de Quervain, D. J., Buchmann, A., Mustovic, H., Wollmer, M. A., Schmidt, C. F., et al. (2007). Better memory and neural efficiency in young apolipoprotein E epsilon4 carriers. *Cerebral Cortex, 17*(8), 1934–1947.

Montagne, A., Barnes, S. R., Sweeney, M. D., Halliday, M. R., Sagare, A. P., Zhao, Z., et al. (2015). Blood-brain barrier breakdown in the aging human hippocampus. *Neuron, 85*(2), 296–302.

Moorhouse, P., & Rockwood, K. (2008). Vascular cognitive impairment: Current concepts and clinical developments. *Lancet Neurology, 7*(3), 246–255.

Mosconi, L., Berti, V., Glodzik, L., Pupi, A., De Santi, S., & de Leon, M. J. (2010). Pre-clinical detection of Alzheimer's disease using FDG-PET, with or without amyloid imaging. *Journal of Alzheimer's Disease, 20*(3), 843–854.

Mosconi, L., Mistur, R., Switalski, R., Tsui, W. H., Glodzik, L., Li, Y., et al. (2009). FDG-PET changes in brain glucose metabolism from normal cognition to pathologically verified Alzheimer's disease. *European Journal of Nuclear Medicine and Molecular Imaging, 36*(5), 811–822.

Nagele, R. G., Wegiel, J., Venkataraman, V., Imaki, H., & Wang, K. C. (2004). Contribution of glial cells to the development of amyloid plaques in Alzheimer's disease. *Neurobiology of Aging, 25*(5), 663–674.

Nakamura, T., Watanabe, A., Fujino, T., Hosono, T., & Michikawa, M. (2009). Apolipoprotein E4 (1-272) fragment is associated with mitochondrial proteins and affects mitochondrial function in neuronal cells. *Molecular Neurodegeneration, 4*, 35.

Nichol, K., Deeny, S. P., Seif, J., Camaclang, K., & Cotman, C. W. (2009). Exercise improves cognition and hippocampal plasticity in APOE epsilon4 mice. *Alzheimer's & Dementia, 5*(4), 287–294.

Nishitsuji, K., Hosono, T., Nakamura, T., Bu, G., & Michikawa, M. (2011). Apolipoprotein E regulates the integrity of tight junctions in an isoform-dependent manner in an *in vitro* blood-brain barrier model. *The Journal of Biological Chemistry, 286*(20), 17536–17542.

Oikawa, N., Hatsuta, H., Murayama, S., Suzuki, A., & Yanagisawa, K. (2014). Influence of APOE genotype and the presence of Alzheimer's pathology on synaptic membrane lipids of human brains. *Journal of Neuroscience Research, 92*(5), 641–650.

Ong, Q. R., Chan, E. S., Lim, M. L., Cole, G. M., & Wong, B. S. (2014). Reduced phosphorylation of brain insulin receptor substrate and Akt proteins in apolipoprotein-E4 targeted replacement mice. *Scientific Reports, 4*, 3754.

Ostergaard, L., Aamand, R., Gutierrez-Jimenez, E., Ho, Y. C., Blicher, J. U., Madsen, S. M., et al. (2013). The capillary dysfunction hypothesis of Alzheimer's disease. *Neurobiology of Aging, 34*(4), 1018–1031.

Overk, C. R., & Masliah, E. (2014). Pathogenesis of synaptic degeneration in Alzheimer's disease and Lewy body disease. *Biochemical Pharmacology, 88*(4), 508–516.

Ownby, R. L. (2010). Neuroinflammation and cognitive aging. *Current Psychiatry Reports, 12*(1), 39–45.

Peila, R., Rodriguez, B. L., & Launer, L. J. (2002). Type 2 diabetes, APOE gene, and the risk for dementia and related pathologies: The Honolulu-Asia aging study. *Diabetes, 51*(4), 1256–1262.

Praetorius, M., Thorvaldsson, V., Hassing, L. B., & Johansson, B. (2013). Substantial effects of apolipoprotein E epsilon4 on memory decline in very old age: Longitudinal findings from a population-based sample. *Neurobiology of Aging, 34*(12), 2734–2739.

Raber, J., Wong, D., Buttini, M., Orth, M., Bellosta, S., Pitas, R. E., et al. (1998). Isoform-specific effects of human apolipoprotein E on brain function revealed in ApoE knockout mice: Increased susceptibility of females. *Proceedings of the National Academy of Sciences of the United States of America, 95*(18), 10914–10919.

Rall, S. C., Weisgraber, K. H., & Mahley, R. W. (1982). Human apolipoprotein E. The complete amino acid sequence. *The Journal of Biological Chemistry, 257*(8), 4171–4178.

Reger, M. A., Watson, G. S., Green, P. S., Baker, L. D., Cholerton, B., Fishel, M. A., et al. (2008). Intranasal insulin administration dose-dependently modulates verbal memory and plasma amyloid-beta in memory-impaired older adults. *Journal of Alzheimer's Disease, 13*(3), 323–331.

Reger, M. A., Watson, G. S., Green, P. S., Wilkinson, C. W., Baker, L. D., Cholerton, B., et al. (2008). Intranasal insulin improves cognition and modulates beta-amyloid in early AD. *Neurology, 70*(6), 440–448.

Reinvang, I., Winjevoll, I. L., Rootwelt, H., & Espeseth, T. (2010). Working memory deficits in healthy APOE epsilon 4 carriers. *Neuropsychologia, 48*(2), 566–573.

Riddell, D. R., Zhou, H., Comery, T. A., Kouranova, E., Lo, C. F., Warwick, H. K., et al. (2007). The LXR agonist TO901317 selectively lowers hippocampal Abeta42 and improves memory in the Tg2576 mouse model of Alzheimer's disease. *Molecular and Cellular Neurosciences, 34*(4), 621–628.

Rockwood, K., Wentzel, C., Hachinski, V., Hogan, D. B., MacKnight, C., & McDowell, I. (2000). Prevalence and outcomes of vascular cognitive impairment. Vascular cognitive impairment investigators of the canadian study of health and aging. *Neurology, 54*(2), 447–451.

Ruitenberg, A., den Heijer, T., Bakker, S. L., van Swieten, J. C., Koudstaal, P. J., Hofman, A., et al. (2005). Cerebral hypoperfusion and clinical onset of dementia: The Rotterdam Study. *Annals of Neurology, 57*(6), 789–794.

Sagare, A. P., Bell, R. D., Zhao, Z., Ma, Q., Winkler, E. A., Ramanathan, A., et al. (2013). Pericyte loss influences Alzheimer-like neurodegeneration in mice. *Nature Communications, 4,* 2932.

Saido, T., & Leissring, M. A. (2012). Proteolytic degradation of amyloid beta-protein. *Cold Spring Harbor Perspectives in Medicine, 2*(6), a006379.

Salloway, S., Sperling, R., Fox, N. C., Blennow, K., Klunk, W., Raskind, M., et al. (2014). Two phase 3 trials of bapineuzumab in mild-to-moderate Alzheimer's disease. *The New England Journal of Medicine, 370*(4), 322–333.

Scarmeas, N., Habeck, C. G., Stern, Y., & Anderson, K. E. (2003). APOE genotype and cerebral blood flow in healthy young individuals. *JAMA, 290*(12), 1581–1582.

Schachter, F., Faure-Delanef, L., Guenot, F., Rouger, H., Froguel, P., Lesueur-Ginot, L., et al. (1994). Genetic associations with human longevity at the APOE and ACE loci. *Nature Genetics, 6*(1), 29–32.

Schiepers, O. J., Harris, S. E., Gow, A. J., Pattie, A., Brett, C. E., Starr, J. M., et al. (2012). APOE E4 status predicts age-related cognitive decline in the ninth decade: Longitudinal follow-up of the lothian birth cohort 1921. *Molecular Psychiatry, 17*(3), 315–324.

Shiiki, T., Ohtsuki, S., Kurihara, A., Naganuma, H., Nishimura, K., Tachikawa, M., et al. (2004). Brain insulin impairs amyloid-beta(1-40) clearance from the brain. *The Journal of Neuroscience, 24*(43), 9632–9637.

Steen, E., Terry, B. M., Rivera, E. J., Cannon, J. L., Neely, T. R., Tavares, R., et al. (2005). Impaired insulin and insulin-like growth factor expression and signaling mechanisms in Alzheimer's disease—Is this type 3 diabetes? *Journal of Alzheimer's Disease, 7*(1), 63–80.

Strittmatter, W. J., Saunders, A. M., Goedert, M., Weisgraber, K. H., Dong, L. M., Jakes, R., et al. (1994). Isoform-specific interactions of apolipoprotein E with microtubule-associated protein tau: Implications for Alzheimer disease. *Proceedings of the National Academy of Sciences of the United States of America, 91*(23), 11183–11186.

Szekely, C. A., Breitner, J. C., Fitzpatrick, A. L., Rea, T. D., Psaty, B. M., Kuller, L. H., et al. (2008). NSAID use and dementia risk in the cardiovascular health study: Role of APOE and NSAID type. *Neurology, 70*(1), 17–24.

Talbot, K., Wang, H. Y., Kazi, H., Han, L. Y., Bakshi, K. P., Stucky, A., et al. (2012). Demonstrated brain insulin resistance in Alzheimer's disease patients is associated with IGF-1 resistance, IRS-1 dysregulation, and cognitive decline. *The Journal of Clinical Investigation, 122*(4), 1316–1338.

Terwel, D., Steffensen, K. R., Verghese, P. B., Kummer, M. P., Gustafsson, J. A., Holtzman, D. M., et al. (2011). Critical role of astroglial apolipoprotein E and liver X receptor-alpha expression for microglial Abeta phagocytosis. *The Journal of Neuroscience, 31*(19), 7049–7059.

Tesseur, I., Van Dorpe, J., Spittaels, K., Van den Haute, C., Moechars, D., & Van Leuven, F. (2000). Expression of human apolipoprotein E4 in neurons causes hyperphosphorylation of protein tau in the brains of transgenic mice. *The American Journal of Pathology, 156*(3), 951–964.

Thambisetty, M., Beason-Held, L., An, Y., Kraut, M. A., & Resnick, S. M. (2010). APOE epsilon4 genotype and longitudinal changes in cerebral blood flow in normal aging. *Archives of Neurology, 67*(1), 93–98.

Trommer, B. L., Shah, C., Yun, S. H., Gamkrelidze, G., Pasternak, E. S., Ye, G. L., et al. (2004). ApoE isoform affects LTP in human targeted replacement mice. *Neuroreport, 15*(17), 2655–2658.

Valla, J., Berndt, J. D., & Gonzalez-Lima, F. (2001). Energy hypometabolism in posterior cingulate cortex of Alzheimer's patients: Superficial laminar cytochrome oxidase associated with disease duration. *The Journal of Neuroscience, 21*(13), 4923–4930.

Valla, J., Yaari, R., Wolf, A. B., Kusne, Y., Beach, T. G., Roher, A. E., et al. (2010). Reduced posterior cingulate mitochondrial activity in expired young adult carriers of the APOE epsilon4 allele, the major late-onset alzheimer's susceptibility gene. *Journal of Alzheimer's Disease, 22*(1), 307–313.

Vanmierlo, T., Rutten, K., Dederen, J., Bloks, V. W., van Vark-van der Zee, L. C., Kuipers, F., et al. (2011). Liver X receptor activation restores memory in aged AD mice without reducing amyloid. *Neurobiology of Aging, 32*(7), 1262–1272.

Vemuri, P., Lesnick, T. G., Przybelski, S. A., Knopman, D. S., Preboske, G. M., Kantarci, K., et al. (2015). Vascular and amyloid pathologies are independent predictors of cognitive decline in normal elderly. *Brain, 138*(Pt 3), 761–771.

Vemuri, P., Lesnick, T. G., Przybelski, S. A., Machulda, M., Knopman, D. S., Mielke, M. M., et al. (2014). Association of lifetime intellectual enrichment with cognitive decline in the older population. *JAMA Neurology, 71*(8), 1017–1024.

Verges, D. K., Restivo, J. L., Goebel, W. D., Holtzman, D. M., & Cirrito, J. R. (2011). Opposing synaptic regulation of amyloid-beta metabolism by NMDA receptors *in vivo*. *The Journal of Neuroscience, 31*(31), 11328–11337.

Vlad, S. C., Miller, D. R., Kowall, N. W., & Felson, D. T. (2008). Protective effects of NSAIDs on the development of Alzheimer disease. *Neurology, 70*(19), 1672–1677.

Wang, C., Wilson, W. A., Moore, S. D., Mace, B. E., Maeda, N., Schmechel, D. E., et al. (2005). Human apoE4-targeted replacement mice display synaptic deficits in the absence of neuropathology. *Neurobiology of Disease, 18*(2), 390–398.

Weisgraber, K. H., Rall, S. C., & Mahley, R. W. (1981). Human E apoprotein heterogeneity. Cysteine-arginine interchanges in the amino acid sequence of the apo-E isoforms. *The Journal of Biological Chemistry, 256*(17), 9077–9083.

Wierenga, C. E., Clark, L. R., Dev, S. I., Shin, D. D., Jurick, S. M., Rissman, R. A., et al. (2013). Interaction of age and APOE genotype on cerebral blood flow at rest. *Journal of Alzheimer's Disease, 34*(4), 921–935.

Wilson, R. S., Bienias, J. L., Berry-Kravis, E., Evans, D. A., & Bennett, D. A. (2002). The apolipoprotein E epsilon 2 allele and decline in episodic memory. *Journal of Neurology, Neurosurgery, and Psychiatry, 73*(6), 672–677.

Wisdom, N. M., Callahan, J. L., & Hawkins, K. A. (2011). The effects of apolipoprotein E on non-impaired cognitive functioning: A meta-analysis. *Neurobiology of Aging, 32*(1), 63–74.

Xu, Q., Bernardo, A., Walker, D., Kanegawa, T., Mahley, R. W., & Huang, Y. (2006). Profile and regulation of apolipoprotein E (ApoE) expression in the CNS in mice with targeting of green fluorescent protein gene to the ApoE locus. *The Journal of Neuroscience, 26*(19), 4985–4994.

Xu, W. L., Qiu, C. X., Wahlin, A., Winblad, B., & Fratiglioni, L. (2004). Diabetes mellitus and risk of dementia in the Kungsholmen project: A 6-year follow-up study. *Neurology, 63*(7), 1181–1186.

Yu, Y. W., Lin, C. H., Chen, S. P., Hong, C. J., & Tsai, S. J. (2000). Intelligence and event-related potentials for young female human volunteer apolipoprotein E epsilon4 and non-epsilon4 carriers. *Neuroscience Letters, 294*(3), 179–181.

Zehnder, A. E., Blasi, S., Berres, M., Monsch, A. U., Stahelin, H. B., & Spiegel, R. (2009). Impact of APOE status on cognitive maintenance in healthy elderly persons. *International Journal of Geriatric Psychiatry, 24*(2), 132–141.

Zhang, F., & Jiang, L. (2015). Neuroinflammation in Alzheimer's disease. *Neuropsychiatric Disease and Treatment, 11*, 243–256.

Zhu, Y., Kodvawala, A., & Hui, D. Y. (2010). Apolipoprotein E inhibits Toll-like receptor (TLR)-3- and TLR-4-mediated macrophage activation through distinct mechanisms. *The Biochemical Journal, 428*(1), 47–54.

Zhu, Y., Nwabuisi-Heath, E., Dumanis, S. B., Tai, L. M., Yu, C., Rebeck, G. W., et al. (2012). APOE genotype alters glial activation and loss of synaptic markers in mice. *Glia, 60*(4), 559–569.

Zlokovic, B. V. (2011). Neurovascular pathways to neurodegeneration in Alzheimer's disease and other disorders. *Nature Reviews Neuroscience, 12*(12), 723–738.

Zlokovic, B. V. (2013). Cerebrovascular effects of apolipoprotein e: Implications for Alzheimer disease. *JAMA Neurology, 70*(4), 440–444.

Lifestyle and Alzheimer's Disease: The Role of Environmental Factors in Disease Development

Nancy Bartolotti and Orly Lazarov

Department of Anatomy and Cell Biology, College of Medicine, University of Illinois at Chicago, Chicago, IL, United States

O U T L I N E

Introduction	198
Epidemiological Studies	201
Cognitive Reserve and Education	*201*
Brain Reserve	*204*
Cognitive Reserve	*205*
The Benefits of Cognitive Complexity Following the Onset of Dementia	207
Physical Activity and Exercise	209
Chemical Exposure and AD Risk	211
Metals	212
Air Pollution and Tobacco Smoke	213
Nutrition and the Microbiome	214
Nutritional Elements that Decrease AD Risk	*214*
Nutrition and Increased Risk for AD	*216*
Microbiome	*217*

Genes, Environment and Alzheimer's Disease.
DOI: http://dx.doi.org/10.1016/B978-0-12-802851-3.00007-3

197

Sleep and Circadian Rhythm	218
Socialization	219
Conclusion	219
References	220

INTRODUCTION

Environmental factors profoundly influence brain structure and function. Lifestyle elements such as nutrition, physical activity, exposure to chemicals, social interaction, and participation in cognitively demanding activities significantly impact the structure and function of the brain. It is now abundantly clear that this effect is profound to the extent of modulating susceptibility to or inducing neurodegenerative disease. The role of the environment is particularly significant in neurodegenerative diseases that are not clearly caused by a single genetic alteration, such as Late Onset Alzheimer's disease (LOAD). The primary risk factor for LOAD is advanced age. Age-dependent lifestyle changes often include reduced mobility, fewer opportunities for socializing, infrequent participation in cognitively stimulating activities, and poor nutrition. Here we examine evidence that lifestyle and environmental factors affect the risk for the development of AD.

In the early 2000s, several pioneering studies provided the first experimental evidence that environmental factors are disease modulators. These experiments modified the living environment of mouse models of Alzheimer's disease (Ambree et al., 2006; Billings, Green, McGaugh, & LaFerla, 2007; Costa et al., 2007; Herring et al., 2009; Hu et al., 2010; Jankowsky et al., 2005; Lazarov et al., 2005; Levi, Jongen-Relo, Feldon, Roses, & Michaelson, 2003; Nichol, Deeny, Seif, Camaclang, & Cotman, 2009; Nichol et al., 2008; Parachikova, Nichol, & Cotman, 2008; Russo-Neustadt, Beard, & Cotman, 1999; Wolf et al., 2006). Experiments included a paradigm known as environmental enrichment (EE, Figure 7.1), learning tasks, or exercising paradigms. EE typically consists of a larger living space than the standard cage, social interactions, novel objects and nesting materials, as well as running wheels for physical exercise (Figure 7.1). This paradigm enhances many aspects of plasticity in the brain of wild-type mice, as well as improves learning and memory (for recent reviews, see Hirase & Shinohara, 2014; Simpson & Kelly, 2011). Experience of mice expressing familial Alzheimer's disease linked mutant variants (FAD mice) in EE resulted in reduced neuropathology and enhanced neuroplasticity, including enhanced and rescued hippocampal neurogenesis and long term potentiation (LTP), reduced oligomeric $A\beta$ and tau hyperphosphorylation, along with upregulation of critical components of anterograde

FIGURE 7.1　The enriched environment experimental paradigm. In the enriched environment, mice are housed in a large cage with novel stimuli such as tunnels and huts. Running wheels are provided for physical activity. Exposure to an enriched environment such as the one pictured here has been shown to enhance brain plasticity and cognitive function in rodents.

axonal transport (for example, see Hu et al., 2010). Specifically, several studies reported decreased amyloid pathology following EE (Ambree et al., 2006; Berchtold, Chinn, Chou, Kesslak, & Cotman, 2005; Billings et al., 2007; Costa et al., 2007; Arne Herring et al., 2011; Hu et al., 2010; Lazarov et al., 2005; Mirochnic, Wolf, Staufenbiel, & Kempermann, 2009; Nichol et al., 2009; Nichol et al., 2008; Parachikova et al., 2008; Perreau, Adlard, Anderson, & Cotman, 2005; Russo-Neustadt et al., 1999).

Importantly, mice that did not use the environment (e.g., did not run on the running wheel) showed no change in levels of pathology in their brains (Lazarov et al., 2005). In the APPswe/PS1ΔE9 mouse model of AD, EE was shown not only to decrease amyloid-related pathology, but also to increase levels of neprilysin, a protein involved in the degradation of Aβ (Lazarov et al., 2005). In support of that, aging-dependent decline in neprilysin activity (Hellstrom-Lindahl, Ravid, & Nordberg, 2008; Wang, Iwata, Hama, Saido, & Dickson, 2003) can be mitigated by experience of aging mice in EE (Mainardi et al., 2014). Importantly, improvements in learning and memory were observed in FAD mice following EE, even when levels of amyloid deposition were found to be upregulated (Jankowsky et al., 2005), indicating that the protective effect of EE surpasses the detrimental effect of amyloid deposition.

Taken together, these experiments suggest that a complex experience that involves multiple stimuli, such as exercise, novelty, and exploration, dramatically affects brain plasticity and mitigates cognitive dysfunction in mouse models of FAD. Later experiments have unraveled, at least in part, the signaling pathways underlying EE (Berchtold et al., 2005; Billings et al., 2007; Hu, Long, Pigino, Brady, & Lazarov, 2013; Lazarov et al., 2005; Perreau et al., 2005). Since then, numerous studies have described the effect of environmental components on the development of AD in rodents and humans. We describe and consider the evidence that supports or defers the link between lifestyle and environmental factors and the development of cognitive deficits and AD, and discuss potential mechanisms underlying these effects (for summary see Table 7.1).

TABLE 7.1   Summary of Discussed Environmental Factors and Their Influence on AD Risk

Environmental factor	Potential mechanism for influencing AD risk ⬇ Decreases AD risk ⬆ Increases AD risk	
	Humans	*Rodents*
Cognitive stimulation (Humans: reading, writing, puzzle solving, education, bilingualism. Rodents: environmental enrichment)	⬇ Increased brain volume ⬇ Increased neuronal density ⬇ Increased hippocampal activation ⬇ Enhanced connectivity (Cognitive Reserve) ⬇ Reduced AD pathology, particularly amyloid pathology	⬇ Enhanced neurogenesis ⬇ Increased neprilysin activity (increased Aβ clearance) ⬇ Reduced GSK3-β activity (reduced tau hyperphosphorylation) ⬇ Enhanced plasticity ⬇ Enhanced LTP ⬇ Reduced AD pathology (Aβ and phosphorylated tau)
Physical activity	⬇ Increased hippocampal activity and volume	⬇ Increased hippocampal activity and volume ⬇ Increased neurotrophins ⬇ Reduced myelin degeneration ⬇ Reduced AD pathology ⬇ Reduced oxidative stress ⬇ Increased neurogenesis ⬆ Increased amyloid pathology
Pesticides (DDT, chlorpyrifos, rotenone)	⬆ Extent of risk and mechanism unclear	⬆ Increased APP ⬆ Increased Aβ ⬆ Increased reactive oxygen species ⬆ Degeneration of cholinergic neurons
Metals (copper, lead, and aluminum are increased in AD, selenium and manganese are decreased in AD, zinc is unclear)	⬇ Zinc associated with decreased plaque density	⬆ Copper decreases neprilysin activity (reduced Aβ clearance) ⬇ Zinc reduces copper absorption ⬆ Zinc increases APP cleavage and Aβ deposition ⬆ Alterations in metal homeostasis may result in a more oxidizing environment in the brain

(Continued)

TABLE 7.1 (Continued)

Environmental factor	Potential mechanism for influencing AD risk ↓ Decreases AD risk ↑ Increases AD risk	
	Humans	*Rodents*
Pollution	↑ Increased inflammation ↑ Increased Aβ and tau pathology ↑ Increased neurodegeneration ↑ Increased blood-brain barrier damage	↑ Increased AD pathology
Good Nutrition (high in antioxidants, fruits, vegetables, cocoa, vitamins, folate, caffeine)	↓ Reduced oxidative stress via antioxidants ↓ Reduced AD pathology	↓ Reduced oxidative stress via antioxidants ↓ Increased neurotrophins ↓ Reduced AD pathology ↓ Increased neprilysin activity ↓ Reduced GSK3-β activity (reduced tau hyperphosphorylation) ↓ Increased acetylcholine ↓ Reduced activity of acetylcholinesterase ↓ Altered presenilin activity ↓ Promoting microbiome health
Bad Nutrition (high in animal products, fat, cholesterol, sugar)	↑ Increased diabetes risk ↑ Decreased vascular health	↑ Exacerbate Aβ and tau pathology ↑ Reduced cholinergic neurons ↑ Disrupting microbiome health
Sleep	See Chapter 10	See Chapter 10
Socialization	↓ Cognitive reserve?	

Potential ways by which an environmental factor can affect the development of the disease or the risk—effect on the genetics—APP, PS1, effect on oxidative stress, mitochondria, blood circulation/elimination of toxins, oxygen level, upregulation of neurotrophic factors, upregulation of neurogenesis, synaptic components, etc.

EPIDEMIOLOGICAL STUDIES

Cognitive Reserve and Education

Compelling evidence for the contribution of environmental factors to AD risk comes from studies like the ones performed in American religious orders (Bennett, 2006). In these studies, the cognitive ability of hundreds of individuals was tested regularly throughout life, and the brains examined following death. Religious orders are a particularly informative observational population since the lifestyles of the members are similarly regulated, thus facilitating a more focused analysis of the effect of a single environmental variable. One factor that was repeatedly found to be associated with risk of AD was the extent of cognitive complexity experienced during life (Bennett, Schneider, Wilson, Bienias, & Arnold, 2005b). For example, the density of ideas in writing samples of nuns during early and midlife was associated with a reduced risk for AD, superior cognitive functioning, and decreased levels of pathological markers for AD (Mortimer, 2012; Riley, Snowdon, Desrosiers, & Markesbery, 2005; Snowdon et al., 1996). The extent of idea density in writing samples may indicate a lifetime of cognitive complexity, in part resulting from higher levels of education (Mitzner & Kemper, 2003). Indeed, in one study of nuns, lower education level was correlated with earlier onset of cognitive impairment (Tyas, Snowdon, Desrosiers, Riley, & Markesbery, 2007). Similarly, a recent meta analysis of the literature on education and risk for AD determined that higher education reduced the risk of incidence of AD (Meng & D'Arcy, 2012). Interestingly, even low levels of education appear to exert a protective effect when compared to no formal education (Farfel et al., 2013). Other studies also describe the inverse correlation between education level and severity of cognitive impairments in spite of the presence of AD-related pathology (Bennett, Schneider, Tang, Arnold, & Wilson, 2006; Bennett et al., 2003; Roe, Xiong, Miller, & Morris, 2007).

In addition to education, other environmental factors can offer cognitive complexity during life and confer a benefit to the aging brain. For example, many studies show that lifetime bilinguals have a delayed average onset for AD (Akbaraly et al., 2009; Alladi et al., 2013; Freedman et al., 2014; Rovio et al., 2005; Wilson, Scherr, Schneider, Tang, & Bennett, 2007). For this reason, bilingualism is often now considered an environmental factor (Gold, 2014). Participation in other cognitively demanding activities such as reading, writing, and playing an instrument or games has also been associated with a reduced risk for AD (Sattler, Toro, Schönknecht, & Schröder, 2012; Wilson, Boyle, Yang, James, & Bennett, 2015; Wilson et al., 2002). This observation is still manifest when preclinical depression, which might result in a decreased interest in participating in leisure activities, is accounted for (Verghese et al., 2003). Like religious orders, studies of twins aid in the identification of important environmental factors, since the extent of genetic and environmental variability is smaller. In studies of twins, the twin with greater involvement in leisure activities, or with a more cognitively complex job, had a lower risk for AD (Andel et al., 2005; Crowe et al., 2003; Potter, Helms, Burke, Steffens, & Plassman, 2007). Similarly, in a monozygotic twin study, the twin with greater involvement in cognitive leisure activities had a reduced risk for AD, particularly when the twins were carriers for APOE4, a genetic risk factor for AD (Carlson et al., 2008). These studies offer compelling evidence that participation in a cognitively complex environment can modulate AD risk independent of genetic contributions to that risk.

Little is known about the mechanism of protection conferred by a lifetime of cognitive activity. Deciphering this mechanism in the human brain is particularly challenging, simply because of the limited ways to detect and measure cellular processes in live individuals. One way to measure the status of AD pathology is by examining the levels of Aβ and tau in the cerebral spinal fluid (CSF), which is thought to be reflective of Aβ and tau pathology in the brain (Wang et al., 2014). Longitudinal studies have shown that greater education results in a slower decline in CSF $A\beta_{42}$ (declining $A\beta_{42}$ in the CSF is indicative of worsening brain pathology), and in postmortem analysis individuals with higher levels of education had lower levels of Aβ in the brain compared to their less educated, but still cognitively intact age-matched counterparts (Lo & Jagust, 2013; Yasuno et al., 2014). In addition, higher levels of self-reported cognitive activity (such as reading and writing) in early life was associated with reduced levels of Pittsburgh Compound B (PiB), a radioactive tracer for amyloid plaques, in later life (Landau et al., 2012). These studies offer evidence that a lifetime of cognitively complex activities may reduce the occurrence of amyloid pathology.

Similarly, in a postmortem analysis, level of education was found to temper the effects of amyloidosis, meaning that individuals with higher education had greater cognitive function when matched to an individual with similar levels of amyloid (Bennett, Schneider, Wilson, Bienias, & Arnold, 2005a). A recent meta analysis of the literature on education and risk for AD determined that the AD pathology was greater than the cognitive performance during life would have otherwise suggested (Meng & D'Arcy, 2012). As in the higher-educated brain, the pathological state of the bilingual brain observed postmortem is often far more severe than the symptoms would have suggested during life, strengthening the argument that an active brain can enhance cognitive functioning even in the presence of AD pathology (Schweizer, Ware, Fischer, Craik, & Bialystok, 2012). Finally, other cognitively stimulating activities performed regularly during life such as reading and playing games have also been shown to be associated with enhanced cognitive function even in the presence of AD pathology in a longitudinal study (Negash et al., 2013).

One interpretation of these findings is that a lifetime of cognitive complexity builds up a "reserve" that allows the brain to better cope with the insults of pathology, requiring a more advanced pathological state before cognitive decline becomes apparent. This reserve may exist in two forms, brain reserve or cognitive reserve. Brain reserve typically refers to greater physical content of the brain, which means that the individual starts with more brain matter before decay begins, so more decay needs to happen before cognitive symptoms are observed. Cognitive reserve usually refers to increased efficiency or alternate neural pathways that allow for intact cognitive functioning in spite of advanced decay. In this section we will examine the evidence for the protective effects of brain and cognitive reserve on AD risk.

Brain Reserve

Brain reserve hypothetically decreases risk for AD by increasing the quantity of brain tissue, which could mean that a greater extent of degeneration would need to occur before symptoms are manifested (Steffener & Stern, 2012). This could potentially be achieved by greater brain volume or high neuronal density. Greater brain volume has previously been associated with a reduction of AD symptoms (Mori et al., 1997). Individuals with higher education levels have greater gray matter volumes of areas that are typically affected by AD, such as the entorhinal cortex, suggesting that these areas are built up during complex experiences like education and are therefore more resistant to degeneration after onset of AD pathology (Serra et al., 2011). The hippocampus is one of the primary brain structures involved in learning and memory, and lower hippocampal volume has been associated with impaired recall (Mortimer, Gosche, Riley, Markesbery, & Snowdon, 2004). Exercises requiring periods of sustained learning have been associated with persisting changes in gray matter, particularly in the hippocampus (Draganski et al., 2006). Further, individuals who report a lifetime of greater cognitive complexity also have a reduced rate of hippocampal degeneration during aging (Valenzuela, Sachdev, Wen, Chen, & Brodaty, 2008). Older individuals who remain cognitively intact exhibit hippocampal activation similar to that of young individuals, indicating that preserved hippocampal functioning may mean the difference between normal aging and dementia (Persson, Kalpouzos, Nilsson, Ryberg, & Nyberg, 2011). Therefore, a lifetime of cognitive complexity may protect against cognitive decline by preventing hippocampal degeneration and thus preserving function.

Brain reserve can also be theoretically achieved through increasing neuronal or gray matter density. Learning a second language is associated with increased gray matter density (Mechelli et al., 2004). Similarly, participating in musical training is also known to increase gray matter density in areas of the brain important for memory and higher-level cognitive processes (James et al., 2014). These results are meaningful because a higher neuronal density has been linked to a slower cognitive decline, so activities that enhance neuronal density may slow the progression of Alzheimer's symptoms (Wilson, et al., 2013). White matter (myelin) also degenerates during aging, but cognitively active individuals such as lifetime bilinguals maintain their white matter for a longer time (Luk, Bialystok, Craik, & Grady, 2011). Cognitive training has also been shown to increase myelination, even in older adults, resulting in enhanced connectivity (Engvig et al., 2012; Lövdén et al., 2010; Takeuchi et al., 2010). Additional support for this result comes from experiments in middle age rats, in which EE can increase myelination and improve cognitive performance (Qiu et al., 2012). These studies provide evidence that cognitive complexity can increase the amount and density of brain matter, potentially requiring a greater amount of damage before reaching a point where symptoms of dementia are apparent.

Cognitive Reserve

While cognitive complexity may build up a structural brain reserve, individuals who have experienced a lifetime of cognitive complexity sometimes show extensive brain decay at time of death and yet are still able to remain cognitively intact, suggesting that their brains were somehow better able to deal with the degeneration (Iacono et al., 2009). These observations led to the formulation of the cognitive reserve hypothesis (Stern, 2002). In this hypothesis, a lifetime of cognitive complexity builds up a reserve of alternate neural connections and greater efficiency in brain processing, acting as an architectural and metabolic buffer for the brain, making it less affected by neurodegeneration (Foubert-Samier et al., 2012; Yaakov Stern, 2009). Cognitive reserve may account for why older individuals with higher levels of education are able to perform better at cognitive tasks in spite of comparable reductions in gray matter (Steffener et al., 2014). Similarly, when white matter begins to degenerate during the early stages of AD, experiences associated with greater cognitive complexity such as education and bilingualism may also protect against this degeneration and allow normal functioning to occur for longer before cognitive decline becomes apparent (Brickman et al., 2011; Gold, Johnson, & Powell, 2013; Molinuevo et al., 2014; Schweizer et al., 2012). Indeed, while higher levels of self-reported cognitive activity (such as reading and writing) in early life were associated with reduced levels of PiB, higher levels of self-reported cognitive activity in late life were not significantly associated with lower levels of PiB, suggesting that late life enrichment may not be able to reverse amyloid pathology, but may still improve cognitive function (Landau et al., 2012). In a recent longitudinal study, higher cognitive activity in youth and higher cognitive activity in old age were both independently associated with greater cognitive functioning in old age, regardless of brain pathology, suggesting again that cognitive activity can neutralize the detrimental effects of AD pathology (Wilson, Boyle, et al., 2013). Evidence for cognitive reserve is also observed in mouse models. In the Tg2576 mouse model of AD, exposure to EE prior to the onset of pathology had a lasting effect into aging, mitigating the effects of AD pathology and slowing cognitive decline (Verret et al., 2013).

It has been previously shown that during AD, alternate pathways are recruited in the brain during a task, compared to cognitively intact age-matched controls, indicating that the brain may be trying to redirect processing around pathology-damaged connections (Stern et al., 2000). Multiple studies have shown that a greater cognitive reserve resulted in a protection of cognitive function, even when the extent of amyloid pathology, measured in the CSF and by PiB, would have otherwise predicted cognitive impairments (Dumurgier et al., 2010; Rentz et al., 2010; Roe et al., 2008; Soldan et al., 2013; Sole-Padulles et al., 2011; Yaffe et al., 2011). An individual who has experienced a lifetime of cognitive complexity may have cultivated a system of alternate connections, and may subsequently be in a better position to make functionally relevant alternate connections when pathology begins to damage the brain. Indeed, an analysis of metabolic brain usage in highly educated pre-AD individuals showed higher levels of activation in certain parts of the brain, compared to their poorly educated cohorts, indicating that they were better able to utilize their brain resources and recruit alternate pathways (Morbelli et al., 2013; Perneczky et al., 2006). It has also been hypothesized that bilingual individuals may make more efficient use of their neural resources, allowing them to function normally for a longer period of time (Guzmán-Vélez & Tranel, 2015). Interestingly, the benefits of bilingualism are greater in low-educated populations compared to high-educated populations, suggesting that reserve is limited, and there may be multiple ways to acquire a cognitive reserve (Gollan, Salmon, Montoya, & Galasko, 2011).

Higher education and greater cognitive complexity can also mitigate the increased risk of AD from carrying the APOE4 allele. The APOE gene is one of the genetic factors associated with an increased risk of LOAD. Of the three human isoforms (APOE2, APOE3, and APOE4) carrying one copy of the APOE4 allele increases the lifetime risk of AD from 10–15% to about 20–30%, while carrying two copies of the APOE4 allele increases risk to 50–60% (Genin et al., 2011). In one study, individuals with high levels of cognitive reserve and carrying the APOE4 allele had a comparable AD risk to individuals without a copy of APOE4 (Ferrari et al., 2014). In addition, APOE4 carriers with a high cognitive reserve had a later onset for AD (Ferrari et al., 2013). Prior to the onset of symptoms, APOE4 carriers with a greater cognitive reserve also had reduced PiB uptake, indicating reduced amyloidosis (Wirth, Villeneuve, La Joie, Marks, & Jagust, 2014). Finally, cognitive reserve resulting from high levels of education has been shown to be just as protective in APOE4 carriers as in non-APOE4 carriers (Garibotto et al., 2012). These results suggest that cultivating a cognitive reserve throughout life may be particularly beneficial for individuals at greater genetic risk for AD.

In addition to slowing the onset of cognitive impairment, the cognitive reserve may also delay the conversion from mild-cognitive impairment

(MCI) into AD dementia. One recent study showed that while the initial AD-related decline began around the same time in individuals with either high or low cognitive reserve, the conversion to dementia was delayed by approximately seven years in the high cognitive reserve group (Amieva et al., 2014). In a postmortem analysis, level of education was found to temper the effects of amyloidosis (greater education, greater cognitive function in spite of amyloidosis), but not the effects of neurofibrillary tangles (Bennett et al., 2005a).

The hyperphosphorylation of tau and the formation of neurofibrillary tangles is another hallmark of AD pathology and likely indicates a later and more severe disease state (Sperling et al., 2011). This may suggest that cognitive reserve can offer protection at earlier stages of pathology (Abner et al., 2011; Soldan et al., 2013; Wilson et al., 2004). In line with this idea, it has been shown that while cognitive reserve confers a protective effect at all stages of amyloid pathology, cognitive reserve is less protective when levels of tau and phosphorylated tau are high in the CSF, a phenomenon that may occur in the more advanced stages of the disease (Soldan et al., 2013). However, even when levels of tau and phosphorylated tau are high in the CSF, the time to conversion to dementia is modified by level of education, indicating that even in an advanced pathological state the cognitive reserve may be slowing disease progression (Roe et al., 2011). Interestingly, evidence from mouse models suggests that EE can modify tau pathology. In the 3xTG-AD mouse model of AD, mice that spent time learning a memory task after the onset of pathology exhibited improved memory and reduced severity amyloid and tau pathology, as well as reduced GSK3-β activity, a kinase that may be responsible for the hyperphosphorylation of tau in AD (Billings et al., 2007). Similarly, in the APPswe/PS1ΔE9 mouse model of AD, EE reduced hyperphosphorylated tau in the hippocampus (Hu et al., 2010). Therefore, even though cognitive reserve may be most effective in preventing the onset of AD or mitigating the effects of the earlier stages of pathology, engaging in cognitively complex experiences may still offer some benefit to individuals with a more severe form of AD by slowing the rate of cognitive decline.

THE BENEFITS OF COGNITIVE COMPLEXITY FOLLOWING THE ONSET OF DEMENTIA

Following the onset of dementia, cognitive complexity may still offer some benefit in slowing decline of cognitive functioning and improving quality of life (Liberati, Raffone, & Olivetti Belardinelli, 2012). Participation in cognitively engaging activities can be particularly effective in slowing the progression of cognitive impairment early after the onset of symptoms (Treiber et al., 2011). A review of clinical trials examining the effect of cognitive stimulation therapy (CST) showed that regular participation in activities such as solving puzzles, playing games, and participating in social activities can improve functioning in people with moderate dementia (Woods, Aguirre, Spector, & Orrell, 2012). CST seems to be particularly beneficial for memory and language-related functioning (Hall, Orrell, Stott, & Spector, 2013). Cognitive training for two months improved memory and increased activation in the hippocampus of persons with MCI (Rosen, Sugiura, Kramer, Whitfield-Gabrieli, & Gabrieli, 2011). CST (via a mneumonic training task) increased activity in the hippocampus of MCI patients during both encoding and retrieval of the memories (Hampstead, Stringer, Stilla, Giddens, & Sathian, 2012). Another review showed that CST could improve memory and general cognition in people with AD (Olazarán et al., 2010). Indeed, CST has been shown to improve cognition, and analysis by fMRI showed changes in the neural circuitry that seem to indicate the AD brain retains some elements of plasticity (Baglio et al., 2015; Belleville et al., 2011). Additional evidence for this plasticity comes from event-related potential (ERP; which measures electrical activity in the brain) studies on individuals with AD who show alterations in ERP response following cognitive training (Spironelli, Bergamaschi, Mondini, Villani, & Angrilli, 2013). These effects seem to be specific to CST, as CST improves memory function even relative to other therapeutic interventions such as occupational therapy (Mapelli, Di Rosa, Nocita, & Sava, 2013). Small clinical trials have shown that CST and involvement in artistic activities stabilized cognition and improved quality of life for both patients and caregivers (Maci et al., 2012; Viola et al., 2011). The benefits of CST appear to be long lasting, and may extend at least as long as 10 years following the therapy (Luttenberger, Hofner, & Graessel, 2012). CST may be particularly beneficial when combined with typical pharmacological interventions used for AD. In combination with a cholinesterase inhibitor, CST resulted in enhanced cognition compared to the drug alone, indicating that cognitive stimulation may be a valuable component of a multiapproach therapy for AD (Matsuda et al., 2010; Onder et al., 2005; Orrell et al., 2014).

CST can be achieved by multiple methods. One possible form of CST is studying a language later in life, which could offer some value as a form of neuroprotection, since this task involves a complex pathway, perhaps more so than other cognitively demanding tasks (Antoniou, Gunasekera, & Wong, 2013). Another way to enhance the effectiveness of CST could be to use music. Patients with AD experience better memory formation when information is sung rather than spoken, indicating a cognitive benefit may exist for musical training (Simmons-Stern, Budson, & Ally, 2010). One challenge for CST for treatment of dementia is that it can be difficult to teach the task. Video game training shows some promise as a means of cognitive stimulation therapy. Research groups are working on making video games intuitive to learn for people suffering from dementia (Boulay, Benveniste, Boespflug, Jouvelot, & Rigaud, 2011). It may even be possible to provide CST of this kind over the Internet, increasing availability and access (Tarraga et al., 2006). In summary, current evidence suggests that access to cognitively stimulating materials such as puzzles, games, and memory exercises may improve quality of life in the already demented (Bharwani, Parikh, Lawhorne, VanVlymen, & Bharwani, 2012).

PHYSICAL ACTIVITY AND EXERCISE

Physical activity has been proposed as one way to potentially modulate AD risk and progression due to the increase in hippocampal volume and activity in humans, and even in older, healthy humans who are more physically active (Niemann, Godde, & Voelcker-Rehage, 2014; Shah et al., 2014; Varma, Chuang, Harris, Tan, & Carlson, 2015). In rodents too, access to a running wheel has profound effects on areas of the brain commonly affected in AD, such as the hippocampus (for recent review, see Hooghiemstra, Eggermont, Scheltens, van der Flier, & Scherder, 2012). In the 3xTG-AD mouse model of AD, mice that had access to a running wheel had decreased pathology and reduced cognitive impairment (Garcia-Mesa et al., 2011). Also in the 3xTg-AD model following the onset of pathology, access to a running wheel rescued deficits in proteins important for synaptic function normally reduced in this mouse model, such as NR2B PSD-95, synaptophysin, GDNF, and SIRT1 (Revilla et al., 2014). Similar results were observed in the Tg2576 mouse model of AD in which access to a running wheel lessened AD-related brain pathology and improved performance in a memory task (Nichol, Parachikova, & Cotman, 2007; Parachikova et al., 2008; Yuede et al., 2009). In the APPswe/PS1ΔE9 mouse model of AD, physical activity also improved spatial memory and reduced AD-related pathology (Tapia-Rojas, Aranguiz, Varela-Nallar, & Inestrosa, 2015). In a mouse model of APOE4 carriers, access to a running wheel improved performance on a memory task to resemble the performance by mice not carrying APOE4 (Nichol et al., 2009). Voluntary running may also ameliorate tau-related pathology, as experiments in the THY-Tau22 mouse model of AD have shown (Belarbi et al., 2011). However, other evidence suggests that exercise may be less effective as a therapy in more advanced stages of tau-related pathology (Ohia-Nwoko, Montazari, Lau, & Eriksen, 2014). The molecular mechanism behind this benefit is still under investigation, but may include reducing oxidative stress, preventing degeneration, increasing neurogenesis, or enhancing neuroprotective factors and signaling pathways involved in learning and memory (Dao, Zagaar, & Alkadhi, 2014; Herring et al., 2010; Mirochnic et al., 2009). Importantly, the molecular pathways that underlie memory improvements in AD may be different from those underlying memory improvements in healthy individuals, and further studies will be required to identify the pathways modulated by physical activity in AD specifically (Rao et al., 2015).

However, some evidence suggests that physical activity may not improve AD-related pathology in mouse models. For example, one study found no beneficial effect of wheel running on memory performance in the TgCRND8 mouse model of AD following the onset of pathology, and even observed a worsening of amyloid pathology in the wheel running group (Richter et al., 2008). Another study showed that initiating treadmill running in APP/PS1 mice after the onset of pathology improved memory, but did not improve brain amyloid pathology (Zhao, Liu, Zhang, & Tong, 2015). Treadmill running in the APP/PS1 mouse model did prevent degeneration of myelin in the hippocampus, offering one potential mechanism for the protective effect observed on memory (Chao et al., 2015). Some of the discrepancy in the literature regarding the effect of physical activity on AD may arise from the use of different mouse models of AD, as well as when in the disease course and for how long the wheels are introduced. However, early data in human trials is also conflicting. Preliminary evidence from a cross-sectional study suggested that greater physical activity may reduce risk for AD (Okonkwo et al., 2014). Another report suggests that greater physical activity results in a lower risk for AD (Nikolaos Scarmeas et al., 2009). One clinical trial showed that greater participation in physical activity reduced risk for MCI or AD (Schlosser Covell et al., 2015). However, other studies do not find a correlation between reduced physical activity and dementia risk (Paganini-Hill, Kawas, & Corrada, 2015). Therefore it is not clear whether in humans the effect of physical activity on cognitive performance is as great as the effect of cognitive leisure activities, such as reading and writing (Lautenschlager et al., 2008; Scarmeas, Levy, Tang, Manly, & Stern, 2001).

Although this is not an easy task to address in mice and its translatability may be questionable, this issue is, to some extent, controversial in the mouse too. Thus, some studies in mouse models of FAD have shown that voluntary wheel running does or does not (Nichol et al., 2007; Nichol et al., 2008; Wolf et al., 2006) improve memory or increase markers of brain plasticity such as neurogenesis in a mouse model of AD, whereas total enrichment does enhance memory and plasticity in the same mouse model of AD (Wolf et al., 2006). Similar results were observed in another study, where physical activity was not enough to improve pathology, but cognitive stimulation in addition to physical activity did improve memory function (Cracchiolo et al., 2007). For this reason, physical activity may be most effective when combined with multiple types of leisure activities (Karp et al., 2006). Indeed, physical activity may enhance the effects of cognitive stimulation therapy (Thiel et al., 2012). A recently published clinical trial showed that targeting physical activity, diet, and cognitive training together could be a promising therapeutic approach for improving cognition in individuals with AD (Ngandu et al., 2015). Other clinical trials investigating the effect of physical activity on AD are currently underway and more work will need to be done to determine the effects of physical activity on AD risk and progression, either alone or in conjunction with other interventions (Hardman, Kennedy, Macpherson, Scholey, & Pipingas, 2015; Yu et al., 2014).

CHEMICAL EXPOSURE AND AD RISK

It is thought that pesticides may have many effects on the brain. Thus, exposure to pesticides may be another environmental factor influencing AD risk (for review see Casida & Durkin, 2013). A challenge when studying the effect of pesticide exposure on the development of AD is that pesticides may cause comorbidities that could perhaps in turn influence the development of AD, or cause death before AD has a chance to develop. Some studies have suggested that exposure to pesticides increases risk for AD (Hayden et al., 2010; Parrón, Requena, Hernández, & Alarcón, 2011; Singh et al., 2013; Tyas et al., 2007). For example, exposure to dichlorodiphenyltrichloroethane (DDT), a ubiquitous mid-twentieth century pesticide, has recently been linked to an increased risk for AD (Richardson et al., 2009). Dichlorodiphenyldichloroethane (DDE; a derivative of DDT) was found to be higher in the serum of individuals with AD, and increased levels of DDE in the serum were correlated with increased levels of DDE in the brain. In addition, DDT concentrations that correlate to levels of what is considered high exposure in humans have been shown to increase the level of APP in cultured neurons, offering a potential mechanistic link to AD risk (Richardson et al., 2014). Although DDT use is infrequent in the United States, it is still used in some countries, and food from these countries is imported to countries not using DDT, making DDT exposure a continuing global concern (Eskenazi et al., 2009). Pesticides other than DDT and DDE are also under investigation. The pesticide chlorpyrifos appears to increase $A\beta$ in the brains of the Tg2576 mouse model of AD (Salazar et al., 2011). Rotenone, a commonly used pesticide and insecticide, may cause degeneration of cholinergic neurons (Ullrich & Humpel, 2009). It has also been suggested that pesticides may be contributing to AD pathology by creating reactive oxygen species in the brain (Leuner et al., 2012). These preliminary experiments indicate that more work should be done to determine the contribution of pesticide exposure to AD risk and disease progression.

Genetic factors may play a critical role in susceptibility to chemically-induced risk for AD and should be considered in future experiments. The presence of the APOE4 allele is thought to worsen the cognitive impairments due to DDE (Richardson et al., 2014). Two other genes, CYP2D6 and GSTP1, have recently been identified to interact with certain pesticide and metal products, suggesting that individuals with these particular polymorphisms may perhaps be at increased risk for pesticide or metal-induced AD (Singh, Banerjee, Bala, Basu, & Chhillar, 2014).

It is important to note that the data for pesticide exposure on AD risk is still in the early stages and the results are occasionally conflicting (Tanner, Goldman, Ross, & Grate, 2014). For example, some studies do not find an association between organochlorine pesticides and increased AD risk, while other studies suggest an association does exist (Medehouenou et al., 2014; Singh et al., 2013; Tanner et al., 2014). Therefore, it is important to continue to research the role of pesticides in AD to determine which compounds increase risk for AD, especially since pesticides may still be an important part of improving health in developing countries. Longitudinal studies should help assess the effects of pesticide exposure, which otherwise may not be immediately apparent (Baldi et al., 2003).

METALS

In addition to pesticides, exposure to metals in the environment may modulate AD risk. Metals seem to be particularly important in regulating amyloid pathology in AD. For example, copper, lead, and aluminum have been shown to be increased in AD brains, while zinc, selenium, and manganese are decreased (Gonzalez-Dominguez, Garcia-Barrera, & Gomez-Ariza, 2014). High levels of copper in the brain have been linked to AD, and increased copper exposure appears to worsen memory function (Pal, Siotto, Prasad, & Squitti, 2015; Yu et al., 2015). Presenilin has been suggested to play a role in copper uptake from the diet, thus dysfunction of presenilin may lead to defective copper uptake (Southon et al., 2013). Other studies have shown that copper interacts with $A\beta$ and disrupts its homeostasis (Hou & Zagorski, 2006; Singh et al., 2013). In addition, copper decreases the activity of neprilysin, which is important for the degradation of $A\beta$ (Li et al., 2010). One means of exposure to high levels of copper may be through the presence of copper in the soil, and thus food. In that regard, higher levels of copper and iron in the soil have been associated with increased AD severity (Shen, Yu, Zhang, Xie, & Jiang, 2014). It has been suggested that a diet with low copper may be one treatment strategy for the prevention of AD (Squitti, Siotto, & Polimanti, 2014).

Interestingly, in one study mice modeling AD were fed zinc and subsequently showed reduced levels of amyloidosis and copper, suggesting that zinc may prevent absorption of copper from the diet (Harris et al., 2014). Some studies suggest that zinc is deficient in AD, and that supplementing zinc into the diet, along with other nutrients, may ameliorate symptoms (Loef, von Stillfried, & Walach, 2012). However, in a mouse line modeling APOE4 carriers, additional zinc in the diet worsened performance on a spatial memory task, indicating that additional research on zinc and AD risk is necessary (Flinn, Bozzelli, Adlard, & Railey, 2014). Indeed, some studies have shown that zinc levels tend to be higher in postmortem AD brain, and a positive correlation may exist between levels of amyloid and zinc (Religa et al., 2006). In addition, zinc has been shown to increase APP cleavage and $A\beta$ deposition in the brain of a mouse model of AD (Wang et al., 2010). In a nun study, serum levels of zinc within the normal range were associated with lower plaque density, suggesting that zinc may be an important modulator of APP (Tully, Snowdon, & Markesbery, 1995). Ultimately, imbalance in metals may lead to a more oxidating environment in the brain, which may in turn exacerbate AD symptoms (Stelmashook et al., 2014). Therefore, it may be most critical to maintain an optimal balance of metals in the brain in AD so that metals are present when needed, but not exceeding an optimal level and resulting in an oxidating environment. More research will need to be done to unravel the role of metals in the development of the disease and to determine the dose-dependent effect of metal exposure.

AIR POLLUTION AND TOBACCO SMOKE

Recently, the effects of air pollution (which consists of particulate matter and ozone) on cognitive decline have received increased attention. Individuals who live in an area with high air pollution have been shown to have higher postmortem levels of inflammatory markers and greater levels of $A\beta_{42}$ (Calderon-Garciduenas et al., 2004). It has also been shown that prolonged exposure to air pollution increases risk of AD (Jung, Lin, & Hwang, 2015). Similarly, individuals who are exposed long-term to high pollution levels experience accelerated cognitive decline (Weuve et al., 2012). Alarmingly, one study showed that children raised in areas with high levels of pollution exhibit abnormalities in brain volume and function (Calderón-Garcidueñas, Torres-Jardón, Kulesza, Park, & D'Angiulli, 2014). For example, in a group of accidental death postmortem analysis, nearly half of children from urban areas exhibited plaque and tangle pathology, compared to none of the children growing up in low pollution areas (Calderón-Garcidueñas et al., 2012). Interestingly, consuming dark cocoa decreased endothelin 1 (which is increased following exposure to pollution) in the hippocampus of urban children, suggesting that the negative effects of pollution may be countered by positive nutritional interventions (Calderón-Garcidueñas et al., 2013). In addition to air pollution, exposure to tobacco smoke may increase risk for AD. Individuals who smoke have an increased risk for AD (Anstey, von Sanden, Salim, & O'Kearney, 2007; Cataldo, Prochaska, &

Glantz, 2010). Even second-hand or "environmental" or "passive" smoke exposure may increase the risk for AD (Chen, 2012). In rats, exposure to air with tobacco smoke appeared to increase the aging of the brain and the expression of AD pathology (Ho et al., 2012). The mechanism for the effect of pollution on AD risk is still under investigation, however there is some evidence that pollution may be interfering with the integrity of the blood–brain barrier (Calderón-Garcidueñas et al., 2015). Studies on air pollution and AD risk are still in the early stages, and more studies are warranted for the determination of the effect of air quality on the risk for AD.

NUTRITION AND THE MICROBIOME

Another way the composition of the environment may impact brain function is through nutrition. Certain nutritional factors may alter the oxidative environment of the brain, while others may alter the composition of the endogenous bacteria in the gut, which can have profound repercussions on learning and memory. In this section we review recent literature concerning the effects of diet and AD risk.

Nutritional Elements that Decrease AD Risk

Much of the work on nutrition and AD risk has focused on the effects of individual nutritional elements. For example, one study showed that fruits such as black currants and bilberries lessen AD pathology in the APP/PS1 mouse model of AD (Vepsäläinen et al., 2013). Similarly, addition of pomegranate extract to the diet of another mouse model of AD improved performance on memory tasks (Subash et al., 2015). In a cell culture study, cocoa powder extract was able to promote brain-derived neurotrophic factor (BDNF) signalling, even in the presence of Aβ (Cimini et al., 2013). An extract from green tea, epigallocatechin gallate (EGCG), has been shown to improve memory in a mouse model of AD (Walker et al., 2015). Long-term treatment with resveratol, a nutritional element from red wine, was found to decrease cognitive impairments and AD pathology in a mouse model of AD (Porquet et al., 2013). Resveratol derivatives have been shown to rescue Abeta-induced impairments in LTP (Wang et al., 2014). Taken together, these studies show that individual nutritional elements have the potential to influence AD risk.

The benefits of the foods described above are typically attributed to their antioxidant properties, and indeed, nutrients that act as antioxidants, such as vitamin E, beta carotene, and vitamin C can lower the risk of AD (Li, Shen, & Ji, 2012). However, some nutritional elements may be directly interacting with pathological components. EGCG and curcumin have been shown to have antiamyloidogenic properties, including the ability to regulate neprilysin activity, likely resulting in greater clearance of Aβ (Hyung et al., 2013; Melzig & Janka, 2003; Wang et al., 2014). Other compounds such as resveratrol may modulate tau pathology by regulating the activity of GSK3-β, an enzyme that can lead to the hyperphosphroylation of tau (Varamini, Sikalidis, & Bradford, 2014). Nutritional elements from apples may also directly interact with the molecular components of AD (Hyson, 2011). Apple juice concentrate has been shown to increase availability of acetylcholine, a neurotransmitter reduced in AD (Chan, Graves, & Shea, 2006). Nutrition can also modulate the activity of presenilin, a protein that is mutated in some forms of familial, early-onset AD. While deficits in folate and vitamin E enhance presenilin activity and increase levels of Aβ, apple juice can reduce this overactivity (Chan & Shea, 2006, 2007, 2009). Interestingly, in a religious order study, those with higher levels of folate in the blood were more likely to be cognitively intact in spite of the presence of AD pathology (Snowdon, Tully, Smith, Riley, & Markesbery, 2000; Wang et al., 2012). Therefore, increasing folate through apple consumption may enhance cognitive function by regulating presenilin activity and rescuing amyloid-induced degeneration. Another nutritional element, caffeine, has also been shown to decrease cognitive impairments and Abeta levels in mouse models of AD (Arendash et al., 2006; Chu et al., 2012; Han, Jia, Li, Yang, & Min, 2013; Laurent et al., 2014). Additional benefits of caffeine include increased absorption of zinc and inhibition of acetylcholinesterase, an enzyme that may be increased in AD and responsible for causing the reduction in acetylcholine (Chang & Ho, 2014; Pohanka & Dobes, 2013).

Although studying individual foods and nutritional elements can be illuminating, it is likely more relevant to study patterns of eating (Eskelinen, Ngandu, Tuomilehto, Soininen, & Kivipelto, 2011). Studies on the Mediterranean diet have indicated that following a diet high in fruits, vegetables, extra virgin olive oil, and low in meat products and sugar are beneficial in lowering the risk of AD and slowing the rate of cognitive decline in humans and in mouse models of AD (Grossi et al., 2013, 2014; Gu, Nieves, Stern, Luchsinger, & Scarmeas, 2010; Lourida et al., 2013; Ozawa et al., 2013; Shah, 2013; Singh et al., 2014; Vassallo & Scerri, 2013). In twin studies, a nonsignificant trend toward decreased risk of AD was observed in the twin with a diet higher in fruits and vegetables (Gustaw-Rothenberg, 2009; Hughes et al., 2010). These observations have led to the development of multinutrient interventions in mouse models of AD, which have been shown to improve learning and memory and decrease AD-related pathology (Jansen et al., 2013; Jansen et al., 2014; van Wijk et al., 2014; Wiesmann et al., 2013).

While a diet consisting primarily of fish and vegetables may decrease risk of AD, these lifestyle factors may instead be indicative of an overall healthier lifestyle that reduces risk for AD. Thus, more targeted research should be done on dietary factors to determine whether they are the causative factors behind the observed modulation of AD risk (Barberger-Gateau et al., 2007). That may also help resolve some conflicting evidence concerning some dietary interventions. For example, a meta-analysis of antioxidant consumption found that antioxidants do not delay or prevent AD in humans (Crichton, Bryan, & Murphy, 2013; Polidori & Nelles, 2014). A similar result was found in a meta-analysis of the Mediterranean diet (Otaegui-Arrazola, Amiano, Elbusto, Urdaneta, & Martínez-Lage, 2014). Another study even showed that in a mouse model of AD a diet heavy in fish and vegetables exacerbated memory impairments (Parrott, Winocur, Bazinet, Ma, & Greenwood, 2015). Conflicting evidence also exists concerning the risk of high cholesterol on AD. Two 30-year longitudinal studies reached opposite conclusions regarding the risk of high cholesterol as a risk for AD (Mielke et al., 2010; Solomon, Kivipelto, Wolozin, Zhou, & Whitmer, 2009). Therefore, more work should be done on dietary patterns and risk for AD, particularly in the context of other factors known to modify risk for AD. In the meantime, the most recent recommendations for a diet aimed at minimizing risk for AD include reducing consumption of foods high in saturated fats (particularly from dairy and meat); increasing consumption of vegetables and legumes; limiting consumption of metals such as copper, iron, and aluminum; and incorporating regular aerobic exercise (Barnard et al., 2014; Shea & Remington, 2015).

Nutrition and Increased Risk for AD

Dietary foods and patterns that have been shown to reduce risk for AD may also be doing so by counteracting the effects that a poor diet has on increasing risk for AD. Diets high in cholesterol and fat may be particularly problematic. High-fat diets increase risk for Type 2 diabetes and negatively affect cardiovascular health. Diabetes and poor cardiovascular health are factors that may increase risk for AD and indeed, poor vascular health was associated with more severe cognitive decline in a nun study (Reitz & Mayeux, 2014; Snowdon et al., 1997). Diets high in fat have also been shown to worsen behavior impairments and Aβ pathology in a mouse model of AD (Barron, Rosario, Elteriefi, & Pike, 2013). A diet high in cholesterol has been shown to lead to impaired spatial memory, decreased numbers of cholinergic neurons (a neuronal population particularly vulnerable in AD), and increased $A\beta_{42}$ and phosphorylated tau in the brains of rats and in a mouse model of AD (Ehrlich & Humpel, 2012; Park et al., 2013). Indeed, a high-fat diet also increased GSK3-β activity, which correlated with increases in phosphorylated tau in the hippocampus, offering a mechanism for how high-fat diets may be exacerbating AD pathology and cognitive decline (Bhat & Thirumangalakudi, 2013).

It is particularly important to consider high cholesterol diets in the presence of APOE4, since this gene is involved in lipid metabolism (Lim,

Kowgier, Yu, Buchman, & Bennett, 2013). In a mouse model of an APOE4 carrier, high carbohydrate diets led to increased memory impairments and reductions in BDNF (Maioli et al., 2012). Also, while eating fatty fish may decrease the risk for AD in many individuals, the effect is not as pronounced if the person is a carrier of APOE4 (Huang et al., 2005). Clinical studies on fatty acid supplementation in AD are still in early stages and appear to have some beneficial effects, but it will be important to consider the contribution of the APOE4 allele on such interventions (Faxen-Irving et al., 2013; Shinto et al., 2014). Interestingly, while a high-fat diet increased AD pathology in a mouse model of AD, experience in an enriched environment reversed these effects, nicely demonstrating the interaction between environmental components in modulating AD (Maesako, Uemura, Kubota, Kuzuya, Sasaki, Asada, et al., 2012). Similarly, physical exercise was also able to counter the effects of a high-fat diet, possibly by increasing the degradation of Aβ (Maesako, Uemura, Kubota, Kuzuya, Sasaki, Hayashida, et al., 2012). In addition, caffeine can prevent cognitive impairments in mice fed a high-fat diet and increased BDNF in the hippocampus (Moy & McNay, 2013). These experiments again demonstrate that the environment includes many factors that modulate risk for AD, and positive factors such as exercise and cognitive complexity can counteract negative factors such as poor diet.

Microbiome

Another way nutrients from the diet may be impacting the brain is by interacting with the resident gut microbiota. A recent study showed that a diet high in fat resulted in memory impairments in mice, with associated changes in the microbiome (Jorgensen et al., 2014). Interestingly, caffeine, which as we previously discussed can reverse the effects of a high-fat diet, may interact with certain gut microbes, suggesting that a healthy microbiome may mediate some of the effects of caffeine on the treatment of AD (Chang & Ho, 2014).

Recent studies have shown that a healthy microbiome may be a critical part of cognitive functioning. The composition of the microbiota has been shown to modify learning and memory (Li, Dowd, Scurlock, Acosta-Martinez, & Lyte, 2009). Certain strains of microbiota have been shown to improve learning and memory in mice (Matthews & Jenks, 2013). Interestingly, certain microbiota profiles can be correlated with performance on memory tests (Jorgensen et al., 2014). The microbiome has been shown to modify many aspects of brain signaling including GABA, NMDA, and BDNF (for review see Bhattacharjee & Lukiw, 2013). While more work needs to be done on how the microbiome affects alterations in the brain, it is clear that a healthy microbiome contributes to brain health and reduces risk of brain disorders.

The environment, particularly during early life, can shape the microbiome, through stress, diet, antibiotics, or other measures (O'mahony, Hyland, Dinan, & Cryan, 2011). It has been proposed that individuals who suffer from irritable bowl syndrome, likely resulting from a disruption of optimal microbiota functioning, may be at an increased risk for dementia (Daulatzai, 2014). Gut microbiota may also underlie changes in type 2 diabetes, which in turn predispose for AD (Alam, Alam, Kamal, Abuzenadah, & Haque, 2014). Maintaining the integrity of the microbiome may be particularly critical with aging, since increases in oxidative stress that occur during aging have been shown to alter the composition of the microbiome (Duncan & Flint, 2013; Lynch, Jeffery, Cusack, O'Connor, & O'Toole, 2015; Patrignani, Tacconelli, & Bruno, 2014). These aging-related changes in the microbiome could be due to changes in nutrition during aging, but even in controlled mouse studies the microbiome changes with age, as does the efficiency for responding to nutrients (Langille et al., 2014). Another recent hypothesis suggests that aging may favor the growth of a particular type of microbiota that can enhance the inflammation in the central nervous system, contributing to AD pathology (Shoemark & Allen, 2014).

While more work needs to be done, recent studies offer a promising glimpse at the role of probiotics (Desbonnet et al., 2010). In a mouse model in which the microbiome was disrupted and hippocampal deficits were observed, treatment with probiotics rescued the functioning of the hippocampus (Smith et al., 2014). In another study, treatment with probiotics was able to reverse age-related deficits in LTP and increase expression of BDNF (Distrutti et al., 2014). This study is particularly interesting since enhancing BDNF has been proposed as one way to treat AD (Lu, Nagappan, Guan, Nathan, & Wren, 2013). Another study showed that a particular probiotic could restore memory function, as well as inhibit acetylcholinesterase (Xiao et al., 2014). In short, further study on the composition of the microbiome in aging may allow for therapies involving probiotics (Pérez Martínez, Bäuerl, & Collado, 2014). It is also critical that more research is done on the particular strains of beneficial bacteria, and that the public be made aware of the state of the research on probiotics, since not all probiotics may be helpful and some may even be harmful (Slashinski, McCurdy, Achenbaum, Whitney, & McGuire, 2012).

SLEEP AND CIRCADIAN RHYTHM

In a religious order study, sleep disturbances were linked to an increased risk of AD and cognitive decline (Lim et al., 2013; Lim, et al., 2013). In the general population, sleep disturbances have also been linked to an increased risk of AD (Benedict et al., 2014). However, it is not clear if sleep disturbances are causative of AD, or an early symptom of cognitive dysfunction. For a more detailed discussion of sleep and AD risk, see Chapter 10.

SOCIALIZATION

Social interactions may be another critical environmental element modulating AD risk. One study has shown that low participation in social activities in late life may increase the risk of dementia (Saczynski et al., 2006). Living alone and having fewer social connections may also increase risk (Fratiglioni, Wang, Ericsson, Maytan, & Winblad, 2000). In particular, feelings of loneliness may be the key aspect of low socialization responsible for cognitive decline (Holwerda et al., 2014). In a longitudinal study, individuals who reported feelings of loneliness had double the risk for AD. These individuals also had a faster rate of cognitive decline, in spite of comparable levels of AD pathology (Wilson, Krueger, et al., 2007). Participating in social activities that are considered an integral part of religion, such as regularly attending church, may help to slow cognitive decline, even in AD (Coin et al., 2010; Corsentino, Collins, Sachs-Ericsson, & Blazer, 2009; Hill, Burdette, Angel, Angel, & Series, 2006; Kaufman, Anaki, Binns, & Freedman, 2007; Reyes-Ortiz et al., 2008). One recent longitudinal study showed that attending religious service correlated with a reduced risk for AD (Paganini-Hill et al., 2015). Virtual socialization may also confer a protective effect as one study recently demonstrated that older individuals who regularly use the Internet had better cognitive health (James, Boyle, Yu, & Bennett, 2013). As in the case of cognitive reserve, social connections may help to mitigate the effects of AD pathology (Bennett et al., 2006; Wilson et al., 2005).

CONCLUSION

Here we have discussed the importance of a stimulating, cognitively complex environment in preventing the onset of AD, through the mechanisms of cognitive reserve or through direct interactions with the molecular pathology. We have also considered the contribution of environmental factors such as exposure to pollution and metals on AD risk. From the evidence presented, we can conclude that a cognitively stimulating environment, including higher education, bilingualism, and participation in cognitively challenging leisure activities such as reading, writing, solving puzzles, and playing games can delay the onset of AD. Physical activity may serve as a valuable way to promote brain health and stave off cognitive decline and AD pathology. A diet high in fruits and vegetables and low in meat and sugar may also confer preventative benefits for AD, by improving overall health and by providing important antioxidants and factors that modulate AD pathology. Finally, environmental manipulations like music training, language learning, video game training, and social interactions can improve quality of life and lessen the severity of cognitive symptoms in individuals with mild and moderate AD, and should be considered as a valuable contribution to a treatment plan with pharmaceutical interventions. Considering multiple environmental factors, such as physical and mental activity, as well as occupational hazards and nutrition may lead to more effective preventative and therapeutic strategies for AD.

References

Abner, E. L., Kryscio, R. J., Schmitt, F. A., Santacruz, K. S., Jicha, G. A., Lin, Y., ... Nelson, P. T. (2011). "End-stage" neurofibrillary tangle pathology in preclinical Alzheimer's disease: Fact or fiction? *Journal of Alzheimer's Disease: JAD*, *25*(3), 445–453. http://dx.doi.org/10.3233/JAD-2011-101980.

Akbaraly, T. N., Portet, F., Fustinoni, S., Dartigues, J. F., Artero, S., Rouaud, O., ... Berr, C. (2009). Leisure activities and the risk of dementia in the elderly: Results from the Three-City Study. *Neurology*, *73*(11), 854–861.

Alam, M. Z., Alam, Q., Kamal, M. A., Abuzenadah, A. M., & Haque, A. (2014). A possible link of gut microbiota alteration in type 2 diabetes and Alzheimer's disease pathogenic-ity: An update. *CNS & Neurological Disorders Drug Targets*, *13*(3), 383–390.

Alladi, S., Bak, T. H., Duggirala, V., Surampudi, B., Shailaja, M., Shukla, A. K., ... Kaul, S. (2013). Bilingualism delays age at onset of dementia, independent of education and immigration status. *Neurology*, *81*(22), 1938–1944.

Ambree, O., Leimer, U., Herring, A., Gortz, N., Sachser, N., Heneka, M. T., ... Keyvani, K. (2006). Reduction of amyloid angiopathy and A beta plaque burden after enriched housing in TgCRND8 mice—Involvement of multiple pathways. *American Journal of Pathology*, *169*(2), 544–552. http://dx.doi.org/10.2353/ajpath.2006.051107.

Amieva, H., Mokri, H., Le Goff, M., Meillon, C., Jacqmin-Gadda, H., Foubert-Samier, A., ... Dartigues, J.-F. (2014). Compensatory mechanisms in higher-educated subjects with Alzheimer's disease: A study of 20 years of cognitive decline. *Brain: A Journal of Neurology*, *137*(Pt 4), 1167–1175.

Andel, R., Crowe, M., Pedersen, N. L., Mortimer, J., Crimmins, E., Johansson, B., et al. (2005). Complexity of work and risk of Alzheimer's disease: A population-based study of Swedish twins. *Journal of Gereontology: Psychological Sciences*, *60*(5), 251–258.

Anstey, K. J., von Sanden, C., Salim, A., & O'Kearney, R. (2007). Smoking as a risk factor for dementia and cognitive decline: A meta-analysis of prospective studies. *American Journal of Epidemiology*, *166*(4), 367–378. http://dx.doi.org/10.1093/aje/kwm116.

Antoniou, M., Gunasekera, G. M., & Wong, P. C. M. (2013). Foreign language training as cognitive therapy for age-related cognitive decline: A hypothesis for future research. *Neuroscience and Biobehavioral Reviews*, *37*(10 Pt 2), 2689–2698.

Arendash, G. W., Schleif, W., Rezai-Zadeh, K., Jackson, E. K., Zacharia, L. C., Cracchiolo, J. R., ... Tan, J. (2006). Caffeine protects Alzheimer's mice against cognitive impairment and reduces brain beta-amyloid production. *Neuroscience*, *142*(4), 941–952.

Baglio, F., Griffanti, L., Saibene, F. L., Ricci, C., Alberoni, M., Critelli, R., ... Farina, E. (2015). Multistimulation group therapy in Alzheimer's disease promotes changes in brain functioning. *Neurorehabilitation and Neural Repair*, *29*(1), 13–24.

Baldi, I., Lebailly, P., Mohammed-Brahim, B., Letenneur, L., Dartigues, J. F., & Brochard, P. (2003). Neurodegenerative diseases and exposure to pesticides in the elderly. *American Journal of Epidemiology, 157*(5), 409–414.

Barberger-Gateau, P., Raffaitin, C., Letenneur, L., Berr, C., Tzourio, C., Dartigues, J. F., & Alpérovitch, A. (2007). Dietary patterns and risk of dementia: The Three-City cohort study. *Neurology, 69*(20), 1921–1930.

Barnard, N. D., Bush, A. I., Ceccarelli, A., Cooper, J., de Jager, C. A., Erickson, K. I., & Fraser, G. (2014). Dietary and lifestyle guidelines for the prevention of Alzheimer's disease. *Neurobiology of Aging, 35*(Suppl), S74–S78. 2 SRC – GoogleScholar. http://dx.doi.org/10.1016/j.neurobiolaging.2014.03.033.

Barron, A. M., Rosario, E. R., Elteriefi, R., & Pike, C. J. (2013). Sex-specific effects of high fat diet on indices of metabolic syndrome in 3xTg-AD mice: Implications for Alzheimer's disease. *PLoS One, 8*(10), e78554.

Belarbi, K., Burnouf, S., Fernandez-Gomez, F. J., Laurent, C., Lestavel, S., Figeac, M., … Blum, D. (2011). Beneficial effects of exercise in a transgenic mouse model of Alzheimer's disease-like Tau pathology. *Neurobiology of Disease, 43*(2), 486–494. http://dx.doi.org/10.1016/j.nbd.2011.04.022.

Belleville, S., Clément, F., Mellah, S., Gilbert, B., Fontaine, F., & Gauthier, S. (2011). Training-related brain plasticity in subjects at risk of developing Alzheimer's disease. *Brain: A Journal of Neurology, 134*(Pt 6), 1623–1634.

Benedict, C., Byberg, L., Cedernaes, J., Hogenkamp, P. S., Giedratis, V., Kilander, L., … Schioth, H. B. (2014). Self-reported sleep disturbance is associated with Alzheimer's disease risk in men. *Alzheimer's & Dementia: The Journal of the Alzheimer's Association, 11*(9), 1090–1097.

Bennett, D. A. (2006). Postmortem indices linking risk factors to cognition: Results from the religious order study and the memory and aging project. *Alzheimer Disease and Associated Disorders, 20*(3 Suppl 2), S63–S68.

Bennett, D. A., Schneider, J. A., Tang, Y. X., Arnold, S. E., & Wilson, R. S. (2006). The effect of social networks on the relation between Alzheimer's disease pathology and level of cognitive function in old people: A longitudinal cohort study. *Lancet Neurology, 5*(5), 406–412. http://dx.doi:10.1016/S1474-4422(06)70417-3.

Bennett, D. A., Schneider, J. A., Wilson, R. S., Bienias, J. L., & Arnold, S. E. (2005a). Education modifies the association of amyloid but not tangles with cognitive function. *Neurology, 65*(6), 953–955.

Bennett, D. A., Schneider, J. A., Wilson, R. S., Bienias, J. L., & Arnold, S. E. (2005b). Education modifies the association of amyloid but not tangles with cognitive function. *Neurology, 65*(6), 953–955. http://dx.doi.org/10.1212/01.wnl.0000176286.17192.69.

Bennett, D. A., Wilson, R. S., Schneider, J. A., Evans, D. A., Mendes de Leon, C. F., Arnold, S. E., … Bienias, J. L. (2003). Education modifies the relation of AD pathology to level of cognitive function in older persons. *Neurology, 60*(12), 1909–1915.

Berchtold, N. C., Chinn, G., Chou, M., Kesslak, J. P., & Cotman, C. W. (2005). Exercise primes a molecular memory for brain-derived neurotrophic factor protein induction in the rat hippocampus. *Neuroscience, 133*(3), 853–861. http://dx.doi.org/10.1016/j.neuroscience.2005.03.026.

Bharwani, G., Parikh, P. J., Lawhorne, L. W., VanVlymen, E., & Bharwani, M. (2012). Individualized behavior management program for Alzheimer's/dementia residents using behavior-based ergonomic therapies. *American Journal of Alzheimer's Disease and Other Dementias, 27*(3), 188–195.

Bhat, N. R., & Thirumangalakudi, L. (2013). Increased tau phosphorylation and impaired brain insulin/IGF signaling in mice fed a high fat/high cholesterol diet. *Journal of Alzheimer's Disease: JAD, 36*(4), 781–789.

Bhattacharjee, S., & Lukiw, W. J. (2013). Alzheimer's disease and the microbiome. *Frontiers in Cellular Neuroscience, 7*, 153.

Billings, L. M., Green, K. N., McGaugh, J. L., & LaFerla, F. M. (2007). Learning decreases A beta*56 and tau pathology and ameliorates behavioral decline in 3xTg-AD mice. *The Journal of Neuroscience: The Official Journal of the Society for Neuroscience, 27*(4), 751–761.

Boulay, M., Benveniste, S., Boespflug, S., Jouvelot, P., & Rigaud, A.-S. (2011). A pilot usability study of MINWii, a music therapy game for demented patients. *Technology and Health Care: Official Journal of the European Society for Engineering and Medicine, 19*(4), 233–246.

Brickman, A. M., Siedlecki, K. L., Muraskin, J., Manly, J. J., Luchsinger, J. A., Yeung, L.-K., ... Stern, Y. (2011). White matter hyperintensities and cognition: Testing the reserve hypothesis. *Neurobiology of Aging, 32*(9), 1588–1598.

Calderón-Garcidueñas, L., Kavanaugh, M., Block, M., D'Angiulli, A., Delgado-Chávez, R., Torres-Jardón, R., ... Diaz, P. (2012). Neuroinflammation, hyperphosphorylated tau, diffuse amyloid plaques, and down-regulation of the cellular prion protein in air pollution exposed children and young adults. *Journal of Alzheimer's Disease: JAD, 28*(1), 93–107.

Calderón-Garcidueñas, L., Mora-Tiscareño, A., Franco-Lira, M., Cross, J. V., Engle, R., Aragón-Flores, M., ... D'Angiulli, A. (2013). Flavonol-rich dark cocoa significantly decreases plasma endothelin-1 and improves cognition in urban children. *Frontiers in Pharmacology, 4*, 104.

Calderon-Garciduenas, L., Reed, W., Maronpot, R. R., Henriquez-Roldan, C., Delgado-Chavez, R., Calderon-Garciduenas, A., ... Swenberg, J. A. (2004). Brain inflammation and Alzheimer's-like pathology in individuals exposed to severe air pollution. *Toxicologic Pathology, 32*(6), 650–658. http://dx.doi.org/10.1080/01926230490520232.

Calderón-Garcidueñas, L., Torres-Jardón, R., Kulesza, R. J., Park, S. -B., & D'Angiulli, A. (2014). Air pollution and detrimental effects on children's brain. The need for a multidisciplinary approach to the issue complexity and challenges. *Frontiers in Human Neuroscience, 8*, 613.

Calderón-Garcidueñas, L., Vojdani, A., Blaurock-Busch, E., Busch, Y., Friedle, A., Franco-Lira, M., ... D'Angiulli, A. (2015). Air pollution and children: Neural and tight junction antibodies and combustion metals, the role of barrier breakdown and brain immunity in neurodegeneration. *Journal of Alzheimer's Disease: JAD, 43*(3), 1039–1058.

Carlson, M. C., Helms, M. J., Steffens, D. C., Burke, J. R., Potter, G. G., & Plassman, B. L. (2008). Midlife activity predicts risk of dementia in older male twin pairs. *Alzheimer's & Dementia: The Journal of the Alzheimer's Association, 4*(5), 324–331.

Casida, J. E., & Durkin, K. A. (2013). Neuroactive insecticides: Targets, selectivity, resistance, and secondary effects. *Annual Review of Entomology, 58*, 99–117.

Cataldo, J. K., Prochaska, J. J., & Glantz, S. A. (2010). Cigarette smoking is a risk factor for Alzheimer's disease: An analysis controlling for tobacco industry affiliation. *Journal of Alzheimer's Disease: JAD, 19*(2), 465–480. http://dx.doi.org/10.3233/JAD-2010-1240.

Chan, A., Graves, V., & Shea, T. B. (2006). Apple juice concentrate maintains acetylcholine levels following dietary compromise. *Journal of Alzheimer's Disease: JAD, 9*(3), 287–291.

Chan, A., & Shea, T. B. (2006). Supplementation with apple juice attenuates presenilin-1 overexpression during dietary and genetically-induced oxidative stress. *Journal of Alzheimer's Disease: JAD, 10*(4), 353–358.

Chan, A., & Shea, T. B. (2007). Folate deprivation increases presenilin expression, gammasecretase activity, and Abeta levels in murine brain: Potentiation by ApoE deficiency and alleviation by dietary S-adenosyl methionine. *Journal of Neurochemistry, 102*(3), 753–760.

Chan, A., & Shea, T. B. (2009). Dietary supplementation with apple juice decreases endogenous amyloid-beta levels in murine brain. *Journal of Alzheimer's Disease: JAD, 16*(1), 167–171.

Chang, K. L., & Ho, P. C. (2014). Gas chromatography time-of-flight mass spectrometry (GC-TOF-MS)-based metabolomics for comparison of caffeinated and decaffeinated coffee and its implications for Alzheimer's disease. *PLoS One, 9*(8), e104621.

Chao, F., Zhang, L., Luo, Y., Xiao, Q., Lv, F., He, Q., ... Tang, Y. (2015). Running exercise reduces myelinated fiber loss in the dentate gyrus of the hippocampus in APP/PS1 transgenic mice. *Current Alzheimer Research, 12*(4), 377–383.

Chen, R. (2012). Association of environmental tobacco smoke with dementia and Alzheimer's disease among never smokers. *Alzheimer's & Dementia: The Journal of the Alzheimer's Association*, 8(6), 590–595. http://dx.doi.org/10.1016/j.jalz.2011.09.231.

Chu, Y.-F., Chang, W.-H., Black, R. M., Liu, J.-R., Sompol, P., Chen, Y., … Cheng, I. H. (2012). Crude caffeine reduces memory impairment and amyloid β(1-42) levels in an Alzheimer's mouse model. *Food Chemistry*, 135(3), 2095–2102.

Cimini, A., Gentile, R., D'Angelo, B., Benedetti, E., Cristiano, L., Avantaggiati, M. L., … Desideri, G. (2013). Cocoa powder triggers neuroprotective and preventive effects in a human Alzheimer's disease model by modulating BDNF signaling pathway. *Journal of Cellular Biochemistry*, 114(10), 2209–2220.

Coin, A., Perissinotto, E., Najjar, M., Girardi, A., Inelmen, E. M., Enzi, G., … Sergi, G. (2010). Does religiosity protect against cognitive and behavioral decline in Alzheimer's dementia? *Current Alzheimer Research*, 7(5), 445–452.

Corsentino, E. A., Collins, N., Sachs-Ericsson, N., & Blazer, D. G. (2009). Religious attendance reduces cognitive decline among older women with high levels of depressive symptoms. *The Journals of Gerontology Series A, Biological Sciences and Medical Sciences*, 64(12), 1283–1289.

Costa, D. A., Craechiolo, J. R., Bachstetter, A. D., Hughes, T. F., Bales, K. R., Paul, S. M., … Potter, H. (2007). Enrichment improves cognition in AD mice by amyloid-related and unrelated mechanisms. *Neurobiology of Aging*, 28(6), 831–844. http://dx.doi.org/10.1016/j.neurobiolaging.2006.04.009.

Cracchiolo, J. R., Mori, T., Nazian, S. J., Tan, J., Potter, H., & Arendash, G. W. (2007). Enhanced cognitive activity-over and above social or physical activity is required to protect Alzheimer's mice against cognitive impairment, reduce A beta deposition, and increase synaptic immunoreactivity. *Neurobiology of Learning and Memory*, 88(3), 277–294. http://dx.doi.org/10.1016/J.Nlm.2007.07.007.

Crichton, G. E., Bryan, J., & Murphy, K. J. (2013). Dietary antioxidants, cognitive function and dementia—A systematic review. *Plant Foods for Human Nutrition (Dordrecht, Netherlands)*, 68(3), 279–292.

Crowe, M., Andel, R., Pedersen, N. L., Johansson, B., Gatz, M., & Series, B. (2003). Does participation in leisure activities lead to reduced risk of Alzheimer's disease? A prospective study of Swedish twins. *Journal of Gerontology: Psychological Sciences*, 58(5), 249–255.

Dao, A. T., Zagaar, M. A., & Alkadhi, K. A. (2014). Moderate treadmill exercise protects synaptic plasticity of the dentate gyrus and related signaling cascade in a rat model of Alzheimer's disease. *Molecular Neurobiology*, 52(3), 1067–1076.

Daulatzai, M. A. (2014). Chronic functional bowel syndrome enhances gut-brain axis dysfunction, neuroinflammation, cognitive impairment, and vulnerability to dementia. *Neurochemical Research*, 39(4), 624–644.

Desbonnet, L., Garrett, L., Clarke, G., Kiely, B., Cryan, J. F., & Dinan, T. G. (2010). Effects of the probiotic Bifidobacterium infantis in the maternal separation model of depression. *Neuroscience*, 170(4), 1179–1188.

Distrutti, E., O'Reilly, J.-A., McDonald, C., Cipriani, S., Renga, B., Lynch, M. A., & Fiorucci, S. (2014). Modulation of intestinal microbiota by the probiotic VSL#3 resets brain gene expression and ameliorates the age-related deficit in LTP. *PLoS One*, 9(9), e106503.

Draganski, B., Gaser, C., Kempermann, G., Kuhn, H. G., Winkler, J., Büchel, C., & May, A. (2006). Temporal and spatial dynamics of brain structure changes during extensive learning. *The Journal of Neuroscience: The Official Journal of the Society for Neuroscience*, 26(23), 6314–6317.

Dumurgier, J., Paquet, C., Benisty, S., Kiffel, C., Lidy, C., Mouton-Liger, F., … Hugon, J. (2010). Inverse association between CSF Aβ 42 levels and years of education in mild form of Alzheimer's disease: The cognitive reserve theory. *Neurobiology of Disease*, 40(2), 456–459.

Duncan, S. H., & Flint, H. J. (2013). Probiotics and prebiotics and health in ageing populations. *Maturitas*, 75(1), 44–50.

Ehrlich, D., & Humpel, C. (2012). Chronic vascular risk factors (cholesterol, homocysteine, ethanol) impair spatial memory, decline cholinergic neurons and induce blood-brain barrier leakage in rats in vivo. *Journal of the Neurological Sciences, 322*(1–2), 92–95.

Engvig, A., Fjell, A. M., Westlye, L. T., Moberget, T., Sundseth, Ø., Larsen, V. A., & Walhovd, K. B. (2012). Memory training impacts short-term changes in aging white matter: A longitudinal diffusion tensor imaging study. *Human Brain Mapping, 33*(10), 2390–2406.

Eskelinen, M. H., Ngandu, T., Tuomilehto, J., Soininen, H., & Kivipelto, M. (2011). Midlife healthy-diet index and late-life dementia and Alzheimer's disease. *Dementia and Geriatric Cognitive Disorders Extra, 1*(1), 103–112.

Eskenazi, B., Chevrier, J., Rosas, L. G., Anderson, H. A., Bornman, M. S., Bouwman, H., … Stapleton, D. (2009). The Pine River statement: Human health consequences of DDT use. *Environmental Health Perspectives, 117*(9), 1359–1367.

Farfel, J. M., Nitrini, R., Suemoto, C. K., Grinberg, L. T., Ferretti, R. E. L., Leite, R. E. P., … Brazilian Aging Brain Study Group, (2013). Very low levels of education and cognitive reserve: A clinicopathologic study. *Neurology, 81*(7), 650–657.

Faxen-Irving, G., Freund-Levi, Y., Eriksdotter-Jonhagen, M., Basun, H., Hjorth, E., Palmblad, J., … Wahlund, L. O. (2013). Effects on transthyretin in plasma and cerebrospinal fluid by DHA-rich n—3 fatty acid supplementation in patients with Alzheimer's disease: The OmegAD study. *Journal of Alzheimer's Disease: JAD, 36*(1), 1–6. http://dx.doi.org/10.3233/JAD-121828.

Ferrari, C., Nacmias, B., Bagnoli, S., Piaceri, I., Lombardi, G., Pradella, S., … Sorbi, S. (2014). Imaging and cognitive reserve studies predict dementia in presymptomatic Alzheimer's disease subjects. *Neuro-Degenerative Diseases, 13*(2–3), 157–159. http://dx.doi.org/10.1159/000353690.

Ferrari, C., Xu, W.-L., Wang, H.-X., Winblad, B., Sorbi, S., Qiu, C., & Fratiglioni, L. (2013). How can elderly apolipoprotein E ε4 carriers remain free from dementia? *Neurobiology of Aging, 34*(1), 13–21.

Flinn, J. M., Bozzelli, P. L., Adlard, P. A., & Railey, A. M. (2014). Spatial memory deficits in a mouse model of late-onset Alzheimer's disease are caused by zinc supplementation and correlate with amyloid-beta levels. *Frontiers in Aging Neuroscience, 6*, 174.

Foubert-Samier, A., Catheline, G., Amieva, H., Dilharreguy, B., Helmer, C., Allard, M., & Dartigues, J.-F. (2012). Education, occupation, leisure activities, and brain reserve: A population-based study. *Neurobiology of Aging, 33*(2) 423.e415-425.

Fratiglioni, L., Wang, H. X., Ericsson, K., Maytan, M., & Winblad, B. (2000). Influence of social network on occurrence of dementia: A community-based longitudinal study. *Lancet, 355*(9212), 1315–1319.

Freedman, M., Alladi, S., Chertkow, H., Bialystok, E., Craik, F. I. M., Phillips, N. A., … Bak, T. H. (2014). Delaying onset of dementia: Are two languages enough? *Behavioural Neurology, 2014*, 808137.

Garcia-Mesa, Y., Lopez-Ramos, J. C., Gimenez-Llort, L., Revilla, S., Guerra, R., Gruart, A., … Sanfeliu, C. (2011). Physical exercise protects against Alzheimer's disease in 3xTg-AD mice. *Journal of Alzheimer's Disease: JAD, 24*(3), 421–454. http://dx.doi.org/10.3233/JAD-2011-101635.

Garibotto, V., Borroni, B., Sorbi, S., Cappa, S. F., Padovani, A., & Perani, D. (2012). Education and occupation provide reserve in both ApoE ε4 carrier and noncarrier patients with probable Alzheimer's disease. *Neurological Sciences: Official Journal of the Italian Neurological Society and of the Italian Society of Clinical Neurophysiology, 33*(5), 1037–1042.

Genin, E., Hannequin, D., Wallon, D., Sleegers, K., Hiltunen, M., Combarros, O., … Campion, D. (2011). APOE and Alzheimer disease: A major gene with semi-dominant inheritance. *Molecular Psychiatry, 16*(9), 903–907. http://dx.doi.org/10.1038/mp.2011.52.

Gold, B. T. (2014). Lifelong bilingualism and neural reserve against Alzheimer's disease: A review of findings and potential mechanisms. *Behavioural Brain Research, 281C*, 9–15. http://dx.doi.org/10.1016/j.bbr.2014.12.006.

Gold, B. T., Johnson, N. F., & Powell, D. K. (2013). Lifelong bilingualism contributes to cognitive reserve against white matter integrity declines in aging. *Neuropsychologia, 51*(13), 2841–2846.

Gollan, T. H., Salmon, D. P., Montoya, R. I., & Galasko, D. R. (2011). Degree of bilingualism predicts age of diagnosis of Alzheimer's disease in low-education but not in highly educated Hispanics. *Neuropsychologia, 49*(14), 3826–3830.

Gonzalez-Dominguez, R., Garcia-Barrera, T., & Gomez-Ariza, J. L. (2014). Homeostasis of metals in the progression of Alzheimer's disease. *Biometals, 27*(3), 539–549. http://dx.doi.org/10.1007/s10534-014-9728-5.

Grossi, C., Ed Dami, T., Rigacci, S., Stefani, M., Luccarini, I., & Casamenti, F. (2014). Employing Alzheimer disease animal models for translational research: Focus on dietary components. *Neuro-Degenerative Diseases, 13*(2–3), 131–134. http://dx.doi.org/10.1159/000355461.

Grossi, C., Rigacci, S., Ambrosini, S., Ed Dami, T., Luccarini, I., Traini, C., … Stefani, M. (2013). The polyphenol oleuropein aglycone protects TgCRND8 mice against Ass plaque pathology. *PLoS One, 8*(8), e71702.http://dx.doi.org/10.1371/journal.pone.0071702.

Gu, Y., Nieves, J. W., Stern, Y., Luchsinger, J. A., & Scarmeas, N. (2010). Food combination and Alzheimer disease risk: A protective diet. *Archives of Neurology, 67*(6), 699–706.

Gustaw-Rothenberg, K. (2009). Dietary patterns associated with Alzheimer's disease: Population based study. *International Journal of Environmental Research and Public Health, 6*(4), 1335–1340.

Guzmán-Vélez, E., & Tranel, D. (2015). Does bilingualism contribute to cognitive reserve? Cognitive and neural perspectives. *Neuropsychology, 29*(1), 139–150.

Hall, L., Orrell, M., Stott, J., & Spector, A. (2013). Cognitive stimulation therapy (CST): Neuropsychological mechanisms of change. *International Psychogeriatrics, 25*(3), 479–489.

Hampstead, B. M., Stringer, A. Y., Stilla, R. F., Giddens, M., & Sathian, K. (2012). Mnemonic strategy training partially restores hippocampal activity in patients with mild cognitive impairment. *Hippocampus, 22*(8), 1652–1658.

Han, K., Jia, N., Li, J., Yang, L., & Min, L.-Q. (2013). Chronic caffeine treatment reverses memory impairment and the expression of brain BNDF and TrkB in the PS1/APP double transgenic mouse model of Alzheimer's disease. *Molecular Medicine Reports, 8*(3), 737–740.

Hardman, R. J., Kennedy, G., Macpherson, H., Scholey, A. B., & Pipingas, A. (2015). A randomised controlled trial investigating the effects of Mediterranean diet and aerobic exercise on cognition in cognitively healthy older people living independently within aged care facilities: The Lifestyle Intervention in Independent Living Aged Care (LIILAC) study protocol [ACTRN12614001133628]. *Nutrition Journal, 14*(1), 53. http://dx.doi.org/10.1186/s12937-015-0042-z.

Harris, C. J., Voss, K., Murchison, C., Ralle, M., Frahler, K., Carter, R., … Quinn, J. F. (2014). Oral zinc reduces amyloid burden in Tg2576 mice. *Journal of Alzheimer's Disease: JAD, 41*(1), 179–192.

Hayden, K. M., Norton, M. C., Darcey, D., Ostbye, T., Zandi, P. P., Breitner, J. C. S., … Cache County Study Investigators. (2010). Occupational exposure to pesticides increases the risk of incident AD: The Cache County study. *Neurology, 74*(19), 1524–1530.

Hellstrom-Lindahl, E., Ravid, R., & Nordberg, A. (2008). Age-dependent decline of neprilysin in Alzheimer's disease and normal brain: Inverse correlation with A beta levels. *Neurobiology of Aging, 29*(2), 210–221. http://dx.doi.org/10.1016/j.neurobiolaging.2006.10.010.

Herring, A., Ambree, O., Tomm, M., Habermann, H., Sachser, N., Paulus, W., & Keyvani, K. (2009). Environmental enrichment enhances cellular plasticity in transgenic mice with Alzheimer-like pathology. *Experimental Neurology, 216*(1), 184–192. http://dx.doi.org/10.1016/j.expneurol.2008.11.027.

Herring, A., Blome, M., Ambree, O., Sachser, N., Paulus, W., & Keyvani, K. (2010). Reduction of cerebral oxidative stress following environmental enrichment in mice with

Alzheimer-like pathology. *Brain Pathology (Zurich, Switzerland)*, 20(1), 166–175. http:// dx.doi.org/10.1111/j.1750-3639.2008.00257.x.

Herring, A., Lewejohann, L., Panzer, A.-L., Donath, A., Kröll, O., Sachser, N., ... Keyvani, K. (2011). Preventive and therapeutic types of environmental enrichment counteract beta amyloid pathology by different molecular mechanisms. *Neurobiology of Disease*, 42(3), 530–538.

Hill, T. D., Burdette, A. M., Angel, J. L., Angel, R. J., & Series, B. (2006). Religious attendance and cognitive functioning among older Mexican Americans. *Journal of Gerontology: Psychological Sciences*, 61(1), 3–9.

Hirase, H., & Shinohara, Y. (2014). Transformation of cortical and hippocampal neural circuit by environmental enrichment. *Neuroscience*, 280, 282–298.

Ho, Y. S., Yang, X., Yeung, S. C., Chiu, K., Lau, C. F., Tsang, A. W., ... Chang, R. C. (2012). Cigarette smoking accelerated brain aging and induced pre-Alzheimer-like neuropathology in rats. *PLoS One*, 7(5), e36752. http://dx.doi.org/10.1371/journal.pone.0036752.

Holwerda, T. J., Deeg, D. J. H., Beekman, A. T. F., van Tilburg, T. G., Stek, M. L., Jonker, C., & Schoevers, R. A. (2014). Feelings of loneliness, but not social isolation, predict dementia onset: Results from the Amsterdam Study of the Elderly (AMSTEL). *Journal of Neurology, Neurosurgery, and Psychiatry*, 85(2), 135–142.

Hooghiemstra, A. M., Eggermont, L. H. P., Scheltens, P., van der Flier, W. M., & Scherder, E. J. A. (2012). Exercise and early-onset Alzheimer's disease: Theoretical considerations. *Dementia and Geriatric Cognitive Disorders Extra*, 2, 132–145.

Hou, L., & Zagorski, M. G. (2006). NMR reveals anomalous copper(II) binding to the amyloid A beta peptide of Alzheimer's disease. *Journal of the American Chemical Society*, 128(29), 9260–9261.

Hu, Y. S., Long, N., Pigino, G., Brady, S. T., & Lazarov, O. (2013). Molecular mechanisms of environmental enrichment: Impairments in Akt/GSK3 beta, neurotrophin-3 and CREB signaling. *PLoS One*, 8(5), e64460. http://dx.doi.org/10.1371/journal.pone.0064460.

Hu, Y. -S., Xu, P., Pigino, G., Brady, S. T., Larson, J., & Lazarov, O. (2010). Complex environment experience rescues impaired neurogenesis, enhances synaptic plasticity, and attenuates neuropathology in familial Alzheimer's disease-linked APPswe/PS1DeltaE9 mice. *FASEB Journal: Official Publication of the Federation of American Societies for Experimental Biology*, 24(6), 1667–1681.

Huang, T. L., Zandi, P. P., Tucker, K. L., Fitzpatrick, A. L., Kuller, L. H., Fried, L. P., ... Carlson, M. C. (2005). Benefits of fatty fish on dementia risk are stronger for those without APOE epsilon4. *Neurology*, 65(9), 1409–1414.

Hughes, T. F., Andel, R., Small, B. J., Borenstein, A. R., Mortimer, J. A., Wolk, A., ... Gatz, M. (2010). Midlife fruit and vegetable consumption and risk of dementia in later life in Swedish twins. *The American Journal of Geriatric Psychiatry: Official Journal of the American Association for Geriatric Psychiatry*, 18(5), 413–420.

Hyson, D. A. (2011). A comprehensive review of apples and apple components and their relationship to human health. *Advances in Nutrition (Bethesda, Md.)*, 2(5), 408–420.

Hyung, S.-J., DeToma, A. S., Brender, J. R., Lee, S., Vivekanandan, S., Kochi, A., ... Lim, M. H. (2013). Insights into antiamyloidogenic properties of the green tea extract (-)-epigallocatechin-3-gallate toward metal-associated amyloid-β species. *Proceedings of the National Academy of Sciences of the United States of America*, 110(10), 3743–3748.

Iacono, D., Markesbery, W. R., Gross, M., Pletnikova, O., Rudow, G., Zandi, P., & Troncoso, J. C. (2009). The Nun Study: Clinically silent AD, neuronal hypertrophy, and linguistic skills in early life. *Neurology*, 73(9), 665–673.

James, B. D., Boyle, P. A., Yu, L., & Bennett, D. A. (2013). Internet use and decision making in community-based older adults. *Frontiers in Psychology*, 4, 605.

James, C. E., Oechslin, M. S., Van De Ville, D., Hauert, C.-A., Descloux, C., & Lazeyras, F. (2014). Musical training intensity yields opposite effects on grey matter density in cognitive versus sensorimotor networks. *Brain Structure & Function, 219*(1), 353–366.

Jankowsky, J. L., Melnikova, T., Fadale, D. J., Xu, G. M., Slunt, H. H., Gonzales, V., ... Savonenko, A. V. (2005). Environmental enrichment mitigates cognitive deficits in a mouse model of Alzheimer's disease. *The Journal of Neuroscience: The Official Journal of the Society for Neuroscience, 25*(21), 5217–5224.

Jansen, D., Zerbi, V., Arnoldussen, I. A. C., Wiesmann, M., Rijpma, A., Fang, X. T., ... Kiliaan, A. J. (2013). Effects of specific multi-nutrient enriched diets on cerebral metabolism, cognition and neuropathology in AβPPswe-PS1dE9 mice. *PLoS One, 8*(9), e75393.

Jansen, D., Zerbi, V., Janssen, C. I. F., van Rooij, D., Zinnhardt, B., Dederen, P. J., ... Kiliaan, A. J. (2014). Impact of a multi-nutrient diet on cognition, brain metabolism, hemodynamics, and plasticity in apoE4 carrier and apoE knockout mice. *Brain Structure & Function, 219*(5), 1841–1868.

Jorgensen, B. P., Hansen, J. T., Krych, L., Larsen, C., Klein, A. B., Nielsen, D. S., ... Sorensen, D. B. (2014). A possible link between food and mood: Dietary impact on gut microbiota and behavior in BALB/c mice. *PLoS One, 9*(8), e103398. http://dx.doi.org/10.1371/journal.pone.0103398.

Jung, C.-R., Lin, Y.-T., & Hwang, B.-F. (2015). Ozone, particulate matter, and newly diagnosed Alzheimer's disease: A population-based cohort study in Taiwan. *Journal of Alzheimer's Disease: JAD, 44*(2), 573–584.

Karp, A., Paillard-Borg, S., Wang, H.-X., Silverstein, M., Winblad, B., & Fratiglioni, L. (2006). Mental, physical and social components in leisure activities equally contribute to decrease dementia risk. *Dementia and Geriatric Cognitive Disorders, 21*(2), 65–73.

Kaufman, Y., Anaki, D., Binns, M., & Freedman, M. (2007). Cognitive decline in Alzheimer disease: Impact of spirituality, religiosity, and QOL. *Neurology, 68*(18), 1509–1514.

Landau, S. M., Marks, S. M., Mormino, E. C., Rabinovici, G. D., Oh, H., O'Neil, J. P., ... Jagust, W. J. (2012). Association of lifetime cognitive engagement and low β-amyloid deposition. *Archives of Neurology, 69*(5), 623–629.

Langille, M. G., Meehan, C. J., Koenig, J. E., Dhanani, A. S., Rose, R. A., Howlett, S. E., & Beiko, R. G. (2014). Microbial shifts in the aging mouse gut. *Microbiome, 2*(1), 50.

Laurent, C., Eddarkaoui, S., Derisbourg, M., Leboucher, A., Demeyer, D., Carrier, S., ... Blum, D. (2014). Beneficial effects of caffeine in a transgenic model of Alzheimer's disease-like tau pathology. *Neurobiology of Aging, 35*(7), 2079–2090.

Lautenschlager, N. T., Cox, K. L., Flicker, L., Foster, J. K., van Bockxmeer, F. M., Xiao, J., ... Almeida, O. P. (2008). Effect of physical activity on cognitive function in older adults at risk for Alzheimer disease: A randomized trial. *JAMA, 300*(9), 1027–1037.

Lazarov, O., Robinson, J., Tang, Y. P., Hairston, I. S., Korade-Mirnics, Z., Lee, V. M., ... Sisodia, S. S. (2005). Environmental enrichment reduces Abeta levels and amyloid deposition in transgenic mice. *Cell, 120*(5), 701–713. http://dx.doi.org/10.1016/j.cell.2005.01.015.

Leuner, K., Schütt, T., Kurz, C., Eckert, S. H., Schiller, C., Occhipinti, A., ... Müller, W. E. (2012). Mitochondrion-derived reactive oxygen species lead to enhanced amyloid beta formation. *Antioxidants & Redox Signaling, 16*(12), 1421–1433.

Levi, O., Jongen-Relo, A. L., Feldon, J., Roses, A. D., & Michaelson, D. M. (2003). ApoE4 impairs hippocampal plasticity isoform-specifically and blocks the environmental stimulation of synaptogenesis and memory. *Neurobiology of Disease, 13*(3), 273–282.

Li, F.-J., Shen, L., & Ji, H.-F. (2012). Dietary intakes of vitamin E, vitamin C, and β-carotene and risk of Alzheimer's disease: A meta-analysis. *Journal of Alzheimer's disease: JAD, 31*(2), 253–258.

Li, M., Sun, M., Liu, Y., Yu, J., Yang, H., Fan, D., & Chui, D. (2010). Copper downregulates neprilysin activity through modulation of neprilysin degradation. *Journal of Alzheimer's Disease: JAD, 19*(1), 161–169.

Li, W., Dowd, S. E., Scurlock, B., Acosta-Martinez, V., & Lyte, M. (2009). Memory and learning behavior in mice is temporally associated with diet-induced alterations in gut bacteria. *Physiology & Behavior, 96*(4–5), 557–567.

Liberati, G., Raffone, A., & Olivetti Belardinelli, M. (2012). Cognitive reserve and its implications for rehabilitation and Alzheimer's disease. *Cognitive Processing, 13*(1), 1–12.

Lim, A. S. P., Kowgier, M., Yu, L., Buchman, A. S., & Bennett, D. A. (2013). Sleep fragmentation and the risk of incident Alzheimer's disease and cognitive decline in older persons. *Sleep, 36*(7), 1027–1032.

Lim, A. S. P., Yu, L., Kowgier, M., Schneider, J. A., Buchman, A. S., & Bennett, D. A. (2013). Modification of the relationship of the apolipoprotein E ε4 allele to the risk of Alzheimer disease and neurofibrillary tangle density by sleep. *JAMA Neurology, 70*(12), 1544–1551.

Lim, W. L. F., Lam, S. M., Shui, G., Mondal, A., Ong, D., Duan, X., … Martins, R. N. (2013). Effects of a high-fat, high-cholesterol diet on brain lipid profiles in apolipoprotein E ε3 and ε4 knock-in mice. *Neurobiology of Aging, 34*(9), 2217–2224.

Lo, R. Y., & Jagust, W. J. (2013). Effect of cognitive reserve markers on Alzheimer pathologic progression. *Alzheimer Disease and Associated Disorders, 27*(4), 343–350. http://dx.doi.org/10.1097/WAD.0b013e3182900b2b.

Loef, M., von Stillfried, N., & Walach, H. (2012). Zinc diet and Alzheimer's disease: A systematic review. *Nutritional Neuroscience, 15*(5), 2–12.

Lourida, I., Soni, M., Thompson-Coon, J., Purandare, N., Lang, I. A., Ukoumunne, O. C., & Llewellyn, D. J. (2013). Mediterranean diet, cognitive function, and dementia: A systematic review. *Epidemiology (Cambridge, Mass.), 24*(4), 479–489.

Lövdén, M., Bodammer, N. C., Kühn, S., Kaufmann, J., Schütze, H., Tempelmann, C., … Lindenberger, U. (2010). Experience-dependent plasticity of white-matter microstructure extends into old age. *Neuropsychologia, 48*(13), 3878–3883.

Lu, B., Nagappan, G., Guan, X., Nathan, P. J., & Wren, P. (2013). BDNF-based synaptic repair as a disease-modifying strategy for neurodegenerative diseases. *Nature Reviews Neuroscience, 14*(6), 401–416.

Luk, G., Bialystok, E., Craik, F. I. M., & Grady, C. L. (2011). Lifelong bilingualism maintains white matter integrity in older adults. *The Journal of Neuroscience: The Official Journal of the Society for Neuroscience, 31*(46), 16808–16813.

Luttenberger, K., Hofner, B., & Graessel, E. (2012). Are the effects of a non-drug multimodal activation therapy of dementia sustainable? Follow-up study 10 months after completion of a randomised controlled trial. *BMC Neurology, 12*, 151.

Lynch, D. B., Jeffery, I. B., Cusack, S., O'Connor, E. M., & O'Toole, P. W. (2015). Diet-microbiota-health interactions in older subjects: Implications for healthy aging. *Interdisciplinary Topics in Gerontology, 40*, 141–154.

Maci, T., Pira, F. L., Quattrocchi, G., Nuovo, S. D., Perciavalle, V., & Zappia, M. (2012). Physical and cognitive stimulation in Alzheimer disease. the GAIA Project: A pilot study. *American Journal of Alzheimer's Disease and Other Dementias, 27*(2), 107–113.

Maesako, M., Uemura, K., Kubota, M., Kuzuya, A., Sasaki, K., Asada, M., … Kinoshita, A. (2012). Environmental enrichment ameliorated high-fat diet-induced Aβ deposition and memory deficit in APP transgenic mice. *Neurobiology of Aging, 33*(5) 1011.e1011-1023.

Maesako, M., Uemura, K., Kubota, M., Kuzuya, A., Sasaki, K., Hayashida, N., … Kinoshita, A. (2012). Exercise is more effective than diet control in preventing high fat diet-induced β-amyloid deposition and memory deficit in amyloid precursor protein transgenic mice. *The Journal of Biological Chemistry, 287*(27), 23024–23033.

Mainardi, M., Di Garbo, A., Caleo, M., Berardi, N., Sale, A., & Maffei, L. (2014). Environmental enrichment strengthens corticocortical interactions and reduces amyloid-beta oligomers in aged mice. *Frontiers in Aging Neuroscience, 6*, 1. http://dx.doi.org/10.3389/fnagi.2014.00001.

Maioli, S., Puerta, E., Merino-Serrais, P., Fusari, L., Gil-Bea, F., Rimondini, R., & Cedazo-Minguez, A. (2012). Combination of apolipoprotein E4 and high carbohydrate diet reduces hippocampal BDNF and arc levels and impairs memory in young mice. *Journal of Alzheimer's Disease: JAD, 32*(2), 341–355.

Mapelli, D., Di Rosa, E., Nocita, R., & Sava, D. (2013). Cognitive stimulation in patients with dementia: Randomized controlled trial. *Dementia and Geriatric Cognitive Disorders Extra, 3*(1), 263–271.

Matsuda, O., Shido, E., Hashikai, A., Shibuya, H., Kouno, M., Hara, C., & Saito, M. (2010). Short-term effect of combined drug therapy and cognitive stimulation therapy on the cognitive function of Alzheimer's disease. *Psychogeriatrics: The Official Journal of the Japanese Psychogeriatric Society, 10*(4), 167–172.

Matthews, D. M., & Jenks, S. M. (2013). Ingestion of Mycobacterium vaccae decreases anxiety-related behavior and improves learning in mice. *Behavioural Processes, 96*, 27–35.

Mechelli, A., Crinion, J. T., Noppeney, U., O'Doherty, J., Ashburner, J., Frackowiak, R. S., & Price, C. J. (2004). Neurolinguistics: Structural plasticity in the bilingual brain. *Nature, 431*(7010), 757.

Medehouenou, T. C. M., Ayotte, P., Carmichael, P.-H., Kröger, E., Verreault, R., Lindsay, J., … Laurin, D. (2014). Plasma polychlorinated biphenyl and organochlorine pesticide concentrations in dementia: The Canadian Study of Health and Aging. *Environment International, 69*, 141–147.

Melzig, M. F., & Janka, M. (2003). Enhancement of neutral endopeptidase activity in SK-N-SH cells by green tea extract. *Phytomedicine: International Journal of Phytotherapy and Phytopharmacology, 10*(6–7), 494–498.

Meng, X., & D'Arcy, C. (2012). Education and dementia in the context of the cognitive reserve hypothesis: A systematic review with meta-analyses and qualitative analyses. *PLoS One, 7*(6), e38268. http://dx.doi.org/10.1371/journal.pone.0038268.

Mielke, M. M., Zandi, P. P., Shao, H., Waern, M., Östling, S., Guo, X., … Gustafson, D. R. (2010). The 32-year relationship between cholesterol and dementia from midlife to late life. *Neurology, 75*(21), 1888–1895.

Mirochnic, S., Wolf, S., Staufenbiel, M., & Kempermann, G. (2009). Age effects on the regulation of adult hippocampal neurogenesis by physical activity and environmental enrichment in the APP23 mouse model of Alzheimer disease. *Hippocampus, 19*(10), 1008–1018. http://dx.doi.org/10.1002/hipo.20560.

Mitzner, T. L., & Kemper, S. (2003). Oral and written language in late adulthood: Findings from the Nun Study. *Experimental Aging Research, 29*(4), 457–474. http://dx.doi.org/10.1080/03610730303698.

Molinuevo, J. L., Ripolles, P., Simó, M., Lladó, A., Olives, J., Balasa, M., … Rami, L. (2014). White matter changes in preclinical Alzheimer's disease: A magnetic resonance imaging-diffusion tensor imaging study on cognitively normal older people with positive amyloid β protein 42 levels. *Neurobiology of Aging, 35*(12), 2671–2680.

Morbelli, S., Perneczky, R., Drzezga, A., Frisoni, G. B., Caroli, A., van Berckel, B. N. M., … Nobili, F. (2013). Metabolic networks underlying cognitive reserve in prodromal Alzheimer disease: A European Alzheimer disease consortium project. *Journal of Nuclear Medicine: Official Publication, Society of Nuclear Medicine, 54*(6), 894–902.

Mori, E., Hirono, N., Yamashita, H., Imamura, T., Ikejiri, Y., Ikeda, M., & Kitagaki, H. (1997). *Premorbid brain size as a determinant of reserve capacity against intellectual decline in Alzheimer's disease* (Vol. 154).

Mortimer, J. A. (2012). The Nun Study: Risk factors for pathology and clinical-pathologic correlations. *Current Alzheimer Research, 9*(6), 621–627.

Mortimer, J. A., Gosche, K. M., Riley, K. P., Markesbery, W. R., & Snowdon, D. A. (2004). Delayed recall, hippocampal volume and Alzheimer neuropathology: Findings from the Nun Study. *Neurology, 62*(3), 428–432.

Moy, G. A., & McNay, E. C. (2013). Caffeine prevents weight gain and cognitive impairment caused by a high-fat diet while elevating hippocampal BDNF. *Physiology & Behavior, 109*, 69–74.

Negash, S., Wilson, R. S., Leurgans, S. E., Wolk, D. A., Schneider, J. A., Buchman, A. S., et al. (2013). Resilient brain aging: Characterization of discordance between Alzheimer's disease pathology and cognition. *Current Alzheimer Research, 10*(8), 844–851.

Ngandu, T., Lehtisalo, J., Solomon, A., Levalahti, E., Ahtiluoto, S., Antikainen, R., … Kivipelto, M. (2015). A 2-year multidomain intervention of diet, exercise, cognitive training, and vascular risk monitoring versus control to prevent cognitive decline in at-risk elderly people (FINGER): A randomised controlled trial. *Lancet, 385*(9984), 2255–2263.

Nichol, K., Deeny, S. P., Seif, J., Camaclang, K., & Cotman, C. W. (2009). Exercise improves cognition and hippocampal plasticity in APOE epsilon4 mice. *Alzheimer's & Dementia: The Journal of the Alzheimer's Association, 5*(4), 287–294. http://dx.doi.org/10.1016/j.jalz.2009.02.006.

Nichol, K. E., Parachikova, A. I., & Cotman, C. W. (2007). Three weeks of running wheel exposure improves cognitive performance in the aged Tg2576 mouse. *Behavioural Brain Research, 184*(2), 124–132. http://dx.doi.org/10.1016/j.bbr.2007.06.027.

Nichol, K. E., Poon, W. W., Parachikova, A. I., Cribbs, D. H., Glabe, C. G., & Cotman, C. W. (2008). Exercise alters the immune profile in Tg2576 Alzheimer mice toward a response coincident with improved cognitive performance and decreased amyloid. *Journal of Neuroinflammation, 5*, 13. http://dx.doi.org/10.1186/1742-2094-5-13.

Niemann, C., Godde, B., & Voelcker-Rehage, C. (2014). Not only cardiovascular, but also coordinative exercise increases hippocampal volume in older adults. *Frontiers in Aging Neuroscience, 6*, 170. http://dx.doi.org/10.3389/fnagi.2014.00170.

Ohia-Nwoko, O., Montazari, S., Lau, Y. S., & Eriksen, J. L. (2014). Long-term treadmill exercise attenuates tau pathology in P301S tau transgenic mice. *Molecular Neurodegeneration, 9*, 54. http://dx.doi.org/10.1186/1750-1326-9-54.

Okonkwo, O. C., Schultz, S. A., Oh, J. M., Larson, J., Edwards, D., Cook, D., … Sager, M. A. (2014). Physical activity attenuates age-related biomarker alterations in preclinical AD. *Neurology, 83*(19), 1753–1760. http://dx.doi.org/10.1212/WNL.0000000000000964.

Olazarán, J., Reisberg, B., Clare, L., Cruz, I., Peña-Casanova, J., Del Ser, T., … Muñiz, R. (2010). Nonpharmacological therapies in Alzheimer's disease: A systematic review of efficacy. *Dementia and Geriatric Cognitive Disorders, 30*(2), 161–178.

O'mahony, S. M., Hyland, N. P., Dinan, T. G., & Cryan, J. F. (2011). Maternal separation as a model of brain-gut axis dysfunction. *Psychopharmacology, 214*(1), 71–88. http://dx.doi.org/10.1007/S00213-010-2010-9.

Onder, G., Zanetti, O., Giacobini, E., Frisoni, G. B., Bartorelli, L., Carbone, G., … Bernabei, R. (2005). Reality orientation therapy combined with cholinesterase inhibitors in Alzheimer's disease: Randomised controlled trial. *The British Journal of Psychiatry: The Journal of Mental Science, 187*, 450–455.

Orrell, M., Aguirre, E., Spector, A., Hoare, Z., Woods, R. T., Streater, A., … Russell, I. (2014). Maintenance cognitive stimulation therapy for dementia: Single-blind, multicentre, pragmatic randomised controlled trial. *The British Journal of Psychiatry: The Journal of Mental Science, 204*(6), 454–461.

Otaegui-Arrazola, A., Amiano, P., Elbusto, A., Urdaneta, E., & Martínez-Lage, P. (2014). Diet, cognition, and Alzheimer's disease: Food for thought. *European Journal of Nutrition, 53*(1), 1–23.

Ozawa, M., Ninomiya, T., Ohara, T., Doi, Y., Uchida, K., Shirota, T., … Kiyohara, Y. (2013). Dietary patterns and risk of dementia in an elderly Japanese population: The Hisayama Study. *The American Journal of Clinical Nutrition, 97*(5), 1076–1082.

Paganini-Hill, A., Kawas, C. H., & Corrada, M. M. (2015). Lifestyle factors and dementia in the oldest-old: The 90+ study. *Alzheimer Disease and Associated Disorders*. epub ahead of print.

Pal, A., Siotto, M., Prasad, R., & Squitti, R. (2015). Towards a unified vision of copper involvement in Alzheimer's disease: A review connecting basic, experimental, and clinical research. *Journal of Alzheimer's Disease: JAD, 44*(2), 343–354.

Parachikova, A., Nichol, K. E., & Cotman, C. W. (2008). Short-term exercise in aged Tg2576 mice alters neuroinflammation and improves cognition. *Neurobiology of Disease, 30*(1), 121–129. http://dx.doi.org/10.1016/j.nbd.2007.12.008.

Park, S. H., Kim, J. H., Choi, K. H., Jang, Y. J., Bae, S. S., Choi, B. T., & Shin, H. K. (2013). Hypercholesterolemia accelerates amyloid β-induced cognitive deficits. *International Journal of Molecular Medicine, 31*(3), 577–582.

Parrón, T., Requena, M., Hernández, A. F., & Alarcón, R. (2011). Association between environmental exposure to pesticides and neurodegenerative diseases. *Toxicology and Applied Pharmacology, 256*(3), 379–385.

Parrott, M. D., Winocur, G., Bazinet, R. P., Ma, D. W. L., & Greenwood, C. E. (2015). Whole-food diet worsened cognitive dysfunction in an Alzheimer's disease mouse model. *Neurobiology of Aging, 36*(1), 90–99.

Patrignani, P., Tacconelli, S., & Bruno, A. (2014). Gut microbiota, host gene expression, and aging. *Journal of Clinical Gastroenterology, 48*(Suppl 1), S28–S31. http://dx.doi.org/10.1097/MCG.0000000000000229.

Pérez Martínez, G., Bäuerl, C., & Collado, M. C. (2014). Understanding gut microbiota in elderly's health will enable intervention through probiotics. *Beneficial Microbes, 5*(3), 235–246.

Perneczky, R., Drzezga, A., Diehl-Schmid, J., Schmid, G., Wohlschläger, A., Kars, S., … Kurz, A. (2006). Schooling mediates brain reserve in Alzheimer's disease: Findings of fluoro-deoxy-glucose-positron emission tomography. *Journal of Neurology, Neurosurgery, and Psychiatry, 77*(9), 1060–1063.

Perreau, V. M., Adlard, P. A., Anderson, A. J., & Cotman, C. W. (2005). Exercise-induced gene expression changes in the rat spinal cord. *Gene Expression, 12*(2), 107–121.

Persson, J., Kalpouzos, G., Nilsson, L.-G., Ryberg, M., & Nyberg, L. (2011). Preserved hippocampus activation in normal aging as revealed by fMRI. *Hippocampus, 21*(7), 753–766.

Pohanka, M., & Dobes, P. (2013). Caffeine inhibits acetylcholinesterase, but not butyrylcholinesterase. *International Journal of Molecular Sciences, 14*(5), 9873–9882.

Polidori, M. C., & Nelles, G. (2014). Antioxidant clinical trials in mild cognitive impairment and Alzheimer's disease-challenges and perspectives. *Current Pharmaceutical Design, 20*(18), 3083–3092.

Porquet, D., Casadesús, G., Bayod, S., Vicente, A., Canudas, A. M., Vilaplana, J., … del Valle, J. (2013). Dietary resveratrol prevents Alzheimer's markers and increases life span in SAMP8. *Age (Dordrecht, Netherlands), 35*(5), 1851–1865.

Potter, G. G., Helms, M. J., Burke, J. R., Steffens, D. C., & Plassman, B. L. (2007). Job demands and dementia risk among male twin pairs. *Alzheimer's & Dementia: The Journal of the Alzheimer's Association, 3*(3), 192–199.

Qiu, X., Huang, C.-X., Lu, W., Yang, S., Li, C., Shi, X.-Y., … Tang, Y. (2012). Effects of a 4 month enriched environment on the hippocampus and the myelinated fibers in the hippocampus of middle-aged rats. *Brain Research, 1465*, 26–33.

Rao, S. K., Ross, J. M., Harrison, F. E., Bernardo, A., Reiserer, R. S., Reiserer, R. S., … McDonald, M. P. (2015). Differential proteomic and behavioral effects of long-term voluntary exercise in wild-type and APP-overexpressing transgenics. *Neurobiology of Disease, 78*, 45–55. http://dx.doi.org/10.1016/j.nbd.2015.03.018.

Reitz, C., & Mayeux, R. (2014). Alzheimer disease: Epidemiology, diagnostic criteria, risk factors and biomarkers. *Biochemical Pharmacology, 88*(4), 640–651.

Religa, D., Strozyk, D., Cherny, R. A., Volitakis, I., Haroutunian, V., Winblad, B., … Bush, A. I. (2006). Elevated cortical zinc in Alzheimer disease. *Neurology, 67*(1), 69–75.

Rentz, D. M., Locascio, J. J., Becker, J. A., Moran, E. K., Eng, E., Buckner, R. L., … Johnson, K. A. (2010). Cognition, reserve, and amyloid deposition in normal aging. *Annals of Neurology, 67*(3), 353–364.

Revilla, S., Suñol, C., García-Mesa, Y., Giménez-Llort, L., Sanfeliu, C., & Cristòfol, R. (2014). Physical exercise improves synaptic dysfunction and recovers the loss of survival factors in 3xTg-AD mouse brain. *Neuropharmacology, 81*, 55–63.

Reyes-Ortiz, C. A., Berges, I. M., Raji, M. A., Koenig, H. G., Kuo, Y. F., & Markides, K. S. (2008). Church attendance mediates the association between depressive symptoms and cognitive functioning among older Mexican Americans. *Journals of Gerontology Series A, Biological Sciences and Medical Sciences, 63*(5), 480–486.

Richardson, J. R., Roy, A., Shalat, S. L., von Stein, R. T., Hossain, M. M., Buckley, B., … German, D. C. (2014). Elevated serum pesticide levels and risk for Alzheimer disease. *JAMA Neurology, 71*(3), 284–290.

Richardson, J. R., Shalat, S. L., Buckley, B., Winnik, B., O'Suilleabhain, P., Diaz-Arrastia, R., … German, D. C. (2009). Elevated serum pesticide levels and risk of Parkinson disease. *Archives of Neurology, 66*(7), 870–875.

Richter, H., Ambree, O., Lewejohann, L., Herring, A., Keyvani, K., Paulus, W., … Sachser, N. (2008). Wheel-running in a transgenic mouse model of Alzheimer's disease: Protection or symptom? *Behavioural Brain Research, 190*(1), 74–84. http://dx.doi.org/10.1016/j.bbr.2008.02.005.

Riley, K. P., Snowdon, D. A., Desrosiers, M. F., & Markesbery, W. R. (2005). Early life linguistic ability, late life cognitive function, and neuropathology: Findings from the Nun Study. *Neurobiology of Aging, 26*(3), 341–347.

Roe, C. M., Fagan, A. M., Grant, E. A., Marcus, D. S., Benzinger, T. L. S., Mintun, M. A., … Morris, J. C. (2011). Cerebrospinal fluid biomarkers, education, brain volume, and future cognition. *Archives of Neurology, 68*(9), 1145–1151.

Roe, C. M., Mintun, M. A., D'Angelo, G., Xiong, C., Grant, E. A., & Morris, J. C. (2008). Alzheimer disease and cognitive reserve: Variation of education effect with carbon 11-labeled Pittsburgh Compound B uptake. *Archives of Neurology, 65*(11), 1467–1471.

Roe, C. M., Xiong, C., Miller, J. P., & Morris, J. C. (2007). Education and Alzheimer disease without dementia: Support for the cognitive reserve hypothesis. *Neurology, 68*(3), 223–228.

Rosen, A. C., Sugiura, L., Kramer, J. H., Whitfield-Gabrieli, S., & Gabrieli, J. D. (2011). Cognitive training changes hippocampal function in mild cognitive impairment: A pilot study. *Journal of Alzheimers Disease: JAD, 26*(Suppl), 349–357. 3 SRC – GoogleScholar. http://dx.doi.org/10.3233/jad-2011-0009.

Rovio, S., Kåreholt, I., Helkala, E.-L., Viitanen, M., Winblad, B., Tuomilehto, J., … Kivipelto, M. (2005). Leisure-time physical activity at midlife and the risk of dementia and Alzheimer's disease. *The Lancet Neurology, 4*(11), 705–711.

Russo-Neustadt, A., Beard, R. C., & Cotman, C. W. (1999). Exercise, antidepressant medications, and enhanced brain derived neurotrophic factor expression. *Neuropsychopharmacology, 21*(5), 679–682. doi:10.1016/S0893-133x(99)00059-7.

Saczynski, J. S., Pfeifer, L. A., Masaki, K., Korf, E. S. C., Laurin, D., White, L., & Launer, L. J. (2006). The effect of social engagement on incident dementia: The Honolulu-Asia Aging Study. *American Journal of Epidemiology, 163*(5), 433–440.

Salazar, J. G., Ribes, D., Cabre, M., Domingo, J. L., Sanchez-Santed, F., & Colomina, M. T. (2011). Amyloid beta peptide levels increase in brain of AbetaPP Swedish mice after exposure to chlorpyrifos. *Current Alzheimer Research, 8*(7), 732–740.

Sattler, C., Toro, P., Schönknecht, P., & Schröder, J. (2012). Cognitive activity, education and socioeconomic status as preventive factors for mild cognitive impairment and Alzheimer's disease. *Psychiatry Research, 196*(1), 90–95.

Scarmeas, N., Levy, G., Tang, M. X., Manly, J., & Stern, Y. (2001). Influence of leisure activity on the incidence of Alzheimer's disease. *Neurology, 57*(12), 2236–2242.

Scarmeas, N., Luchsinger, J. A., Schupf, N., Brickman, A. M., Cosentino, S., Tang, M. X., & Stern, Y. (2009). Physical activity, diet, and risk of Alzheimer disease. *JAMA, 302*(6), 627–637.

Schlosser Covell, G. E., Hoffman-Snyder, C. R., Wellik, K. E., Woodruff, B. K., Geda, Y. E., Caselli, R. J., ... Wingerchuk, D. M. (2015). Physical activity level and future risk of mild cognitive impairment or dementia: A critically appraised topic. *Neurologist, 19*(3), 89–91. http://dx.doi.org/10.1097/NRL.0000000000000013.

Schweizer, T. A., Ware, J., Fischer, C. E., Craik, F. I. M., & Bialystok, E. (2012). Bilingualism as a contributor to cognitive reserve: Evidence from brain atrophy in Alzheimer's disease. *Cortex: A Journal Devoted to the Study of the Nervous System and Behavior, 48*(8), 991–996.

Serra, L., Cercignani, M., Petrosini, L., Basile, B., Perri, R., Fadda, L., ... Bozzali, M. (2011). Neuroanatomical correlates of cognitive reserve in Alzheimer disease. *Rejuvenation Research, 14*(2), 143–151.

Shah, R. (2013). The role of nutrition and diet in Alzheimer disease: A systematic review. *Journal of the American Medical Directors Association, 14*(6), 398–402.

Shah, T., Verdile, G., Sohrabi, H., Campbell, A., Putland, E., Cheetham, C., ... Martins, R. N. (2014). A combination of physical activity and computerized brain training improves verbal memory and increases cerebral glucose metabolism in the elderly. *Translational Psychiatry, 4*, e487. http://dx.doi.org/10.1038/tp.2014.122.

Shea, T. B., & Remington, R. (2015). Nutritional supplementation for Alzheimer's disease? *Current Opinion in Psychiatry, 28*(2), 141–147.

Shen, X.-L., Yu, J.-H., Zhang, D.-F., Xie, J.-X., & Jiang, H. (2014). Positive relationship between mortality from Alzheimer's disease and soil metal concentration in mainland China. *Journal of Alzheimer's Disease: JAD, 42*(3), 893–900.

Shinto, L., Quinn, J., Montine, T., Dodge, H. H., Woodward, W., Baldauf-Wagner, S., ... Kaye, J. (2014). A randomized placebo-controlled pilot trial of omega-3 fatty acids and alpha lipoic acid in Alzheimer's disease. *Journal of Alzheimer's Disease: JAD, 38*(1), 111–120.

Shoemark, D. K., & Allen, S. J. (2014). The microbiome and disease: Reviewing the links between the oral microbiome, aging, and Alzheimer's disease. *Journal of Alzheimer's Disease: JAD, 43*, 725–738. 3 SRC—GoogleScholar. http://dx.doi.org/10.3233/jad-141170.

Simmons-Stern, N. R., Budson, A. E., & Ally, B. A. (2010). Music as a memory enhancer in patients with Alzheimer's disease. *Neuropsychologia, 48*(10), 3164–3167.

Simpson, J., & Kelly, J. P. (2011). The impact of environmental enrichment in laboratory rats— Behavioural and neurochemical aspects. *Behavioural Brain Research, 222*(1), 246–264.

Singh, B., Parsaik, A. K., Mielke, M. M., Erwin, P. J., Knopman, D. S., Petersen, R. C., & Roberts, R. O. (2014). Association of mediterranean diet with mild cognitive impairment and Alzheimer's disease: A systematic review and meta-analysis. *Journal of Alzheimer's Disease: JAD, 39*(2), 271–282.

Singh, I., Sagare, A. P., Coma, M., Perlmutter, D., Gelein, R., Bell, R. D., ... Deane, R. (2013). Low levels of copper disrupt brain amyloid-β homeostasis by altering its production and clearance. *Proceedings of the National Academy of Sciences of the United States of America, 110*(36), 14771–14776.

Singh, N., Chhillar, N., Banerjee, B., Bala, K., Basu, M., & Mustafa, M. (2013). Organochlorine pesticide levels and risk of Alzheimer's disease in north Indian population. *Human & Experimental Toxicology, 32*(1), 24–30.

Singh, N. K., Banerjee, B. D., Bala, K., Basu, M., & Chhillar, N. (2014). Polymorphism in cytochrome P450 2D6, glutathione S-transferases Pi 1 genes, and organochlorine pesticides in Alzheimer disease: A case–control study in north Indian population. *Journal of Geriatric Psychiatry and Neurology, 27*(2), 119–127. http://dx.doi.org/10.1177/0891988714522698.

Slashinski, M. J., McCurdy, S. A., Achenbaum, L. S., Whitney, S. N., & McGuire, A. L. (2012). "Snake-oil," "quack medicine," and "industrially cultured organisms:" Biovalue and the commercialization of human microbiome research. *BMC Medical Ethics, 13*, 28.

Smith, C. J., Emge, J. R., Berzins, K., Lung, L., Khamishon, R., Shah, P., ... Gareau, M. G. (2014). Probiotics normalize the gut-brain-microbiota axis in immunodeficient mice. *American Journal of Physiology Gastrointestinal and Liver Physiology, 307*(8), G793–G802.

Snowdon, D. A., Greiner, L. H., Mortimer, J. A., Riley, K. P., Greiner, P. A., & Markesbery, W. R. (1997). Brain infarction and the clinical expression of Alzheimer disease. The Nun Study. *JAMA, 277*(10), 813–817.

Snowdon, D. A., Kemper, S. J., Mortimer, J. A., Greiner, L. H., Wekstein, D. R., & Markesbery, W. R. (1996). Linguistic ability in early life and cognitive function and Alzheimer's disease in late life. Findings from the Nun Study. *JAMA, 275*(7), 528–532.

Snowdon, D. A., Tully, C. L., Smith, C. D., Riley, K. P., & Markesbery, W. R. (2000). Serum folate and the severity of atrophy of the neocortex in Alzheimer disease: Findings from the Nun Study. *American Journal of Clinical Nutrition, 71*(4), 993–998.

Soldan, A., Pettigrew, C., Li, S., Wang, M. -C., Moghekar, A., Selnes, O. A., ... Team, B. R. (2013). Relationship of cognitive reserve and cerebrospinal fluid biomarkers to the emergence of clinical symptoms in preclinical Alzheimer's disease. *Neurobiology of Aging, 34*(12), 2827–2834.

Sole-Padulles, C., Llado, A., Bartres-Faz, D., Fortea, J., Sanchez-Valle, R., Bosch, B., ... Rami, L. (2011). Association between cerebrospinal fluid tau and brain atrophy is not related to clinical severity in the Alzheimer's disease continuum. *Psychiatry Research, 192*(3), 140–146. http://dx.doi.org/10.1016/j.psychresns.2010.12.001.

Solomon, A., Kivipelto, M., Wolozin, B., Zhou, J., & Whitmer, R. A. (2009). Midlife serum cholesterol and increased risk of Alzheimer's and vascular dementia three decades later. *Dementia and Geriatric Cognitive Disorders, 28*(1), 75–80.

Southon, A., Greenough, M. A., Ganio, G., Bush, A. I., Burke, R., & Camakaris, J. (2013). Presenilin promotes dietary copper uptake. *PLoS One, 8*(5), e62811.

Sperling, R. A., Aisen, P. S., Beckett, L. A., Bennett, D. A., Craft, S., Fagan, A. M., ... Phelps, C. H. (2011). Toward defining the preclinical stages of Alzheimer's disease: Recommendations from the National Institute on Aging-Alzheimer's Association workgroups on diagnostic guidelines for Alzheimer's disease. *Alzheimer's & Dementia: The Journal of the Alzheimer's Association, 7*(3), 280–292.

Spironelli, C., Bergamaschi, S., Mondini, S., Villani, D., & Angrilli, A. (2013). Functional plasticity in Alzheimer's disease: Effect of cognitive training on language-related ERP components. *Neuropsychologia, 51*(8), 1638–1648.

Squitti, R., Siotto, M., & Polimanti, R. (2014). Low-copper diet as a preventive strategy for Alzheimer's disease. *Neurobiology of Aging, 35*(Suppl), S40–S50. 2 SRC—GoogleScholar. http://dx.doi.org/10.1016/j.neurobiolaging.2014.02.031.

Steffener, J., Barulli, D., Habeck, C., O'Shea, D., Razlighi, Q., & Stern, Y. (2014). The role of education and verbal abilities in altering the effect of age-related gray matter differences on cognition. *PLoS One, 9*(3), e91196.

Steffener, J., & Stern, Y. (2012). Exploring the neural basis of cognitive reserve in aging. *Biochimica et Biophysica Acta, 1822*(3), 467–473.

Stelmashook, E. V., Isaev, N. K., Genrikhs, E. E., Amelkina, G. A., Khaspekov, L. G., Skrebitsky, V. G., & Illarioshkin, S. N. (2014). Role of zinc and copper ions in the pathogenetic mechanisms of Alzheimer's and Parkinson's diseases. *Biochemistry Biokhimiia, 79*(5), 391–396.

Stern, Y. (2002). What is cognitive reserve? Theory and research application of the reserve concept. *Journal of the International Neuropsychological Society: JINS, 8*(3), 448–460.

Stern, Y. (2009). Cognitive reserve. *Neuropsychologia, 47*(10), 2015–2028.

Stern, Y., Moeller, J. R., Anderson, K. E., Luber, B., Zubin, N. R., DiMauro, A. A., et al. (2000). Different brain networks mediate task performance in normal aging and AD: Defining compensation. *Neurology, 15*(9), 1291–1297.

Subash, S., Braidy, N., Essa, M. M., Zayana, A. -B., Ragini, V., Al-Adawi, S., ... Guillemin, G. J. (2015). Long-term (15 mo) dietary supplementation with pomegranates from Oman attenuates cognitive and behavioral deficits in a transgenic mice model of Alzheimer's disease. *Nutrition (Burbank, Los Angeles County, Calif.), 31*(1), 223–229.

Takeuchi, H., Sekiguchi, A., Taki, Y., Yokoyama, S., Yomogida, Y., Komuro, N., ... Kawashima, R. (2010). Training of working memory impacts structural connectivity. *The Journal of Neuroscience: The Official Journal of the Society for Neuroscience, 30*(9), 3297–3303.

Tanner, C. M., Goldman, S. M., Ross, G. W., & Grate, S. J. (2014). The disease intersection of susceptibility and exposure: Chemical exposures and neurodegenerative disease risk. *Alzheimer's & Dementia: The Journal of the Alzheimer's Association, 10*(3 Suppl), S213–S225. http://dx.doi.org/10.1016/j.jalz.2014.04.014.

Tapia-Rojas, C., Aranguiz, F., Varela-Nallar, L., & Inestrosa, N. C. (2015). Voluntary running attenuates memory loss, decreases neuropathological changes and induces neurogenesis in a mouse model of Alzheimer's disease. *Brain Pathology (Zurich, Switzerland).* epub ahead of print.

Tarraga, L., Boada, M., Modinos, G., Espinosa, A., Diego, S., Morera, A., ... Becker, J. T. (2006). A randomised pilot study to assess the efficacy of an interactive, multimedia tool of cognitive stimulation in Alzheimer's disease. *Journal of Neurology, Neurosurgery, and Psychiatry, 77*(10), 1116–1121. http://dx.doi.org/10.1136/jnnp.2005.086074.

Thiel, C., Vogt, L., Tesky, V. A., Meroth, L., Jakob, M., Sahlender, S., ... Banzer, W. (2012). Cognitive intervention response is related to habitual physical activity in older adults. *Aging Clinical and Experimental Research, 24*(1), 47–55.

Treiber, K. A., Carlson, M. C., Corcoran, C., Norton, M. C., Breitner, J. C. S., Piercy, K. W., ... Tschanz, J. T. (2011). Cognitive stimulation and cognitive and functional decline in Alzheimer's disease: The cache county dementia progression study. *The Journals of Gerontology Series B, Psychological Sciences and Social Sciences, 66*(4), 416–425.

Tully, C. L., Snowdon, D. A., & Markesbery, W. R. (1995). Serum zinc, senile plaques, and neurofibrillary tangles: Findings from the Nun Study. *Neuroreport, 6*(16), 2105–2108.

Tyas, S. L., Snowdon, D. A., Desrosiers, M. F., Riley, K. P., & Markesbery, W. R. (2007). Healthy ageing in the Nun Study: Definition and neuropathologic correlates. *Age and Ageing, 36*(6), 650–655. http://dx.doi.org/10.1093/ageing/afm120.

Ullrich, C., & Humpel, C. (2009). Rotenone induces cell death of cholinergic neurons in an organotypic co-culture brain slice model. *Neurochemical Research, 34*(12), 2147–2153.

Valenzuela, M. J., Sachdev, P., Wen, W., Chen, X., & Brodaty, H. (2008). Lifespan mental activity predicts diminished rate of hippocampal atrophy. *PLoS One, 3*(7), e2598. http://dx.doi.org/10.1371/journal.pone.0002598.

van Wijk, N., Broersen, L. M., de Wilde, M. C., Hageman, R. J., Groenendijk, M., Sijben, J. W., & Kamphuis, P. J. (2014). Targeting synaptic dysfunction in Alzheimer's disease by administering a specific nutrient combination. *Journal of Alzheimer's Disease: JAD, 38*(3), 459–479. http://dx.doi.org/10.3233/JAD-130998.

Varamini, B., Sikalidis, A. K., & Bradford, K. L. (2014). Resveratrol increases cerebral glycogen synthase kinase phosphorylation as well as protein levels of drebrin and transthyretin in mice: An exploratory study. *International Journal of Food Sciences and Nutrition, 65*(1), 89–96.

Varma, V. R., Chuang, Y. F., Harris, G. C., Tan, E. J., & Carlson, M. C. (2015). Low-intensity daily walking activity is associated with hippocampal volume in older adults. *Hippocampus, 25*(5), 605–615. http://dx.doi.org/10.1002/Hipo.22397.

Vassallo, N., & Scerri, C. (2013). Mediterranean diet and dementia of the Alzheimer type. *Current Aging Science, 6*(2), 150–162.

Vepsäläinen, S., Koivisto, H., Pekkarinen, E., Mäkinen, P., Dobson, G., McDougall, G. J., ... Hiltunen, M. (2013). Anthocyanin-enriched bilberry and blackcurrant extracts modulate amyloid precursor protein processing and alleviate behavioral abnormalities in the APP/PS1 mouse model of Alzheimer's disease. *The Journal of Nutritional Biochemistry, 24*(1), 360–370.

Verghese, J., Lipton, R. B., Katz, M. J., Hall, C. B., Derby, C. A., Kuslansky, G., ... Buschke, H. (2003). Leisure activities and the risk of dementia in the elderly. *The New England Journal of Medicine, 348*(25), 2508–2516.

Verret, L., Krezymon, A., Halley, H., Trouche, S., Zerwas, M., Lazouret, M., ... Rampon, C. (2013). Transient enriched housing before amyloidosis onset sustains cognitive improvement in Tg2576 mice. *Neurobiology of Aging, 34*(1), 211–225.

Viola, L. F., Nunes, P. V., Yassuda, M. S., Aprahamian, I., Santos, F. S., Santos, G. D., et al. (2011). Effects of a multidisciplinary cognitive rehabilitation program for patients with mild Alzheimer's disease. *Clinics (Sao Paulo), 66*(8), 1395–1400.

Walker, J. M., Klakotskaia, D., Ajit, D., Weisman, G. A., Wood, W. G., Sun, G. Y., ... Schachtman, T. R. (2015). Beneficial effects of dietary EGCG and voluntary exercise on behavior in an Alzheimer's disease mouse model. *Journal of Alzheimer's Disease: JAD, 44*(2), 561–572.

Wang, C. Y., Wang, T., Zheng, W., Zhao, B. L., Danscher, G., Chen, Y. H., & Wang, Z. Y. (2010). Zinc overload enhances APP cleavage and Abeta deposition in the Alzheimer mouse brain. *PLoS One, 5*(12), e15349. http://dx.doi.org/10.1371/journal.pone.0015349.

Wang, D. S., Iwata, N., Hama, E., Saido, T. C., & Dickson, D. W. (2003). Oxidized neprilysin in aging and Alzheimer's disease brains. *Biochemical and Biophysical Research Communications, 310*(1), 236–241.

Wang, H., Odegaard, A., Thyagarajan, B., Hayes, J., Cruz, K. S., Derosiers, M. F., ... Gross, M. D. (2012). Blood folate is associated with asymptomatic or partially symptomatic Alzheimer's disease in the Nun Study. *Journal of Alzheimer's Disease: JAD, 28*(3), 637–645.

Wang, J., Bi, W., Cheng, A., Freire, D., Vempati, P., Zhao, W., ... Pasinetti, G. M. (2014). Targeting multiple pathogenic mechanisms with polyphenols for the treatment of Alzheimer's disease—Experimental approach and therapeutic implications. *Frontiers in Aging Neuroscience, 6*, 42. http://dx.doi.org/10.3389/fnagi.2014.00042.

Wang, P., Su, C., Li, R., Wang, H., Ren, Y., Sun, H., ... Jiang, S. (2014). Mechanisms and effects of curcumin on spatial learning and memory improvement in APPswe/PS1dE9 mice. *Journal of Neuroscience Research, 92*(2), 218–231.

Weuve, J., Puett, R. C., Schwartz, J., Yanosky, J. D., Laden, F., & Grodstein, F. (2012). Exposure to particulate air pollution and cognitive decline in older women. *Archives of Internal Medicine, 172*(3), 219–227.

Wiesmann, M., Jansen, D., Zerbi, V., Broersen, L. M., Garthe, A., & Kiliaan, A. J. (2013). Improved spatial learning strategy and memory in aged Alzheimer AβPPswe/PS1dE9 mice on a multi-nutrient diet. *Journal of Alzheimer's Disease: JAD, 37*(1), 233–245.

Wilson, R. S., Boyle, P. A., Yang, J., James, B. D., & Bennett, D. A. (2015). Early life instruction in foreign language and music and incidence of mild cognitive impairment. *Neuropsychology, 29*(2), 292–302. http://dx.doi.org/10.1037/neu0000129.

Wilson, R. S., Boyle, P. A., Yu, L., Barnes, L. L., Schneider, J. A., & Bennett, D. A. (2013). Lifespan cognitive activity, neuropathologic burden, and cognitive aging. *Neurology, 81*(4), 314–321. http://dx.doi.org/10.1212/WNL.0b013e31829c5e8a.

Wilson, R. S., Krueger, K. R., Arnold, S. E., Schneider, J. A., Kelly, J. F., Barnes, L. L., ... Bennett, D. A. (2007). Loneliness and risk of Alzheimer disease. *Archives of General Psychiatry, 64*(2), 234–240. http://dx.doi.org/10.1001/archpsyc.64.2.234.

Wilson, R. S., Li, Y., Aggarwal, N. T., Barnes, L. L., McCann, J. J., Gilley, D. W., & Evans, D. A. (2004). Education and the course of cognitive decline in Alzheimer disease. *Neurology, 63*(7), 1198–1202.

Wilson, R. S., Mendes De Leon, C. F., Barnes, L. L., Schneider, J. A., Bienias, J. L., Evans, D. A., & Bennett, D. A. (2002). Participation in cognitively stimulating activities and risk of incident Alzheimer disease. *JAMA, 287*(6), 742–748.

Wilson, R. S., Nag, S., Boyle, P. A., Hizel, L. P., Yu, L., Buchman, A. S., ... Bennett, D. A. (2013). Neural reserve, neuronal density in the locus ceruleus, and cognitive decline. *Neurology, 80*(13), 1202–1208. http://dx.doi.org/10.1212/WNL.0b013e3182897103.

Wilson, R. S., Scherr, P. A., Hoganson, G., Bienias, J. L., Evans, D. A., & Bennett, D. A. (2005). Early life socioeconomic status and late life risk of Alzheimer's disease. *Neuroepidemiology, 25*(1), 8–14. http://dx.doi.org/10.1159/000085307.

Wilson, R. S., Scherr, P. A., Schneider, J. A., Tang, Y., & Bennett, D. A. (2007). Relation of cognitive activity to risk of developing Alzheimer disease. *Neurology, 69*(20), 1911–1920.

Wirth, M., Villeneuve, S., La Joie, R., Marks, S. M., & Jagust, W. J. (2014). Gene-environment interactions: Lifetime cognitive activity, APOE genotype, and β-amyloid burden. *The Journal of Neuroscience: The Official Journal of the Society for Neuroscience, 34*(25), 8612–8617.

Wolf, S. A., Kronenberg, G., Lehmann, K., Blankenship, A., Overall, R., Staufenbiel, M., & Kempermann, G. (2006). Cognitive and physical activity differently modulate disease progression in the amyloid precursor protein (APP)-23 model of Alzheimer's disease. *Biological Psychiatry, 60*(12), 1314–1323. http://dx.doi.org/10.1016/j.biopsych.2006.04.004.

Woods, B., Aguirre, E., Spector, A. E., & Orrell, M. (2012). Cognitive stimulation to improve cognitive functioning in people with dementia. *The Cochrane Database of Systematic Reviews, 2*, CD005562.

Xiao, J., Li, S., Sui, Y., Wu, Q., Li, X., Xie, B., ... Sun, Z. (2014). Lactobacillus casei-01 facilitates the ameliorative effects of proanthocyanidins extracted from lotus seedpod on learning and memory impairment in scopolamine-induced amnesia mice. *PLoS One, 9*(11), e112773.

Yaffe, K., Weston, A., Graff-Radford, N. R., Satterfield, S., Simonsick, E. M., Younkin, S. G., ... Harris, T. B. (2011). Association of plasma beta-amyloid level and cognitive reserve with subsequent cognitive decline. *JAMA, 305*(3), 261–266.

Yasuno, F., Kazui, H., Morita, N., Kajimoto, K., Ihara, M., Taguchi, A., ... Kishimoto, T. (2014). Low amyloid-β deposition correlates with high education in cognitively normal older adults: A pilot study. *International Journal of Geriatric Psychiatry, 30*(9), 919–926.

Yu, F., Bronas, U. G., Konety, S., Nelson, N. W., Dysken, M., Jack, C., Jr., ... Smith, G. (2014). Effects of aerobic exercise on cognition and hippocampal volume in Alzheimer's disease: Study protocol of a randomized controlled trial (The FIT-AD trial). *Trials, 15*, 394. http://dx.doi.org/10.1186/1745-6215-15-394.

Yu, J., Luo, X., Xu, H., Ma, Q., Yuan, J., Li, X., ... Yang, X. (2015). Identification of the key molecules involved in chronic copper exposure-aggravated memory impairment in transgenic mice of Alzheimer's disease using proteomic analysis. *Journal of Alzheimer's Disease: JAD, 44*(2), 455–469.

Yuede, C. M., Zimmerman, S. D., Dong, H., Kling, M. J., Bero, A. W., Holtzman, D. M., ... Csernansky, J. G. (2009). Effects of voluntary and forced exercise on plaque deposition, hippocampal volume, and behavior in the Tg2576 mouse model of Alzheimer's disease. *Neurobiology of Disease, 35*(3), 426–432. http://dx.doi.org/10.1016/j.nbd.2009.06.002.

Zhao, G., Liu, H. L., Zhang, H., & Tong, X. J. (2015). Treadmill exercise enhances synaptic plasticity, but does not alter beta-amyloid deposition in hippocampi of aged APP/PS1 transgenic mice. *Neuroscience, 298*, 357–366. http://dx.doi.org/10.1016/j.neuroscience.2015.04.038.

8

Role of BACE1 in Cognitive Function, from Alzheimer's Disease to Traumatic Brain Injury

Sylvia Lombardo and Giuseppina Tesco

Alzheimer's Disease Research Laboratory, Department of Neuroscience, Tufts University School of Medicine, Boston, MA, United States

OUTLINE

List of Abbreviations	239
Introduction	240
BACE1	241
BACE1 Null Mice	242
An Overview of BACE1 Substrates	244
BACE1 Transgenic Models	245
BACE1 in AD	247
BACE1 Is a Stress-Induced Protease Elevated in AD and Acute Brain Injuries	253
BACE Inhibition, a Therapeutic Approach for AD	256
References	258

List of Abbreviations

AD Alzheimer's disease
Aβ amyloid-β
APP amyloid precursor protein 1

Genes, Environment and Alzheimer's Disease.
DOI: http://dx.doi.org/10.1016/B978-0-12-802851-3.00008-5

APLP1 amyloid precursor-like protein 1
APPswe APP Swedish mutation
APLP2 amyloid precursor-like protein 2
APP-CTFs APP-C-terminal fragments
CTF carboxyl-terminal fragment
BAC bacterial artificial chromosome
BACE1 β-site APP cleaving enzyme 1
BBB blood–brain barrier
CaMKII Cam Kinase II
CHL1 cell-adhesion molecule L1-like
FAD familial Alzheimer's disease
hAPP human APP
hBACE1 human BACE1
Jag1 Jagged 1
KI knockin
KO knockout
NRG Neuregulin 1
PrP prion protein
PSGL-1 P-selectin glycoprotein ligand-1
TACE tumor necrosis factor-α converting enzyme
VEGFR1 vascular endothelial growth factor receptor 1

INTRODUCTION

Alzheimer's disease (AD) is a devastating neurodegenerative disorder that results in loss of memory and alteration of cognitive function, and eventually leads to dementia. AD is the most diffuse form of dementia, representing 70% of all cases, and its incidence is expected to increase over the next years (Alzheimer's Association, 2015). A key neuropathological event in AD etiology is the cerebral accumulation of an ~4 kDa peptide termed amyloid-β (Aβ), the principle component of senile plaques. Amyloid plaques are formed by aggregates of Aβ-peptides, 37–43 amino-acid fragments (predominantly $A\beta_{40}$ and $A\beta_{42}$) derived by serial proteolysis of the amyloid precursor protein (APP) by β- and γ-secretase. APP more commonly undergoes a nonamyloidogenic processing by α-secretase that cleaves in the middle of the β-amyloid domain (De Strooper and Annaert, 2000). β-secretase has been identified as a novel membrane-tethered member of the aspartyl proteases, termed β-site APP cleaving enzyme 1 (BACE1) (Hong et al., 2000; Hussain et al., 1999; Lin et al., 2000; Sinha et al., 1999; Vassar et al., 1999; Yan et al., 1999), while candidate α-secretases include ADAM 9, 10, and 17 (TACE, tumor necrosis factor-α converting enzyme) (Buxbaum et al., 1998; Lammich et al., 1999). APP proteolysis by β- and α-secretases results in the production of secreted APP polypeptides (β- or α-APPs, respectively) along with membrane-associated C99 and C83 APP-C-terminal fragments (APP-CTFs), respectively. The C99 and C83

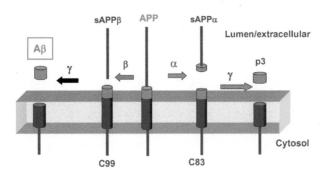

FIGURE 8.1   Proteolysis of the APP by α-, β-, and γ-secretase and relative fragments obtained.

APP-CTFs can then serve as substrates for γ-secretase resulting in the production of Aβ or p3, respectively (Figure 8.1).

BACE1

BACE1 is an N-glycosylated type 1 transmembrane protein that undergoes constitutive N-terminal processing in the Golgi apparatus. The ectodomain contains four glycosylation sites and two signature sequences typically associated with aspartyl proteases (DT/SGT/S). BACE1 is targeted through the secretory pathway to the plasma membrane where it can be internalized to endosomes (Citron, 2004). The BACE1-carboxyl-terminal fragment (CTF) contains a specific di-leucine sorting signal ([495]DDISLL[500]) (Huse, Pijak, Leslie, Lee, & Doms, 2000; Koh, von Arnim, Hyman, Tanzi, & Tesco, 2005; Pastorino, Ikin, Nairn, Pursnani, & Buxbaum, 2002) and an ubiquitination site at K501 (Kang, Cameron, Piazza, Walker, & Tesco, 2010). GGA1, 2, and 3, a family of monomeric clathrin adaptors, have been shown to bind to the BACE1 [496]DISLL[500] motif via their VHS domain and the phosphorylation of BACE1-S498 appears to increase their binding (von Arnim et al., 2004; He et al., 2003; Shiba et al., 2004; Wahle et al., 2005; Walter et al., 2001). RNAi-mediated depletion of GGA results in BACE1 accumulation in early endosomes (He, Li, Chang, & Tang, 2005; Kang et al., 2010). Our previous studies have shown that BACE1 is degraded via the lysosomal pathway (Koh et al., 2005) and that depletion of GGA3 results in increased BACE1 levels and activity owing to its impaired lysosomal trafficking and degradation (Kang et al., 2010; Tesco et al., 2007). We reported that GGA3 regulates BACE1 degradation independently of the VHS/di-leucine motif interaction but requires binding to ubiquitin (Kang et al., 2010). More importantly we found that GGA3 levels are significantly decreased and inversely correlated with BACE1 levels in the postmortem temporal cortex of patients with AD (Tesco et al., 2007). Our original findings have been confirmed by two independent studies conducted in AD brain samples of Australian and European origin (Santosa et al., 2011) (US Patent Application 20120276076, Annaert, Wim, et al., 2012). Finally, we have demonstrated the role of GGA3 in the regulation of BACE1 *in vivo* by showing that BACE1 levels are increased in the brain of GGA3 null mice (Walker, Kang, Whalen, Shen, & Tesco, 2012). BACE1 trafficking is regulated by other molecules including reticulons, Rab11 and EHD 1 and 3, and we refer to recent reviews of these topics (Buggia-Prévot and Thinakaran, 2015; Vassar et al., 2014).

BACE1 Null Mice

Shortly after the discovery that BACE1 has β-secretase activity, different BACE1 null mouse lines were generated (Cai et al., 2001; Dominguez et al., 2005; Luo et al., 2001; Roberds et al., 2001). These lines were essential to determine the *in vivo* role of BACE1 and to validate BACE1 as a key enzyme in Aβ generation. The absence of Aβ in BACE1 null mutant mice (BACE1 KO) confirmed that BACE1 is the major β-secretase in the brain and that inhibition of this enzyme could be used as therapeutic strategy for AD. Initially BACE1 KO mice were described as having no deleterious phenotype. Roberds et al. (2001) detected no behavioral differences between BACE1 KO and WT in SHIRPA test. In addition, BACE1 KO displayed normal development, normal brain morphology, and no abnormalities in neuronal or nonneuronal tissues (Cai et al., 2001; Luo et al., 2001; Roberds et al., 2001).

However, following studies identified a multitude of phenotypes in BACE1 KO mice such as memory deficits (Kobayashi et al., 2008; Laird et al., 2005; Ohno et al., 2004, 2006, 2007), increased pain sensitivity (Hu et al., 2006), reduced grip strength (Hu et al., 2006), hyperactivity (Dominguez et al., 2005; Savonenko et al., 2008), low exploratory behavior (Harrison et al., 2003), impaired axon guidance (Cao, Rickenbacher, Rodriguez, Moulia, & Albers, 2012; Hitt et al., 2012; Rajapaksha, Eimer, Bozza, & Vassar, 2011), schizophrenia-like phenotype (Savonenko et al., 2008), and hypomyelination in peripheral and central nervous systems (CNS) (Hu et al., 2006, Hu, Hu, Dai, Trapp, & Yan, 2015; Willem et al., 2006). Moreover, it was shown that BACE1 is expressed in the retina (Xiong et al., 2007) and that BACE1 KO mice display degeneration of the retina with apoptotic neuronal death in the ganglion cell layer (Cai et al., 2012). Cheret et al. (2013) observed coordination defects and severe deficits in formation and maturation of muscle spindles in BACE1 KO mice. Moreover, alteration of neurochemical activity was also detected, like spontaneous seizures (Hitt, Jaramillo, Chetkovich, & Vassar, 2010; Hu et al., 2010; Kobayashi et al., 2008), decrease in 5-HT levels in hippocampus, and decreased dopamine in striatum (Harrison et al., 2003). In addition, deficit in presynaptic function at mossy fiber synapses in CA3 and absence of mossy fiber LTP were reported (Wang, Song, Laird, Wong, & Lee, 2008). Neonatal mortality, growth retardation (Dominguez et al., 2005), and decreased body weight (Meakin et al., 2012) were also observed. However these findings were not consistent between different studies. Indeed one study showed normal body weight in BACE1 KO (Harrison et al., 2003). These differences could be specific to the BACE1 KO mouse model investigated. Interestingly, aged BACE1 KO mice displayed neurodegeneration probably due to chronic epilepsy (Hu et al., 2010). Metabolic abnormalities such as decrease of total body fat (in particular whole-body lipid content) and increased glucose clearance from the peripheral blood circulation were observed in BACE1 KO mice, suggesting that BACE1 plays an important role in insulin signaling pathway (Meakin et al., 2012). Moreover, Hu, He, Luo, Tsubota, and Yan (2013) observed increased astrogenesis and decreased neurogenesis in the hippocampus of BACE1 KO mice during early development (Table 8.1).

TABLE 8.1   Phenotypes Detected in Different Lines of BACE1 KO Mice

Phenotypes of BACE1 KO mice	References
Memory deficits	Kobayashi et al. (2008), Laird et al. (2005), Ohno et al. (2004, 2006, 2007)
Hypomyelination	Hu et al. (2006), Willem et al. (2006)
Increased pain sensitivity	Hu et al. (2006)
Neonatal mortality, growth retardation	Dominguez et al. (2005)
Hyperactivity	Dominguez et al. (2005), Savonenko et al. (2008)
Low exploratory behavior	Harrison et al. (2003)
Spontaneous seizures	Hitt et al. (2010), Hu et al. (2010), Kobayashi et al. (2008)
Neurodegeneration (aged mice)	Hu et al. (2010)
Impaired axon guidance	Cao et al. (2012), Hitt et al. (2012), Rajapaksha et al. (2011)
Schizophrenia-like behavior	Savonenko et al. (2008)
Metabolic abnormalities	Meakin et al. (2012)
Retina degeneration	Cai et al. (2012)
Muscle spindle deficit	Cheret et al. (2013)
Increased astrogenesis and decreased neurogenesis	Hu et al. (2013)
LTP deficit	Wang et al. (2008)

An Overview of BACE1 Substrates

The variety of phenotypes described in the BACE1 null mice is linked to BACE1 multiple substrates. The most known and studied BACE1 substrate is APP, given its role in AD pathogenesis. APP belongs to a family of proteins that includes amyloid precursor-like protein 1 (APLP1) and APLP2 (Wasco et al., 1992). BACE1 cleaves all members of APLP family (Eggert et al., 2004; Li and Südhof, 2004; Pastorino et al., 2004). APP and APLPs share conserved motifs supporting the theory of functional redundancy. However the Aβ sequence is present only in APP. The exact function of APP and APLPs is not clear. Nonetheless, some evidence points to a role of the APP family members in the development of the nervous system. It was demonstrated that the expression of APP increases during neuronal differentiation (Hung, Koo, Haass, & Selkoe, 1992) and its expression is also upregulated following brain injury in adult animal models (Murakami et al., 1998). APP is highly expressed during synaptogenesis (Moya, Benowitz, Schneider, & Allinquant, 1994) and undergoes rapid anterograde axonal transport to synaptic terminals (Lyckman, Confaloni, Thinakaran, Sisodia, & Moya, 1998). In addition, animal models overexpressing wild-type APP exhibit enlarged neurons, pointing to a neurotrophic effect (Oh et al., 2009). These observations suggest a role for APP in cell growth, motility, and survival during development (for a review on APP family, see Shariati and De Strooper, 2013).

Another well-studied BACE1 substrate is Neuregulin 1 (NRG1). Two separated studies showed myelination deficits in central and peripheral nervous systems associated with a reduction in myelin proteins in BACE1 null mutants (Hu et al., 2006; Willem et al., 2006). Indeed, WT mice have high levels of BACE1 in early postnatal stages, which corresponds to the myelination period. Moreover, accumulation of unprocessed NRG1 and hypomyelinated axons were observed in BACE1 KO mice (Willem et al., 2006). Lack of BACE1 processing of NRG1 has been proposed to be responsible for the hyperactivity and schizophrenia endophenotypes (Dominguez et al., 2005; Savonenko et al., 2008), spine density reduction (Savonenko et al., 2008), and deficits in formation and maturation of muscle spindles (Cheret et al., 2013). NRG3, which belongs to the same family of NRG1, is also cleaved by BACE1 (Hu et al., 2008). A dysfunction in Notch-Jagged 1 (Jag1) signaling pathway has been proposed as the mechanism underlying enhanced astrogenesis and reduction of neurogenesis observed in the hippocampus of BACE1 KO mice. Indeed, Jag1 is a BACE1 substrate and its proteolytic cleavage by BACE1 is required for the regulation of Notch signaling (Hu et al., 2013). L1 and its close homolog cell-adhesion molecule L1-like are neural cell adhesion molecules of the immunoglobulin superfamily, which are BACE1 substrates (Hitt et al., 2012; Kuhn et al., 2012; Zhou et al., 2012). These molecules are important for axon guidance and maintenance of neuronal circuits both altered in BACE1 KO mice (Cao et al., 2012; Rajapaksha et al., 2011). Another BACE1 substrate is the vascular endothelial growth factor receptor 1 (VEGFR1). It was postulated that VEGFR1 has a role in the mechanism by which BACE1 inhibition leads to retinal pathology (Cai et al., 2012).

BACE1 cleaves also the β subunit of Na_V1, a voltage-gated sodium channel expressed on the plasma membrane (Kim, Ingano, Carey, Pettingell, & Kovacs, 2005; Wong et al., 2005). This evidence highlights a potential role for BACE1 in regulating sodium currents and it could be the underlying mechanism of the epileptic phenotype observed in BACE1 null mice (Hitt et al., 2010; Hu et al., 2010; Kim et al., 2007; Kobayashi et al., 2008). Additional BACE1 substrates have been identified by quantitative proteomics analysis (Dislich et al., 2015; Hemming, Elias, Gygi, & Selkoe, 2009; Kuhn et al., 2012) but only some of them have been validated (Table 8.2). All together these findings shed some light on BACE1 function but also raise concerns about the possible side effects of therapeutic BACE1 inhibition (for review, see Vassar et al., 2014; Yan and Vassar, 2014).

BACE1 Transgenic Models

In addition to BACE1 null models, several lines of mice overexpressing human BACE1 (hBACE1) were also generated (Table 8.3) to investigate the consequences of BACE1 elevation *in vivo* (Bodendorf et al., 2002; Harrison et al., 2003; Lee et al., 2005; Plucińska et al., 2014).

Bodendorf et al. (2002) generated a transgenic model overexpressing hBACE1 under the control of Thy-1 promoter. BACE1 highest expression was found in the neocortex and hippocampus, while the lowest expression was detected in cerebellum. The overexpression of hBACE1 in neuronal cells of the brain promoted the amyloidogenic processing of APP *in vivo*, indeed total endogenous full-length APP was significantly reduced in hBACE1 transgenic mice (Bodendorf et al., 2002).

Harrison et al. (2003) generated both a BACE1 KO and a transgenic model overexpressing hBACE1. BACE1 transgenic mice showed between a four- and tenfold increase of BACE1 (human and murine) levels compared to WT mice. In this model hBACE1 expression was driven by the Cam Kinase II (CaMKII) promoter. hBACE1 expression was detected in hippocampus, cortex, caudate putamen, and striatum. Behavioral characterization of mice expressing hBACE1 showed a less anxious phenotype, even though none of the results reached statistical significance. In addition, transgenic mice displayed lower body weight and increase in 5-HT turnover in different brain areas (cerebellum, hippocampus, hypothalamus, nucleus accumbens, and caudate striatum). Also observed was a decrease in the level of dopamine in the hypothalamus and nucleus accumbens. It was clear that these mice have a biochemical systemic alteration in which

TABLE 8.2   Validated Substrates of BACE1

Protein name	Function	Localization	References
APP	Multiple functions	Plasma membrane Endosomes	Hussain et al. (1999), Vassar et al. (1999), Sinha et al. (1999), Yan et al. (1999), Lin et al. (2000)
APLP1	Multiple functions	Plasma membrane	Eggert et al. (2004), Li and Südhof (2004)
APLP2	Multiple functions	Plasma membrane	Pastorino et al. (2004), Eggert et al. (2004), Li and Südhof (2004)
NRG1 NRG3	Growth factor	Plasma membrane	Hu et al. (2006), Willem et al. (2006), Hu et al. (2008)
ST6Gal I	Sialytranferase	Golgi apparatus	Kitazume et al. (2001)
PSGL-1	Leuckocyte adhesion	Plasma membrane	Lichtenthaler et al. (2003)
Na_v1	Voltage dependent sodium channel	Plasma membrane	Kim et al. (2005), Wong et al. (2005)
LRP	Endocytosis	Plasma membrane	Arnim et al. (2005)
IL-1R2	Interleukin-1 decoy receptor	Plasma membrane	Kuhn et al. (2007)
Contactin-2	Adhesion molecule	Plasma membrane	Gautam et al. (2014)
Jagged1	Signaling molecule	Plasma membrane	Hu et al. (2013)
CHL1	Adhesion molecule	Plasma membrane	Hitt et al. (2012), Kuhn et al. (2012), Zhou et al. (2012)
L1	Adhesion molecule	Plasma membrane	Hitt et al. (2012), Kuhn et al. (2012), Zhou et al. (2012)
VEGFR1	Growth factor	Plasma membrane	Cai et al. (2012)

serotoninergic and dopaminergic systems are involved. Interestingly, a dysfunction in cholinergic and serotoninergic systems was found in AD patients. However more studies are required to investigate whether this dysfunction is caused by BACE1 overexpression itself or indirectly by increased Aβ levels.

Recently, the first knockin (KI) model expressing hBACE1 was generated (Plucińska et al., 2014). The sequence of hBACE1 was introduced

TABLE 8.3 Relative Phenotypes of Transgenic and Knockin Models of BACE1

Phenotypes of BACE1 transgenic/KI mice	Promoter used to express BACE1	References
Reduced anxiety Lower body weight Higher 5-HT turnover Lower dopamine levels	CaMKII	Harrison et al. (2003)
Lower APP full-length Higher Aβ	Thy-1	Bodendorf et al. (2002)
Delayed habituation Elevated locomotor activity Anxiolytic behavior	CaMKII	Plucińska et al. (2014)

under the control of the CaMKII promoter in the safe HPRT locus by targeted KI. Unlike transgenic models, where the gene integration is random, KI mice are generated with a targeted integration of the transgene in a specific locus. Young transgenic animals displayed normal body weight that was decreasing with age (from 6 months for males and 12 months for females). The behavioral characterization showed delayed habituation, global reduction in locomotor activity linked to age, and an anxiolytic phenotype. No motor impairment or working memory deficit was detected in these mice. BACE1-mediated proteolysis of endogenous APP was found to be increased in both 6- and 12-month-old transgenic mice compared with age-matched WT mice. βAPP-CTF fragment, the direct product of BACE1 cleavage of APP, was also prominently increased in forebrains of transgenic mice. High levels of dodecameric and hexameric Aβ in brain extract of transgenic mice were detected. Interestingly, no monomeric Aβ was detected. Elevated Aβ extraneuronal accumulation was found in CA1, dentate gyrus, and forebrain areas. Astrocyte activation was also observed and used as a marker for inflammatory reactions in 12-month-old transgenic mice. Highest levels of astrogliosis were encountered within the dentate gyrus, but also in CA1 and piriform cortex. In conclusion the overexpression of hBACE1 causes an increased processing of endogenous APP and increased Aβ synthesis, together with systemic behavioral and physiological alterations.

BACE1 in AD

The generation of BACE1 null mice has clearly demonstrated that BACE1 is the major β-secretase and is the limiting enzyme for Aβ generation. According to the amyloid hypothesis, which states that Aβ production and its accumulation initiates a cascade of events causing AD

(Hardy and Higgins, 1992), BACE1 plays a key role in AD pathogenesis. In support of the latter, autosomal dominant mutations responsible for the familial form of AD (FAD) have been found in the APP sequence near the β-secretase site. Two of these mutations are located near the β-secretase site and result in increased Aβ production. These are the Swedish double mutation (K670N/M671L) (Mullan et al., 1992) and the A673V mutation (Di Fede et al., 2009). Conversely, it was reported that a coding mutation on APP sequence close to the β-secretase site (A673T) is protective against AD and reduces Aβ production by 40% *in vitro* (Jonsson et al., 2012). Moreover, several groups have reported that BACE1 levels are increased in postmortem AD brain samples, suggesting that BACE1 elevation may also play a role in sporadic AD (Fukumoto, Cheung, Hyman, & Irizarry, 2002; Holsinger, McLean, Beyreuther, Masters, & Evin, 2002; Li et al., 2004; Tesco et al., 2007; Tyler, Dawbarn, Wilcock, & Allen, 2002; Yang et al., 2003). In order to study the consequences of BACE1 overexpression on AD phenotypes, bigenic mice models coexpressing hBACE1 and human APP (hAPP) were generated (Table 8.4).

TABLE 8.4   Transgenic or BACE1 Mice Crossed with AD Models and the Consequences on Aβ Levels in Each Model

BACE1	AD line used	Effect on Aβ level	References
Tg hBACE1	APP51/22 line APP23 line	↑	Bodendorf et al. (2002)
Tg hBACE1	hAPPSWE	↑	Mohajeri et al. (2004)
Tg hBACE1	APP[V7171] line	↑	Willem et al. (2004)
Tg hBACE1	Tg2576 line	↑	Chiocco et al. (2004)
Tg hBACE1	YAC APP R1.40 line	↑	Chiocco and Lamb (2007)
Tg hBACE1	Tg2576 line	↓	Lee et al. (2005)
Tg hBACE1	hAPP line 41	↓	Rockenstein et al. (2005)
BACE1$^{-/-}$	Tg2576 line	↓	Ohno et al. (2004)
BACE1$^{-/-}$	APPSweΔE9	↓	Laird et al. (2005)
BACE1$^{-/-}$	5XFAD	↓	Ohno et al. (2007)
BACE1$^{+/-}$	APPSweΔE9	↓	Laird et al. (2005)
BACE1$^{+/-}$	5XFAD	↓	Kimura et al. (2010)
BACE1$^{+/-}$	5XFAD	Female ↓ Male =	Sadleir et al. (2015)
BACE1$^{+/-}$	PDAPP	↓	McConlogue et al. (2007)

Bodendorf et al. (2002) generated a double transgenic model expressing hBACE1 (Thy1 promoter) as well as hAPP751 (the APP51/22 line). In this line the reduction of full-length hAPP was accompanied by a significant increase in the carboxyl-terminal metabolites generated by β-secretase and in sAPPβ, the complementary metabolite of APP-C99. Furthermore, the hBACE1 line was crossed with the APP23 mice. This AD model expresses the hAPP Swedish mutation (hAPPswe). In this line, the presence of hBACE1 significantly reduced full-length hAPPswe and elevated levels of the Aβ peptides compared hBACE1/hAPP751 bigenic mice confirming that APPswe is a better substrate for BACE1 than APP wild type (Bodendorf et al., 2002).

Similarly, Mohajeri, Saini, and Nitsch (2004) investigated the effect of hBACE1 (Thy1 promoter) overexpression on the cleavage of hAPPswe (a line expressing APPswe under the control of the PrP promoter). In this bigenic line, Aβ levels were increased and AD-like pathology was accelerated. Indeed, early signs of extracellular Aβ deposition and reactive astrocytes were found in two-month-old bigenic mice, but not in single hAPPswe transgenic mice. Furthermore, well-defined amyloid deposits surrounded by activated astrocytes could be detected at only four months of age in the double transgenic mice. Thus, simultaneous expression of hBACE1 and its substrate (hAPPswe) lead to an accelerated amyloid plaque formation in this model (Mohajeri et al., 2004).

A different bigenic line was generated overexpressing hBACE1 (Thy1 promoter) and APP[V717I] mice (a line expressing APP with the London mutation) (Willem et al., 2004). The single transgenic APP[V717I] mice developed a robust AD-related phenotype with early cognitive and behavioral deficits, followed by widespread amyloid plaques in brain parenchyma. In heterozygous animals for both transgenes the level of mature APP full-length was reduced considerably in the brain of double transgenic mice at age 16 and 22 month of age while the level of immature APP was similar to the one detected in the single transgenic APP[V717I]. This could be due to similar expression of APP but increased turnover of mature APP caused by higher BACE1 activity. Increased levels of total $A\beta_{40}$ and total $A\beta_{42}$ in the brain of double transgenic mice were detected. In addition, in accordance with an increase in Aβ levels, an increase in amyloid plaques was observed in the cerebral cortex. On the other hand a decrease in vascular amyloid deposition was found in double transgenic mice (Willem et al., 2004).

Chiocco et al. (2004) generated a new model with the use of a genomic bacterial artificial chromosome (BAC) clone containing the full-length human BACE1 gene. The hBACE1 mouse was then crossed with the Tg2576 line to obtain a bigenic mouse (BACE1/Tg2576). The transgenic animals express elevated levels of hBACE1, localized primarily to olfactory bulb neurons, cortical layers, and hippocampus. BACE1/Tg2576 were characterized by elevated levels of Aβ species as well as significant increases of Aβ deposits. Interestingly, bigenic mice had significantly higher levels of Aβ in the olfactory bulb and cortex, a profile overlapping with the regions where BACE1 expression is the highest. This suggests that BACE1 overexpression may control Aβ brain-regional deposition pattern and AD-like neuropathology. To further investigate the spatial and temporal pattern of Aβ production throughout postnatal development, Chiocco and Lamb (2007) crossed the human BACE1 BAC transgenic line (Chiocco et al., 2004) with the YAC APP R1.40 transgenic line. BACE1 levels were significantly higher in neonatal than adult mice and this corresponded to a parallel higher Aβ production.

In contrast with these findings other groups found that BACE1 overexpression decreased Aβ deposition despite enhanced β-cleavage of APP in a transgenic mouse model. Lee et al. (2005) generated three lines of transgenic mice expressing different levels of hBACE1 driven by the prion protein promoter. BACE1 transgenic lines were crossed with the Tg2576 mouse model (overexpressing APP with Swedish mutation), obtaining the BACE1/APP line. ELISA quantification of Aβ derived from both exogenous and endogenous APP in young mice (between 5 and 7 months of age) showed decreased Aβ levels. Thus, amyloid plaque pathology was investigated in old BACE1/APP mice (between 14 and 16 months of age). Despite the high levels of BACE1 expression, amyloid burden was reduced in BACE1/APP mice compared to the Tg2576 in cortex and hippocampus. Further experiments showed that BACE1 overexpression altered Aβ deposition via a mechanism related to the spatial location of APP processing. Indeed BACE1 overexpression reduced APP axonal transport, thus inducing β-cleavage earlier in the secretory pathway. Overexpression of both APP and BACE1 could potentially result in aberrant trafficking of either proteins (Lee et al., 2005). Similarly, Rockenstein et al. (2005) generated a bigenic mouse expressing hBACE1 and hAPP line 41, which expresses mutant hAPP751 with both the Swedish and the London mutations under the control of the mThy1 promoter. hBACE1/hAPP41 double transgenic mice had increased levels of APP C-terminal fragments and decreased levels of full-length APP and Aβ. Bigenic hBACE1/hAPP41 mice showed degeneration of neurons in the neocortex and hippocampus and degradation of myelin. These results demonstrate that high levels of BACE1 activity are sufficient to elicit neurodegeneration and neurological decline *in vivo*. In addition, this pathogenic pathway does not involve the accumulation of Aβ, but rather other APP toxic fragments (e.g., APP-CTFs) or misprocessing of other BACE1 substrates. It remains unclear whether the opposite effect on Aβ levels in the different bigenic mouse lines depend on the different mouse genetic background or, most likely, on the levels of BACE1 transgene overexpression. While low and intermediate levels of BACE1 overexpression increase Aβ production, high levels result in decreased Aβ most likely owing to impaired APP trafficking.

In order to determine whether BACE1 inhibition prevents the development of AD phenotypes, several groups have crossed BACE1 null mice with different lines of APP transgenic mice (Table 8.4). Ohno et al. (2004) crossed BACE1 KO with transgenic mice that overexpress human APPswe (Tg2576 line). Spatial memory was investigated: Tg2576 mice displayed a spatial working memory deficit that was rescued to wild-type control level in the bigenic mice. Moreover, Aβ generation was nearly abolished in the bigenic line. It was previously demonstrated that in the Tg2576 line cholinergic modulation of GABAergic transmission is impaired in the cerebral cortex (Zhong et al., 2003); in addition, cholinergic transmission in the hippocampus is a key event in memory formation. It was observed that cholinergic regulation of hippocampal neuronal excitability was rescued in the bigenic mice. This functional recovery associated with the lower production of Aβ can be the cause for memory improvement observed in this line (Ohno et al., 2004). Accordingly, Laird et al. (2005) demonstrated that deletion of BACE1 in APPswe/PS1ΔE9 mice prevents both Aβ deposition and age-associated cognitive abnormalities that occur in this model of AD. Moreover, the authors measured Aβ level in the three different lines BACE1$^{+/+}$/APPswe/PS1ΔE9, BACE1$^{+/-}$/APPswe/PS1ΔE9, and BACE1$^{-/-}$/APPswe/PS1ΔE9 showing that Aβ burden is sensitive to BACE1 dosage. Similarly, McConlogue et al. (2007) generated a different model by crossing BACE1 KO mice with PDAPP (APP with the V717F Indiana mutation) mice. The PDAPP mouse line 109 is characterized by aggressive plaque pathology. The importance of this study is to compare the effects of homo- or heterozygous ablation of BACE1 on Aβ generation and plaque load. As expected, the homozygous ablation of BACE1 reduced plaque and synaptic pathologies in the PDAPP mouse model. Ablation of a single BACE1 allele had only a modest effect on Aβ levels (12% reduction), but was sufficient to reduce plaque and synaptic pathologies in these mice. This small decrease of Aβ in young mice (3 months old) reduced plaque formation in older mice (90% less plaques in 13-month-old mice). Also observed was a drastic reduction of the dystrophic neurites around amyloid plaques. Thus a modest reduction in Aβ could be beneficial in AD patients to lower plaque load and synaptic dysfunction (McConlogue et al., 2007).

The effect of BACE1 deletion was also studied in another AD model, the 5XFAD transgenic line (Ohno et al., 2007). This mouse line expresses five familial AD mutations (APPK670N/M671L, APPI716V, APPV717I, PS1M146L, PS1L286V) and develops cerebral amyloid plaques and gliosis at two months of age, achieves massive Aβ$_{42}$ burdens, and has reduced synaptic markers, increased p25 levels, neuron loss, and memory impairment in the Y-maze (Oakley et al., 2006). Neuronal loss in cortical layer 5 and subiculum starts at nine months of age (Oakley et al., 2006). Moreover, BACE1 levels become elevated in neurons around amyloid plaques and increase concurrently with amyloid burden (Zhao et al.,

2007). BACE1-deficient 5XFAD (BACE1$^{-/-}$/5XFAD) mice did not have amyloid plaques, astrogliosis, or memory deficits found in age-matched 5XFAD mice. Most importantly, neuronal death in the cerebral cortex and subiculum was prevented in the BACE1$^{-/-}$/5XFAD mice. This was the first demonstration of genetic rescue of neuronal loss via ablation of BACE1, and consequently of Aβ, in an AD transgenic mouse model and provides strong evidence that Aβ ultimately kills neurons *in vivo* (Ohno et al., 2007).

To address the question of what degree of BACE1 inhibition is necessary to prevent the development of AD-like phenotypes, Kimura, Devi, and Ohno (2010) investigated the effect of partial deletion of BACE1 in 5XFAD mice (BACE1$^{+/-}$/5XFAD). Immunoblot analysis of hemibrain homogenates demonstrated that BACE1$^{+/-}$ genotype significantly reduced cerebral BACE1 expression (by 50%) in 5XFAD mice in concordance with the reduction of gene copy number. However the full-length APP expression levels were not different between BACE1$^{+/-}$/5XFAD and 5XFAD mice. Levels of sAPPβ and APP-C99 were significantly reduced in BACE1$^{+/-}$/5XFAD mice compared to 5XFAD. The lower expression of BACE1 significantly decreased levels of Aβ$_{40}$ and Aβ$_{42}$ and amyloid burden in hippocampus and anterior cingulate cortex compared to the respective areas of 5XFAD littermate control mice at six months of age. 5XFAD mice show hippocampal synaptic and memory dysfunctions and also exhibit remote memory destabilization caused by cortical Aβ accumulation at four to 6 months of age. In contextual fear conditioning, BACE1$^{+/-}$/5XFAD mice showed significantly higher levels of freezing than the 5XFAD mice. Moreover, 5XFAD mice showed deficits in spatial working memory at 6 months of age that were rescued in BACE1$^{+/-}$/5XFAD. Thus, BACE1$^{+/-}$ gene deletion was efficacious in preventing memory deficits in 5XFAD mice. In addition, 5XFAD mice showed hippocampal synaptic dysfunctions and reductions in basal transmission and LTP at 6 months of age. In BACE1$^{+/-}$/5XFAD the LTP reduction was rescued, but not the impaired basal transmission. Given that BDNF and its receptor TrkB are critically involved in the late phase of LTP, BDNF-TrkB signaling pathways were investigated in BACE1$^{+/-}$/5XFAD mice, in order to address potential mechanisms underlying the LTP maintenance in this line. BDNF levels were significantly reduced in 6-month-old 5XFAD mice compared to wild-type controls while BDNF levels were restored to wild-type levels in BACE1$^{+/-}$/5XFAD mice. Therefore, BACE1$^{+/-}$ deletion improved deficiencies in hippocampal BDNF-TrkB signaling in 5XFAD mice in line with the rescue of their CA1 LTP impairments, in addition to improved memory deficit and lowered Aβ levels (Kimura et al., 2010).

More recently, Sadleir, Eimer, Cole, and Vassar (2015) reported that plaques and Aβ levels were reduced only in BACE1$^{+/-}$/5XFAD females, but not in males. They showed that 5XFAD females have higher levels of

$A\beta_{42}$ and full-length APP than males, most likely owing to the presence of an estrogen response element in the Thy-1 promoter. Thus, the authors hypothesized that the lower level of expression of the APP transgene in 5XFAD males leads to a situation where BACE1 is in excess of APP, so that no $A\beta_{42}$ lowering occurs in BACE1$^{+/-}$/5XFAD mice. In contrast, because of higher transgenic APP levels in 5XFAD females, BACE1 is not in excess over APP, thus resulting in substantial $A\beta_{42}$ lowering with 50% BACE1 reduction (Sadleir et al., 2015). These results are in contrast with previous studies by Kimura et al. (2010) that showed $A\beta$ decrease in BACE1$^{+/-}$/5XFAD mice. However, in the latter paper the authors did not specify the sexes used for their study. Thus, these contrasting results remain to be clarified.

BACE1 Is a Stress-Induced Protease Elevated in AD and Acute Brain Injuries

Several studies have shown that BACE1 protein levels and β-secretase activity are increased in AD brains (Fukumoto et al., 2002; Holsinger et al., 2002; Li et al., 2004; Tesco et al., 2007; Tyler et al., 2002; Yang et al., 2003). Moreover increasing evidence suggests that BACE1 is a stress-induced protease. BACE1 levels increase in cells exposed to oxidative stress (Tamagno et al., 2002, 2003, 2005; Tong et al., 2005) and apoptosis (Tesco et al., 2007) in *in vivo* animal models following traumatic brain injury (TBI) (Blasko et al., 2004; Chen et al., 2004; Loane et al., 2009; Walker et al., 2012; Yu, Zhang, & Chuang, 2012), cerebral ischemia (Tesco et al., 2007; Wen, Onyewuchi, Yang, Liu, & Simpkins, 2004), and impaired energy metabolism (O'Connor et al., 2008; Velliquette, O'Connor, & Vassar, 2005). To date, several transcriptional and posttranscriptional mechanisms have been reported to regulate BACE1 levels (Willem, Lammich, & Haass, 2009). The regulation of BACE1 levels seems to be mediated by molecular mechanisms influenced by both genetic and environmental factors. Thus, BACE1 elevation may be the first step in increasing $A\beta$ and triggering AD pathology, at least in the sporadic cases. The identification of the molecular pathways that regulate BACE1 is expected to shed some light on the mechanisms leading to BACE1 accumulation in AD brains and to lead to the discovery of novel therapeutic targets for the treatment of AD.

We have identified a GGA3-dependent mechanism regulating BACE1 levels and activity (Tesco et al., 2007). We reported that BACE1 protein levels and consequently, β-secretase cleavage of APP, are potentiated during apoptosis. In exploring the mechanism underlying this novel apoptotic event, we found that the elevation in BACE1 following caspase activation is due to posttranslational stabilization owing to a significant impairment in the degradation and turnover of BACE1. Given our previous findings that BACE1 is normally degraded in the lysosomes and since the C-terminal di-leucine motif is required for trafficking to lysosomes, we next investigated the fate of GGA3 during apoptosis. We discovered that GGA3 is a novel caspase 3 substrate that is cleaved during apoptosis. This was shown both *in vitro*, in cell cultures, and *in vivo*, in a rat model of cerebral ischemia. Moreover, we observed that as GGA3 is removed by caspase cleavage, BACE1 is increasing following ischemia. To investigate whether decreased GGA3 protein levels may account for increased BACE1 levels and β-secretase activity in AD brains, we measured levels of BACE1 and GGA3 in a series of AD and nondemented (ND) control brain samples. While BACE1 protein levels were significantly increased in the AD temporal cortex, in contrast, GGA3 protein levels were significantly decreased and were inversely correlated with BACE1 levels in the AD group, but not in the ND control group. We next analyzed BACE1 and GGA3 levels in the cerebellum, which is usually spared of AD pathology, and found that GGA3 levels were decreased in AD subjects. These data suggest that levels of GGA3 are decreased independently of Aβ pathology (e.g., owing to genetic factors) and that subjects with lower levels of GGA3 could be at risk of developing AD. This may be also true for patients with stroke and TBI in which caspase activation occurs even in the chronic period after injury (Endres et al., 1998; Lenzlinger, Marx, Trentz, Kossmann, & Morganti-Kossmann, 2002; Uzan et al., 2006).

More recently, we have found that GGA3 is depleted while BACE1 levels increase in the acute phase post-TBI (Walker et al., 2012). We have demonstrated the role of GGA3 in the regulation of BACE1 *in vivo* by showing that BACE1 levels are increased in the brain of GGA3 null mice and that *Gga3* deletion leads to increased β-secretase activity with aging. We next asked to what extent the deletion of GGA3 affects BACE1 elevation following TBI; we found that head trauma potentiates BACE1 elevation in GGA3 null mice at 48h post-TBI. Consequently, these findings indicate that in addition to the GGA3 mediated posttranslational stabilization of BACE1 other mechanisms also contribute to BACE1 accumulation in the acute phase postinjury. In an effort to find other mechanisms responsible for the BACE1 elevation observed at 48h post-TBI, we discovered that GGA1, a homolog of GGA3, is depleted by caspase cleavage both *in vitro* following apoptosis and *in vivo* at 48h post-TBI. Furthermore, GGA1 silencing potentiates BACE1 elevation induced by GGA3 deletion in neurons *in vitro*. Thus, we conclude that depletion of GGA1 by RNAi-mediated silencing or by caspase activation following TBI potentiates the BACE1 elevation produced by GGA3 deletion *in vitro* and *in vivo*, respectively. Importantly we have shown that decreased levels of GGA1 are associated with the depletion of GGA3 and BACE1 elevation observed in a series of postmortem AD brains. These findings confirm and extend a previous report showing that GGA1 levels were significantly decreased (~40%) in AD brains (Wahle et al., 2006). Collectively, our data indicate that depletion of GGA1 and GGA3 synergistically elevate BACE1 levels and suggest that the BACE1 elevation observed in AD brains is mediated by the concurrent depletion of GGA1 and GGA3 (Walker et al., 2012). To date, several mechanisms have been proposed to explain the increased accumulation of BACE1 in AD brains: depletion of GGA3 and GGA1 (Tesco et al., 2007; Walker et al., 2012); increased phosphorylation of translation factor eIF2α (O'Connor et al., 2008); increased expression of a non coding antisense BACE1 transcript (Faghihi et al., 2008); and decreased expression of the BACE1 regulating microRNAs, miR-29 and miR-107 (Hébert et al., 2008; Wang et al., 2008).

In addition to the posttranslational regulation of BACE1 via caspase mediated depletion of GGA3 and GGA1, BACE1 levels can also be regulated at the transcriptional and translational level following acute brain injuries. A number of known and hypothesized BACE1 regulating factors are acutely altered (usually 3–24 h postinjury) following experimental TBI in rodents. These include activation/upregulation of the well-known transcriptional molecules: STAT1 (Zhao, O'Connor, & Vassar, 2011), STAT3 (Oliva, Kang, Sanchez-Molano, Furones, & Atkins, 2012; Zhao et al., 2011), HIF-1α (Anderson, Sandhir, Hamilton, & Berman, 2009), NF-β (Sanz, Acarin, González, & Castellano, 2002), and TNFα (a potent activator of NF-κβ pathways) (Lotocki, Alonso, Dietrich, & Keane, 2004). Moreover eiF2α phosphorylation is increased in the hippocampus of rodents 24 h postinjury following fluid percussion injury (Singleton, Zhu, Stone, & Povlishock, 2002), while TBI induced by controlled cortical impact in rodents has been shown to acutely decrease the levels of miR-107 and miR-328 (Redell, Liu, & Dash, 2009; Wang et al., 2010). Thus, in addition to GGA3 and GGA1 depletion, any or all of these additional BACE1 regulatory mechanisms may also contribute to the elevation of BACE1 observed in the acute phase postinjury. However, in order to confirm the contribution of these other mechanisms, postinjury levels of BACE1 would need to be analyzed in animals in which these molecules or pathways have been pharmacologically or genetically inhibited. At this stage these studies are currently missing. To date, this is the first study providing evidence for a molecular mechanism of BACE1 elevation following TBI taking advantage of a novel mouse model null for *Gga3*.

In the same study, we also attempted to address the important question of how acute brain injuries (e.g., stroke and head trauma) result in chronic neurodegeneration. We have previously demonstrated that GGA3 regulates BACE1 degradation by trafficking BACE1 to the lysosomes (Kang et al., 2010; Tesco et al., 2007). Thus, GGA3 is expected to play a key role in the disposal of the BACE1 that accumulates during the acute phase postinjury. Consequently, we set out to investigate the effect of GGA3 haploinsufficiency (which best resembles the depletion of GGA3 observed in AD brains) on BACE1 levels in the subacute phase of injury (7 day post-TBI). We found that GGA3 haploinsufficiency results in sustained elevation of BACE1 and Aβ levels in the subacute phase of injury when GGA1 levels are restored.

In conclusion, our data indicate that depletion of GGA1 and GGA3, and most likely additional transcriptional and posttranscriptional mechanisms (Rossner, Sastre, Bourne, & Lichtenthaler, 2006; Vassar, Kovacs, Yan, & Wong, 2009), engender a rapid and robust elevation of BACE1 in the acute phase postinjury. However, the efficient disposal of the acutely accumulated BACE1 solely depends on GGA3 levels in the subacute phase of injury. As a consequence, impaired degradation of BACE1 (e.g., owing to GGA3 haploinsufficiency) represents an attractive molecular mechanism linking acute brain injury to chronic Aβ production and neurodegeneration. Persistent Aβ elevation would be predicted to result in further caspase activation and ensuing GGA1 and GGA3 depletion. According to our findings, this would serve to further elevate BACE1 and Aβ levels leading to a vicious cycle in individuals affected by TBI. As such, our data strongly support the hypothesis that subjects with lower levels of GGA3 may be at increased risk to develop AD following acute brain injury, whether it be stroke, TBI, or some other form of major brain insult. Regulation of BACE1 levels seems to be mediated by molecular mechanisms influenced by both genetic and environmental factors. Thus, BACE1 elevation may be the first step in increasing Aβ and triggering AD pathology, at least in the sporadic cases. The identification of the molecular mechanisms that regulate BACE1 is expected to lead to the discovery of novel therapeutic targets for the treatment and/or prevention of AD.

Although several groups (including ours) have reported that BACE1 levels increase following experimental TBI (Blasko et al., 2004; Chen et al., 2004; Loane et al., 2009; Uryu et al., 2007; Walker et al., 2012; Yu et al., 2012), the effect of BACE1 deletion on brain injury remains controversial. One study reported that BACE1 genetic deletion ameliorates motor and cognitive deficits and reduces cell loss after experimental TBI in mice (Loane et al., 2009). However, another study showed the BACE1 deletion worsened functional outcome after TBI in mice (Mannix et al., 2011). Thus, it remains unclear the extent to which BACE1 inhibition could be a therapeutic treatment for TBI.

BACE Inhibition, a Therapeutic Approach for AD

Aβ synthesis has a central role in the etiology of AD (Hardy and Higgins, 1992), thus regulating Aβ levels could be a straightforward therapeutic strategy. Both β- and γ-secretase are key elements in the regulation of Aβ and are considered promising targets for drug development. While genetic deletion of presenilin-1, the catalytic subunit of the γ-secretase complex, resulted in an embryonic lethal phenotype (De Strooper et al., 1998; Shen et al., 1997), initial studies showed that genetic deletion of BACE1 did not produce any overt phenotype, suggesting that inhibiting BACE1 pharmacologically would be a straightforward strategy for AD treatment (Cai et al., 2001; Dominguez et al., 2005; Luo et al., 2001; Roberds et al., 2001). Many different classes of BACE1 inhibitors have been designed and developed. These inhibitors can be largely classified as either peptidomimetic or nonpeptomimetic inhibitors. Initially, the inhibitors were noncleavable peptide-based modeled after the β-secretase cleavage site of APP. *In vitro*, these peptidomimetic molecules were potent BACE inhibitors. However *in vivo* these inhibitors showed several weaknesses: oral bioavailability, long serum half-life, or blood–brain barrier (BBB) penetration. New classes of inhibitors were generated to solve these problems, and a second and third generation followed. Selective active site small molecule inhibitors of BACE (targeting both BACE1 and BACE2) are currently in clinical trials and show promise for the prevention and/or treatment of AD and for a detailed discussion of these topics we refer to recent reviews (Vassar, 2014; Yan and Vassar, 2014).

Another class of therapeutic strategy for BACE1 inhibition involves immunotherapy approaches to reduce BACE1 processing of APP (Wang, Liu, & Lazarus, 2013). The benefit of using antibodies as therapeutic agents is that they can provide better selectivity. The major issue of using antibodies for CNS targets is due to problems crossing the BBB. An anti-BACE1 antibody that does not bind to the active site but rather binds at an exosite was recently developed (Atwal et al., 2011; Yu et al., 2011). Anti-BACE1 antibody was shown to reduce endogenous BACE1 activity and Aβ production in the human cell line 293-HEK expressing APP and in cultured primary neurons. The structure obtained with x-ray crystallography suggests that anti-BACE1 antibody binds to this exosite and can reduce catalytic efficiency toward APP cleavage. In addition, CNS Aβ levels were reduced in both mice (the Tg2576 line) and monkeys (Atwal et al., 2011). This antibody was ameliorated with the generation of a bispecific antibody, which allowed crossing the BBB. This bispecific antibody achieved a bigger reduction in CNS Aβ levels (Yu et al., 2011).

An attractive alternative approach to the direct inhibition of BACE is represented by the search for therapeutic targets aimed to modulate BACE1 activity. Similarly, γ-secretase modulation is actively pursued given that γ-secretase inhibition not only did not improve but worsened both cognition and function in a phase III clinical trial, most likely owing to mechanism-based side effects (Doody et al., 2013). It has been widely demonstrated that BACE1 undergoes complex regulation at the transcriptional, translational, and posttranslational levels (Rossner et al., 2006; Vassar et al., 2009). Each one of these steps has a role in regulating BACE1 levels and its activity. Thus, another possible approach could be to regulate BACE1 gene expression, its interaction with cellular components, or its degradation (for review see Klaver and Tesco, 2014). For example, the search for therapeutic targets (e.g., GGA3 and GGA1) aimed to restore normal levels of BACE1 is particularly important for conditions associated with BACE1 elevation and increased risk to develop AD (acute brain injuries) and represents an attractive alternative approach to the direct inhibition of BACE1.

Although therapies aimed to reduce β-secretase activity represent a promising treatment for AD, it remains unclear whether long-term BACE inhibition will produce mechanism-based side effects (for reviews see Ghosh, Brindisi, & Tang, 2012; Vassar, 2014; Vassar et al., 2014; Yan and Vassar, 2014). While at least some of the complex neurological abnormalities observed in the BACE1 null mice could be developmental in nature (Yan and Vassar, 2014), recent studies have shown that pharmacological inhibition of BACE1 in adult mice resulted in retinal pathology (Cai et al., 2012), spindle deficits (Cheret et al., 2013), and cognitive deficits (Filser et al., 2015) continuing to raise concerns. Likely the most successful approach to reduce Aβ pathology will be to use a BACE1 inhibitor at a dose that does not produce mechanism-based effects in association with therapies aimed to clear Aβ (e.g., anti-Aβ antibody) (DeMattos et al., 2014; Jacobsen et al., 2014).

Another challenging question that rises from BACE1 inhibitor clinical development concerns the level of BACE1 inhibition required and the stage of AD at which to treat for optimal efficacy. In the future the combination of biomarker analysis and imaging studies will help to determine the appropriate therapeutic approach for each patient.

References

Alzheimer's Association, (2015). Alzheimer's disease facts and figures. *Alzheimers Dementia Journal of Alzheimers Association*, 11, 332–384.

Anderson, J., Sandhir, R., Hamilton, E. S., & Berman, N. E. J. (2009). Impaired expression of neuroprotective molecules in the HIF-1alpha pathway following traumatic brain injury in aged mice. *Journal of Neurotrauma*, 26, 1557–1566.

Annaert, W., et al. (2012). US Patent Application 20120276076.

Arnim, C. A. F., von Kinoshita, A., Peltan, I. D., Tangredi, M. M., Herl, L., Lee, B. M., et al. (2005). The low density Lipoprotein Receptor-related Protein (LRP) is a novel β-secretase (BACE1) substrate. *The Journal of Biological Chemistry*, 280, 17777–17785.

Atwal, J. K., Chen, Y., Chiu, C., Mortensen, D. L., Meilandt, W. J., Liu, Y., et al. (2011). A therapeutic antibody targeting BACE1 inhibits amyloid-β production *in vivo*. *Science Translational Medicine*, 3, 84ra43.

Blasko, I., Beer, R., Bigl, M., Apelt, J., Franz, G., Rudzki, D., et al. (2004). Experimental traumatic brain injury in rats stimulates the expression, production and activity of Alzheimer's disease beta-secretase (BACE1). *Journal of Neural Transmission Vienna Austria*, 1996(111), 523–536.

Bodendorf, U., Danner, S., Fischer, F., Stefani, M., Sturchler-Pierrat, C., Wiederhold, K.-H., et al. (2002). Expression of human beta-secretase in the mouse brain increases the steady-state level of beta-amyloid. *Journal of Neurochemistry*, 80, 799–806.

Buggia-Prévot, V., & Thinakaran, G. (2015). Significance of transcytosis in Alzheimer's disease: BACE1 takes the scenic route to axons. *BioEssays*, 37, 888–898.

Buxbaum, J. D., Liu, K. N., Luo, Y., Slack, J. L., Stocking, K. L., Peschon, J. J., et al. (1998). Evidence that tumor necrosis factor alpha converting enzyme is involved in regulated alpha-secretase cleavage of the Alzheimer amyloid protein precursor. *The Journal of Biological Chemistry, 273*, 27765–27767.

Cai, H., Wang, Y., McCarthy, D., Wen, H., Borchelt, D. R., Price, D. L., et al. (2001). BACE1 is the major β-secretase for generation of Aβ peptides by neurons. *Nature Neuroscience, 4*, 233–234.

Cai, J., Qi, X., Kociok, N., Skosyrski, S., Emilio, A., Ruan, Q., et al. (2012). β-Secretase (BACE1) inhibition causes retinal pathology by vascular dysregulation and accumulation of age pigment. *EMBO Molecular Medicine, 4*, 980–991.

Cao, L., Rickenbacher, G. T., Rodriguez, S., Moulia, T. W., & Albers, M. W. (2012). The precision of axon targeting of mouse olfactory sensory neurons requires the BACE1 protease. *Science Report, 2*, 231.

Chen, X.-H., Siman, R., Iwata, A., Meaney, D. F., Trojanowski, J. Q., & Smith, D. H. (2004). Long-term accumulation of amyloid-beta, beta-secretase, presenilin-1, and caspase-3 in damaged axons following brain trauma. *The American Journal of Pathology, 165*, 357–371.

Cheret, C., Willem, M., Fricker, F. R., Wende, H., Wulf-Goldenberg, A., Tahirovic, S., et al. (2013). Bace1 and Neuregulin-1 cooperate to control formation and maintenance of muscle spindles. *The EMBO Journal, 32*, 2015–2028.

Chiocco, M. J., Kulnane, L. S., Younkin, L., Younkin, S., Evin, G., & Lamb, B. T. (2004). Altered amyloid-beta metabolism and deposition in genomic-based beta-secretase transgenic mice. *The Journal of Biological Chemistry, 279*, 52535–52542.

Chiocco, M. J., & Lamb, B. T. (2007). Spatial and temporal control of age-related APP processing in genomic-based beta-secretase transgenic mice. *Neurobiology of Aging, 28*, 75–84.

Citron, M. (2004). Beta-secretase inhibition for the treatment of Alzheimer's disease—Promise and challenge. *Trends in Pharmacological Sciences, 25*, 92–97.

De Strooper, B., & Annaert, W. (2000). Proteolytic processing and cell biological functions of the amyloid precursor protein. *Journal of Cell Science, 113*(Pt 11), 1857–1870.

De Strooper, B., Saftig, P., Craessaerts, K., Vanderstichele, H., Guhde, G., Annaert, W., et al. (1998). Deficiency of presenilin-1 inhibits the normal cleavage of amyloid precursor protein. *Nature, 391*, 387–390.

DeMattos, R., May, P., Racke, M., Hole, J., Tzaferis, J., Liu, F., et al. (2014). Combination therapy with a plaque-specific Aβ antibody and BACE inhibitor results in dramatic plaque lowering in aged PDAPP transgenic mice. *Alzheimers Dementia Journal of Alzheimers Association, 10*, P149.

Di Fede, G., Catania, M., Morbin, M., Rossi, G., Suardi, S., Mazzoleni, G., et al. (2009). A recessive mutation in the APP gene with dominant-negative effect on amyloidogenesis. *Science, 323*, 1473–1477.

Dislich, B., Wohlrab, F., Bachhuber, T., Mueller, S., Kuhn, P.-H., Hogl, S., et al. (2015). Label-free quantitative proteomics of mouse cerebrospinal fluid detects BACE1 protease substrates *in vivo. Molecular & Cellular Proteomics MCP, 14*(10), 2550–2563.

Dominguez, D., Tournoy, J., Hartmann, D., Huth, T., Cryns, K., Deforce, S., et al. (2005). Phenotypic and biochemical analyses of BACE1- and BACE2-deficient mice. *The Journal of Biological Chemistry, 280*, 30797–30806.

Doody, R. S., Raman, R., Farlow, M., Iwatsubo, T., Vellas, B., Joffe, S., et al. (2013). A phase 3 trial of semagacestat for treatment of Alzheimer's disease. *The New England Journal of Medicine, 369*, 341–350.

Eggert, S., Paliga, K., Soba, P., Evin, G., Masters, C. L., Weidemann, A., et al. (2004). The proteolytic processing of the amyloid precursor protein gene family members APLP-1 and APLP-2 involves alpha-, beta-, gamma-, and epsilon-like cleavages: Modulation of APLP-1 processing by n-glycosylation. *The Journal of Biological Chemistry, 279*, 18146–18156.

Endres, M., Namura, S., Shimizu-Sasamata, M., Waeber, C., Zhang, L., Gómez-Isla, T., et al. (1998). Attenuation of delayed neuronal death after mild focal ischemia in mice by inhibition of the caspase family. *Journal of Cerebral Blood Flow and Metabolism, 18*, 238–247.

Faghihi, M. A., Modarresi, F., Khalil, A. M., Wood, D. E., Sahagan, B. G., Morgan, T. E., et al. (2008). Expression of a noncoding RNA is elevated in Alzheimer's disease and drives rapid feed-forward regulation of beta-secretase. *Nature Medicine, 14*, 723–730.

Filser, S., Ovsepian, S. V., Masana, M., Blazquez-Llorca, L., Brandt Elvang, A., Volbracht, C., et al. (2015). Pharmacological inhibition of BACE1 impairs synaptic plasticity and cognitive functions. *Biological Psychiatry, 77*, 729–739.

Fukumoto, H., Cheung, B. S., Hyman, B. T., & Irizarry, M. C. (2002). Beta-secretase protein and activity are increased in the neocortex in Alzheimer disease. *Archives of Neurology, 59*, 1381–1389.

Gautam, V., D'Avanzo, C., Hebisch, M., Kovacs, D. M., & Kim, D. Y. (2014). BACE1 activity regulates cell surface contactin-2 levels. *Molecular Neurodegeneration, 9*, 4.

Ghosh, A. K., Brindisi, M., & Tang, J. (2012). Developing β-secretase inhibitors for treatment of Alzheimer's disease. *Journal of Neurochemistry, 120*(Suppl. 1), 71–83.

Hardy, J., & Higgins, G. (1992). Alzheimer's disease: The amyloid cascade hypothesis. *Science, 256*, 184–185.

Harrison, S. M., Harper, A. J., Hawkins, J., Duddy, G., Grau, E., Pugh, P. L., et al. (2003). BACE1 (beta-secretase) transgenic and knockout mice: Identification of neurochemical deficits and behavioral changes. *Molecular and Cellular Neurosciences, 24*, 646–655.

He, X., Li, F., Chang, W.-P., & Tang, J. (2005). GGA proteins mediate the recycling pathway of memapsin 2 (BACE). *The Journal of Biological Chemistry, 280*, 11696–11703.

He, X., Zhu, G., Koelsch, G., Rodgers, K. K., Zhang, X. C., & Tang, J. (2003). Biochemical and structural characterization of the interaction of memapsin 2 (beta-secretase) cytosolic domain with the VHS domain of GGA proteins. *Biochemistry (Mosc.), 42*, 12174–12180.

Hébert, S. S., Horré, K., Nicolaï, L., Papadopoulou, A. S., Mandemakers, W., Silahtaroglu, A. N., et al. (2008). Loss of microRNA cluster miR-29a/b-1 in sporadic Alzheimer's disease correlates with increased BACE1/beta-secretase expression. *Proceedings of the National Academy of Sciences of the United States of America* (Vol. 105), (pp. 6415–6420).

Hemming, M. L., Elias, J. E., Gygi, S. P., & Selkoe, D. J. (2009). Identification of beta-secretase (BACE1) substrates using quantitative proteomics. *PLoS One, 4*, e8477.

Hitt, B., Riordan, S. M., Kukreja, L., Eimer, W. A., Rajapaksha, T. W., & Vassar, R. (2012). β-Site amyloid precursor protein (APP)-cleaving enzyme 1 (BACE1)-deficient mice exhibit a close homolog of L1 (CHL1) loss-of-function phenotype involving axon guidance defects. *The Journal of Biological Chemistry, 287*, 38408–38425.

Hitt, B. D., Jaramillo, T. C., Chetkovich, D. M., & Vassar, R. (2010). BACE1−/− mice exhibit seizure activity that does not correlate with sodium channel level or axonal localization. *Molecular Neurodegeneration, 5*, 31.

Holsinger, R. M. D., McLean, C. A., Beyreuther, K., Masters, C. L., & Evin, G. (2002). Increased expression of the amyloid precursor beta-secretase in Alzheimer's disease. *Annals of Neurology, 51*, 783–786.

Hong, L., Koelsch, G., Lin, X., Wu, S., Terzyan, S., Ghosh, A. K., et al. (2000). Structure of the protease domain of memapsin 2 (beta-secretase) complexed with inhibitor. *Science, 290*, 150–153.

Hu, X., He, W., Diaconu, C., Tang, X., Kidd, G. J., Macklin, W. B., et al. (2008). Genetic deletion of BACE1 in mice affects remyelination of sciatic nerves. *The FASEB Journal, 22*, 2970–2980.

Hu, X., He, W., Luo, X., Tsubota, K. E., & Yan, R. (2013). BACE1 regulates hippocampal astrogenesis via the Jagged1-Notch pathway. *Cell Reports, 4*, 40–49.

Hu, X., Hicks, C. W., He, W., Wong, P., Macklin, W. B., Trapp, B. D., et al. (2006). Bace1 modulates myelination in the central and peripheral nervous system. *Nature Neuroscience, 9*, 1520–1525.

Hu, X., Hu, J., Dai, L., Trapp, B., & Yan, R. (2015). Axonal and Schwann cell BACE1 is equally required for remyelination of peripheral nerves. *The Journal of Neuroscience, 35*, 3806–3814.

Hu, X., Zhou, X., He, W., Yang, J., Xiong, W., Wong, P., et al. (2010). BACE1 deficiency causes altered neuronal activity and neurodegeneration. *The Journal of Neuroscience, 30*, 8819–8829.

Hung, A. Y., Koo, E. H., Haass, C., & Selkoe, D. J. (1992). Increased expression of beta-amyloid precursor protein during neuronal differentiation is not accompanied by secretory cleavage. *Proceedings of the National Academy of Sciences* (Vol. 89), (pp. 9439–9443).

Huse, J. T., Pijak, D. S., Leslie, G. J., Lee, V. M., & Doms, R. W. (2000). Maturation and endosomal targeting of beta-site amyloid precursor protein-cleaving enzyme. The Alzheimer's disease beta-secretase. *The Journal of Biological Chemistry, 275*, 33729–33737.

Hussain, I., Powell, D., Howlett, D. R., Tew, D. G., Meek, T. D., Chapman, C., et al. (1999). Identification of a novel aspartic protease (Asp 2) as beta-secretase. *Molecular and Cellular Neurosciences, 14*, 419–427.

Jacobsen, H., Ozmen, L., Caruso, A., Narquizian, R., Hilpert, H., Jacobsen, B., et al. (2014). Combined treatment with a BACE inhibitor and anti-Aβ antibody gantenerumab enhances amyloid reduction in APPLondon mice. *The Journal Neuroscience, 34*, 11621–11630.

Jonsson, T., Atwal, J. K., Steinberg, S., Snaedal, J., Jonsson, P. V., Bjornsson, S., et al. (2012). A mutation in APP protects against Alzheimer's disease and age-related cognitive decline. *Nature, 488*, 96–99.

Kang, E. L., Cameron, A. N., Piazza, F., Walker, K. R., & Tesco, G. (2010). Ubiquitin regulates GGA3-mediated degradation of BACE1. *The Journal of Biological Chemistry, 285*, 24108–24119.

Kim, D. Y., Carey, B. W., Wang, H., Ingano, L. A. M., Binshtok, A. M., Wertz, M. H., et al. (2007). BACE1 regulates voltage-gated sodium channels and neuronal activity. *Nature Cell Biology, 9*, 755–764.

Kim, D. Y., Ingano, L. A. M., Carey, B. W., Pettingell, W. H., & Kovacs, D. M. (2005). Presenilin/γ-Secretase-mediated cleavage of the voltage-gated sodium channel β2-Subunit regulates cell adhesion and migration. *The Journal of Biological Chemistry, 280*, 23251–23261.

Kimura, R., Devi, L., & Ohno, M. (2010). Partial reduction of BACE1 improves synaptic plasticity, recent and remote memories in Alzheimer's disease transgenic mice. *Journal of Neurochemistry, 113*, 248–261.

Kitazume, S., Tachida, Y., Oka, R., Shirotani, K., Saido, T. C., & Hashimoto, Y. (2001). Alzheimer's beta-secretase, beta-site amyloid precursor protein-cleaving enzyme, is responsible for cleavage secretion of a Golgi-resident sialyltransferase. *Proceedings of the National Academy of Sciences of the United States of America, 98*, 13554–13559.

Klaver, D. W., & Tesco, G. (2014). Modulation of BACE1 activity as a potential therapeutic strategy for treating Alzheimer's disease. In *Frontiers in drug design & discovery* (pp. 478–517). Sharjah, UAE: Bentham Science Publishers.

Kobayashi, D., Zeller, M., Cole, T., Buttini, M., McConlogue, L., Sinha, S., et al. (2008). BACE1 gene deletion: Impact on behavioral function in a model of Alzheimer's disease. *Neurobiology of Aging, 29*, 861–873.

Koh, Y. H., von Arnim, C. A. F., Hyman, B. T., Tanzi, R. E., & Tesco, G. (2005). BACE is degraded via the lysosomal pathway. *The Journal of Biological Chemistry, 280*, 32499–32504.

Kuhn, P. -H., Marjaux, E., Imhof, A., De Strooper, B., Haass, C., & Lichtenthaler, S. F. (2007). Regulated intramembrane proteolysis of the interleukin-1 receptor II by alpha-, beta-, and gamma-secretase. *The Journal of Biological Chemistry, 282*, 11982–11995.

Kuhn, P.-H., Koroniak, K., Hogl, S., Colombo, A., Zeitschel, U., Willem, M., et al. (2012). Secretome protein enrichment identifies physiological BACE1 protease substrates in neurons. *The EMBO Journal, 31*, 3157–3168.

Laird, F. M., Cai, H., Savonenko, A. V., Farah, M. H., He, K., Melnikova, T., et al. (2005). BACE1, a major determinant of selective vulnerability of the brain to amyloid-β amyloidogenesis, is essential for cognitive, emotional, and synaptic functions. *The Journal of Neuroscience, 25*, 11693–11709.

Lammich, S., Kojro, E., Postina, R., Gilbert, S., Pfeiffer, R., Jasionowski, M., et al. (1999). Constitutive and regulated alpha-secretase cleavage of Alzheimer's amyloid precursor protein by a disintegrin metalloprotease. *Proceedings of the National Academy of Sciences of the United States of America* (Vol. 96), (pp. 3922–3927).

Lee, E. B., Zhang, B., Liu, K., Greenbaum, E. A., Doms, R. W., Trojanowski, J. Q., et al. (2005). BACE overexpression alters the subcellular processing of APP and inhibits Abeta deposition *in vivo*. *The Journal of Cell Biology, 168*, 291–302.

Lenzlinger, P. M., Marx, A., Trentz, O., Kossmann, T., & Morganti-Kossmann, M. C. (2002). Prolonged intrathecal release of soluble Fas following severe traumatic brain injury in humans. *Journal of Neuroimmunology, 122*, 167–174.

Li, Q., & Südhof, T. C. (2004). Cleavage of amyloid-beta precursor protein and amyloid-beta precursor-like protein by BACE 1. *The Journal of Biological Chemistry, 279*, 10542–10550.

Li, R., Lindholm, K., Yang, L.-B., Yue, X., Citron, M., Yan, R., et al. (2004). Amyloid beta peptide load is correlated with increased beta-secretase activity in sporadic Alzheimer's disease patients. *Proceedings of the National Academy of Sciences of the United States of America* (Vol. 101), (pp. 3632–3637).

Lichtenthaler, S. F., Dominguez, D., Westmeyer, G. G., Reiss, K., Haass, C., Saftig, P., et al. (2003). The cell adhesion protein p-selectin glycoprotein ligand-1 is a substrate for the aspartyl protease BACE1. *The Journal of Biological Chemistry, 278*, 48713–48719.

Lin, X., Koelsch, G., Wu, S., Downs, D., Dashti, A., & Tang, J. (2000). Human aspartic protease memapsin 2 cleaves the beta-secretase site of beta-amyloid precursor protein. *Proceedings of the National Academy of Sciences of the United States of America* (Vol. 97), (pp. 1456–1460).

Loane, D. J., Pocivavsek, A., Moussa, C. E.-H., Thompson, R., Matsuoka, Y., Faden, A. I., et al. (2009). Amyloid precursor protein secretases as therapeutic targets for traumatic brain injury. *Nature Medicine, 15*, 377–379.

Lotocki, G., Alonso, O. F., Dietrich, W. D., & Keane, R. W. (2004). Tumor necrosis factor receptor 1 and its signaling intermediates are recruited to lipid rafts in the traumatized brain. *The Journal of Neuroscience, 24*, 11010–11016.

Luo, Y., Bolon, B., Kahn, S., Bennett, B. D., Babu-Khan, S., Denis, P., et al. (2001). Mice deficient in BACE1, the Alzheimer's β-secretase, have normal phenotype and abolished β-amyloid generation. *Nature Neuroscience, 4*, 231–232.

Lyckman, A. W., Confaloni, A. M., Thinakaran, G., Sisodia, S. S., & Moya, K. L. (1998). Post-translational processing and turnover kinetics of presynaptically targeted amyloid precursor superfamily proteins in the central nervous system. *The Journal of Biological Chemistry, 273*, 11100–11106.

Mannix, R. C., Zhang, J., Park, J., Zhang, X., Bilal, K., Walker, K., et al. (2011). Age-dependent effect of apolipoprotein E4 on functional outcome after controlled cortical impact in mice. *Journal of Cereberal Blood Flow and Metabolism, 31*, 351–361.

McConlogue, L., Buttini, M., Anderson, J. P., Brigham, E. F., Chen, K. S., Freedman, S. B., et al. (2007). Partial reduction of BACE1 has dramatic effects on Alzheimer plaque and synaptic pathology in APP transgenic mice. *The Journal of Biological Chemistry, 282*, 26326–26334.

Meakin, P. J., Harper, A. J., Hamilton, D. L., Gallagher, J., McNeilly, A. D., Burgess, L. A., et al. (2012). Reduction in BACE1 decreases body weight, protects against diet-induced obesity and enhances insulin sensitivity in mice. *The Biochemical Journal, 441*, 285–296.

Mohajeri, M. H., Saini, K. D., & Nitsch, R. M. (2004). Transgenic BACE expression in mouse neurons accelerates amyloid plaque pathology. *Journal of Neural Transmission Vienna Austria, 1996*(111), 413–425.

Moya, K. L., Benowitz, L. I., Schneider, G. E., & Allinquant, B. (1994). The amyloid precursor protein is developmentally regulated and correlated with synaptogenesis. *Developmental Biology, 161*, 597–603.

Mullan, M., Crawford, F., Axelman, K., Houlden, H., Lilius, L., Winblad, B., et al. (1992). A pathogenic mutation for probable Alzheimer's disease in the APP gene at the N-terminus of β-amyloid. *Nature Genetics, 1*, 345–347.

Murakami, N., Yamaki, T., Iwamoto, Y., Sakakibara, T., Kobori, N., Fushiki, S., et al. (1998). Experimental brain injury induces expression of amyloid precursor protein, which may be related to neuronal loss in the hippocampus. *Journal of Neurotrauma, 15*, 993–1003.

Oakley, H., Cole, S. L., Logan, S., Maus, E., Shao, P., Craft, J., et al. (2006). Intraneuronal beta-amyloid aggregates, neurodegeneration, and neuron loss in transgenic mice with five familial Alzheimer's disease mutations: Potential factors in amyloid plaque formation. *The Journal of Neuroscience, 26*, 10129–10140.

O'Connor, T., Sadleir, K. R., Maus, E., Velliquette, R. A., Zhao, J., Cole, S. L., et al. (2008). Phosphorylation of the translation initiation factor eIF2alpha increases BACE1 levels and promotes amyloidogenesis. *Neuron, 60*, 988–1009.

Oh, E. S., Savonenko, A. V., King, J. F., Fangmark Tucker, S. M., Rudow, G. L., Xu, G., et al. (2009). Amyloid precursor protein increases cortical neuron size in transgenic mice. *Neurobiology of Aging, 30*, 1238–1244.

Ohno, M., Chang, L., Tseng, W., Oakley, H., Citron, M., Klein, W. L., et al. (2006). Temporal memory deficits in Alzheimer's mouse models: Rescue by genetic deletion of BACE1. *The European Journal of Neuroscience, 23*, 251–260.

Ohno, M., Cole, S. L., Yasvoina, M., Zhao, J., Citron, M., Berry, R., et al. (2007). BACE1 gene deletion prevents neuron loss and memory deficits in 5XFAD APP/PS1 transgenic mice. *Neurobiology of Disease, 26*, 134–145.

Ohno, M., Sametsky, E. A., Younkin, L. H., Oakley, H., Younkin, S. G., Citron, M., et al. (2004). BACE1 deficiency rescues memory deficits and cholinergic dysfunction in a mouse model of Alzheimer's disease. *Neuron, 41*, 27–33.

Oliva, A. A., Kang, Y., Sanchez-Molano, J., Furones, C., & Atkins, C. M. (2012). STAT3 signaling after traumatic brain injury. *Journal of Neurochemistry, 120*, 710–720.

Pastorino, L., Ikin, A. F., Lamprianou, S., Vacaresse, N., Revelli, J. P., Platt, K., et al. (2004). BACE (beta-secretase) modulates the processing of APLP2 *in vivo*. *Molecular and Cellular Neurosciences, 25*, 642–649.

Pastorino, L., Ikin, A. F., Nairn, A. C., Pursnani, A., & Buxbaum, J. D. (2002). The carboxyl-terminus of BACE contains a sorting signal that regulates BACE trafficking but not the formation of total A(beta). *Molecular and Cellular Neurosciences, 19*, 175–185.

Plucińska, K., Crouch, B., Koss, D., Robinson, L., Siebrecht, M., Riedel, G., et al. (2014). Knock-in of human BACE1 cleaves murine APP and reiterates Alzheimer-like phenotypes. *The Journal of Neuroscience, 34*, 10710–10728.

Rajapaksha, T. W., Eimer, W. A., Bozza, T. C., & Vassar, R. (2011). The Alzheimer's β-secretase enzyme BACE1 is required for accurate axon guidance of olfactory sensory neurons and normal glomerulus formation in the olfactory bulb. *Molecular Neurodegeneration, 6*, 88.

Redell, J. B., Liu, Y., & Dash, P. K. (2009). Traumatic brain injury alters expression of hippocampal microRNAs: Potential regulators of multiple pathophysiological processes. *Journal of Neuroscience Research, 87*, 1435–1448.

Roberds, S. L., Anderson, J., Basi, G., Bienkowski, M. J., Branstetter, D. G., Chen, K. S., et al. (2001). BACE knockout mice are healthy despite lacking the primary β-secretase activity in brain: Implications for Alzheimer's disease therapeutics. *Human Molecular Genetics, 10*, 1317–1324.

Rockenstein, E., Mante, M., Alford, M., Adame, A., Crews, L., Hashimoto, M., et al. (2005). High beta-secretase activity elicits neurodegeneration in transgenic mice despite reductions in amyloid-beta levels: Implications for the treatment of Alzheimer disease. *The Journal of Biological Chemistry, 280*, 32957–32967.

Rossner, S., Sastre, M., Bourne, K., & Lichtenthaler, S. F. (2006). Transcriptional and translational regulation of BACE1 expression—Implications for Alzheimer's disease. *Progress in Neurobiology, 79*, 95–111.

Sadleir, K. R., Eimer, W. A., Cole, S. L., & Vassar, R. (2015). Aβ reduction in BACE1 heterozygous null 5XFAD mice is associated with transgenic APP level. *Molecular Neurodegeneration, 10*, 1.

Santosa, C., Rasche, S., Barakat, A., Bellingham, S. A., Ho, M., Tan, J., et al. (2011). Decreased expression of GGA3 protein in Alzheimer's disease frontal cortex and increased co-distribution of BACE with the amyloid precursor protein. *Neurobiology of Disease, 43,* 176–183.

Sanz, O., Acarin, L., González, B., & Castellano, B. (2002). NF-kappa B and I-kappa B alpha expression following traumatic brain injury to the immature rat brain. *Journal of Neuroscience Research, 67,* 772–780.

Savonenko, A. V., Melnikova, T., Laird, F. M., Stewart, K.-A., Price, D. L., & Wong, P. C. (2008). Alteration of BACE1-dependent NRG1/ErbB4 signaling and schizophrenia-like phenotypes in BACE1-null mice. *Proceedings of the National Academy of Sciences of the United States of America* (Vol. 105), (pp. 5585–5590).

Shariati, S. A. M., & De Strooper, B. (2013). Redundancy and divergence in the amyloid precursor protein family. *FEBS Letters, 587,* 2036–2045.

Shen, J., Bronson, R. T., Chen, D. F., Xia, W., Selkoe, D. J., & Tonegawa, S. (1997). Skeletal and CNS defects in presenilin-1-deficient mice. *Cell, 89,* 629–639.

Shiba, T., Kametaka, S., Kawasaki, M., Shibata, M., Waguri, S., Uchiyama, Y., et al. (2004). Insights into the phosphoregulation of beta-secretase sorting signal by the VHS domain of GGA1. *Traffic Copenhagen Denmark, 5,* 437–448.

Singleton, R. H., Zhu, J., Stone, J. R., & Povlishock, J. T. (2002). Traumatically induced axotomy adjacent to the soma does not result in acute neuronal death. *The Journal Neuroscience, 22,* 791–802.

Sinha, S., Anderson, J. P., Barbour, R., Basi, G. S., Caccavello, R., Davis, D., et al. (1999). Purification and cloning of amyloid precursor protein beta-secretase from human brain. *Nature, 402,* 537–540.

Tamagno, E., Bardini, P., Obbili, A., Vitali, A., Borghi, R., Zaccheo, D., et al. (2002). Oxidative stress increases expression and activity of BACE in NT2 neurons. *Neurobiology of Disease, 10,* 279–288.

Tamagno, E., Guglielmotto, M., Bardini, P., Santoro, G., Davit, A., Di Simone, D., et al. (2003). Dehydroepiandrosterone reduces expression and activity of BACE in NT2 neurons exposed to oxidative stress. *Neurobiology of Disease, 14,* 291–301.

Tamagno, E., Parola, M., Bardini, P., Piccini, A., Borghi, R., Guglielmotto, M., et al. (2005). Beta-site APP cleaving enzyme up-regulation induced by 4-hydroxynonenal is mediated by stress-activated protein kinases pathways. *Journal of Neurochemistry, 92,* 628–636.

Tesco, G., Koh, Y. H., Kang, E. L., Cameron, A. N., Das, S., Sena-Esteves, M., et al. (2007). Depletion of GGA3 stabilizes BACE and enhances beta-secretase activity. *Neuron, 54,* 721–737.

Tong, Y., Zhou, W., Fung, V., Christensen, M. A., Qing, H., Sun, X., et al. (2005). Oxidative stress potentiates BACE1 gene expression and Abeta generation. *Journal of Neural Transmission Vienna Austria, 1996*(112), 455–469.

Tyler, S. J., Dawbarn, D., Wilcock, G. K., & Allen, S. J. (2002). alpha- and beta-secretase: Profound changes in Alzheimer's disease. *Biochemical and Biophysical Research Communications, 299,* 373–376.

Uryu, K., Chen, X.-H., Martinez, D., Browne, K. D., Johnson, V. E., Graham, D. I., et al. (2007). Multiple proteins implicated in neurodegenerative diseases accumulate in axons after brain trauma in humans. *Experimental Neurology, 208,* 185–192.

Uzan, M., Erman, H., Tanriverdi, T., Sanus, G. Z., Kafadar, A., & Uzun, H. (2006). Evaluation of apoptosis in cerebrospinal fluid of patients with severe head injury. *Acta Neurochirurgica (Wien), 148,* 1157–1164; discussion.

Vassar, R. (2014). BACE1 inhibitor drugs in clinical trials for Alzheimer's disease. *Alzheimers Research Therapy, 6,* 89.

Vassar, R., Bennett, B. D., Babu-Khan, S., Kahn, S., Mendiaz, E. A., Denis, P., et al. (1999). Beta-secretase cleavage of Alzheimer's amyloid precursor protein by the transmembrane aspartic protease BACE. *Science, 286,* 735–741.

Vassar, R., Kovacs, D. M., Yan, R., & Wong, P. C. (2009). The beta-secretase enzyme BACE in health and Alzheimer's disease: Regulation, cell biology, function, and therapeutic potential. *The Journal of Neuroscience, 29,* 12787–12794.

Vassar, R., Kuhn, P.-H., Haass, C., Kennedy, M. E., Rajendran, L., Wong, P. C., et al. (2014). Function, therapeutic potential and cell biology of BACE proteases: Current status and future prospects. *Journal of Neurochemistry, 130,* 4–28.

Velliquette, R. A., O'Connor, T., & Vassar, R. (2005). Energy inhibition elevates beta-secretase levels and activity and is potentially amyloidogenic in APP transgenic mice: Possible early events in Alzheimer's disease pathogenesis. *The Journal of Neuroscience, 25,* 10874–10883.

von Arnim, C. A. F., Tangredi, M. M., Peltan, I. D., Lee, B. M., Irizarry, M. C., Kinoshita, A., et al. (2004). Demonstration of BACE (beta-secretase) phosphorylation and its interaction with GGA1 in cells by fluorescence-lifetime imaging microscopy. *Journal of Cell Science, 117,* 5437–5445.

Wahle, T., Prager, K., Raffler, N., Haass, C., Famulok, M., & Walter, J. (2005). GGA proteins regulate retrograde transport of BACE1 from endosomes to the trans-Golgi network. *Molecular and Cellular Neurosciences, 29,* 453–461.

Wahle, T., Thal, D. R., Sastre, M., Rentmeister, A., Bogdanovic, N., Famulok, M., et al. (2006). GGA1 is expressed in the human brain and affects the generation of amyloid beta-peptide. *The Journal of Neuroscience, 26,* 12838–12846.

Walker, K. R., Kang, E. L., Whalen, M. J., Shen, Y., & Tesco, G. (2012). Depletion of GGA1 and GGA3 mediates postinjury elevation of BACE1. *The Journal of Neuroscience, 32,* 10423–10437.

Walter, J., Fluhrer, R., Hartung, B., Willem, M., Kaether, C., Capell, A., et al. (2001). Phosphorylation regulates intracellular trafficking of beta-secretase. *The Journal of Biological Chemistry, 276,* 14634–14641.

Wang, H., Song, L., Laird, F., Wong, P. C., & Lee, H.-K. (2008). BACE1 knock-outs display deficits in activity-dependent potentiation of synaptic transmission at mossy fiber to CA3 synapses in the hippocampus. *The Journal of Neuroscience, 28,* 8677–8681.

Wang, W., Liu, Y., & Lazarus, R. A. (2013). Allosteric inhibition of BACE1 by an exosite-binding antibody. *Current Opinion in Structural Biology, 23,* 797–805.

Wang, W.-X., Wilfred, B. R., Madathil, S. K., Tang, G., Hu, Y., Dimayuga, J., et al. (2010). miR-107 regulates granulin/progranulin with implications for traumatic brain injury and neurodegenerative disease. *The American Journal of Pathology, 177,* 334–345.

Wasco, W., Bupp, K., Magendantz, M., Gusella, J. F., Tanzi, R. E., & Solomon, F. (1992). Identification of a mouse brain cDNA that encodes a protein related to the Alzheimer disease-associated amyloid beta protein precursor. *Proceedings of the National Academy of Sciences* (Vol. 89), (pp. 10758–10762).

Wen, Y., Onyewuchi, O., Yang, S., Liu, R., & Simpkins, J. W. (2004). Increased beta-secretase activity and expression in rats following transient cerebral ischemia. *Brain Research, 1009,* 1–8.

Willem, M., Dewachter, I., Smyth, N., Van Dooren, T., Borghgraef, P., Haass, C., et al. (2004). beta-site amyloid precursor protein cleaving enzyme 1 increases amyloid deposition in brain parenchyma but reduces cerebrovascular amyloid angiopathy in aging BACE x APP[V717I] double-transgenic mice. *The American Journal of Pathology, 165,* 1621–1631.

Willem, M., Garratt, A. N., Novak, B., Citron, M., Kaufmann, S., Rittger, A., et al. (2006). Control of peripheral nerve myelination by the ß-secretase BACE1. *Science, 314,* 664–666.

Willem, M., Lammich, S., & Haass, C. (2009). Function, regulation and therapeutic properties of beta-secretase (BACE1). *Seminars in Cell & Developmental Biology, 20,* 175–182.

Wong, H.-K., Sakurai, T., Oyama, F., Kaneko, K., Wada, K., Miyazaki, H., et al. (2005). Beta Subunits of voltage-gated sodium channels are novel substrates of beta-site amyloid precursor protein-cleaving enzyme (BACE1) and gamma-secretase. *The Journal of Biological Chemistry, 280,* 23009–23017.

Xiong, K., Cai, H., Luo, X.-G., Struble, R. G., Clough, R. W., & Yan, X.-X. (2007). Mitochondrial respiratory inhibition and oxidative stress elevate beta-secretase (BACE1) proteins and activitiy *in vivo* in the rat retina. *Experimental Brain Research, 181*, 435–446.

Yan, R., Bienkowski, M. J., Shuck, M. E., Miao, H., Tory, M. C., Pauley, A. M., et al. (1999). Membrane-anchored aspartyl protease with Alzheimer's disease beta-secretase activity. *Nature, 402*, 533–537.

Yan, R., & Vassar, R. (2014). Targeting the β secretase BACE1 for Alzheimer's disease therapy. *Lancet Neurology, 13*, 319–329.

Yang, L.-B., Lindholm, K., Yan, R., Citron, M., Xia, W., Yang, X.-L., et al. (2003). Elevated beta-secretase expression and enzymatic activity detected in sporadic Alzheimer disease. *Nature Medicine, 9*, 3–4.

Yu, F., Zhang, Y., & Chuang, D.-M. (2012). Lithium reduces BACE1 overexpression, β amyloid accumulation, and spatial learning deficits in mice with traumatic brain injury. *Journal of Neurotrauma, 29*, 2342–2351.

Yu, Y. J., Zhang, Y., Kenrick, M., Hoyte, K., Luk, W., Lu, Y., et al. (2011). Boosting brain uptake of a therapeutic antibody by reducing its affinity for a transcytosis target. *Science Translational Medicine, 3*, 84ra44.

Zhao, J., Fu, Y., Yasvoina, M., Shao, P., Hitt, B., O'Connor, T., et al. (2007). Beta-site amyloid precursor protein cleaving enzyme 1 levels become elevated in neurons around amyloid plaques: Implications for Alzheimer's disease pathogenesis. *The Journal of Neuroscience, 27*, 3639–3649.

Zhao, J., O'Connor, T., & Vassar, R. (2011). The contribution of activated astrocytes to Aβ production: Implications for Alzheimer's disease pathogenesis. *Journal of Neuroinflammation, 8*, 150.

Zhong, P., Gu, Z., Wang, X., Jiang, H., Feng, J., & Yan, Z. (2003). Impaired modulation of GABAergic transmission by muscarinic receptors in a mouse transgenic model of Alzheimer's disease. *The Journal of Biological Chemistry, 278*, 26888–26896.

Zhou, L., Barão, S., Laga, M., Bockstael, K., Borgers, M., Gijsen, H., et al. (2012). The neural cell adhesion molecules L1 and CHL1 are cleaved by BACE1 protease *in vivo*. *The Journal of Biological Chemistry, 287*, 25927–25940.

CHAPTER

9

Traumatic Brain Injury and Rationale for a Neuropsychological Diagnosis of Diffuse Axonal Injury

Amanda R. Rabinowitz and Douglas H. Smith

Center for Brain Injury and Repair, Department of Neurosurgery, University of Pennsylvania, Philadelphia, PA, United States

OUTLINE

Introduction	268
Common TBI-Related Cognitive Deficits	271
Mechanisms of Cognitive Disturbance Following TBI	273
Diffuse Axonal Injury	274
DAI as a Mechanism of Cognitive Dysfunction Following TBI	274
Neurochemical and Neurometabolic Changes	277
Neurometabolic Changes as a Mechanism of Cognitive Dysfunction Following TBI	277
Focal Injuries	279
Focal Brain Injury as a Mechanism of Cognitive Dysfunction Following TBI	279
Neurodegenerative Processes	281

Genes, Environment and Alzheimer's Disease.
DOI: http://dx.doi.org/10.1016/B978-0-12-802851-3.00009-7
267

Neurodegenerative Processes as a Mechanism of
    Cognitive Dysfunction Following TBI   282

Conclusions   283

References   285

INTRODUCTION

Cognitive dysfunction is the hallmark clinical manifestation of moderate and severe traumatic brain injury (TBI) (Schretlen & Shapiro, 2003), often exacting profound effects on patients' lives as the predominant source of disability (Selassie et al., 2008). Moreover, emerging evidence shows that persisting cognitive dysfunction also occurs in approximately 15% of individuals with mild TBI (mTBI) or "concussion" (Boake, McCauley, & Levin, 2005; Røe, Sveen, Alvsåker, & Bautz-Holter, 2009). However, the anatomic substrates specifically related to TBI-related cognitive deficits have not been well characterized.

This diagnostic shortfall of TBI may reflect the complicated and heterogeneous nature of the neuropathologies well characterized in moderate and severe TBI. Indeed, combinations of focal injuries such as contusion or hematoma are commonly found in concert with diffuse changes such as brain swelling, ischemia, and diffuse axonal injury (DAI). Therefore, attribution of clinical signs to specific neuropathologies has posed a significant challenge. Notably, the neuropathologies of mTBI are not well characterized, primarily due to the typically nonmortal nature of the injury. Notwithstanding, a singular clinical neuropathological study of mTBI (Blumbergs, Jones, & North, 1989) joined by multiple advanced neuroimaging studies (Bazarian et al., 2005; Levin, Wilde, & Troyanskaya, 2010; Wilde, McCauley, & Hunter, 2008) and animal neuropathological studies (Browne, Chen, Meaney, & Smith, 2011) of mTBI show an absence of overt focal changes and implicate DAI as the predominant neuropathological feature. As such, this observation provides a unique opportunity to examine the role of DAI in the neurocognitive consequences of mTBI.

There are compelling reasons to focus on DAI and its subsequent neurochemical and neurometabolic cascades. DAI is the most common neuropathological feature across all levels of TBI severity. Curiously, for mTBI, in the *absence* of conventional radiological findings, DAI has been a diagnosis of exclusion for patients with a history of head impact and symptoms of concussion. In addition, multiple recent studies employing advanced neuroimaging methods have shown white matter abnormalities in mTBI patients (Bazarian et al., 2007; Huisman, Schwamm, & Schaefer,

2004; Inglese, Makani, & Johnson, 2005; Levin, Wilde, & Chu, 2008; Mayer, Ling, & Mannell, 2010; Wilde et al., 2008). Growing evidence suggests that DAI triggers acute neurochemical and neurometabolic changes (Büki & Povlishock, 2006; Povlishock & Katz, 2005) that may lead to long-term cognitive dysfunction in some patients, and even neurodegenerative processes (Johnson, Stewart, & Smith, 2013) implicated in the development of dementia (Gardner et al., 2014). Importantly, DAI is a promising therapeutic target and preclinical studies have uncovered candidate interventions that appear to attenuate some of its features (Smith, Hicks, & Povlishock, 2013). Although more research is needed before any DAI therapy is ready for clinical trial, clinical studies are doomed to fail without a reliable clinical diagnosis of DAI to identify the TBI patients most likely to benefit from treatment.

It is surprising that there is no current neuropsychological diagnosis for DAI, given that DAI is the presumed neuropathological substrate of mTBI-related cognitive deficits and is thought to contribute to cognitive dysfunction following moderate and severe TBI (Bigler & Maxwell, 2011; Cicerone, Levin, Malec, Stuss, & Whyte, 2006; Inglese et al., 2005; Kraus et al., 2007). It is not uncommon for a neuropsychologist to diagnose a patient with frontal lobe dysfunction or medial temporal lobe amnesia; however, most practitioners remain agnostic about neuropathological etiology of TBI-related cognitive deficits. Recent decades have seen significant advances in the pathological and behavioral characterizations of TBI, and research has begun to develop links between pathology and functional outcomes in TBI survivors. Neuropsychology as a field is uniquely poised to translate these findings into behavioral tools for DAI diagnosis and prognosis. However, in order to take on this important translational research, the field must confront a number of conceptual and methodological challenges.

One of the challenges in this regard is the complexity and heterogeneity of TBI. TBI can have multiple pathological features in addition to DAI, including cerebral contusion, hemorrhage, inflammation, and neurochemical and neurometabolic changes. Diagnosis of focal pathologies is relatively straightforward. These features are readily visualized with standard clinical imaging techniques, including magnetic resonance imaging and computed tomography (CT). Additionally, the behavioral and cognitive profiles of focal injury are relatively well defined and easily assessable using standard neuropsychological test batteries. DAI, on the other hand, is invisible to standard clinical neuroimaging, and measures of DAI-related cognitive deficits have yet to be validated. However, recent advances in neuroimaging and biomarker research offer promising non-invasive *in vivo* methods for identifying DAI, and these measures will enable researchers to develop sensitive and specific neuropsychological probes of this pathology.

This chapter will review the current literature on the cognitive, behavioral, and neuropathological effects of TBI, while also proposing a paradigm shift in the neuropsychological approach to this disorder. We argue that research on cognition in TBI must shift from merely describing deficits to diagnosing their underlying causes. To this end, we will briefly review what is known about TBI pathophysiology, focusing on the acute and long-term effects of DAI, and discuss the important advances in biomarker and neuroimaging research that have begun to bridge the gap between pathology and function. We also propose a model that posits potential relationships between DAI and both acute and long-term cognitive effects of TBI (Figure 9.1). This model is theoretical in nature, and although some of the links are supported by existing research, more work must be done to rigorously evaluate the relationships we propose.

Attempts to classify TBI subtypes and guide prognosis have typically relied on four major indicators of injury severity—loss of consciousness (LOC), posttraumatic amnesia (PTA), positive findings on CT, and depth of coma as measured by the Glasgow Coma Scale (GCS). According to consensus definitions, moderate and severe TBI are characterized by LOC

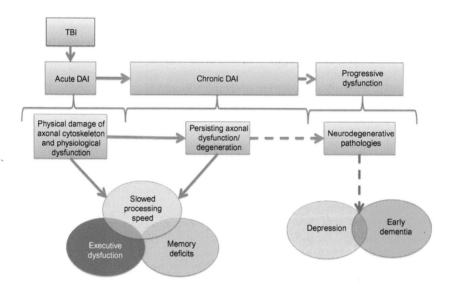

FIGURE 9.1   This figure depicts a theoretical model of the relationship between diffuse axonal injury (DAI) and neurocognitive deficits at acute and chronic stages of TBI. Damage to axonal cytoskeletons disrupts transport along the axon and causes chemical and neurometabolic changes. These pathological cascades may result in persisting axonal dysfunction and degeneration over time. Neurodegenerative pathologies, such as amyloid-beta plaque formation and tauopathy are initiated in some cases. Cognitive and neurobehavioral deficits manifest at various stages after TBI, and may arise due to general cognitive slowing, or directly from dysfunction in the specific white-matter pathways underlying those functions.

greater than 30 min and/or PTA persisting for at least 24 h. GCS scores of eight or less (coma) qualify as severe TBI, whereas GCS scores between 9 and 12 qualify as moderate TBI (purposeful responses in at least one domain). MTBI has been considered a more transient injury, not typically associated with structural damage to the brain. Diagnosis of mTBI is based on a period of confusion, disorientation, or amnesia for less than 24 h following a head injury, up to 30 min of LOC, and a score of 13 or greater on the GCS (purposeful responses in all domains, and normal response in at least one domain) (Carroll, Cassidy, & Peloso, 2004). Notably, GCS is of little utility in measuring the severity of mTBI, as many mTBI patients score at the top of this scale.

Although mTBI is typically associated with lower levels of disability and less severe cognitive dysfunction, there is great heterogeneity among TBI patients. Hence, it is not uncommon for a physician to find two very different TBI patients on her caseload: one with moderate TBI who recovers within months, and another with a supposedly mild injury who suffers from symptoms and cognitive deficits that persist for a year or more. Traditional indicators of TBI severity—LOC, PTA, GCS, and CT findings—have failed to explain this heterogeneity in outcomes. The field is in need of measures that are at once more sensitive to mTBI-related impairments, and also specific to DAI. Cognitive functioning is promising in this regard.

COMMON TBI-RELATED COGNITIVE DEFICITS

Cognitive dysfunction is prevalent following TBI and is the leading cause of disability among TBI survivors (Selassie et al., 2008). Research using standard neuropsychological tests suggests that cognitive deficits associated with mTBI typically resolve within 3–6 months in about 80–85% of patients (Belanger, Curtiss, Demery, Lebowitz, & Vanderploeg, 2005; Vanderploeg, Curtiss, & Belanger, 2005). There appears to be a subgroup of mTBI patients with persistent cognitive complaints. The prevalence of complaints is unclear based on the current literature, but current estimates suggest that 15–30% of mTBI patients fall into this group (Boake et al., 2005; Røe et al., 2009). Athletes with sports-related concussion appear to recover more quickly, showing evidence of cognitive recovery within days or weeks of injury (Belanger & Vanderploeg, 2005; McCrea, 2001). Interestingly, female athletes appear to sustain concussions at a higher rate (Covassin, Swanik, & Sachs, 2003) and experience worse postconcussion outcomes (Broshek et al., 2005) than their male counterparts. Moderate and severe TBI are associated with more severe and persistent cognitive deficits (Dikmen, Machamer, Winn, & Temkin,

1995) with approximately 65% of moderate to severe TBI patients reporting long-term cognitive problems (Whiteneck, Gerhart, & Cusick, 2004). Heterogeneity in cognitive outcomes is also evident among survivors of severe TBI, with some patients demonstrating good recovery at 1 year postinjury, and others exhibiting marked intellectual impairment for 3–5 years postinjury (Levin, Grossman, Rose, & Teasdale, 1979).

TBI patients may show impairments in multiple cognitive domains. Slowed processing speed and increased reaction time (RT) are commonly associated with TBI across all severity levels (Axelrod, Fichtenberg, Liethen, Czarnota, & Stucky, 2001; Mathias, Beall, & Bigler, 2004; Mathias & Wheaton, 2007; Ponsford & Kinsella 1992; Spikman, van Zomeren, & Deelman, 1996; Timmerman & Brouwer, 1999). Memory deficits are also prevalent in TBI survivors (Vakil, 2005), and the results of a meta-analytic review suggest that delayed memory impairment is the most robust acute cognitive finding among patients with mTBI (Belanger et al., 2005). However, some have suggested that memory problems associated with TBI are primarily due to deficits in executive functioning (Cicerone et al., 2006; Raskin & Rearick, 1996; Vakil, 2005). Executive dysfunction is common in TBI patients at all levels of severity (Cicerone et al., 2006; Stuss & Gow, 1992).

Executive functioning is a broad cognitive domain, which includes a number of higher-level cognitive skills, predominantly localized to the frontal lobes. The primary purpose of the executive system is to organize goal-directed behavior (Fuster, 1999). Executive functions include such diverse cognitive tasks as attention shifting, memory acquisition and retrieval, planning, judgment, and cognitive aspects of decision-making. Each of these skills is negatively influenced by slowed speed of processing, hence, weaknesses in these areas are not surprising in the context of overall cognitive slowing (Nelson, Yoash-Gantz, Pickett, & Campbell, 2009). Emotional and behavioral regulation is also included under the general rubric of executive functioning, and these functions include response inhibition (impulsivity), motivation, and emotional aspects of decision-making (Kerr & Zelazo, 2004). Deficits in attention shifting and executive aspects of memory are prevalent across levels of TBI severity (Brooks, Fos, Greve, & Hammond, 1999; Vakil, 2005). However, impulsivity, decreased motivation, and reduced judgment and decision-making abilities are typically observed only in patients with moderate to severe TBI (Hart, Giovannetti, Montgomery, & Schwartz, 1998; Prigatano 2005; Prigatano, 2009). Neuroimaging research suggests that these deficits are associated with focal damage to areas of the frontal lobes underlying these functions (MacDonald, Flashman, & Saykin, 2002; Prigatano, 2009).

There are other cognitive deficits that are mostly observed in patients with moderate to severe TBI that may be associated with more focal injuries to the neuroanatomical areas that support those functions. For example, deficits of awareness (such as failure to recognize cognitive and physical limitations) (Sherer et al., 1998), language impairment (Prigatano, 2009; Sohlberg & Mateer, 1989), and visual and spatial deficits (Ellis & Zahn, 1985; Newcombe, 1982) may be present, and these deficits are associated with neuroimaging findings suggesting damage to corresponding neuroanatomical regions. In some severe TBI cases, global intellectual functioning may be decreased (Levin et al., 1979).

An intriguing line of research from recent years shows that there may be subtle forms of cognitive dysfunction that do not manifest on traditional neuropsychological tests. For example, studies have shown that TBI patients demonstrate greater variability in RT, as compared to healthy uninjured controlled participants (Stuss, Pogue, Buckle, & Bondar, 1994). RT for tasks that involve the transfer of information across the cerebral hemispheres is also sensitive to TBI in some samples (Mathias et al., 2004; Mathias, Bigler, & Jones, 2004). Deficits of oculomotor control, as measured on tests of visual tracking and speeded eye-movements, have been associated with concussion and TBI in several studies (Galetta, Barrett, & Allen, 2011; Galetta, Brandes, & Maki, 2011; Heitger et al., 2006; Heitger et al., 2009; Maruta, Suh, Niogi, Mukherjee, & Ghajar, 2010). Notably, all of these tasks rely on the efficient coordination of multiple anatomically distinct brain regions connected by white matter pathways (i.e., networks). These are precisely the types of functions that are, theoretically, most likely to be disturbed by DAI.

MECHANISMS OF COGNITIVE DISTURBANCE FOLLOWING TBI

In recent years researchers have begun to outline frameworks for the functional neuroanatomy underlying TBI-related cognitive dysfunction (Bigler, 2008; Cicerone et al., 2006). TBI occurs when the brain undergoes rapid acceleration and deceleration forces as a result of physical trauma to the head or body. Focal cortical contusions and subcortical lesions are sometimes present in moderate to serve TBIs. These are caused either by force applied to the skull and transmitted to the brain, or by acceleration/deceleration forces that jostle the brain against boney protuberances of the adjacent skull. DAI is the most common pathological feature of TBI across all levels of severity. Because of its great mass and gyrencephalic organization, the brain can literally pull itself apart when disturbed by physical trauma. Immediately following head-trauma, biomechanical forces lead to disruption of neuronal membranes, axonal stretching, and the opening of voltage-gated ion channels. In some patients, these acute pathophysiological changes may initiate persistent axonal degeneration and other neurodegenerative processes such as amyloid beta (AB) plaque formation and accumulation of tau proteins (Johnson et al., 2013).

DIFFUSE AXONAL INJURY

Direct mechanical injury to axons can lead to a distributed pattern of neuronal injury referred to as DAI. DAI is the result of rapid deformation of axons, which damage the axonal cytoskeleton (Povlishock & Kontos, 1992; Smith & Meaney, 2000; Smith, Meaney, & Shull, 2003) and can directly disrupt axonal transport, leading to swelling. Eventually, gradual disorganization of the structural components of the axon can lead to secondary axonal disconnection (Smith et al., 2003). During this process, the swellings are often seen as tortuous varicosities along the length of the axon or at the disconnected axon terminals known as axon bulbs.

Disturbance of the axonal membrane initiates a neurochemical and neurometabolic cascade of events that disrupts pathway functioning, and can cause further progressive neuronal injury. Mechanical injury perturbs the structure and function of voltage gated sodium channels (Iwata, Stys, & Wolf, 2004; Wolf, Stys, Lusardi, Meaney, & Smith, 2001; Yuen, Browne, Iwata, & Smith, 2009), which indirectly leads to massive calcium influx into axons and associated activation of proteases. Concomitant with calcium influx, mitochondria swell and degrade (Pettus & Povlishock, 1996). Some of these events may be transient and reversible, and hence, amenable to intervention (Smith et al., 2013). However, these disruptions can cause progressive axonal degeneration over time (Yuen et al., 2009).

Rather than strictly "diffuse" in nature, DAI is better characterized as "multifocal." Because of the organization of the human brain, its relationship to the skull, and the physical properties of its major compartments (white matter, gray matter, and cerebral spinal fluid), certain brain regions and white matter pathways are more vulnerable to injury than others (Johnson, Stewart, & Smith, 2013). Specifically, the upper brainstem, medial temporal lobes, pituitary hypothalamic axis, basal forebrain, and cross-hemispheric white matter pathways such as the corpus callosum and anterior commissure have been posited as particularly susceptible to DAI (Bigler, 2007; Bigler, 2008).

DAI AS A MECHANISM OF COGNITIVE DYSFUNCTION FOLLOWING TBI

There is an increasing awareness that human cognition relies on the coherent functioning of anatomically distinct brain regions connected by white matter pathways (i.e., networks) (Van Essen et al., 2013). Complex cognitive behavior arises from the dynamic interplay of multiple brain regions. Hence, axonal injury, and the resultant degradation of white matter connections, decreases the efficiency of information transfer. Dependent on the extent and distribution of damage, multiple sensory, motor, and cognitive functions can be affected. Complex functions involving the integration of information from multiple brain regions may be the most vulnerable to inefficiencies. However, because these complex tasks also tend to be the most amenable to compensatory strategies (due to adaptive redundancies in the cognitive and neuroanatomical organization of higher order cognitive functions), the assessment of these deficits is a challenging endeavor. Consequently, for many domains, DAI affects the efficiency of task performance, rather than its accuracy (Hillary, Medaglia, & Gates, 2011). Improved assessment of DAI-related cognitive changes may necessitate "unmasking" deficits by modulating task difficulty, or otherwise increasing demands on neural networks.

Speeded processing relies on the efficient transfer of information across functionally connected regions of the brain. Hence, processing speed is dependent on the integrity of the brain's white matter pathways. DAI to the brain's vulnerable long-coursing axons will slow response speed, particularly when a task involves integration across multiple cortical regions and must cross the cerebral hemispheres (Mathias et al., 2004). Slowed speed of processing is one of the most common cognitive complaints following TBI, and deficits have been demonstrated in numerous research studies (Bate, Mathias, & Crawford, 2001a, 2001b; Mathias et al., 2004; Mathias & Wheaton, 2007; Ponsford & Kinsella, 1992; Spikman et al., 1996; Timmerman & Brouwer, 1999).

Processing speed is typically referred to as though it were a unitary construct, however, research suggests that the various tasks used to assess processing speed may tap into distinct cognitive processes, which may have different neurophysiological etiologies (Chiaravalloti, Christodoulou, Demaree, & DeLuca, 2003). For example, authors have distinguished between simple and complex processing speed (Chiaravalloti et al., 2003) and simple versus choice RT (Michiels, de Gucht, Cluydts, & Fischler, 1999). Electrophysiological studies suggest that TBI-related processing speed deficits occur at both early and late stages of task performance (Bashore & Ridderinkhof, 2002).

Because most neuropsychological tests rely on briefly presented stimuli and/or a speeded response, slowed processing speed can depress performance on tests designed to tap other cognitive domains. Research suggests that slowed processing speed may underlie apparent deficits in memory (Chiaravalloti et al., 2003; Demaree, DeLuca, Gaudino, & Diamond, 1999) and executive functioning (Demaree et al., 1999; Nelson et al., 2009). This phenomenon has been observed in TBI patients (Nelson et al., 2009) as well as other patient populations (Chiaravalloti et al., 2003; Demaree et al., 1999). Of note, because of its essential role in multiple cognitive domains, slowed processing speed not only influences neuropsychological test performance, but also can lead to errors in everyday memory and executive functioning. Hence, speed of processing does not merely confound the assessment of other cognitive domains, but also serves as an essential component of many cognitive functions.

Evidence in support of DAI as the anatomical substrate underlying slowed processing speed comes from one study showing that the number of damaged white matter structures was positive correlated with cognitive RT, such that those TBI patients with greater neuroimaging evidence of DAI were slower than those with relatively less evidence of white matter damage (Niogi, Mukherjee, & Ghajar, 2008). Interestingly, another study from this group suggests that the relationship between slowed processing speed and white matter integrity can be localized to the pathways underlying execution of a specific cognitive task. Specifically, Niogi, Mukherjee, & Ghajar (2008) found that decreased white matter integrity of the bilateral uncinate fasciculus—the pathway connecting the prefrontal cortex with the parahippocampal gyrus—was associated with deficits in memory, whereas decreased white matter integrity of the left anterior corona radiata—with synapses in the anterior cingulate gyrus— was associated with deficits in attentional control. This latter study is unique in its attempts to identify both the pathophysiological etiology and the anatomical location of TBI-related cognitive dysfunction. More research in this vein will be critical for the future of DAI diagnosis and prognosis.

There is also evidence suggesting that DAI is the mechanism underlying TBI-related executive dysfunction. One study examined 20 mTBI patients within 2 weeks of injury, and found that impaired executive functioning was associated with neuroimaging evidence of axonal injury involving the dorsolateral prefrontal cortex (Lipton, Gulko, & Zimmerman, 2009).

The evolving course of DAI provides plausible explanations for the clinical progression of TBI symptoms, including acute potentially reversible cognitive dysfunction, persisting deficits, and progressive neurobehavioral changes that are associated with remote history of TBI (Guskiewicz, Marshall, & Bailes, 2005; Guskiewicz, Marshall, & Bailes, 2007). Based on the current understanding of this disease course (Povlishock & Katz, 2005), we propose a model of cognitive and behavioral effects of DAI (Figure 9.1). We suggest that the neurochemical and neurometabolic effects of axonal injury lead to transient pathway disruption, which underlie the acute neurocognitive deficits that tend to improve over time. Primary and secondary disconnection of axons, on the other hand, leads to a permanent decrease in pathway efficiency, which would account for persistent cognitive dysfunction. Finally, there is emerging evidence from neuropathological studies and animal models suggesting that DAI initiates neurodegenerative processes akin to what is observed in Alzheimer's disease (AD) (Johnson, Stewart, & Smith, 2012; Johnson et al., 2013). These changes may contribute to progressive delayed onset neurobehavioral outcomes, including depression and early dementia.

NEUROCHEMICAL AND NEUROMETABOLIC CHANGES

In parallel to the direct mechanical injury to axons, uncontrolled ion fluxes across axonal membranes cause excitatory neurotransmitters, such as glutamate, to bind to cell receptors, causing further depolarization as potassium leaves and calcium enters neurons. In order to restore membrane potential, the sodium potassium pump goes into overdrive, transporting potassium into and sodium out of the cell. This is a costly process, requiring increased use of adenosine triphosphate—the neuron's currency for electrochemical transport. Research suggests that glucose metabolism dramatically increases to fuel this process, which is followed by a chronic reduction in glucose metabolism (Bergsneider, Hovda, & McArthur, 2001; Vespa, McArthur, & O'Phelan, 2003). Concurrent with calcium influx, mitochondria undergo swelling, fragmentation, and eventual death, which eliminates cells' ability to provide the energy required for maintaining axons (Büki, Okonkwo, Wang, & Povlishock, 2000).

At the same time, cerebral blood flow decreases, as much as 50% of normal volume in experimental animal models (Doberstein, Velarde, Badie, Hovda, & Becker, 1992; Velarde, Fisher, Hovda, Adelson, & Becker, 1992; Yamakami & McIntosh, 1989; Yuan, Prough, Smith, & Dewitt, 1988). Hence, TBI creates a state of hypermetabolism and hyperglycolisis while cerebral blood flow is decreased and glucose availability in the brain is diminished. This has been referred to as an "energy crisis" in the brain (Giza & Hovda, 2001). The following period of metabolic depression may last as long as 30 days postinjury, and this phenomenon has been posited as the cause of neurologic deficits following TBI (Bergsneider et al., 2001).

Findings from animal studies suggest that chemical and metabolic levels return to normal within 10 days following injury (Hovda, 1996). Clinical studies of TBI patients suggest a similar pattern of hyperglycolisis followed by metabolic depression, however, the time course of these changes appears to be protracted, relative to that characterized in the animal literature. Metabolic recovery can take weeks to months following a moderate to severe TBI (Bergsneider et al., 2001). The time course of metabolic recovery following human mTBI has yet to be well characterized.

NEUROMETABOLIC CHANGES AS A MECHANISM OF COGNITIVE DYSFUNCTION FOLLOWING TBI

Although acute metabolic changes have been associated with cognitive deficits in experimentally injured animals (Hovda, 1996), there is little direct evidence for a link between metabolic disturbance and cognitive deficits from studies in human TBI. However, some studies have demonstrated an association between metabolic changes and patients' general clinical status. The Glasgow Outcome Scale-extended (GOSe) and the Disability Rating Scale (DRS) are clinician-rating scales that measure patients' level of independent functioning following TBI. Cognitive functioning is not directly assessed by these instruments, but rather inferred based on the patient's ability to carry out activities of daily life, such as grooming, feeding, and transportation. Because these scales are only rough proxies of cognitive status, these measures are not sensitive to more subtle deficits associated with mTBI. Studies of patients with moderate to severe TBI have shown a relationship between glucose metabolism and clinician ratings of patient outcome. Bergsneider and colleagues showed a relationship between global cerebral metabolic rate and DRS scores in a sample of mostly severe TBI patients measured up to 16 months postinjury (Bergsneider et al., 2001). Another study conducted by that group demonstrated a correlation between low extracellular glucose and 6-month GOSe scores (Vespa et al., 2003).

Fewer studies have explored relationships between metabolic markers and performance-based measures of cognitive functioning. One study of TBI patients with DAI examined correlations between cerebral metabolism and performance on neuropsychological tests of IQ, memory, processing speed, and executive functioning. Interestingly, only full scale IQ (FSIQ), typically considered less sensitive to TBI than the other domains assessed, demonstrated a significant correlation with regional cerebral metabolism (rCM). FSIQ was positively correlated with rCM in the right cingulate gyrus and bilateral medial frontal gyrus (Kato et al., 2007).

Very few studies have examined cerebral glucose metabolism and cognitive functioning in patients with mTBI. One retrospective study of 20 patients with mild to moderate TBI studied 1–5 years postinjury found an association between neuropsychological test performance and frontal and temporal cerebral glucose metabolism (Gross, Kling, Henry, Herndon, & Lavretsky, 1996). In another study 13 veterans with blast mTBI and 12 cognitively normal community volunteers underwent positron emission tomography (PET) imaging and neuropsychological testing. Injured veterans demonstrated regional hypometabolism in cortical and subcortical brain regions, as well as subtle impairments in processing speed, verbal fluency, attention, and working memory, suggesting that blast mTBI-related cognitive dysfunction is associated with cerebral metabolic changes (Peskind, Petrie, & Cross, 2011).

Although the extant research is suggestive of a relationship between cerebral hypometabolism and cognitive dysfunction, its role as an underlying mechanism of cognitive deficits remains unclear. First, in most cases of moderate to severe TBI and in a subset of mTBI cases, cognitive deficits persist beyond metabolic recovery. Hence, although metabolic changes may contribute to TBI-related cognitive dysfunction, they are unlikely to account for long-term cognitive deficits. Additionally, the relationship between DAI and regional hypometabolism has yet to be fully explicated. Neuronal damage may cause failure of the brain's cognitive networks, hence leading to decreased neuronal activity and, in turn, diffuse glucose hypometabolism (Kato et al., 2007).

FOCAL INJURIES

Focal injuries can occur when blunt head trauma causes cortical contusion, hemorrhage, ischemia, or brainstem damage. These features tend to be absent in cases of mTBI, but characterize a significant proportion of patients with moderate to severe injuries (Adams et al., 2011; Gentry, Godersky, & Thompson, 1988). Clinical and experimental evidence has shown that the frontal and temporal lobes are susceptible to focal injury, due to their adjacency to boney protrusions of the human skull (Ommaya & Gennarelli, 1974). The importance of frontal lobe injury to the clinical presentation of TBI has been appreciated for decades (Jefferson, 1933), and recent research confirms that the frontal and temporal lobes are particularly vulnerable to traumatic injuries (Bigler & Maxwell, 2011; Levin & Kraus, 1994). Numerous autopsy studies of TBI patients have documented high frequency of injury, hemorrhage, or contusion in frontal and anterior temporal regions (Adams, Graham, & Gennarelli, 1985; Courville, 1937; Nevin, 1967).

FOCAL BRAIN INJURY AS A MECHANISM OF COGNITIVE DYSFUNCTION FOLLOWING TBI

The frontal lobes are home to a number of high-level cognitive and behavioral functions, including motivation, self-monitoring, and the control and direction of lower level cognitive abilities. A patient's ability to rapidly produce words according to a given rule (verbal fluency) is often used as an assessment of frontal lobe function. Studies have shown that this ability is depressed following TBI (Henry & Crawford, 2004). Verbal fluency deficits are associated with lesions to the frontal lobe, however, not all patients with frontal lobe lesions demonstrate impairments (Jurado, Mataro, Verger, Bartumeus, & Junque, 2000), suggesting that redundancy in the cognitive and neuroanatomical architecture of the brain allows some patients to compensate for their lesions. Other executive deficits have also been associated with focal lesions to the frontal lobes, including judgment and decision making, working memory, mental flexibility, and attentional control (Stuss & Gow, 1992). Changes in personality consistent with a "frontal syndrome" have also been noted. Specifically, TBI patients demonstrate decreased metacognitive capacity, which refers to the ability to accurately evaluate one's own abilities, monitor impulses, and formulate realistic plans (Levin, Mattis, & Ruff, 1987).

Importantly, there is evidence that the observed "frontal dysfunction" following TBI may not be solely due to focal damage, but rather, could also be the consequence of diffuse damage to the rich network of reciprocal connections between the frontal lobe and other brain areas (Anderson, Bigler, & Blatter, 1995; Stuss & Gow, 1992). This hypothesis is supported by research demonstrating correlations between white matter abnormalities and performance on standard neuropsychological tests of executive functioning (Kraus et al., 2007; Lipton et al., 2009). This research suggests that the efficacy of the brain's functional networks can be significantly reduced due to white matter changes, even in the absence of focal gray matter damage.

Perhaps no cognitive function has been discussed and written of as much as memory. Although we refer to memory as though it were one cognitive function, it is a complex multicomponent construct that operates across all sensory modalities. In the memory literature, theorists have proposed more than 100 different types of memory (Tulving, 2002). However, most agree that memory functions can be divided into two long-term storage and retrieval systems: the explicit (declarative) system, and the implicit (procedural) memory system (Baddeley, 2002; Milner, Squire, & Kandel, 1998; Squire & Knowlton, 2000). Explicit memory—memory for information, objects, and events—is the system that is most likely to be influenced by injury or disease. It is clear that there are distinct stages of the explicit memory process that account for the learning and reproduction of new information. A clinically useful framework for conceptualizing dysfunctional memory proposes a three-stage process (McGaugh, 1966; Parkin, 2001). *Registration memory* refers to the selecting and recording process by which perceptions enter the memory system. This is followed by the *short-term memory* stage, which includes *working memory*—the temporary storage and processing system used for manipulating information over a limited time frame (Baddeley, 1986; Baddeley, 2002). Information stored in short-term memory may be encoded into *long-term memory* (i.e., learned), and retained for durations greater than a few minutes without ongoing rehearsal.

Memory impairment is one of the most reliable neuropsychological deficits reported in TBI patients (Belanger et al., 2005; Binder, 1986; Levin et al., 1987; Richardson, 1984). Medial temporal structures are intimately implicated in memory function, suggesting that focal damage to these regions may underlie memory deficits. Indeed, autopsy studies of patients with fatal injuries (Adams et al., 1985; Courville, 1937; Nevin, 1967) as well as neuroimaging studies in mTBI survivors (Umile, Sandel, Alavi, Terry, &

Plotkin, 2002) suggest a high frequency of focal injury to the temporal lobe associated with TBI. However, Bigler and colleagues conducted a study excluding patients with focal temporal injury, and found that TBI generally was associated with subcortical white matter atrophy in the temporal lobe. Interestingly, they did not find evidence of gray matter atrophy in this group of TBI patients. Taken together, these findings suggest that DAI affects medial temporal structures, leading to white matter and hippocampal atrophy, even in the absence of focal injury to the temporal gray matter (Bigler, Andersob, & Blatter, 2002).

Careful assessment of memory function can allow a clinician to distinguish between deficits at each of these processing stages, and compare memory functioning across sensory (e.g., auditory versus visual) and content (e.g., verbal versus spatial) modalities. The neuroanatomy of long-term memory function is well established (Zola-Morgan & Squire, 1992), and much has been learned about the functional networks that underlie working memory (Schlösser, Wagner, & Sauer, 2006). Just as patterns of language deficits have been used to localize the anatomical source of aphasias (Goodglass & Kaplan, 1972), it is possible that patterns of memory deficits could be used to diagnose the pathophysiology of TBI-related memory impairment (McAllister, Flashman, McDonald, & Saykin, 2006; Vakil, 2005). Focal injury to medial temporal structures is only one potential source of memory problems following TBI. In many cases, changes to the connectivity of the working memory network—including frontal, parietal, and temporal structures—is the likely cause of TBI-related memory impairment (Christodoulou, DeLuca, & Ricker, 2001; McAllister et al., 2006).

NEURODEGENERATIVE PROCESSES

TBI has been recognized as a risk factor for neurodegeneration since dementia pugilistic, or "punch drunk syndrome," was first characterized in boxers (Martland, 1928). Recently, this syndrome has come to be known as chronic traumatic encephalopathy (CTE), and has been reported in athletes exposed to repetitive head trauma as a result of participation in contact sports such as American football (McKee, Cantu, & Nowinski, 2009; Omalu, DeKosky, & Hamilton, 2006; Omalu et al., 2005; Smith, Johnson, & Stewart, 2013). CTE is diagnosed at autopsy according to abnormal intracellular accumulations of hyperphosphorylated tau protein, known as neurofibrillary tangles (NFTs) (Corsellis, Bruton, & Freeman-Browne, 1973; Forman & Trojanowski, 2004; Geddes, Vowles, Nicoll, & Revesz, 1999; Roberts, Allsop, & Bruton, 1990; Tokuda, Ikeda, Yanagisawa, Ihara, & Glenner, 1991). In addition to NFTs, extracellular AB plaques and vascular AB deposits have also been discovered in the brains of CTE cases (Roberts et al., 1990; Tokuda et al., 1991), and AB plaques have been identified in up to 30% of TBI victims who died acutely following a single injury (Chen, Johnson, Uryu, Trojanowski, & Smith, 2009; DeKosky, Abrahamson, & Ciallella, 2007; Gentleman, Greenberg, &, Savage 1997; Ikonomovic, Uryu, & Abrahamson, 2004; Roberts et al., 1994; Smith et al., 2003; Smith et al., 2003; Uryu, Chen, & Martinez, 2007). Axonal damage (i.e., DAI) has been identified as a potential source of AB in animal models and clinical studies of TBI (Johnson et al., 2013).

NEURODEGENERATIVE PROCESSES AS A MECHANISM OF COGNITIVE DYSFUNCTION FOLLOWING TBI

Much of what is known about the long-term neurobehavioral outcomes of TBI comes from studies of retired athletes who have been exposed to repetitive head trauma. In order to evaluate the relationship between head trauma and protracted neurobehavioral outcomes, Guskiewicz and colleagues gathered data from a large sample of retired professional football players. They found associations between recurrent concussion and late-life depression (Guskiewicz et al., 2007), and between recurrent concussion and clinically diagnosed mild cognitive impairment (MCI) (which is considered the preclinical presentation of AD). Earlier onset of AD was also associated with repetitive head injury. Furthermore, a history of multiple head injuries predicted self-reported memory complaints and memory complaints reported by a spouse or close relative (Guskiewicz et al., 2005). A more recent study of older adults demonstrated that history of a single TBI conferred risk for diagnosis of dementia (Gardner et al., 2014). Epidemiological studies such as these strongly suggest a link between head trauma and later-life cognitive and emotional problems; however, the extent to which these symptoms are caused by CTE can only be addressed with neuropathological data.

McKee et al. (2009) have published neuropathological studies reporting tauopathy in individuals with a history of head trauma. Based on postmortem data from 85 brains of athletes, military personnel, and civilians with repetitive head trauma, and a control sample of 18 brains from cognitively intact individuals without a history of head trauma, this group proposed a stage theory of CTE progression (McKee, Stein, & Nowinski, 2013). Stage 1 begins with headache and decreased attention/concentration. Stage II is associated with the onset of depression, explosivity, and short-term memory impairment. Symptoms at stage III include executive dysfunction and cognitive impairment. During the final stage, stage IV, the authors propose the progression of dementia, word-finding difficulties, and aggressive behavior (McKee et al., 2013). Of note, this stage theory is based on the retrospective reports of patients' loved ones. The authors duly note the likelihood of bias in their sample—families of individuals demonstrating symptoms are most likely to participate in brain donation studies. Additionally, individuals exposed to repetitive trauma who did not develop cognitive and behavioral symptoms would make a more suitable control group for correlating neuropathology with clinical symptoms. Longitudinal prospective studies are needed to elucidate the neuropathological and clinical course of CTE.

The recent development of a PET neuroimaging technique (FDDNP-PET) allows, for the first time, the measurement of tau tangle and amyloid plaque deposition in living brains (Shoghi-Jadid, Small, & Agdeppa, 2002). FDDNP-PET has been used to investigate a small sample of retired NFL players with complaints of cognitive and mood symptoms. These researchers found that the retired players had significantly higher FDDNP signals, as compared to control samples, in the caudate, putamen, thalamus, subthalamus, midbrain, cerebellar white matter, and the amygdala (Small, Kepe, & Siddarth, 2013). The results of this small study are very promising, but large-scale longitudinal studies are needed to establish tau and AB pathology as the underlying cause of late-life neurobehavioral dysfunction associated with repetitive head trauma, and to evaluate the utility of FDDNP-PET as a diagnostic tool.

Of note, there are other brain changes related to trauma that may account for long-term changes in cognition and behavior. For example, one study found neuroimaging evidence of white matter abnormalities and decreased regional cerebral blood flow associated with cognitive impairment in retired football players with history of head trauma. The cognitive domains affected were similar to those implicated in MCI and AD—naming, word finding, and memory (Hart, Kraut, & Womack, 2013). These findings do not preclude the role of tau and AB in TBI-related neurodegeneration; however, longitudinal studies employing multimodal neuroimaging and serial neuropsychological assessment would be needed to explicate the time-course of clinical and neuropathological changes following head trauma.

CONCLUSIONS

There is much uncertainty in the current state of TBI care. Clinicians have little to offer their patients by way of prognosis and treatment. This is a frustrating reality for patients and providers alike. TBI is a complex and heterogeneous injury. No two patients are alike with regard to the constellation and course of their cognitive and neurobehavioral symptoms. Likewise, each case has its own idiosyncratic pathophysiological profile—which can include focal brain damage, DAI, neurometabolic changes, and, at later stages of the disease, neurodegenerative processes. Each of these pathological mechanisms contributes to cognitive dysfunction. Careful assessment of cognition could provide insight into the underlying brain pathology relevant to a specific patient, hence clarifying diagnosis and prognosis. Furthermore, elucidating the pathological mechanisms that lead to dysfunction and disability in TBI patients is a necessary first step in designing clinical trials that target treatments to those patients who are most likely to benefit.

DAI is a promising therapeutic target (Smith et al., 2013); however, a reliable clinical diagnosis of DAI is needed in order to design clinical trials of experimental treatments. It is only logical that cognitive functioning could serve as the basis for diagnosis of DAI (Smith et al., 2013), and interestingly, promising tests of pathway functioning already exist (Galetta et al., 2011; Galetta, Brandes, & Maki, 2011; Heitger et al., 2006; Heitger et al., 2009; Mathias, Beall, & Bigler, 2004; Mathias et al., 2004; Maruta et al., 2010). However, in order to capitalize on cognition as a window into the injured brain, more research must be done.

Recent years have seen advances in the development of *in vivo* measures of trauma-induced neuropathology. For decades, focal contusion and hemorrhage have been readily assessed by standard clinical neuroimaging. DAI, on the other hand, is a "stealth pathology" that, until recently, has been invisible to standard imaging techniques. New advanced neuroimaging techniques have been developed that now allow for the assessment of white matter changes that may reflect DAI (Bazarian et al., 2007; Kraus et al., 2007; Wilde et al., 2008). Furthermore, serum biomarkers of neuronal injury have been discovered that are sensitive to white matter changes and cognitive dysfunction in CT-negative mTBI survivors (Siman, Giovannone, & Hanten, 2013). Additionally, the development of new ligands for PET imaging allows for the examination of neurodegenerative processes in living TBI patients (Small et al., 2013).

The next crucial step is thorough examination of the relationships between *in vivo* measures of pathology and cognitive and behavioral functioning. Much of the extant research has correlated neuroimaging or neuropathological data with a general level of dysfunction of disability (Kesler, Adams, & Bigler, 2000). Clinician rating scales and retrospective chart reviews provide blunt measures of cognitive functioning at best. Research shows that patients with scores indicating good recovery still manifest cognitive impairment when assessed with more sensitive neuropsychological tests (Stuss, Ely & Hugenholtz, 1985) and these scales have little applicability to mTBI. A finer level of analysis is necessary to detect the cognitive deficits due to DAI. Future work should examine relationships between *in vivo* measures of DAI and well-defined profiles of cognitive dysfunction.

In addition to improved understanding of TBI pathophysiology, a more refined understanding of TBI-related cognitive dysfunction is needed. Standard protocols for the neuropsychological assessment of TBI tend to focus on tests that have high sensitivity to TBI-related cognitive impairments. Unfortunately, many of these highly sensitive tasks lack specificity to the distinct pathological processes that underlie TBI-related cognitive dysfunction. For example, it has been noted that the currently available tests of executive functioning lack the specificity to differentiate between cognitive processes (Cicerone et al., 2006). Further confounding matters, many tests are highly sensitive to other common head-injury sequelae, such as depression (Christensen, Griffiths, MacKinnon, & Jacomb, 1997), fatigue (Ziino & Ponsford, 2006), and pain (Hart, Martelli, & Zasler, 2000). Tasks that fail to distinguish between pathological etiologies are of limited diagnostic and prognostic utility. A more refined articulation of the cognitive domains affected by DAI may involve the development of new tests and novel applications of existing measures. Some promising experimental measures already exist, including theoretically compelling measures of RT (Mathias et al., 2004; Mathias et al., 2004; Stuss et al., 1994), oculomotor control (Galetta et al., 2011; Galetta et al., 2011; Heitger et al., 2006; Heitger et al., 2009; Maruta et al., 2010), and executive functioning. For these paradigms to become clinically useful, appropriate normative data must be collected.

TBI is a messy injury, with great heterogeneity and complexity in both its clinical presentation and its underlying pathophysiology. Large-scale longitudinal studies combining multimodal neuroimaging, serum biomarkers, and *specific* measures of cognition could provide clarity regarding the sequence of pathological changes and cognitive symptoms initiated by DAI. Recent advances in TBI research have already begun to elucidate the causal relationships between axonal injury, neurometabolic changes, and neurodegenerative processes. With more research correlating these measures of pathology with TBI-related cognitive deficits, neuropsychologists could potentially use cognitive performance to diagnose DAI and predict its course. This would represent a considerable advance in the neuropsychological approach to TBI—a paradigm shift that would take advantage of the unique expertise offered by neuropsychologists, enabling them to make a valuable contribution to the clinical management of this challenging population.

References

Adams, J., Graham, D., & Gennarelli, T. (1985). Contemporary neuropathological considerations regarding brain damage in head injury. *Central Nervous System Trauma Status Report*, 65–77.

Adams, J. H., Jennett, B., Murray, L. S., Teasdale, G. M., Gennarelli, T. A., & Graham, D. I. (2011). Neuropathological findings in disabled survivors of a head injury. *Journal of Neurotrauma*, 28(5), 701–709.

Anderson, C. V., Bigler, E. D., & Blatter, D. D. (1995). Frontal lobe lesions, diffuse damage, and neuropsychological functioning in traumatic brain-injured patients. *Journal of Clinical and Experimental Neuropsychology, 17*(6), 900–908.

Axelrod, B. N., Fichtenberg, N. L., Liethen, P. C., Czarnota, M. A., & Stucky, K. (2001). Performance characteristics of postacute traumatic brain injury patients on the WAIS-III and WMS-III. *The Clinical Neuropsychologist, 15*(4), 516–520.

Baddeley, A. D. (1986). *Working memory Oxford.* Oxford, United Kingdom: Oxford University Press.

Baddeley A. D. (2002). The psychology of memory. In A. D. Baddeley, M. Kopelman, & A. Wilson (Eds.), *The essential handbook of memory disorder for clinicians* (pp. 1–13). Hoboken, NJ: John Wiley & Sons.

Bashore, T. R., & Ridderinkhof, K. R. (2002). Older age, traumatic brain injury, and cognitive slowing: Some convergent and divergent findings. *Psychological Bulletin, 128*(1), 151.

Bate, A. J., Mathias, J. L., & Crawford, J. R. (2001a). The covert orienting of visual attention following severe traumatic brain injury. *Journal of Clinical and Experimental Neuropsychology, 23*(3), 386–398.

Bate, A. J., Mathias, J. L., & Crawford, J. R. (2001b). Performance on the test of everyday attention and standard tests of attention following severe traumatic brain injury. *The Clinical Neuropsychologist, 15*(3), 405–422.

Bazarian, J. J., Mcclung, J., Shah, M. N., Ting Cheng, Y., Flesher, W., & Kraus, J. (2005). Mild traumatic brain injury in the United States, 1998–2000. *Brain Injury, 19*(2), 85–91.

Bazarian, J. J., Zhong, J., Blyth, B., Zhu, T., Kavcic, V., & Peterson, D. (2007). Diffusion tensor imaging detects clinically important axonal damage after mild traumatic brain injury: A pilot study. *Journal of Neurotrauma, 24*(9), 1447–1459.

Belanger, H. G., Curtiss, G., Demery, J. A., Lebowitz, B. K., & Vanderploeg, R. D. (2005). Factors moderating neuropsychological outcomes following mild traumatic brain injury: A meta-analysis. *Journal of the International Neuropsychological Society, 11*(03), 215–227.

Belanger, H. G., & Vanderploeg, R. D. (2005). The neuropsychological impact of sports-related concussion: A meta-analysis. *Journal of the International Neuropsychological Society, 11*(4), 345–357.

Bergsneider, M., Hovda, D. A., McArthur, D. L., et al. (2001). Metabolic recovery following human traumatic brain injury based on FDG-PET: Time course and relationship to neurological disability. *The Journal of Head Trauma Rehabilitation, 16*(2), 135–148.

Bigler, E. D. (2007). Anterior and middle cranial fossa in traumatic brain injury: Relevant neuroanatomy and neuropathology in the study of neuropsychological outcome. *Neuropsychology, 21*(5), 515.

Bigler, E. D. (2008). Neuropsychology and clinical neuroscience of persistent post-concussive syndrome. *Journal of the International Neuropsychological Society, 14*(01), 1–22.

Bigler, E. D., Andersob, C. V., & Blatter, D. D. (2002). Temporal lobe morphology in normal aging and traumatic brain injury. *American Journal of Neuroradiology, 23*(2), 255–266.

Bigler, E. D., & Maxwell, W. L. (2011). Neuroimaging and neuropathology of TBI. *NeuroRehabilitation, 28*(2), 63–74.

Binder, L. M. (1986). Persisting symptoms after mild head injury: A review of the postconcussive syndrome. *Journal of Clinical and Experimental Neuropsychology, 8*(4), 323–346.

Blumbergs, P. C., Jones, N. R., & North, J. B. (1989). Diffuse axonal injury in head trauma. *Journal of Neurology, Neurosurgery, and Psychiatry, 52*(7), 838.

Boake, C., McCauley, S. R., Levin, H. S., et al. (2005). Diagnostic criteria for postconcussional syndrome after mild to moderate traumatic brain injury. *The Journal of Neuropsychiatry and Clinical Neurosciences, 17*(3), 350–356.

Brooks, J., Fos, L. A., Greve, K. W., & Hammond, J. S. (1999). Assessment of executive function in patients with mild traumatic brain injury. *The Journal of Trauma, 46*(1), 159–163.

Broshek, D. K., Kaushik, T., Freeman, J. R., Erlanger, D., Webbe, F., & Barth, J. T. (2005). Sex differences in outcome following sports-related concussion. *Journal of Neurosurgery, 102*(5), 856–863.

Browne, K. D., Chen, X. H., Meaney, D. F., & Smith, D. H. (2011). Mild traumatic brain injury and diffuse axonal injury in swine. *Journal of Neurotrauma, 28*(9), 1747–1755.

Büki, A., Okonkwo, D. O., Wang, K. K., & Povlishock, J. T. (2000). Cytochrome c release and caspase activation in traumatic axonal injury. *The Journal of Neuroscience, 20*(8), 2825–2834.

Büki, A., & Povlishock, J. (2006). All roads lead to disconnection?–Traumatic axonal injury revisited. *Acta Neurochirurgica, 148*(2), 181–194.

Carroll, L., Cassidy, J. D., Peloso, P., et al. (2004). Prognosis for mild traumatic brain injury: Results of the WHO collaborating centre task force on mild traumatic brain injury. *Journal of Rehabilitation Medicine, 36*(0), 84–105.

Chen, X. H., Johnson, V. E., Uryu, K., Trojanowski, J. Q., & Smith, D. H. (2009). A lack of amyloid β plaques despite persistent accumulation of amyloid β in axons of long-term survivors of traumatic brain injury. *Brain Pathology, 19*(2), 214–223.

Chiaravalloti, N. D., Christodoulou, C., Demaree, H. A., & DeLuca, J. (2003). Differentiating simple versus complex processing speed: Influence on new learning and memory performance. *Journal of Clinical and Experimental Neuropsychology, 25*(4), 489–501.

Christensen, H., Griffiths, K., MacKinnon, A., & Jacomb, P. (1997). A quantitative review of cognitive deficits in depression and Alzheimer-type dementia. *Journal of the International Neuropsychological Society, 3*(06), 631–651.

Christodoulou, C., DeLuca, J., Ricker, J., et al. (2001). Functional magnetic resonance imaging of working memory impairment after traumatic brain injury. *Journal of Neurology, Neurosurgery, and Psychiatry, 71*(2), 161–168.

Cicerone, K. D., Levin, H. S., Malec, J. F., Stuss, D. T., & Whyte, J. (2006). Cognitive rehabilitation interventions for executive function: Moving from bench to bedside in patients with traumatic brain injury. *Journal of Cognitive Neuroscience, 18*(7), 1212–1222.

Corsellis, J., Bruton, C., & Freeman-Browne, D. (1973). The aftermath of boxing. *Psychological Medicine, 3*(03), 270–303.

Courville, C. B. (1937). *Pathology of the central nervous system: A study based upon a survey of lesions found in a series of fifteen thousand autopsies.* Mountain View, CA: Pacific Press Publishing Association.

Covassin, T., Swanik, C. B., & Sachs, M. L. (2003). Sex differences and the incidence of concussions among collegiate athletes. *Journal of Athletic Training, 38*(3), 238.

DeKosky, S. T., Abrahamson, E. E., Ciallella, J. R., et al. (2007). Association of increased cortical soluble $A\beta_{42}$ levels with diffuse plaques after severe brain injury in humans. *Archives of Neurology, 64*(4), 541–544.

Demaree, H. A., DeLuca, J., Gaudino, E. A., & Diamond, B. J. (1999). Speed of information processing as a key deficit in multiple sclerosis: Implications for rehabilitation. *Journal of Neurology, Neurosurgery, and Psychiatry, 67*(5), 661–663.

Dikmen, S. S., Machamer, J. E., Winn, H. R., & Temkin, N. R. (1995). Neuropsychological outcome at 1-year post head injury. *Neuropsychology, 9*(1), 80.

Doberstein, C., Velarde, F., Badie, H., Hovda, D., & Becker, D. (1992). Changes in local cerebral blood flow following concussive brain injury. *Society for Neuroscience, 18*, 175.

Ellis, D. W., & Zahn, B. S. (1985). Psychological functioning after severe closed head injury. *Journal of Personality Assessment, 49*(2), 125–128.

Forman M. S., Trojanowski J. Q., Lee V. M.-Y. (2004). Hereditary tauopathies and idiopathic frontotemporal dementias. In: M. Esiri, V. M.-Y. Lee, J. Q. Trojanowski (Eds.), *The neuropathology of dementia* (pp. 257–288). Cambridge: Cambridge University Press.

Fuster, J. (1999). Synopsis of function and dysfunction of the frontal lobe. *Acta Psychiatrica Scandinavica, 99*(s395), 51–57.

Galetta, K., Barrett, J., Allen, M., et al. (2011). The King-Devick test as a determinant of head trauma and concussion in boxers and MMA fighters. *Neurology, 76*(17), 1456–1462.

Galetta, K. M., Brandes, L. E., Maki, K., et al. (2011). The King–Devick test and sports-related concussion: Study of a rapid visual screening tool in a collegiate cohort. *Journal of the Neurological Sciences, 309*(1), 34–39.

Gardner, R. C., Burke, J. F., Nettiksimmons, J., Kaup, A., Barnes, D. E., & Yaffe, K. (2014). Dementia risk after traumatic brain injury vs nonbrain trauma: The role of age and severity. *JAMA Neurology*.

Geddes, J., Vowles, G., Nicoll, J., & Revesz, T. (1999). Neuronal cytoskeletal changes are an early consequence of repetitive head injury. *Acta Neuropathologica*, 98(2), 171–178.

Gentleman, S. M., Greenberg, B. D., Savage, M. J., et al. (1997). A β42 is the predominant form of amyloid b-protein in the brains of short-term survivors of head injury. *Neuroreport*, 8(6), 1519–1522.

Gentry, L. R., Godersky, J. C., & Thompson, B. (1988). MR imaging of head trauma: Review of the distribution and radiopathologic features of traumatic lesions. *American Journal of Neuroradiology*, 9(1), 101–110.

Giza, C. C., & Hovda, D. A. (2001). The neurometabolic cascade of concussion. *Journal of Athletic Training*, 36(3), 228.

Goodglass, H., & Kaplan, E. (1972). *The assessment of aphasia and related disorders.* Philadelphia, PA: Lea & Febiger.

Gross, H., Kling, A., Henry, G., Herndon, C., & Lavretsky, H. (1996). Local cerebral glucose metabolism in patients with long-term behavioral and cognitive deficits following mild traumatic brain injury. *The Journal of Neuropsychiatry and Clinical Neurosciences*, 8(3), 324–334.

Guskiewicz, K. M., Marshall, S. W., Bailes, J., et al. (2005). Association between recurrent concussion and late-life cognitive impairment in retired professional football players. *Neurosurgery*, 57(4), 719.

Guskiewicz, K. M., Marshall, S. W., Bailes, J., et al. (2007). Recurrent concussion and risk of depression in retired professional football players. *Medicine and Science in Sports and Exercise*, 39(6), 903.

Hart, J., Kraut, M. A., Womack, K. B., et al. (2013). Neuroimaging of cognitive dysfunction and depression in aging retired national football league players: A cross-sectional study. *JAMA Neurology*, 70(3), 326–335.

Hart, R. P., Martelli, M. F., & Zasler, N. D. (2000). Chronic pain and neuropsychological functioning. *Neuropsychology Review*, 10(3), 131–149.

Hart, T., Giovannetti, T., Montgomery, M. W., & Schwartz, M. F. (1998). Awareness of errors in naturalistic action after traumatic brain injury. *The Journal of Head Trauma Rehabilitation*, 13(5), 16–28.

Heitger, M. H., Jones, R. D., Dalrymple-Alford, J. C., Frampton, C. M., Ardagh, M. W., & Anderson, T. J. (2006). Motor deficits and recovery during the first year following mild closed head injury. *Brain Injury*, 20(8), 807–824.

Heitger, M. H., Jones, R. D., Macleod, A., Snell, D. L., Frampton, C. M., & Anderson, T. J. (2009). Impaired eye movements in post-concussion syndrome indicate suboptimal brain function beyond the influence of depression, malingering or intellectual ability. *Brain*, 132(10), 2850–2870.

Henry, J. D., & Crawford, J. R. (2004). A meta-analytic review of verbal fluency performance in patients with traumatic brain injury. *Neuropsychology*, 18(4), 621.

Hillary, F. G., Medaglia, J. D., Gates, K., et al. (2011). Examining working memory task acquisition in a disrupted neural network. *Brain*, 134(5), 1555–1570.

Hovda, D. (1996). Metabolic dysfunction. In *Neurotrauma* (pp. 1459–1478). New York, NY: McGraw-Hill Inc.

Huisman, T. A., Schwamm, L. H., Schaefer, P. W., et al. (2004). Diffusion tensor imaging as potential biomarker of white matter injury in diffuse axonal injury. *American Journal of Neuroradiology*, 25(3), 370–376.

Ikonomovic, M. D., Uryu, K., Abrahamson, E. E., et al. (2004). Alzheimer's pathology in human temporal cortex surgically excised after severe brain injury. *Experimental Neurology*, 190(1), 192–203.

Inglese, M., Makani, S., Johnson, G., et al. (2005). Diffuse axonal injury in mild traumatic brain injury: A diffusion tensor imaging study. *Journal of Neurosurgery, 103*(2), 298–303.

Iwata, A., Stys, P. K., Wolf, J. A., et al. (2004). Traumatic axonal injury induces proteolytic cleavage of the voltage-gated sodium channels modulated by tetrodotoxin and protease inhibitors. *The Journal of Neuroscience, 24*(19), 4605–4613.

Jefferson, G. (1933). Remarks on the treatment of acute head injuries. *British Medical Journal, 2*(3800), 807.

Johnson, V. E., Stewart, W., & Smith, D. H. (2012). Widespread tau and amyloid-beta pathology many years after a single traumatic brain injury in humans. *Brain Pathology, 22*(2), 142–149.

Johnson, V. E., Stewart, W., & Smith, D. H. (2013). Axonal pathology in traumatic brain injury. *Experimental Neurology, 246,* 35–43.

Jurado, M. A., Mataro, M., Verger, K., Bartumeus, F., & Junque, C. (2000). Phonemic and semantic fluencies in traumatic brain injury patients with focal frontal lesions. *Brain Injury, 14*(9), 789–795.

Kato, T., Nakayama, N., Yasokawa, Y., Okumura, A., Shinoda, J., & Iwama, T. (2007). Statistical image analysis of cerebral glucose metabolism in patients with cognitive impairment following diffuse traumatic brain injury. *Journal of Neurotrauma, 24*(6), 919–926.

Kerr, A., & Zelazo, P. D. (2004). Development of "hot" executive function: The children's gambling task. *Brain and Cognition, 55*(1), 148–157.

Kesler, S. R., Adams, H. F., & Bigler, E. D. (2000). SPECT, MR and quantitative MR imaging: Correlates with neuropsychological and psychological outcome in traumatic brain injury. *Brain Injury, 14*(10), 851–857.

Kraus, M. F., Susmaras, T., Caughlin, B. P., Walker, C. J., Sweeney, J. A., & Little, D. M. (2007). White matter integrity and cognition in chronic traumatic brain injury: A diffusion tensor imaging study. *Brain, 130*(10), 2508–2519.

Levin, H., & Kraus, M. F. (1994). The frontal lobes and traumatic brain injury. *The Journal of Neuropsychiatry and Clinical Neurosciences, 6,* 443–454.

Levin, H. S., Grossman, R. G., Rose, J. E., & Teasdale, G. (1979). Long-term neuropsychological outcome of closed head injury. *Journal of Neurosurgery, 50,* 412–422.

Levin, H. S., Mattis, S., Ruff, R. M., et al. (1987). Neurobehavioral outcome following minor head injury: A three-center study. *Journal of Neurosurgery, 66*(2), 234–243.

Levin, H. S., Wilde, E., Troyanskaya, M., et al. (2010). Diffusion tensor imaging of mild to moderate blast-related traumatic brain injury and its sequelae. *Journal of Neurotrauma, 27*(4), 683–694.

Levin, H. S., Wilde, E. A., Chu, Z., et al. (2008). Diffusion tensor imaging in relation to cognitive and functional outcome of traumatic brain injury in children. *The Journal of Head Trauma Rehabilitation, 23*(4), 197.

Lipton, M. L., Gulko, E., Zimmerman, M. E., et al. (2009). Diffusion-tensor imaging implicates prefrontal axonal injury in executive function impairment following very mild traumatic brain injury1. *Radiology, 252*(3), 816–824.

MacDonald, B. C., Flashman, L. A., & Saykin, A. J. (2002). Executive dysfunction following traumatic brain injury: Neural substrates and treatment strategies. *NeuroRehabilitation, 17,* 333–344.

Martland, H. (1928). Dementia pugilistica. *Journal of the American Medical Association, 91,* 1103–1107.

Maruta, J., Suh, M., Niogi, S. N., Mukherjee, P., & Ghajar, J. (2010). Visual tracking synchronization as a metric for concussion screening. *The Journal of Head Trauma Rehabilitation, 25*(4), 293–305.

Mathias, J. L., Beall, J. A., & Bigler, E. D. (2004). Neuropsychological and information processing deficits following mild traumatic brain injury. *Journal of the International Neuropsychological Society, 10*(02), 286–297.

Mathias, J. L., Bigler, E. D., Jones, N. R., et al. (2004). Neuropsychological and information processing performance and its relationship to white matter changes following moderate and severe traumatic brain injury: A preliminary study. *Applied Neuropsychology*, *11*(3), 134–152.

Mathias, J. L., & Wheaton, P. (2007). Changes in attention and information-processing speed following severe traumatic brain injury: A meta-analytic review. *Neuropsychology*, *21*(2), 212.

Mayer, A., Ling, J., Mannell, M., et al. (2010). A prospective diffusion tensor imaging study in mild traumatic brain injury. *Neurology*, *74*(8), 643–650.

McAllister, T. W., Flashman, L. A., McDonald, B. C., & Saykin, A. J. (2006). Mechanisms of working memory dysfunction after mild and moderate TBI: Evidence from functional MRI and neurogenetics. *Journal of Neurotrauma*, *23*(10), 1450–1467.

McCrea, M. (2001). Standardized mental status assessment of sports concussion. *Clinical Journal of Sport Medicine*, *11*(3), 176.

McGaugh, J. L. (1966). Time-dependent processes in memory storage. *Science*, *153*(3742), 1351–1358.

McKee, A. C., Cantu, R. C., Nowinski, C. J., et al. (2009). Chronic traumatic encephalopathy in athletes: Progressive tauopathy following repetitive head injury. *Journal of Neuropathology and Experimental Neurology*, *68*(7), 709.

McKee, A. C., Stein, T. D., Nowinski, C. J., et al. (2013). The spectrum of disease in chronic traumatic encephalopathy. *Brain*, *136*(1), 43–64.

Michiels, V., de Gucht, V., Cluydts, R., & Fischler, B. (1999). Attention and information processing efficiency in patients with chronic fatigue syndrome. *Journal of Clinical and Experimental Neuropsychology*, *21*(5), 709–729.

Milner, B., Squire, L. R., & Kandel, E. R. (1998). Cognitive neuroscience and the study of memory. *Neuron*, *20*(3), 445–468.

Nelson, L. A., Yoash-Gantz, R. E., Pickett, T. C., & Campbell, T. A. (2009). Relationship between processing speed and executive functioning performance among OEF/OIF veterans: Implications for postdeployment rehabilitation. *The Journal of Head Trauma Rehabilitation*, *24*(1), 32–40.

Nevin, N. C. (1967). Neuropathological changes in the white matter following head injury. *Journal of Neuropathology and Experimental Neurology*, *26*(1), 77–84.

Newcombe, F. (1982). The psychological consequences of closed head injury: Assessment and rehabilitation. *Injury*, *14*(2), 111–136.

Niogi, S., Mukherjee, P., Ghajar, J., et al. (2008). Extent of microstructural white matter injury in postconcussive syndrome correlates with impaired cognitive reaction time: A 3T diffusion tensor imaging study of mild traumatic brain injury. *American Journal of Neuroradiology*, *29*(5), 967–973.

Niogi, S. N., Mukherjee, P., Ghajar, J., et al. (2008). Structural dissociation of attentional control and memory in adults with and without mild traumatic brain injury. *Brain*, *131*(12), 3209–3221.

Omalu, B. I., DeKosky, S. T., Hamilton, R. L., et al. (2006). Chronic traumatic encephalopathy in a national football league player: Part II. *Neurosurgery*, *59*(5), 1086–1093.

Omalu, B. I., DeKosky, S. T., Minster, R. L., Kamboh, M. I., Hamilton, R. L., & Wecht, C. H. (2005). Chronic traumatic encephalopathy in a national football league player. *Neurosurgery*, *57*(1), 128–134.

Ommaya, A. K., & Gennarelli, T. (1974). Cerebral concussion and traumatic unconsciousness correlation of experimental and clinical observations on blunt head injuries. *Brain*, *97*(4), 633–654.

Parkin, A. J. (2001). The structure and mechanisms of memory. In R. Brenda (Ed.), *The handbook of cognitive neuropsychology: What deficits reveal about the human mind* (pp. 399–422). Philadelphia, PA: Psychology Press.

Peskind, E. R., Petrie, E. C., Cross, D. J., et al. (2011). Cerebrocerebellar hypometabolism associated with repetitive blast exposure mild traumatic brain injury in 12 Iraq war veterans with persistent post-concussive symptoms. *Neuroimage, 54*, S76–S82.

Pettus, E. H., & Povlishock, J. T. (1996). Characterization of a distinct set of intra-axonal ultrastructural changes associated with traumatically induced alteration in axolemmal permeability. *Brain Research, 722*(1), 1–11.

Ponsford, J., & Kinsella, G. (1992). Attentional deficits following closed-head injury. *Journal of Clinical and Experimental Neuropsychology, 14*(5), 822–838.

Povlishock, J., & Kontos, H. (1992). The role of oxygen radicals in the pathobiology of traumatic brain injury. *Human Cell, 5*(4), 345–353.

Povlishock, J. T., & Katz, D. I. (2005). Update of neuropathology and neurological recovery after traumatic brain injury. *The Journal of Head Trauma Rehabilitation, 20*(1), 76–94.

Prigatano, G. (2005). Impaired self-awareness after moderately severe to severe traumatic brain injury. In K. R. H. von Wilde (Ed.), *Re-engineering of the damaged brain and spinal cord* (pp. 39–42). New York, NY: Springer.

Prigatano, G. P. (2009). Anosognosia: Clinical and ethical considerations. *Current Opinion in Neurology, 22*(6), 606–611.

Raskin, S. A., & Rearick, E. (1996). Verbal fluency in individuals with mild traumatic brain injury. *Neuropsychology, 10*(3), 416.

Richardson, J. T. (1984). The effects of closed head injury upon intrusions and confusions in free recall. *Cortex, 20*(3), 413–420.

Roberts, G., Gentleman, S., Lynch, A., Murray, L., Landon, M., & Graham, D. (1994). Beta amyloid protein deposition in the brain after severe head injury: Implications for the pathogenesis of Alzheimer's disease. *Journal of Neurology, Neurosurgery, and Psychiatry, 57*(4), 419–425.

Roberts, G. W., Allsop, D., & Bruton, C. (1990). The occult aftermath of boxing. *Journal of Neurology, Neurosurgery, and Psychiatry, 53*(5), 373–378.

Røe, C., Sveen, U., Alvsåker, K., & Bautz-Holter, E. (2009). Post-concussion symptoms after mild traumatic brain injury: Influence of demographic factors and injury severity in a 1-year cohort study. *Disability and Rehabilitation, 31*(15), 1235–1243.

Schlösser, R., Wagner, G., & Sauer, H. (2006). Assessing the working memory network: Studies with functional magnetic resonance imaging and structural equation modeling. *Neuroscience, 139*(1), 91–103.

Schretlen, D. J., & Shapiro, A. M. (2003). A quantitative review of the effects of traumatic brain injury on cognitive functioning. *International Review of Psychiatry, 15*(4), 341–349.

Selassie, A. W., Zaloshnja, E., Langlois, J. A., Miller, T., Jones, P., & Steiner, C. (2008). Incidence of long-term disability following traumatic brain injury hospitalization, United States, 2003. *The Journal of Head Trauma Rehabilitation, 23*(2), 123–131.

Sherer, M., Boake, C., Levin, E., Silver, B. V., Ringholz, G., & High, W. M. (1998). Characteristics of impaired awareness after traumatic brain injury. *Journal of the International Neuropsychological Society, 4*, 380–387.

Shoghi-Jadid, K., Small, G. W., Agdeppa, E. D., et al. (2002). Localization of neurofibrillary tangles and beta-amyloid plaques in the brains of living patients with Alzheimer disease. *The American Journal of Geriatric Psychiatry, 10*(1), 24–35.

Siman, R., Giovannone N., Hanten G., et al. (2013). Evidence that the blood biomarker SNTF predicts brain imaging changes and persistent cognitive dysfunction in mild TBI patients. *Frontiers of Neurology. 4*, 190.

Small, G. W., Kepe, V., Siddarth, P., et al. (2013). PET scanning of brain tau in retired national football league players: Preliminary findings. *The American Journal of Geriatric Psychiatry, 21*(2), 138–144.

Smith, D. H., Chen, X. -h, Iwata, A., & Graham, D. I. (2003). Amyloid β accumulation in axons after traumatic brain injury in humans. *Journal of Neurosurgery, 98*(5), 1072–1077.

Smith, D. H., Hicks, R., & Povlishock, J. T. (2013). Therapy development for diffuse axonal injury. *Journal of Neurotrauma*, 30(5), 307–323.

Smith, D. H., Johnson, V. E., & Stewart, W. (2013). Chronic neuropathologies of single and repetitive TBI: substrates of dementia? *Nature Reviews Neurology*, 9(4), 211–221.

Smith, D. H., & Meaney, D. F. (2000). Axonal damage in traumatic brain injury. *The Neuroscientist*, 6(6), 483–495.

Smith, D. H., Meaney, D. F., & Shull, W. H. (2003). Diffuse axonal injury in head trauma. *The Journal of Head Trauma Rehabilitation*, 18(4), 307.

Sohlberg, M. M., & Mateer, C. A. (1989). The assessment of cognitive-communicative functions in head injury. *Topics in Language Disorders*, 9(2), 15–33.

Spikman, J. M., van Zomeren, A. H., & Deelman, B. G. (1996). Deficits of attention after closed-head injury: Slowness only? *Journal of Clinical and Experimental Neuropsychology*, 18(5), 755–767.

Squire, L. R., & Knowlton, B. J. (2000). The medial temporal lobe, the hippocampus, and the memory systems of the brain. *The New Cognitive Neurosciences*, 2, 756–776.

Stuss, D., Ely, P., & Hugenholtz, H., et al. (1985). Subtle neuropsychological deficits in patients with good recovery after closed head injury. *Neurosurgery*, 17(1), 41–47.

Stuss, D. T., & Gow, C. A. (1992). "Frontal dysfunction" after traumatic brain injury. *Cognitive and Behavioral Neurology*, 5(4), 272–282.

Stuss, D. T., Pogue, J., Buckle, L., & Bondar, J. (1994). Characterization of stability of performance in patients with traumatic brain injury: Variability and consistency on reaction time tests. *Neuropsychology*, 8(3), 316.

Timmerman, M., & Brouwer, W. (1999). Slow information processing after very severe closed head injury: Impaired access to declarative knowledge and intact application and acquisition of procedural knowledge. *Neuropsychologia*, 37(4), 467–478.

Tokuda, T., Ikeda, S., Yanagisawa, N., Ihara, Y., & Glenner, G. (1991). Re-examination of ex-boxers' brains using immunohistochemistry with antibodies to amyloid β-protein and tau protein. *Acta Neuropathologica*, 82(4), 280–285.

Tulving, E. (2002). Episodic memory: From mind to brain. *Annual Review of Psychology*, 53(1), 1–25.

Umile, E. M., Sandel, M. E., Alavi, A., Terry, C. M., & Plotkin, R. C. (2002). Dynamic imaging in mild traumatic brain injury: Support for the theory of medial temporal vulnerability. *Archives of Physical Medicine and Rehabilitation*, 83(11), 1506–1513.

Uryu, K., Chen, X. -H., Martinez, D., et al. (2007). Multiple proteins implicated in neurodegenerative diseases accumulate in axons after brain trauma in humans. *Experimental Neurology*, 208(2), 185–192.

Vakil, E. (2005). The effect of moderate to severe traumatic brain injury (TBI) on different aspects of memory: A selective review. *Journal of Clinical and Experimental Neuropsychology*, 27(8), 977–1021.

Vanderploeg, R. D., Curtiss, G., & Belanger, H. G. (2005). Long-term neuropsychological outcomes following mild traumatic brain injury. *Journal of the International Neuropsychological Society*, 11(03), 228–236.

Van Essen, D. C., Smith, S. M., Barch, D. M., Behrens, T. E., Yacoub, E., & Ugurbil, K. (2013). The WU-Minn human connectome project: An overview. *NeuroImage*, 80, 62–79.

Velarde, F., Fisher, D., Hovda, D., Adelson, P., & Becker, D. (1992). Fluid percussion injury induces prolonged changes in cerebral blood flow. *Journal of Neurotrauma*, 9, 402.

Vespa, P. M., McArthur, D., O'Phelan, K., et al. (2003). Persistently low extracellular glucose correlates with poor outcome 6 months after human traumatic brain injury despite a lack of increased lactate & colon; A microdialysis study. *Journal of Cerebral Blood Flow and Metabolism*, 23(7), 865–877.

Whiteneck, G. G., Gerhart, K. A., & Cusick, C. P. (2004). Identifying environmental factors that influence the outcomes of people with traumatic brain injury. *The Journal of Head Trauma Rehabilitation*, 19(3), 191–204.

Wilde, E., McCauley, S., Hunter, J., et al. (2008). Diffusion tensor imaging of acute mild traumatic brain injury in adolescents. *Neurology, 70*(12), 948–955.

Wolf, J. A., Stys, P. K., Lusardi, T., Meaney, D., & Smith, D. H. (2001). Traumatic axonal injury induces calcium influx modulated by tetrodotoxin-sensitive sodium channels. *The Journal of Neuroscience, 21*(6), 1923–1930.

Yamakami, I., & McIntosh, T. K. (1989). Effects of traumatic brain injury on regional cerebral blood flow in rats as measured with radiolabeled microspheres. *Journal of Cerebral Blood Flow and Metabolism, 9*(1), 117–124.

Yuan, X. Q., Prough, D. S., Smith, T. L., & Dewitt, D. S. (1988). The effects of traumatic brain injury on regional cerebral blood flow in rats. *Journal of Neurotrauma, 5*(4), 289–301.

Yuen, T. J., Browne, K. D., Iwata, A., & Smith, D. H. (2009). Sodium channelopathy induced by mild axonal trauma worsens outcome after a repeat injury. *Journal of Neuroscience Research, 87*(16), 3620–3625.

Ziino, C., & Ponsford, J. (2006). Selective attention deficits and subjective fatigue following traumatic brain injury. *Neuropsychology, 20*(3), 383.

Zola-Morgan, S., & Squire, L. R. (1992). The components of the medial temporal lobe memory system. In L. R. Squire & N. Butters (Eds.), *Neuropsychology of memory* (2nd ed). New York, NY: Guilford.

Alzheimer's Disease and the Sleep–Wake Cycle

Adam W. Bero[1,2] and Li-Huei Tsai[1,3]

[1]Department of Brain and Cognitive Sciences, Picower Institute for Learning and Memory, Massachusetts Institute of Technology, Cambridge, MA, United States [2]Department of Neuroscience, Merck Research Laboratories, Boston, MA, United States [3]Broad Institute of Harvard University and Massachusetts Institute of Technology, Cambridge, MA, United States

OUTLINE

Introduction	295
To Sleep, Perchance to Learn	297
Sleep Disturbances and Cognitive Decline	300
Sleep and Aβ Pathology: A Pathogenic Loop	301
Summary and Closing Remarks	308
References	309

INTRODUCTION

Alzheimer's disease (AD) is the most common type of dementia and the sixth leading cause of death in the United States (Thies & Bleiler, 2013). Histopathologically, AD is characterized by deposition of toxic proteinaceous aggregates that are widely hypothesized to result in the progressive deterioration of neural circuits that subserve memory and cognition. Despite intensive efforts in the academic, biotechnological, and pharmaceutical research communities, no treatment that delays the onset

Genes, Environment and Alzheimer's Disease.
DOI: http://dx.doi.org/10.1016/B978-0-12-802851-3.00010-3

or slows the progression of AD is currently available. As the number of aged individuals in the United States is projected to increase dramatically in the coming decades, the number of domestic AD cases is estimated to triple by 2050 and exceed $1 trillion in annual costs (Thies & Bleiler, 2013). Therefore, a deeper understanding of the mechanisms of AD pathogenesis is urgently needed to accelerate therapeutic discovery and avoid a public health crisis.

The search for genetic risk factors for AD has revealed that approximately 1% of AD cases are inherited in a fully penetrant, autosomal dominant fashion in which dementia manifests between 30 and 60 years of age (familial AD, fAD). Missense mutations in the coding sequence of the gene encoding the amyloid precursor protein (APP), from which the amyloid-β (Aβ) peptide is derived, were the first mutations identified to segregate with fAD (Goate et al., 1991), and several mutations in APP, PSEN1, and PSEN2 have subsequently been found to be sufficient to cause fAD (Holtzman, Morris, & Goate, 2011). In addition, it is now well established that the apolipoprotein E (APOE) ε4 allele is the greatest genetic risk factor for the more common, late-onset form of AD (LOAD) (Corder et al., 1993, 1994; Saunders et al., 1993). However, as approximately 50% of individuals with LOAD harbor an APOE ε4 allele, additional genetic variants likely conspire to modulate risk for AD. Accordingly, recent large-scale genomics efforts have unveiled several additional LOAD susceptibility loci, including ABCA7, CD33, CLU, CR1, and PICALM (Lambert et al., 2013), further refining our understanding of the genetic architecture of AD pathogenesis.

As gene–environment interactions constitute an integral component of a wide spectrum of complex human diseases, environmental factors are likely to significantly influence risk for AD. While the spectrum of environmental risk factors for AD remains incompletely understood, epidemiological studies indicate that traumatic brain injury is associated with a greater risk for AD (Barnes et al., 2014; Lye & Shores, 2000; Tran, LaFerla, Holtzman, & Brody, 2011; Tran, Sanchez, Esparza, & Brody, 2011), while education level and engagement in physical exercise are associated with reduced AD risk (Adlard, Perreau, Pop, & Cotman, 2005; Flicker, 2010). Indeed, dissection of the mechanisms by which these and other environmental factors modulate AD risk is an active area of investigation. Intriguingly, sleep has recently emerged as a potent environmental modulator of Aβ pathophysiology and AD risk. Specifically, recent data suggest the existence of a pathogenic partnership between sleep and AD, wherein Aβ accumulation and aggregation disrupt sleep quality, and disrupted sleep quality further exacerbates Aβ aggregation and increases risk for AD (Figure 10.1). Therefore, sleep disruption may represent an early biomarker of AD pathology in the brain and may serve as a quantitative, functional endophenotype whose assessment may inform AD diagnosis

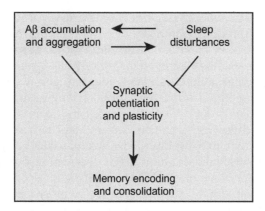

FIGURE 10.1 Potential feed-forward cycle of sleep disturbances and Aβ aggregation in Alzheimer's disease (AD). An extensive literature demonstrates that Aβ aggregation and disruptions in the sleep–wake cycle can independently impair synaptic plasticity and memory. Intriguingly, recent data suggest that Aβ pathology can disrupt the sleep–wake cycle, thereby further increasing Aβ aggregation propensity and risk for cognitive decline associated with AD.

and response to therapeutic intervention. In this chapter, we describe the critical role of sleep in memory processing, discuss the relationship between sleep disturbances and cognitive decline, and review the evidence for a bidirectional relationship between sleep disturbances and cerebral Aβ pathology.

TO SLEEP, PERCHANCE TO LEARN

Sleep is ubiquitous across the animal kingdom. Indeed, sleep-like states have been identified across distantly related species, from fruit flies to humans (Campbell & Tobler, 1984). Given the significant evolutionary pressure to select against consequences associated with sleep, including preclusion of reproduction and vulnerability to predation, sleep must confer critical functional advantages. While a unifying theory of sleep function remains elusive, a large body of literature demonstrates that sleep is a critical regulator of synaptic plasticity events necessary for long-term memory encoding and consolidation.

Sleep is required for learning and memory across species. For instance, episodic memory encoding capacity in humans decreases across a waking interval (Mander, Santhanam, Saletin, & Walker, 2011), and sleep deprivation prior to memory encoding impairs recency discrimination of previously presented visual stimuli (Harrison & Horne, 2000; Walker & Stickgold, 2006) and impairs performance in both free recall (Drummond et al., 2000) and recognition tests of verbal memory (Walker & Stickgold,

2006). Similarly, studies using a variety of learning paradigms have demonstrated that sleep deprivation significantly disrupts memory encoding in animal models. For example, sleep deprivation in mice immediately prior to training on the Morris water maze task significantly decreases memory retention measured 24 h later, suggesting that sleep deprivation prior to encoding impairs the formation of long-term spatial memory (Beaulieu & Godbout, 2000; Guan, Peng, & Fang, 2004). Moreover, learning is impaired following 6 or 12 h of sleep deprivation and at the conclusion of a normal waking day in Drosophila melanogaster (Seugnet, Suzuki, Vine, Gottschalk, & Shaw, 2008). Interestingly, selective disruption of rapid eye movement (REM; paradoxical) sleep is sufficient to impair performance on multiple declarative memory tasks, including the Morris water maze, contextual fear conditioning, passive avoidance, and taste aversion (McGrath & Cohen, 1978; Smith, 1985). Together, these data suggest that sleep deprivation prior to learning significantly impairs memory encoding, and identify paradoxical sleep as a critical regulator of sleep-mediated learning capacity. Further exploring the role of sleep in memory processing, a robust literature also demonstrates a necessary role for postlearning sleep in the consolidation of newly formed memories. For instance, while human subjects exhibit improved performance on a visual discrimination learning task following a night of normal sleep, such improvement is abolished by posttraining sleep deprivation (Karni, Tanne, Rubenstein, Askenasy, & Sagi, 1994). Interestingly, sleep-dependent memory consolidation may be time-limited as sleep deprivation-induced memory deficits persist following two full nights of postdeprivation recovery sleep (Stickgold, James, & Hobson, 2000). In accord with human data, sleep deprivation following learning impairs consolidation of spatial memory in rats (Smith & Rose, 1996, 1997), contextual fear memory in mice (Graves, Heller, Pack, & Abel, 2003), and courtship memory in Drosophila (Ganguly-Fitzgerald, Donlea, & Shaw, 2006). Conversely, inducing sleep on-demand following memory encoding facilitates long-term memory consolidation (Donlea, Thimgan, Suzuki, Gottschalk, & Shaw, 2011). Taken together, these results demonstrate that sleep represents a key determinant of memory encoding capacity and postlearning memory consolidation across species.

Given the central role of sleep in memory function, a plethora of studies have investigated the systems, circuit, cellular, and molecular-level mechanisms by which sleep regulates learning and memory. In this regard, human positron emission tomography and functional magnetic resonance imaging data demonstrate that non–rapid-eye-movement sleep is associated with a significant decrease in cerebral glucose utilization relative to wakefulness (Buchsbaum et al., 1989; Maquet et al., 1990) and that sleep is associated with decreased functional connectivity across information processing hubs in the neocortex (Horovitz et al., 2009; Larson-Prior et al.,

2011; Samann et al., 2011). Intriguingly however, functional connectivity between medial temporal regions and frontal cortex increases at the conclusion of a waking day, perhaps reflecting postlearning memory consolidation processes (Shannon et al., 2013). Indeed, electrophysiological and metabolic measures of synaptic activity indicate that while wakefulness is associated with increased slope and amplitude of local field potentials (Vyazovskiy, Cirelli, Pfister-Genskow, Faraguna, & Tononi, 2008), increased neuronal activity (Bero et al., 2011; Vyazovskiy et al., 2009), increased miniature excitatory postsynaptic current frequency and amplitude (Liu, Faraguna, Cirelli, Tononi, & Gao, 2010), and enhanced glucose utilization (Vyazovskiy, Cirelli, Tononi, & Tobler, 2008) relative to sleep, recordings of neuronal ensembles in the cortex and hippocampus demonstrate that neurons that fire together during spatial memory encoding tend to fire together during subsequent sleep, suggesting that sleep may function to consolidate information learned during the waking state into neural memory stores (Ji & Wilson, 2007; Louie & Wilson, 2001; Wilson & McNaughton, 1994). Interestingly, a recent study suggests that such "replay" can be facilitated through the presentation of task-related sensory cues, suggesting that sleep-dependent memory consolidation can be enhanced by external stimulation (Bendor & Wilson, 2012).

On the molecular level, candidate-based gene expression data in animal models suggest that the expression of synaptic plasticity-related genes, including *Arc* and *BDNF*, is increased during wakefulness relative to sleep (Cirelli & Tononi, 2000a). These data were subsequently confirmed and extended by genomewide microarray expression data, which revealed that wakefulness is associated with increased expression of genes regulating synaptic plasticity, transcription, energy metabolism, and excitatory neurotransmission, including those encoding Arc, Bdnf, Ngf, and Glut1. Conversely, sleep is associated with increased expression of genes involved in the negative regulation of transcription and inhibitory neurotransmission (Cirelli, Gutierrez, & Tononi, 2004; Cirelli & Tononi, 2000b). In accord with these gene expression data, levels of glutamatergic AMPA receptor expression and phosphorylation are upregulated in synaptoneurosomes prepared from the cerebral cortex and hippocampus following naturally occurring or enforced wakefulness relative to sleep (Vyazovskiy, Cirelli, Pfister-Genskow, Faraguna & Tononi, 2008). Indeed, longitudinal *in vivo* multiphoton microscopy data suggest that wakefulness results in a net increase in dendritic spines in the mouse neocortex, while sleep is associated with a net reduction in spine density (Maret, Faraguna, Nelson, Cirelli, & Tononi, 2011).

Interestingly, similar sleep-dependent alterations in the expression of synaptic markers and synaptic strength have been found to occur in Drosophila (Bushey, Tononi, & Cirelli, 2011; Gilestro, Tononi, & Cirelli, 2009). Together, these data suggest that synaptic plasticity events occurring during wakefulness elicit a net increase in synaptic potentiation, and that renormalizing of synaptic weight to an energetically sustainable baseline may represent a principal function of sleep (Tononi & Cirelli, 2006). More generally, these data indicate that the sleep–wake cycle is critical for neuroplasticity that underlies both the initial encoding and subsequent consolidation of long-term memory.

SLEEP DISTURBANCES AND COGNITIVE DECLINE

Impairments in memory and executive function are early clinical manifestations of AD. In accord with the extensive literature demonstrating that the sleep–wake cycle is a central regulator of memory and cognition, recent findings suggest that sleep disturbances may be closely associated with AD and potentially represent a harbinger of cognitive decline. For instance, human epidemiological data suggest that ≤5h (Tworoger, Lee, Schernhammer, & Grodstein, 2006) and ≥11h (Faubel et al., 2009) of sleep per night are associated with impaired cognition, as assessed by adaptations of the mini-mental state examination. Moreover, both subjective and actigraphy-based sleep quality metrics reveal that sleep quality is associated with cognitive impairment. Specifically, cross-sectional (Blackwell et al., 2006) and prospective (Potvin et al., 2012; Yaffe et al., 2011) studies indicate that prolonged wakefulness after sleep onset, reduced sleep efficiency, increased sleep latency, sleep-disordered breathing, and exacerbated daytime dysfunction are associated with impaired cognition. Actigraphically interrogated sleep–wake rhythms further suggest that fragmentation of the daily sleep–wake rhythm is correlated with significant decreases in measures of mental speed, semantic memory, working memory, and executive function (Lim et al., 2012; Oosterman, van Someren, Vogels, Van Harten, & Scherder, 2009).

Consistent with these reports, cross-sectional studies of community-dwelling individuals suggest that 25–40% of individuals with AD experience significant sleep disturbances. These disturbances include difficulty falling asleep, multiple awakenings during the night, early morning awakenings, and disrupted sleep rhythm assessed by reduced activity amplitude and phase delay (Ancoli-Israel et al., 1997; McCurry et al., 1999; Moran et al., 2005; Vitiello, Prinz, Williams, Frommlet, & Ries, 1990). Moreover, a recent prospective study of 737 cognitively normal older adults found that greater levels of sleep fragmentation were associated with 1.5-fold increase in risk of incident AD over a six-year period and a 22% increase in the rate of cognitive decline, after controlling for age, sex, and education (Lim, Kowgier, Yu, Buchman, & Bennett, 2013). Electroencephalographic examination of individuals with amnestic mild cognitive impairment (aMCI) suggests that aMCI patients exhibit fewer sleep spindles and decreased time spent in slow-wave sleep (Westerberg et al., 2012), which may contribute to deficits in memory consolidation (Rasch, Buchel, Gais, & Born, 2007; Tamminen, Payne, Stickgold, Wamsley, & Gaskell, 2010). Furthermore, sleep disturbances in AD are positively correlated with measures of aggressiveness (Moran et al., 2005) and constitute a primary driver of caregiver stress and patient institutionalization (Donaldson, Tarrier, & Burns, 1998; Hope, Keene, Gedling, Fairburn, & Jacoby, 1998; Pollak & Perlick, 1991).

While extant data therefore suggest that AD is associated with sleep abnormalities, it is not known whether sleep disturbances are present in the early stages of AD, prior to the onset of clinical presentation. Indeed, recent reports suggest that the neuropathology of AD begins to develop approximately 15–20 years prior to the emergence of clinical symptoms (Bateman et al., 2012; Fagan et al., 2014; Fleisher et al., 2012; Reiman et al., 2012). To address this issue, Ju and colleagues examined the relationship between cerebral amyloid plaque deposition and actigraphy-based measures of sleep quality in a cohort of cognitively normal volunteers (Ju et al., 2013). As $A\beta_{42}$ concentration in the cerebrospinal fluid (CSF) declines as a function of cerebral amyloid plaque deposition (Fagan et al., 2006), CSF $A\beta_{42}$ levels were used a proxy of $A\beta$ aggregation in the brain. After controlling for age, sex, and *APOE* ε4 carrier status, cerebral amyloid deposition was associated with decreased sleep efficiency and increased frequency of daytime napping (Ju et al., 2013), suggesting that sleep disturbances may precede the clinical manifestation of hallmark AD symptomatology and thus represent an early functional biomarker of AD-associated neuropathology. Furthermore, a cross-sectional study of 70 cognitively normal volunteers recently found that self-reported shorter sleep duration was associated with increased amyloid plaque deposition, as assessed by ^{11}C-Pittsburgh compound B positron emission tomography (Spira et al., 2013). Taken together, these data demonstrate that AD is associated with significant disruptions in the sleep–wake cycle and that such abnormalities may emerge during the preclinical stage of the disease. To dissect the causal basis of these relationships, recent studies in animal models have begun to explore potential mechanistic links between the sleep–wake cycle, $A\beta$ neuropathology, and AD risk.

SLEEP AND Aβ PATHOLOGY: A PATHOGENIC LOOP

Several decades after their initial description by Alois Alzheimer in 1906, intraparenchymal amyloid plaques extracted from postmortem AD brain were found to be principally composed of aggregated forms of $A\beta$ (Glenner & Wong, 1984; Masters et al., 1985). $A\beta$ is a 38–43 amino acid peptide of undescribed function that is derived from the sequential proteolytic processing of APP by β-secretase and γ-secretase (De Strooper et al., 1998; Haass et al., 1992; Seubert et al., 1992; Shoji et al., 1992; Vassar et al., 1999). Alterative APP processing by α-secretase occurs within the Aβ peptide sequence and thereby precludes Aβ generation (Lammich et al., 1999). Under normal conditions, Aβ is produced in neurons and secreted in a soluble form that can be detected in the plasma and CSF of cognitively normal individuals (Fagan et al., 2006; Haass et al., 1992). In AD pathogenesis, Aβ accumulates and aggregates into higher-order species such as soluble oligomers and insoluble amyloid plaques in a concentration-dependent manner (Hasegawa, Yamaguchi, Omata, Gejyo, & Naiki, 1999; Meyer-Luehmann et al., 2003; Yan et al., 2009), and such aggregation is associated with myriad neurotoxic effects, including disrupted brain metabolism (Klunk et al., 2004; Sperling et al., 2009), reduced functional connectivity (Bero et al., 2012; Drzezga et al., 2011; Hedden et al., 2009; Sheline et al., 2010; Supekar Menon, Rubin, Musen, & Greicius, 2008), neuritic dystrophy (Brendza et al., 2005; Meyer-Luehmann et al., 2008), neurofibrillary tangle formation (Bolmont et al., 2007), neuro-inflammation (Bolmont et al., 2008; Meyer-Luehmann et al., 2008), synap-totoxicity (Koffie et al., 2009, 2012; Shankar et al., 2008) and neuronal loss (Jack et al., 2009). As a result, Aβ aggregation is widely hypothesized to represent an initiating event in AD (Hardy & Selkoe, 2002; Selkoe, 2001). Therefore, understanding the mechanisms by which the sleep–wake cycle may modulate cerebral Aβ metabolism may provide critical insight into the pathophysiology of AD.

As discussed herein, converging evidence suggests that wakefulness is associated with increased cerebral glucose metabolism and neuronal activity relative to sleep (Buchsbaum et al., 1989; Liu et al., 2010; Maquet et al., 1990; Vyazovskiy, Cirelli, Pfister-Genskow, et al., 2008; Vyazovskiy, Cirelli, Tononi, et al., 2008; Vyazovskiy et al., 2009). Intriguingly, work over the last decade has identified neuronal activity as a potent regulator of cerebral Aβ concentration (Jagust & Mormino, 2011). For instance, a seminal report from the Malinow laboratory showed that pharmacological manipulation of organotypic hippocampal slices significantly modu-lates the extent of Aβ secretion *in vitro*. Specifically, hippocampal slices prepared from APP transgenic mice exhibited decreased Aβ release when maintained in the presence of tetrodotoxin, a voltage-gated sodium chan-nel blocker. Conversely, hippocampal slices treated with picrotoxin, a GABA$_A$ receptor blocker, exhibited enhanced Aβ release (Kamenetz et al., 2003). To determine whether neuronal activity dynamically modulates Aβ concentration *in vivo*, Cirrito et al. (2003) developed an Aβ microdialysis technique to sample soluble Aβ levels in the brain interstitial fluid (ISF) of awake, behaving mice while performing concurrent electrical or phar-macological manipulation of neuronal activity. Consistent with previous findings indicating that APP is transported from the entorhinal cortex to the hippocampus via anterograde axonal transport (Buxbaum et al., 1998) and that lesions of the perforant pathway prevent Aβ plaque deposition in the hippocampus of APP transgenic mice (Lazarov, Lee, Peterson, & Sisodia, 2002; Sheng, Price, & Koliatsos, 2002), Cirrito et al. (2005) found that, within minutes, acute electrical stimulation of the perforant pathway upregulated ISF Aβ levels in the hippocampus of APP transgenic mice. Using reverse microdialysis, they next showed that local infusion of tetrodotoxin to the hippocampus reduced ISF Aβ levels and neuronal activity as assessed by depth electroencephalographic recording, and that pharmacological inhibition of synaptic vesicle exocytosis using tetanus toxin was sufficient to halt Aβ release. Furthermore, experiments using organotypic hippocampal slice culture demonstrated that pharmacological induction of synaptic vesicle exocytosis stimulated ISF Aβ release even in the absence of action potential firing, suggesting that synaptic activity directly regulates ISF Aβ levels (Cirrito et al., 2005). Consistent with *in vitro* data suggesting that APP localized to the cell surface can be internalized via clathrin-mediated endocytosis (Nordstedt, Caporaso, Thyberg, Gandy, & Greengard, 1993), and that proteolytic processing of APP occurs within early and late endosomes (Lah & Levey, 2000; Vassar et al., 1999), secretion of Aβ at the synapse *in vivo* was subsequently shown to involve clathrin-mediated endocytosis (Cirrito et al., 2008).

A great deal of knowledge has been gained from human neuroimaging experiments in which neural activation induced during task performance is compared to a resting-state control condition. Indeed, such experiments have enabled relation of neural topography to function (Petersen, Fox, Posner, Mintun, & Raichle, 1988). However, a subset of brain areas, including the medial prefrontal, lateral parietal, and posterior cingulate/retrosplenial cortices, exhibit preferential activation during undirected mentation and reduced activity during performance of an externally directed task. In light of this activation profile, this neural network was termed the "default-mode network" (Buckner, Andrews-Hanna, & Schacter, 2008; Raichle et al., 2001; Raichle & Mintun, 2006; Raichle & Snyder, 2007). As resting-state glucose metabolism accounts for the majority of the brain's total energy budget, and given the near proportionality between neuronal energy consumption and neurotransmitter processing (Hyder et al., 2006), neuronal activity within the default-mode network is hypothesized to represent the majority of total brain metabolism. Buckner et al. (2005) made the remarkable discovery that the spatial distribution of resting-state brain activity in cognitively normal humans is highly correlated with the stereotypical region-specific distribution of amyloid plaque deposition in AD, further suggesting a close relationship between local cerebral metabolism and Aβ plaque pathology. Additionally, recent extensions of this work have demonstrated that brain regions that comprise the default-mode network exhibit elevated intrinsic functional connectivity (Bero et al., 2012; Buckner et al., 2009; Lu et al., 2012) and aerobic glycolysis (Vaishnavi et al., 2010; Vlassenko et al., 2010), indicating that cortical "hubs" of neural information processing are particularly vulnerable to developing Aβ plaque pathology. Finally, recent data in animal models demonstrate a close relationship between brain region-specific levels of steady-state neuronal activity and ISF Aβ concentration, as ISF Aβ levels assessed in several brain regions of APP transgenic mice prior to the onset of amyloid plaque deposition are proportional to the level of local neuronal activity and predictive of the degree of subsequent region-specific amyloid plaque deposition (Bero et al., 2011). Moreover, physiological manipulation of neuronal activity is sufficient to bidirectionally regulate local ISF Aβ levels and alter amyloid plaque growth dynamics (Bero et al., 2011), thus implicating regional differences in steady-state neuronal activity as a primary driver of region-specific Aβ pathology in AD.

To begin to explore the mechanisms by which the sleep–wake cycle might be related to the development of Aβ pathology *in vivo*, Kang and colleagues examined cerebral Aβ dynamics across the sleep–wake cycle (Kang et al., 2009). Several months prior to the onset of amyloid plaque deposition, young APP transgenic (Tg2576) and wild-type mice were implanted with a unilateral microdialysis probe in the hippocampus and electroencephalograph/electromyogram electrodes to permit simultaneous examination of ISF Aβ levels and polysomnographic determination of sleep stage. Using this approach, the investigators observed that ISF Aβ levels exhibited diurnal fluctuation: ISF Aβ levels were significantly greater during the dark cycle (when animals were awake) than during the light cycle (when animals were asleep) and that ISF Aβ concentration was highly correlated with wakefulness (Kang et al., 2009).

Interestingly, a recent study found that the diurnal fluctuation of ISF Aβ was closely associated with the diurnal fluctuation of neuronal activity, and that ISF Aβ concentration is highly correlated with neuronal activity across the sleep–wake cycle (Bero et al., 2011), suggesting that the diurnal fluctuation of ISF Aβ may be due, at least in part, to diurnal fluctuations in endogenous neuronal activity. Moreover, acute sleep deprivation of APP transgenic mice during the light phase was sufficient to attenuate the normal reduction in ISF Aβ levels in a stress-independent manner, suggesting that hippocampal Aβ concentration is causally linked to time spent awake (Kang et al., 2009). Exploration of the molecular mechanism by which wakefulness regulates ISF Aβ levels has suggested a critical role for the orexin (hypocretin) system. Orexin is a hypothalamic neuropeptide that is strongly implicated in regulating sleep and arousal states. Indeed, orexin knockout mice exhibit a phenotype that closely resembles human narcolepsy patients (Chemelli et al., 1999). Interestingly, hypothalamic neurons release orexin in a diurnal fashion (Yoshida et al., 2001), and project to the hippocampus (Peyron et al., 1998). Thus, to determine whether the orexin system modulates ISF Aβ concentration *in vivo*, either orexin or an orexin receptor 1/2 antagonist (almorexant) was continuously administered to young APP transgenic mice via intracerebroventricular infusion during the light cycle. Orexin administration significantly increased, while almorexant decreased, ISF Aβ levels, suggesting that the orexin system may represent a central regulator of the diurnal fluctuation of Aβ (Kang et al., 2009). Finally, as the sleep–wake cycle regulates ISF Aβ levels, and ISF Aβ concentration is closely associated with amyloid plaque formation and growth (Yan et al., 2009), Kang and colleagues investigated whether chronic sleep deprivation might accelerate amyloid plaque deposition in APP/PSEN1 bitransgenic mice. Specifically, APP/PSEN1 mice were subjected to 20h of sleep deprivation per day for 21 consecutive days. Remarkably, sleep-restricted mice exhibited significantly increased Aβ plaque deposition across all brain regions examined, including the hippocampus as well as the cingulate, piriform, and entorhinal cortices. Conversely, chronic almorexant treatment significantly decelerated Aβ plaque deposition in these brain regions relative to vehicle-treated controls (Kang et al., 2009). Therefore, given that Aβ is released by neurons in an activity-dependent manner and that the sleep–wake cycle regulates neuronal activity and ISF Aβ levels, the sleep–wake cycle may represent an important regulator of activity-dependent Aβ secretion and concentration-dependent Aβ aggregation.

As Aβ aggregates in a concentration-dependent manner (Hasegawa et al., 1999; Meyer-Luehmann et al., 2003; Yan et al., 2009), alterations in the amount of Aβ synthesis can have profound effects on the development of cerebral Aβ pathology. However, the steady-state concentration of Aβ in the brain, and therefore, the likelihood of cerebral Aβ aggregation, is the result of a balance between the rates of Aβ synthesis and clearance. Indeed, Aβ clearance rate in the central nervous system represents a critical driver of the risk for AD (Mawuenyega et al., 2010). Accordingly, the strongest genetic risk factor for LOAD is the *APOE* ε4 allele (Corder et al., 1993, 1994; Saunders et al., 1993), the presence of which significantly impairs Aβ clearance and increases Aβ plaque deposition (Castellano et al., 2011). Intriguingly, a recent report suggests that sleep may also function to facilitate Aβ clearance from the brain ISF via a novel pathway involving convective exchange with the CSF (Xie et al., 2013). This pathway, termed the "glymphatic" pathway, is characterized by periarterial influx of CSF into the brain ISF, followed by paravenous clearance of ISF metabolites from the parenchyma into the subarachnoid CSF, bloodstream, and cervical lymphatics (Iliff et al., 2012).

To examine whether the sleep–wake cycle might influence ISF Aβ levels via regulation of the glymphatic system, Xie et al. (2013) performed simultaneous two-photon imaging, fluorescent CSF tracer infusion, and polysomnography in mice during wakefulness or sleep to enable simultaneous, real-time assessments of CSF tracer influx into the brain parenchyma and sleep stage. Using this method, the authors made the surprising discovery that periarterial and parenchymal influx of the CSF tracer was decreased by 95% in the waking state relative to sleep (Xie et al., 2013), suggesting that sleep drives convective exchange of CSF and ISF. Interestingly, sleep-induced expansion of CSF influx can be recapitulated by treatment with ketamine/xylazine anesthesia, suggesting that glympathtic CSF influx may be regulated by neuronal activity. Further investigation revealed that sleep is associated with a 60% increase in the volume of the interstitium relative to wakefulness, suggesting that expansion of the ISF during sleep may reduce resistance to convective flow and thereby increase CSF influx into the brain parenchyma (Xie et al., 2013). As a previous report demonstrated that the glymphatic pathway may represent a principal route of Aβ clearance from the brain (Iliff et al., 2012), the authors next tested whether sleep is associated with enhanced Aβ clearance rate. Remarkably, the clearance rate of ^{125}I-Aβ injected into the frontal cortex was enhanced twofold during sleep relative to awake controls, and this effect was recapitulated by ketamine/xylazine anesthesia. Thus, the sleep–wake cycle may regulate cerebral Aβ metabolism via modulation of both Aβ synthesis and clearance rates. Specifically, as cerebral energy metabolism and neuronal activity is reduced during sleep compared to wakefulness (Buchsbaum et al., 1989; Liu et al., 2010; Maquet et al., 1990; Vyazovskiy, Cirelli, Pfister-Genskow, et al., 2008; Vyazovskiy, Cirelli, Tononi, et al., 2008; Vyazovskiy et al., 2009), sleep may decrease activity-dependent Aβ generation and release (Bero et al., 2011; Kang et al., 2009), thereby reducing steady-state soluble Aβ concentration. Additionally, as sleep is associated with increased convective exchange of CSF and ISF via the glymphatic pathway (Xie et al., 2013), sleep also serves to enhance the rate of Aβ clearance from the brain, thus further reducing cerebral Aβ levels during sleep. Notably, since the sleep-induced enhancement in glymphatic pathway-mediated Aβ clearance can be recapitulated by pharmacological agents that reduce neuronal activity (Xie et al., 2013), the dampening in synaptic activity associated with sleep may serve to both reduce activity-dependent Aβ generation and create a permissive cerebral milieu for enhanced convective exchange and accelerated Aβ clearance. Taken together, these data demonstrate that the sleep–wake cycle represents a potent regulator of ISF Aβ dynamics and can thereby influence the progression of cerebral amyloid plaque deposition characteristic of AD.

Given that the sleep-wake cycle can regulate Aβ concentration and amyloid plaque deposition, might Aβ aggregation reciprocally affect the sleep–wake cycle? To address this possibility, several groups have examined sleep–wake regulation in mouse models of cerebral β-amyloidosis. For instance, mice expressing human *APP*, *MAPT*, and *PSEN1* exhibit reduced time spent in NREM sleep and increased wakefulness relative to controls (Platt et al., 2011), and aged APP transgenic mice (PDAPP) exhibit increased wakefulness and reduced REM sleep during the dark period compared to wild-type controls (Huitron-Resendiz et al., 2002). To further explore this issue, Roh and colleagues confirmed and extended previous results obtained from Tg2576 mice (Bero et al., 2011; Kang et al., 2009), showing that ISF Aβ exhibits diurnal fluctuation in APP/PSEN1 bitransgenic mouse brain and that diurnal fluctuation occurs both in the hippocampus and striatum (Roh et al., 2012).

Similar to human AD (Buckner et al., 2005; Klunk et al., 2004) and other mouse models of cerebral β-amyloidosis (Bero et al., 2011), APP/PSEN1 exhibit regional differences in the degree of amyloid plaque deposition. Specifically, significant Aβ plaque deposition is present in the hippocampus of APP/PSEN1 mice by 6 months of age, while Aβ deposition in the striatum is not present until 9 months of age. As Aβ plaques can rapidly sequester soluble Aβ present in the ISF and thus alter Aβ metabolism (Cirrito et al., 2003; Hong et al., 2011), the authors hypothesized that Aβ aggregation would be associated with region-specific disruption of the diurnal fluctuation of ISF Aβ. In accord with this hypothesis, while diurnal fluctuation of ISF Aβ was normal in the hippocampus and striatum of 3-month-old APP/PSEN1, the diurnal fluctuation of Aβ was abolished in the hippocampus of 6-month-old APP/PSEN1 mice, while similar effects were not observed in the striatum until 9 months of age (Roh et al., 2012). Consistent with these data, the amplitude of the diurnal fluctuation of neuronal activity in each brain region was decreased in the presence of amyloid plaque deposition (Roh et al., 2012). Together, these data demonstrate that Aβ plaque deposition is closely associated with local disruption of Aβ dynamics associated with the sleep–wake cycle.

To directly determine whether cerebral Aβ aggregation is associated with disruption of the sleep–wake cycle, Roh and colleagues performed polysomnography in APP/PSEN1 mice at 3, 6, and 9 months of age to permit examination of wakefulness, REM sleep, and NREM sleep as a function of disease severity. Interestingly, 9-month-old APP/PSEN1 mice that harbor robust amyloid plaque deposition exhibited significant reductions in time spent in REM and NREM sleep states during the light phase compared to 3-month-old APP/PSEN1 mice that do not exhibit amyloid plaque deposition, suggesting that Aβ aggregation is closely associated with prolonged wakefulness (Roh et al., 2012). To examine the causal role of Aβ in disruption of the sleep–wake cycle, the authors next performed active immunization of 1.5-month-old presymptomatic APP/ PSEN1 mice via chronic injection of synthetic Aβ peptide. In accord with previous studies (Brody & Holtzman, 2008), Aβ-immunized mice harbored significantly fewer amyloid plaque deposits relative to phosphate-buffered saline-treated controls (Roh et al., 2012). Importantly, active Aβ immunization prevented suppression of the diurnal fluctuation of ISF Aβ and disruption of the sleep–wake cycle in 9-month-old APP/PSEN1 mice (Roh et al., 2012), suggesting that Aβ accumulation and aggregation suppresses the diurnal fluctuation of Aβ concentration and disrupts the sleep–wake cycle.

As data obtained in animal models suggest the existence of a bidirectional relationship between the sleep–wake cycle and cerebral Aβ dynamics, investigators have recently begun to explore the interplay between sleep and Aβ pathology in humans. For instance, to directly determine whether the sleep–wake cycle may regulate cerebral Aβ metabolism in humans, lumbar catheters were placed in the intrathecal space of cognitively normal volunteers to permit longitudinal assessment of Aβ concentration in the CSF. In remarkable accord with data obtained in animal models, Kang et al. (2009) found that CSF Aβ levels gradually increased over the course of a waking day and decreased during subsequent sleep. Moreover, while a night of unrestricted sleep is associated with a significant reduction in CSF Aβ levels in cognitively normal individuals, sleep deprivation prevents this reduction (Ooms et al., 2014). Finally, the amplitude of the diurnal fluctuation of CSF Aβ levels is abolished in the presence of cerebral amyloid deposition (Huang et al., 2012; Roh et al., 2012). Together, these results suggest that diurnal fluctuation of soluble Aβ levels is intrinsic to normal brain physiology, that the sleep–wake cycle directly regulates Aβ metabolism, and that amyloid plaque deposition is closely associated with disturbed sleep across species.

SUMMARY AND CLOSING REMARKS

AD is a devastating neurodegenerative disorder for which no effective treatment is available. As human lifespan, and thus the prevalence of AD, increase, development of treatments that halt or reverse the progression of AD will soon become an economic imperative. While advances in genomics have identified several genetic variants that can influence risk for AD (Lambert et al., 2013), environmental modifiers of AD risk are not well understood. In this context, recent data obtained in humans and animal models have identified the sleep–wake cycle as a novel regulator of cerebral Aβ metabolism (Bero et al., 2011; Kang et al., 2009; Ooms et al., 2014; Xie et al., 2013): Aβ levels are increased during wakefulness and gradually decline during sleep. Moreover, consistent with epidemiological studies indicating that up to 40% of individuals with AD experience sleep disturbances (Moran et al., 2005), a growing literature suggests that Aβ pathology can disturb the sleep–wake cycle and thereby elicit further Aβ aggregation (Huang et al., 2012; Roh et al., 2012). Therefore, a feed-forward relationship between the sleep–wake cycle and Aβ pathology, wherein Aβ aggregation disrupts sleep, and disrupted sleep hastens Aβ aggregation, may significantly increase risk for AD (Ju, Lucey, & Holtzman, 2014). Further investigation will be necessary to determine whether other pathological features of AD, such as tau hyperphosphorylation and neurofibrillary tangle formation, are altered by the sleep–wake cycle. Particularly interesting in this regard is the discovery that tau, primarily a cytoplasmic protein that stabilizes microtubules, is normally released into the brain ISF (Yamada et al., 2011) and is regulated by neuronal activity (Yamada et al., 2014). As discussed in this chapter, the sleep–wake cycle regulates the diurnal fluctuation of neuronal activity (Buchsbaum et al., 1989; Liu et al., 2010; Maquet et al., 1990; Vyazovskiy, Cirelli, Pfister-Genskow, et al., 2008; Vyazovskiy, Cirelli, Tononi, et al., 2008; Vyazovskiy et al., 2009) and the rate of metabolite clearance from the brain (Xie et al., 2013), and may thus represent a central regulator of AD neuropathology. Finally, as recent data suggest that cerebral Aβ deposition is associated with worsened sleep quality in the preclinical stage of AD, compromised sleep quality may represent an antecedent biomarker of AD (Ju et al., 2013; Spira et al., 2013). Further examination will be critical to determine whether sleep may indeed serve as a functional biomarker of AD neuropathology and thus inform AD diagnosis, prognosis, and clinical trial design.

References

Adlard, P. A., Perreau, V. M., Pop, V., & Cotman, C. W. (2005). Voluntary exercise decreases amyloid load in a transgenic model of Alzheimer's disease. *The Journal of Neuroscience, 25*, 4217–4221.

Ancoli-Israel, S., Klauber, M. R., Jones, D. W., Kripke, D. F., Martin, J., Mason, W., et al. (1997). Variations in circadian rhythms of activity, sleep, and light exposure related to dementia in nursing-home patients. *Sleep, 20*, 18–23.

Barnes, D. E., Kaup, A., Kirby, K. A., Byers, A. L., Diaz-Arrastia, R., & Yaffe, K. (2014). Traumatic brain injury and risk of dementia in older veterans. *Neurology, 83*, 312–319.

Bateman, R. J., Xiong, C., Benzinger, T. L., Fagan, A. M., Goate, A., Fox, N. C., et al. (2012). Clinical and biomarker changes in dominantly inherited Alzheimer's disease. *The New England Journal of Medicine, 367*, 795–804.

Beaulieu, I., & Godbout, R. (2000). Spatial learning on the Morris water Maze test after a short-term paradoxical sleep deprivation in the rat. *Brain and Cognition, 43*, 27–31.

Bendor, D., & Wilson, M. A. (2012). Biasing the content of hippocampal replay during sleep. *Nature Neuroscience, 15*, 1439–1444.

Bero, A. W., Bauer, A. Q., Stewart, F. R., White, B. R., Cirrito, J. R., Raichle, M. E., et al. (2012). Bidirectional relationship between functional connectivity and Amyloid-beta deposition in mouse brain. *The Journal of Neuroscience, 32*, 4334–4340.

Bero, A. W., Yan, P., Roh, J. H., Cirrito, J. R., Stewart, F. R., Raichle, M. E., et al. (2011). Neuronal activity regulates the regional vulnerability to amyloid-beta deposition. *Nature Neuroscience, 14*, 750–756.

Blackwell, T., Yaffe, K., Ancoli-Israel, S., Schneider, J. L., Cauley, J. A., Hillier, T. A., et al. (2006). Poor sleep is associated with impaired cognitive function in older women: The study of osteoporotic fractures. *The Journals of Gerontology Series A, Biological Sciences and Medical Sciences, 61*, 405–410.

Bolmont, T., Clavaguera, F., Meyer-Luehmann, M., Herzig, M. C., Radde, R., Staufenbiel, M., et al. (2007). Induction of tau pathology by intracerebral infusion of amyloid-beta–containing brain extract and by amyloid-beta deposition in APP x Tau transgenic mice. *The American Journal of Pathology, 171*, 2012–2020.

Bolmont, T., Haiss, F., Eicke, D., Radde, R., Mathis, C. A., Klunk, W. E., et al. (2008). Dynamics of the microglial/amyloid interaction indicate a role in plaque maintenance. *The Journal of Neuroscience, 28*, 4283–4292.

Brendza, R. P., Bacskai, B. J., Cirrito, J. R., Simmons, K. A., Skoch, J. M., Klunk, W. E., et al. (2005). Anti-Abeta antibody treatment promotes the rapid recovery of amyloid-associated neuritic dystrophy in PDAPP transgenic mice. *The Journal of Clinical Investigation, 115*, 428–433.

Brody, D. L., & Holtzman, D. M. (2008). Active and passive immunotherapy for neurodegenerative disorders. *Annual Review of Neuroscience, 31*, 175–193.

Buchsbaum, M. S., Gillin, J. C., Wu, J., Hazlett, E., Sicotte, N., Dupont, R. M., et al. (1989). Regional cerebral glucose metabolic rate in human sleep assessed by positron emission tomography. *Life Sciences, 45*, 1349–1356.

Buckner, R. L., Andrews-Hanna, J. R., & Schacter, D. L. (2008). The brain's default network: Anatomy, function, and relevance to disease. *Annals of the New York Academy of Sciences, 1124*, 1–38.

Buckner, R. L., Sepulcre, J., Talukdar, T., Krienen, F. M., Liu, H., Hedden, T., et al. (2009). Cortical hubs revealed by intrinsic functional connectivity: Mapping, assessment of stability, and relation to Alzheimer's disease. *The Journal of Neuroscience, 29*, 1860–1873.

Buckner, R. L., Snyder, A. Z., Shannon, B. J., LaRossa, G., Sachs, R., Fotenos, A. F., et al. (2005). Molecular, structural, and functional characterization of Alzheimer's disease: Evidence for a relationship between default activity, amyloid, and memory. *The Journal of Neuroscience, 25*, 7709–7717.

Bushey, D., Tononi, G., & Cirelli, C. (2011). Sleep and synaptic homeostasis: Structural evidence in Drosophila. *Science, 332*, 1576–1581.

Buxbaum, J. D., Thinakaran, G., Koliatsos, V., O'Callahan, J., Slunt, H. H., Price, D. L., et al. (1998). Alzheimer amyloid protein precursor in the rat hippocampus: Transport and processing through the perforant path. *The Journal of Neuroscience, 18*, 9629–9637.

Campbell, S. S., & Tobler, I. (1984). Animal sleep: A review of sleep duration across phylogeny. *Neuroscience and Biobehavioral Reviews, 8*, 269–300.

Castellano, J. M., Kim, J., Stewart, F. R., Jiang, H., DeMattos, R. B., Patterson, B. W., et al. (2011). Human apoE isoforms differentially regulate brain amyloid-beta peptide clearance. *Science Translational Medicine, 3*, 89ra57.

Chemelli, R. M., Willie, J. T., Sinton, C. M., Elmquist, J. K., Scammell, T., Lee, C., et al. (1999). Narcolepsy in orexin knockout mice: Molecular genetics of sleep regulation. *Cell, 98*, 437–451.

Cirelli, C., Gutierrez, C. M., & Tononi, G. (2004). Extensive and divergent effects of sleep and wakefulness on brain gene expression. *Neuron, 41*, 35–43.

Cirelli, C., & Tononi, G. (2000a). Differential expression of plasticity-related genes in waking and sleep and their regulation by the noradrenergic system. *The Journal of Neuroscience, 20*, 9187–9194.

Cirelli, C., & Tononi, G. (2000b). Gene expression in the brain across the sleep-waking cycle. *Brain Research, 885*, 303–321.

Cirrito, J. R., Kang, J. E., Lee, J., Stewart, F. R., Verges, D. K., Silverio, L. M., et al. (2008). Endocytosis is required for synaptic activity-dependent release of amyloid-beta *in vivo*. *Neuron, 58*, 42–51.

Cirrito, J. R., May, P. C., O'Dell, M. A., Taylor, J. W., Parsadanian, M., Cramer, J. W., et al. (2003). *In vivo* assessment of brain interstitial fluid with microdialysis reveals plaque-associated changes in amyloid-beta metabolism and half-life. *The Journal of Neuroscience, 23,* 8844–8853.

Cirrito, J. R., Yamada, K. A., Finn, M. B., Sloviter, R. S., Bales, K. R., May, P. C., et al. (2005). Synaptic activity regulates interstitial fluid amyloid-beta levels *in vivo. Neuron, 48,* 913–922.

Corder, E. H., Saunders, A. M., Risch, N. J., Strittmatter, W. J., Schmechel, D. E., Gaskell, P. C., Jr., et al. (1994). Protective effect of apolipoprotein E type 2 allele for late onset Alzheimer disease. *Nature Genetics, 7,* 180–184.

Corder, E. H., Saunders, A. M., Strittmatter, W. J., Schmechel, D. E., Gaskell, P. C., Small, G. W., et al. (1993). Gene dose of apolipoprotein E type 4 allele and the risk of Alzheimer's disease in late onset families. *Science, 261,* 921–923.

De Strooper, B., Saftig, P., Craessaerts, K., Vanderstichele, H., Guhde, G., Annaert, W., et al. (1998). Deficiency of presenilin-1 inhibits the normal cleavage of amyloid precursor protein. *Nature, 391,* 387–390.

Donaldson, C., Tarrier, N., & Burns, A. (1998). Determinants of carer stress in Alzheimer's disease. *International Journal of Geriatric Psychiatry, 13,* 248–256.

Donlea, J. M., Thimgan, M. S., Suzuki, Y., Gottschalk, L., & Shaw, P. J. (2011). Inducing sleep by remote control facilitates memory consolidation in Drosophila. *Science, 332,* 1571–1576.

Drummond, S. P., Brown, G. G., Gillin, J. C., Stricker, J. L., Wong, E. C., & Buxton, R. B. (2000). Altered brain response to verbal learning following sleep deprivation. *Nature, 403,* 655–657.

Drzezga, A., Becker, J. A., Van Dijk, K. R., Sreenivasan, A., Talukdar, T., Sullivan, C., et al. (2011). Neuronal dysfunction and disconnection of cortical hubs in non-demented subjects with elevated amyloid burden. *Brain, 134,* 1635–1646.

Fagan, A. M., Mintun, M. A., Mach, R. H., Lee, S. Y., Dence, C. S., Shah, A. R., et al. (2006). Inverse relation between *in vivo* amyloid imaging load and cerebrospinal fluid Abeta42 in humans. *Annals of Neurology, 59,* 512–519.

Fagan, A. M., Xiong, C., Jasielec, M. S., Bateman, R. J., Goate, A. M., Benzinger, T. L., et al. (2014). Longitudinal change in CSF biomarkers in autosomal-dominant Alzheimer's disease. *Science Translational Medicine, 6,* 226ra230.

Faubel, R., Lopez-Garcia, E., Guallar-Castillon, P., Graciani, A., Banegas, J. R., & Rodriguez-Artalejo, F. (2009). Usual sleep duration and cognitive function in older adults in Spain. *Journal of Sleep Research, 18,* 427–435.

Fleisher, A. S., Chen, K., Quiroz, Y. T., Jakimovich, L. J., Gomez, M. G., Langois, C. M., et al. (2012). Florbetapir PET analysis of amyloid-beta deposition in the presenilin 1 E280A autosomal dominant Alzheimer's disease kindred: A cross-sectional study. *Lancet Neurology, 11,* 1057–1065.

Flicker, L. (2010). Modifiable lifestyle risk factors for Alzheimer's disease. *Journal of Alzheimer's Disease, 20,* 803–811.

Ganguly-Fitzgerald, I., Donlea, J., & Shaw, P. J. (2006). Waking experience affects sleep need in Drosophila. *Science, 313,* 1775–1781.

Gilestro, G. F., Tononi, G., & Cirelli, C. (2009). Widespread changes in synaptic markers as a function of sleep and wakefulness in Drosophila. *Science, 324,* 109–112.

Glenner, G. G., & Wong, C. W. (1984). Alzheimer's disease: Initial report of the purification and characterization of a novel cerebrovascular amyloid protein. *Biochemical and Biophysical Research Communications, 120,* 885–890.

Goate, A., Chartier-Harlin, M. C., Mullan, M., Brown, J., Crawford, F., Fidani, L., et al. (1991). Segregation of a missense mutation in the amyloid precursor protein gene with familial Alzheimer's disease. *Nature, 349,* 704–706.

Graves, L. A., Heller, E. A., Pack, A. I., & Abel, T. (2003). Sleep deprivation selectively impairs memory consolidation for contextual fear conditioning. *Learning & Memory, 10*, 168–176.

Guan, Z., Peng, X., & Fang, J. (2004). Sleep deprivation impairs spatial memory and decreases extracellular signal-regulated kinase phosphorylation in the hippocampus. *Brain Research, 1018*, 38–47.

Haass, C., Schlossmacher, M. G., Hung, A. Y., Vigo-Pelfrey, C., Mellon, A., Ostaszewski, B. L., et al. (1992). Amyloid beta-peptide is produced by cultured cells during normal metabolism. *Nature, 359*, 322–325.

Hardy, J., & Selkoe, D. J. (2002). The amyloid hypothesis of Alzheimer's disease: Progress and problems on the road to therapeutics. *Science, 297*, 353–356.

Harrison, Y., & Horne, J. A. (2000). Sleep loss and temporal memory. *The Quarterly Journal of Experimental Psychology: Human Experimental Psychology, 53A*, 271–279.

Hasegawa, K., Yamaguchi, I., Omata, S., Gejyo, F., & Naiki, H. (1999). Interaction between A beta(1-42) and A beta(1-40) in Alzheimer's beta-amyloid fibril formation *in vitro*. *Biochemistry, 38*, 15514–15521.

Hedden, T., Van Dijk, K. R., Becker, J. A., Mehta, A., Sperling, R. A., Johnson, K. A., et al. (2009). Disruption of functional connectivity in clinically normal older adults harboring amyloid burden. *The Journal of Neuroscience, 29*, 12686–12694.

Holtzman, D. M., Morris, J. C., & Goate, A. M. (2011). Alzheimer's disease: The challenge of the second century. *Science Translational Medicine, 3*, 77sr71.

Hong, S., Quintero-Monzon, O., Ostaszewski, B. L., Podlisny, D. R., Cavanaugh, W. T., Yang, T., et al. (2011). Dynamic analysis of amyloid beta-protein in behaving mice reveals opposing changes in ISF versus parenchymal Abeta during age-related plaque formation. *The Journal of Neuroscience, 31*, 15861–15869.

Hope, T., Keene, J., Gedling, K., Fairburn, C. G., & Jacoby, R. (1998). Predictors of institutionalization for people with dementia living at home with a carer. *International Journal of Geriatric Psychiatry, 13*, 682–690.

Horovitz, S. G., Braun, A. R., Carr, W. S., Picchioni, D., Balkin, T. J., Fukunaga, M., et al. (2009). Decoupling of the brain's default mode network during deep sleep. *Proceedings of the National Academy of Sciences of the United States of America, 106*, 11376–11381.

Huang, Y., Potter, R., Sigurdson, W., Santacruz, A., Shih, S., Ju, Y. E., et al. (2012). Effects of age and amyloid deposition on Abeta dynamics in the human central nervous system. *Archives of Neurology, 69*, 51–58.

Huitron-Resendiz, S., Sanchez-Alavez, M., Gallegos, R., Berg, G., Crawford, E., Giacchino, J. L., et al. (2002). Age-independent and age-related deficits in visuospatial learning, sleep-wake states, thermoregulation and motor activity in PDAPP mice. *Brain Research, 928*, 126–137.

Hyder, F., Patel, A. B., Gjedde, A., Rothman, D. L., Behar, K. L., & Shulman, R. G. (2006). Neuronal-glial glucose oxidation and glutamatergic-GABAergic function. *Journal of Cerebral Blood Flow and Metabolism, 26*, 865–877.

Iliff, J. J., Wang, M., Liao, Y., Plogg, B. A., Peng, W., Gundersen, G. A., et al. (2012). A paravascular pathway facilitates CSF flow through the brain parenchyma and the clearance of interstitial solutes, including amyloid beta. *Science Translational Medicine, 4*, 147ra111.

Jack, C. R., Jr., et al., Lowe, V. J., Weigand, S. D., Wiste, H. J., Senjem, M. L., Knopman, D. S., et al. (2009). Serial PIB and MRI in normal, mild cognitive impairment and Alzheimer's disease: Implications for sequence of pathological events in Alzheimer's disease. *Brain, 132*, 1355–1365.

Jagust, W. J., & Mormino, E. C. (2011). Lifespan brain activity, beta-amyloid, and Alzheimer's disease. *Trends in Cognitive Sciences, 15*, 520–526.

Ji, D., & Wilson, M. A. (2007). Coordinated memory replay in the visual cortex and hippocampus during sleep. *Nature Neuroscience, 10*, 100–107.

Ju, Y. E., Lucey, B. P., & Holtzman, D. M. (2014). Sleep and Alzheimer disease pathology—A bidirectional relationship. *Nature Reviews Neurology, 10*, 115–119.

Ju, Y. E., McLeland, J. S., Toedebusch, C. D., Xiong, C., Fagan, A. M., Duntley, S. P., et al. (2013). Sleep quality and preclinical Alzheimer disease. *JAMA Neurol, 70*, 587–593.

Kamenetz, F., Tomita, T., Hsieh, H., Seabrook, G., Borchelt, D., Iwatsubo, T., et al. (2003). APP processing and synaptic function. *Neuron, 37*, 925–937.

Kang, J. E., Lim, M. M., Bateman, R. J., Lee, J. J., Smyth, L. P., Cirrito, J. R., et al. (2009). Amyloid-beta dynamics are regulated by orexin and the sleep-wake cycle. *Science, 326*, 1005–1007.

Karni, A., Tanne, D., Rubenstein, B. S., Askenasy, J. J., & Sagi, D. (1994). Dependence on REM sleep of overnight improvement of a perceptual skill. *Science, 265*, 679–682.

Klunk, W. E., Engler, H., Nordberg, A., Wang, Y., Blomqvist, G., Holt, D. P., et al. (2004). Imaging brain amyloid in Alzheimer's disease with Pittsburgh compound-B. *Annals of Neurology, 55*, 306–319.

Koffie, R. M., Hashimoto, T., Tai, H. C., Kay, K. R., Serrano-Pozo, A., Joyner, D., et al. (2012). Apolipoprotein E4 effects in Alzheimer's disease are mediated by synaptotoxic oligomeric amyloid-beta. *Brain, 135*, 2155–2168.

Koffie, R. M., Meyer-Luehmann, M., Hashimoto, T., Adams, K. W., Mielke, M. L., Garcia-Alloza, M., et al. (2009). Oligomeric amyloid beta associates with postsynaptic densities and correlates with excitatory synapse loss near senile plaques. *Proceedings of the National Academy of Sciences of the United States of America, 106*, 4012–4017.

Lah, J. J., & Levey, A. I. (2000). Endogenous presenilin-1 targets to endocytic rather than biosynthetic compartments. *Molecular and Cellular Neurosciences, 16*, 111–126.

Lambert, J. C., Ibrahim-Verbaas, C. A., Harold, D., Naj, A. C., Sims, R., Bellenguez, C., et al. (2013). Meta-analysis of 74,046 individuals identifies 11 new susceptibility loci for Alzheimer's disease. *Nature Genetics, 45*, 1452–1458.

Lammich, S., Kojro, E., Postina, R., Gilbert, S., Pfeiffer, R., Jasionowski, M., et al. (1999). Constitutive and regulated alpha-secretase cleavage of Alzheimer's amyloid precursor protein by a disintegrin metalloprotease. *Proceedings of the National Academy of Sciences of the United States of America, 96*, 3922–3927.

Larson-Prior, L. J., Power, J. D., Vincent, J. L., Nolan, T. S., Coalson, R. S., Zempel, J., et al. (2011). Modulation of the brain's functional network architecture in the transition from wake to sleep. *Progress in Brain Research, 193*, 277–294.

Lazarov, O., Lee, M., Peterson, D. A., & Sisodia, S. S. (2002). Evidence that synaptically released beta-amyloid accumulates as extracellular deposits in the hippocampus of transgenic mice. *The Journal of Neuroscience, 22*, 9785–9793.

Lim, A. S., Kowgier, M., Yu, L., Buchman, A. S., & Bennett, D. A. (2013). Sleep fragmentation and the risk of incident Alzheimer's disease and cognitive decline in older persons. *Sleep, 36*, 1027–1032.

Lim, A. S., Yu, L., Costa, M. D., Leurgans, S. E., Buchman, A. S., Bennett, D. A., et al. (2012). Increased fragmentation of rest-activity patterns is associated with a characteristic pattern of cognitive impairment in older individuals. *Sleep, 35*, 633–640B.

Liu, Z. W., Faraguna, U., Cirelli, C., Tononi, G., & Gao, X. B. (2010). Direct evidence for wake-related increases and sleep-related decreases in synaptic strength in rodent cortex. *The Journal of Neuroscience, 30*, 8671–8675.

Louie, K., & Wilson, M. A. (2001). Temporally structured replay of awake hippocampal ensemble activity during rapid eye movement sleep. *Neuron, 29*, 145–156.

Lu, H., Zou, Q., Gu, H., Raichle, M. E., Stein, E. A., & Yang, Y. (2012). Rat brains also have a default mode network. *Proceedings of the National Academy of Sciences of the United States of America, 109*, 3979–3984.

Lye, T. C., & Shores, E. A. (2000). Traumatic brain injury as a risk factor for Alzheimer's disease: A review. *Neuropsychology Review, 10*, 115–129.

Mander, B. A., Santhanam, S., Saletin, J. M., & Walker, M. P. (2011). Wake deterioration and sleep restoration of human learning. *Current Biology, 21*, R183–184.

Maquet, P., Dive, D., Salmon, E., Sadzot, B., Franco, G., Poirrier, R., et al. (1990). Cerebral glucose utilization during sleep-wake cycle in man determined by positron emission tomography and [18F]2-fluoro-2-deoxy-D-glucose method. *Brain Research, 513*, 136–143.

Maret, S., Faraguna, U., Nelson, A. B., Cirelli, C., & Tononi, G. (2011). Sleep and waking modulate spine turnover in the adolescent mouse cortex. *Nature Neuroscience, 14*, 1418–1420.

Masters, C. L., Simms, G., Weinman, N. A., Multhaup, G., McDonald, B. L., & Beyreuther, K. (1985). Amyloid plaque core protein in Alzheimer disease and down syndrome. *Proceedings of the National Academy of Sciences of the United States of America, 82*, 4245–4249.

Mawuenyega, K. G., Sigurdson, W., Ovod, V., Munsell, L., Kasten, T., Morris, J. C., et al. (2010). Decreased clearance of CNS beta-amyloid in Alzheimer's disease. *Science, 330*, 1774.

McCurry, S. M., Logsdon, R. G., Teri, L., Gibbons, L. E., Kukull, W. A., Bowen, J. D., et al. (1999). Characteristics of sleep disturbance in community-dwelling Alzheimer's disease patients. *Journal of Geriatric Psychiatry and Neurology, 12*, 53–59.

McGrath, M. J., & Cohen, D. B. (1978). REM sleep facilitation of adaptive waking behavior: A review of the literature. *Psychological Bulletin, 85*, 24–57.

Meyer-Luehmann, M., Spires-Jones, T. L., Prada, C., Garcia-Alloza, M., de Calignon, A., Rozkalne, A., et al. (2008). Rapid appearance and local toxicity of amyloid-beta plaques in a mouse model of Alzheimer's disease. *Nature, 451*, 720–724.

Meyer-Luehmann, M., Stalder, M., Herzig, M. C., Kaeser, S. A., Kohler, E., Pfeifer, M., et al. (2003). Extracellular amyloid formation and associated pathology in neural grafts. *Nature Neuroscience, 6*, 370–377.

Moran, M., Lynch, C. A., Walsh, C., Coen, R., Coakley, D., & Lawlor, B. A. (2005). Sleep disturbance in mild to moderate Alzheimer's disease. *Sleep Medicine, 6*, 347–352.

Nordstedt, C., Caporaso, G. L., Thyberg, J., Gandy, S. E., & Greengard, P. (1993). Identification of the Alzheimer beta/A4 amyloid precursor protein in clathrin-coated vesicles purified from PC12 cells. *The Journal of Biological Chemistry, 268*, 608–612.

Ooms, S., Overeem, S., Besse, K., Rikkert, M. O., Verbeek, M., & Claassen, J. A. (2014). Effect of 1 night of total sleep deprivation on cerebrospinal fluid beta-amyloid 42 in healthy middle-aged men: A randomized clinical trial. *JAMA Neurology, 71*(8), 971–977.

Oosterman, J. M., van Someren, E. J., Vogels, R. L., Van Harten, B., & Scherder, E. J. (2009). Fragmentation of the rest-activity rhythm correlates with age-related cognitive deficits. *Journal of Sleep Research, 18*, 129–135.

Petersen, S. E., Fox, P. T., Posner, M. I., Mintun, M., & Raichle, M. E. (1988). Positron emission tomographic studies of the cortical anatomy of single-word processing. *Nature, 331*, 585–589.

Peyron, C., Tighe, D. K., van den Pol, A. N., de Lecea, L., Heller, H. C., Sutcliffe, J. G., et al. (1998). Neurons containing hypocretin (orexin) project to multiple neuronal systems. *The Journal of Neuroscience, 18*, 9996–10015.

Platt, B., Drever, B., Koss, D., Stoppelkamp, S., Jyoti, A., Plano, A., et al. (2011). Abnormal cognition, sleep, EEG and brain metabolism in a novel knock-in Alzheimer mouse, PLB1. *PLoS One, 6*, e27068.

Pollak, C. P., & Perlick, D. (1991). Sleep problems and institutionalization of the elderly. *Journal of Geriatric Psychiatry and Neurology, 4*, 204–210.

Potvin, O., Lorrain, D., Forget, H., Dube, M., Grenier, S., Preville, M., et al. (2012). Sleep quality and 1-year incident cognitive impairment in community-dwelling older adults. *Sleep, 35*, 491–499.

Raichle, M. E., MacLeod, A. M., Snyder, A. Z., Powers, W. J., Gusnard, D. A., & Shulman, G. L. (2001). A default mode of brain function. *Proceedings of the National Academy of Sciences of the United States of America, 98*, 676–682.

Raichle, M. E., & Mintun, M. A. (2006). Brain work and brain imaging. *Annual Review of Neuroscience, 29*, 449–476.

Raichle, M. E., & Snyder, A. Z. (2007). A default mode of brain function: A brief history of an evolving idea. *Neuroimage, 37,* 1083–1090.

Rasch, B., Buchel, C., Gais, S., & Born, J. (2007). Odor cues during slow-wave sleep prompt declarative memory consolidation. *Science, 315,* 1426–1429.

Reiman, E. M., Quiroz, Y. T., Fleisher, A. S., Chen, K., Velez-Pardo, C., Jimenez-Del-Rio, M., et al. (2012). Brain imaging and fluid biomarker analysis in young adults at genetic risk for autosomal dominant Alzheimer's disease in the presenilin 1 E280A kindred: A case-control study. *Lancet Neurology, 11,* 1048–1056.

Roh, J. H., Huang, Y., Bero, A. W., Kasten, T., Stewart, F. R., Bateman, R. J., et al. (2012). Disruption of the sleep-wake cycle and diurnal fluctuation of beta-amyloid in mice with Alzheimer's disease pathology. *Science Translational Medicine, 4,* 150ra122.

Samann, P. G., Wehrle, R., Hoehn, D., Spoormaker, V. I., Peters, H., Tully, C., et al. (2011). Development of the brain's default mode network from wakefulness to slow wave sleep. *Cerebral Cortex, 21,* 2082–2093.

Saunders, A. M., Strittmatter, W. J., Schmechel, D., George-Hyslop, P. H., Pericak-Vance, M. A., Joo, S. H., et al. (1993). Association of apolipoprotein E allele epsilon 4 with late-onset familial and sporadic Alzheimer's disease. *Neurology, 43,* 1467–1472.

Selkoe, D. J. (2001). Alzheimer's disease: Genes, proteins, and therapy. *Physiological Reviews, 81,* 741–766.

Seubert, P., Vigo-Pelfrey, C., Esch, F., Lee, M., Dovey, H., Davis, D., et al. (1992). Isolation and quantification of soluble Alzheimer's beta-peptide from biological fluids. *Nature, 359,* 325–327.

Seugnet, L., Suzuki, Y., Vine, L., Gottschalk, L., & Shaw, P. J. (2008). D1 receptor activation in the mushroom bodies rescues sleep-loss-induced learning impairments in Drosophila. *Current Biology, 18,* 1110–1117.

Shankar, G. M., Li, S., Mehta, T. H., Garcia-Munoz, A., Shepardson, N. E., Smith, I., et al. (2008). Amyloid-beta protein dimers isolated directly from Alzheimer's brains impair synaptic plasticity and memory. *Nature Medicine, 14,* 837–842.

Shannon, B. J., Dosenbach, R. A., Su, Y., Vlassenko, A. G., Larson-Prior, L. J., Nolan, T. S., et al. (2013). Morning-evening variation in human brain metabolism and memory circuits. *Journal of Neurophysiology, 109,* 1444–1456.

Sheline, Y. I., Raichle, M. E., Snyder, A. Z., Morris, J. C., Head, D., Wang, S., et al. (2010). Amyloid plaques disrupt resting state default mode network connectivity in cognitively normal elderly. *Biological Psychiatry, 67,* 584–587.

Sheng, J. G., Price, D. L., & Koliatsos, V. E. (2002). Disruption of corticocortical connections ameliorates amyloid burden in terminal fields in a transgenic model of Abeta amyloidosis. *The Journal of Neuroscience, 22,* 9794–9799.

Shoji, M., Golde, T. E., Ghiso, J., Cheung, T. T., Estus, S., Shaffer, L. M., et al. (1992). Production of the Alzheimer amyloid beta protein by normal proteolytic processing. *Science, 258,* 126–129.

Smith, C. (1985). Sleep states and learning: A review of the animal literature. *Neuroscience and Biobehavioral Reviews, 9,* 157–168.

Smith, C., & Rose, G. M. (1996). Evidence for a paradoxical sleep window for place learning in the Morris water maze. *Physiology & Behavior, 59,* 93–97.

Smith, C., & Rose, G. M. (1997). Posttraining paradoxical sleep in rats is increased after spatial learning in the Morris water maze. *Behavioral Neuroscience, 111,* 1197–1204.

Sperling, R. A., Laviolette, P. S., O'Keefe, K., O'Brien, J., Rentz, D. M., Pihlajamaki, M., et al. (2009). Amyloid deposition is associated with impaired default network function in older persons without dementia. *Neuron, 63,* 178–188.

Spira, A. P., Gamaldo, A. A., An, Y., Wu, M. N., Simonsick, E. M., Bilgel, M., et al. (2013). Self-reported sleep and beta-amyloid deposition in community-dwelling older adults. *JAMA Neurol, 70,* 1537–1543.

Stickgold, R., James, L., & Hobson, J. A. (2000). Visual discrimination learning requires sleep after training. *Nature Neuroscience, 3,* 1237–1238.

Supekar, K., Menon, V., Rubin, D., Musen, M., & Greicius, M. D. (2008). Network analysis of intrinsic functional brain connectivity in Alzheimer's disease. *PLoS Computational Biology, 4,* e1000100.

Tamminen, J., Payne, J. D., Stickgold, R., Wamsley, E. J., & Gaskell, M. G. (2010). Sleep spindle activity is associated with the integration of new memories and existing knowledge. *The Journal of Neuroscience, 30,* 14356–14360.

Thies, W., & Bleiler, L. (2013). 2013 Alzheimer's disease facts and figures. *Alzheimer's & Dementia, 9,* 208–245.

Tononi, G., & Cirelli, C. (2006). Sleep function and synaptic homeostasis. *Sleep Medicine Reviews, 10,* 49–62.

Tran, H. T., LaFerla, F. M., Holtzman, D. M., & Brody, D. L. (2011). Controlled cortical impact traumatic brain injury in 3xTg-AD mice causes acute intra-axonal amyloid-beta accumulation and independently accelerates the development of tau abnormalities. *The Journal of Neuroscience, 31,* 9513–9525.

Tran, H. T., Sanchez, L., Esparza, T. J., & Brody, D. L. (2011). Distinct temporal and anatomical distributions of amyloid-beta and tau abnormalities following controlled cortical impact in transgenic mice. *PLoS One, 6,* e25475.

Tworoger, S. S., Lee, S., Schernhammer, E. S., & Grodstein, F. (2006). The association of self-reported sleep duration, difficulty sleeping, and snoring with cognitive function in older women. *Alzheimer Disease and Associated Disorders, 20,* 41–48.

Vaishnavi, S. N., Vlassenko, A. G., Rundle, M. M., Snyder, A. Z., Mintun, M. A., & Raichle, M. E. (2010). Regional aerobic glycolysis in the human brain. *Proceedings of the National Academy of Sciences of the United States of America, 107,* 17757–17762.

Vassar, R., Bennett, B. D., Babu-Khan, S., Kahn, S., Mendiaz, E. A., Denis, P., et al. (1999). Beta-secretase cleavage of Alzheimer's amyloid precursor protein by the transmembrane aspartic protease BACE. *Science, 286,* 735–741.

Vitiello, M. V., Prinz, P. N., Williams, D. E., Frommlet, M. S., & Ries, R. K. (1990). Sleep disturbances in patients with mild-stage Alzheimer's disease. *Journal of Gerontology, 45,* M131–138.

Vlassenko, A. G., Vaishnavi, S. N., Couture, L., Sacco, D., Shannon, B. J., Mach, R. H., et al. (2010). Spatial correlation between brain aerobic glycolysis and amyloid-beta (Abeta) deposition. *Proceedings of the National Academy of Sciences of the United States of America, 107,* 17763–17767.

Vyazovskiy, V. V., Cirelli, C., Pfister-Genskow, M., Faraguna, U., & Tononi, G. (2008). Molecular and electrophysiological evidence for net synaptic potentiation in wake and depression in sleep. *Nature Neuroscience, 11,* 200–208.

Vyazovskiy, V. V., Cirelli, C., Tononi, G., & Tobler, I. (2008). Cortical metabolic rates as measured by 2-deoxyglucose-uptake are increased after waking and decreased after sleep in mice. *Brain Research Bulletin, 75,* 591–597.

Vyazovskiy, V. V., Olcese, U., Lazimy, Y. M., Faraguna, U., Esser, S. K., Williams, J. C., et al. (2009). Cortical firing and sleep homeostasis. *Neuron, 63,* 865–878.

Walker, M. P., & Stickgold, R. (2006). Sleep, memory, and plasticity. *Annual Review of Psychology, 57,* 139–166.

Westerberg, C. E., Mander, B. A., Florczak, S. M., Weintraub, S., Mesulam, M. M., Zee, P. C., et al. (2012). Concurrent impairments in sleep and memory in amnestic mild cognitive impairment. *Journal of the International Neuropsychological Society, 18,* 490–500.

Wilson, M. A., & McNaughton, B. L. (1994). Reactivation of hippocampal ensemble memories during sleep. *Science, 265,* 676–679.

Xie, L., Kang, H., Xu, Q., Chen, M. J., Liao, Y., Thiyagarajan, M., et al. (2013). Sleep drives metabolite clearance from the adult brain. *Science, 342,* 373–377.

Yaffe, K., Laffan, A. M., Harrison, S. L., Redline, S., Spira, A. P., Ensrud, K. E., et al. (2011). Sleep-disordered breathing, hypoxia, and risk of mild cognitive impairment and dementia in older women. *JAMA, 306*, 613–619.

Yamada, K., Cirrito, J. R., Stewart, F. R., Jiang, H., Finn, M. B., Holmes, B. B., et al. (2011). *In vivo* microdialysis reveals age-dependent decrease of brain interstitial fluid tau levels in P301S human tau transgenic mice. *The Journal of Neuroscience, 31*, 13110–13117.

Yamada, K., Holth, J. K., Liao, F., Stewart, F. R., Mahan, T. E., Jiang, H., et al. (2014). Neuronal activity regulates extracellular tau *in vivo*. *The Journal of Experimental Medicine, 211*, 387–393.

Yan, P., Bero, A. W., Cirrito, J. R., Xiao, Q., Hu, X., Wang, Y., et al. (2009). Characterizing the appearance and growth of amyloid plaques in APP/PS1 mice. *The Journal of Neuroscience, 29*, 10706–10714.

Yoshida, Y., Fujiki, N., Nakajima, T., Ripley, B., Matsumura, H., Yoneda, H., et al. (2001). Fluctuation of extracellular hypocretin-1 (orexin A) levels in the rat in relation to the light-dark cycle and sleep-wake activities. *The European Journal of Neuroscience, 14*, 1075–1081.

CHAPTER

11

Stroke, Cognitive Function, and Alzheimer's Disease

Katherine A. Jackman[1], Toby Cumming[1] and Alyson A. Miller[2]

[1]Melbourne Brain Centre (Austin Campus), The Florey Institute of Neuroscience and Mental Health, The University of Melbourne, Heidelberg, Victoria, Australia [2]School of Medical Sciences, Health Innovations Research Institute, RMIT University, Melbourne, Victoria, Australia

OUTLINE

Introduction	320
Stroke: An Overview	320
Cognitive Impairment and Dementia: AD, VaD, and VCI	324
The Role of the NVU in AD and VCI	327
Contribution of Aβ to Neurovascular Dysfunction in AD	*330*
Neurovascular Dysfunction: Cause or Consequence of Neurodegeneration?	*333*
VCI After Stroke	333
Incidence and Prevalence	*333*
Clinical Presentation of VCI	*334*
Experimental Studies	*335*
AD and Stroke	341
Clinical Findings	*341*
Interactions Between Stroke and AD	*342*
Mechanistic Links	*343*
Prospects for Treatment and Prevention	344
Conclusions	347
References	347

Genes, Environment and Alzheimer's Disease.
DOI: http://dx.doi.org/10.1016/B978-0-12-802851-3.00011-5

319

INTRODUCTION

Stroke and dementia are two of the most serious health issues confronting our world. For many years, the two were considered to be separate entities. Stroke was known to occur as a result of a sudden onset of blocked blood supply to the brain that resulted in focal deficits, while dementia was an atrophic process that lead to cognitive decline over a period of several years. However, it is now apparent that cerebrovascular disease and cognitive impairment are often related. Ischemic pathology frequently coexists with Alzheimer's pathology. Furthermore, vascular factors not only predict stroke risk, but have also been shown to influence risk of developing Alzheimer's disease (AD). Consistent with this relationship, cognitive impairment is frequently observed in stroke survivors. The questions that remain to be answered are the extent and direction of any causal relationship between cerebrovascular disease and dementia, and ultimately, the underlying pathophysiological mechanisms that mediate these effects. The aim of this chapter is to summarize current evidence of the relationship between stroke and cognitive dysfunction, including both AD and vascular cognitive impairment (VCI), making reference to both animal and human studies.

STROKE: AN OVERVIEW

Stroke represents a major cause of death worldwide, ranking number four behind heart disease, cancer, and chronic lower respiratory disease. It is also a leading cause of major long-term disability, with impairments ranging from paralysis and difficulties with mobility to aphasia and cognitive impairment. As such, stroke is a major economic burden on society and has an associated mean lifetime cost estimated at over $140,000 (Taylor et al., 1996). Cognitive dysfunction is frequently observed in stroke survivors, occurring in approximately 20% of patients (Douiri, Rudd, & Wolfe, 2013), and has profound effects on quality of life. It is associated with higher mortality (Tatemichi et al., 1994a), greater rates of poststroke institutionalization (Pasquini, Leys, Rousseaux, Pasquier, & Hénon, 2007), and results in higher health care costs (Claesson, Lindén, Skoog, & Blomstrand, 2005). It is therefore an important target for therapeutic intervention.

The World Health Organization defines stroke as "rapidly developed clinical signs of focal (or global) disturbance of cerebral function, lasting more than 24 h or leading to death, with no apparent cause other than of vascular origin" (Aho et al., 1980). Stroke is divided into two major categories: ischemic, accounting for ~87% of total strokes, and hemorrhagic, which occurs as a result of either intracerebral hemorrhage (ICH) (~10%)

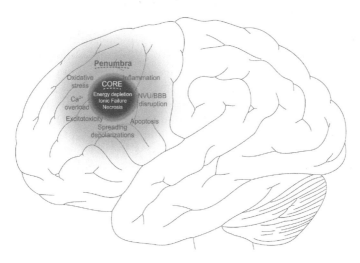

FIGURE 11.1   Pathogenesis of ischemic stroke. Processes of ischemic cell death are complex and interrelated. In the infarct core, death occurs via necrosis as a result of energy failure. Expansion of brain injury into penumbral tissue is mediated by multiple processes, including inflammation, oxidative stress, excitotoxicity and calcium (Ca^{2+}) overload, apoptosis, and blood–brain barrier (BBB) dysfunction.

or subarachnoid hemorrhage (~3%) (Go, Mozaffarian, Roger, & Benjamin, 2014). Subtypes of stroke can be further classified depending on their etiology, with ~50% of ischemic strokes attributed to large-artery atherothrombosis, ~25% intracranial small vessel disease resulting in lacunar stroke, ~20% cardioembolism and the remaining ~5% occurring as a result of rare causes such as vasculitis or arterial dissection (Warlow, Sudlow, Dennis, Wardlaw, & Sandercock, 2003).

The pathogenesis of stroke is extremely complex (Figure 11.1), with the majority of our understanding of these processes coming from animal studies. In tissue of the ischemic core—a region characterized by a severe reduction in cerebral perfusion—cell death is rapid and occurs largely as a result of energy failure and subsequent necrotic cell death (Dirnagl, Iadecola, & Moskowitz, 1999). In contrast, expansion of injury into the surrounding ischemic penumbra—an area of structurally intact but functionally impaired tissue (Astrup, Siesjo, & Symon, 1981)—develops over a period of days after insult and is mediated by multiple mechanisms of cell death. These include glutamate mediated excitotoxicity, calcium dysregulation, mitochondrial dysfunction, cortical spreading depolarizations, and apoptotic cell death (Moskowitz, Lo, & Iadecola, 2010). Oxidative and nitrosative stress play a key role in injury development in this region (Allen & Bayraktutan, 2009). A large body of evidence supports NADPH oxidase (or "Nox oxidases"), in particular the Nox2 oxidase isoform (Jackman et al., 2009; Kunz, Anrather, Zhou, Orio, & Iadecola, 2006; Walder et al., 1997), as a key source of oxidative stress following cerebral ischemia (Chrissobolis, Miller, Drummond, Kemp-Harper, & Sobey, 2011; Kahles & Brandes, 2012). In addition, mitochondria and uncoupled endothelial nitric oxide synthase represent other major sources of ROS in the ischemic brain. There is considerable evidence supporting the importance of inflammation and immune system activation in injury development and expansion after stroke, with key roles of adhesion molecules (e.g., P-selectin, ICAM, VCAM), pro-inflammatory cytokines and chemokines (e.g., IL-6, IL-1β, TNF, MCP-1, RANTES), proteases (e.g., elastase, MMP-2 & -9), and pro-inflammatory enzymes (e.g., COX-2, iNOS), in addition to both the innate and adaptive immune system (Iadecola & Anrather, 2011).

While stroke research has historically focused on understanding the responses of single cells (e.g., neuron, microglia, astrocyte), the importance of their interaction with each other is now particularly apparent and this has shifted the focus of ischemia research toward the combined components of the neurovascular unit (NVU) (del Zoppo, 2010; Dirnagl, 2012). The NVU is a collective term for the structural and functional association between neurons, perivascular astrocytes, vascular smooth muscle cells (pericytes/astrocytes), endothelial cells, and the basal lamina (Figure 11.2) (Stanimirovic & Friedman, 2012). Together, these components of the NVU act to regulate and maintain cerebral perfusion, preserve homeostatic balance in the brain, and control immune regulation. The NVU ensures adequate perfusion of the brain in response to changes in neural activity (functional hyperemia) and arterial pressure (cerebral autoregulation) (Girouard & Iadecola, 2006; Jackman & Iadecola, 2013). Furthermore, it represents the primary site of the blood–brain barrier—endothelial cells form tight junctions and express selective transporters, while vascular myocytes/pericytes, perivascular astrocytes, and the basal lamina provide both structural and functional support—separating the brain from plasma components and thereby preserving its delicate homeostatic balance (Zlokovic, 2008). Cerebral ischemia has devastating effects on both the structure and functionality of the NVU. It impairs cerebral perfusion and vascular reactivity, increases blood–brain barrier permeability, and enhances inflammatory cell infiltration, thereby increasing both the development and exacerbation of brain injury occurring in response to stroke (Jackman & Iadecola, 2013).

A number of animal models of stroke have been described (Durukan & Tatlisumak, 2007; Hossmann, 2008; Krafft et al., 2012; Tajiri et al., 2013). While rodents, in particular rats and mice, are by far the most commonly utilized species due to lower costs associated with purchase, breeding and maintenance, and greater ethical acceptability, larger domesticated animals

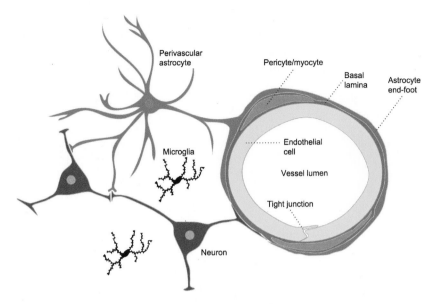

FIGURE 11.2   The neurovascular unit (NVU). The NVU is the structural and functional association between neurons, endothelial cells, perivascular astrocytes, pericytes, and the basal lamina. This functional unit regulates cerebral perfusion and preserves homeostatic balance in the brain.

such as rabbits, cats and dogs, and primates are also studied (Howells et al., 2010). Mice are becoming increasingly popular in experimental stroke research due to an ever-expanding variety of genetically modified strains, allowing researchers to more accurately and precisely evaluate the role of specific proteins and signaling pathways in stroke pathogenesis. Both ischemic and hemorrhagic stroke can be modelled experimentally, with the majority of ischemic stroke models involving occlusion of the middle cerebral artery or its branches (Table 11.1). Intraluminal filament/ endovascular middle cerebral artery occlusion (MCAO) is a particularly popular model, in which the middle cerebral artery (MCA) is blocked at its origin using either a heat-blunted or silicone-coated nylon suture (Jackman, Kunz, & Iadecola, 2011). It has the advantages of being less invasive and inducing reproducible, MCA territory infarcts (Durukan & Tatlisumak, 2007). In addition, it can be used to induce either transient or permanent cerebral ischemia. Other techniques of inducing cerebral ischemia include direct surgical ligation of the MCA, photothrombotic MCA occlusion using a photoreactive dye (e.g., rose bengal), embolic models (both autologous clot and nonclot models) and MCA vasoconstriction with endothelin-1 (Durukan & Tatlisumak, 2007).

TABLE 11.1   Experimental Models of Ischemic Stroke

Stroke model	Benefits	Disadvantages
Intraluminal filament (endovascular) **Filament advanced via the ICA to occlude MCA**	• Less invasive • Reproducible infarcts • Transient or permanent ischemia	• Poststroke hyperthermia with longer ischemic periods • Risk of subarachnoid hemorrhage (less likely with laser Doppler)
Endothelin-1 (ET-1) **Vasoconstriction of the MCA**	• Less invasive/quick • Stroke induced in conscious animals	• Magnitude/duration of ischemia highly variable • Requires craniotomy • Additional, nonvasoactive effects of ET-1 (e.g., astrocytosis)
Direct surgical occlusion MCA ligation/ cauterization/clipping	• Control of the site of occlusion (i.e., proximal vs distal)	• Invasive • No reperfusion with cauterization • Possible brain trauma
Autologous clot embolism **Injected via ICA**	• More closely mimics human embolic stroke • Allows study of thrombolytics	• Highly variable infarct sizes • Uncontrollable reperfusion
Nonclot embolism **Artificial clots (e.g., microspheres injected via ICA or CCA)**	• Slow lesion development (increased therapeutic) window)	• Multiple, scattered infarcts • No reperfusion
Photo-thrombosis **Injection of photoactive dye followed by laser irradiation**	• Produces cortical infarcts of defined location/size • Less invasive	• Infarcts are usually small • Fails to produce typical penumbra

Adapted from: Jackman et al. (2011).
MCA, middle cerebral artery; ICA, internal carotid artery; ET-1, endothelin 1; CCA, common carotid artery.

COGNITIVE IMPAIRMENT AND DEMENTIA: AD, VaD, AND VCI

AD is the most common form of dementia, accounting for up to 60–70% of cases (Ferri et al., 2005). The AD phenotype is defined generally by an impairment in episodic memory in addition to a deficit in at least one additional cognitive domain, however these diagnostic criteria continue to be revised and refined (Dubois et al., 2007; Sarazin, de Souza, Lehéricy, & Dubois, 2012). Pathologically, it is characterized by the accumulation of amyloid beta (Aβ) in the brain parenchyma (amyloid plaques) or blood vessels (cerebral amyloid angiopathy (CAA)) in addition to the presence of neurofibrillary tangles (Querfurth & LaFerla, 2010). Second in prevalence to AD is VaD, being responsible for at least 20% of cases of dementia (Gorelick et al., 2011). The concept of VaD has existed for many years based on the theory that vascular brain injury of many types can lead to global decline in cognitive function. VaD is probably the most commonly used term today for vascular causes of dementia. Recently, however, there has been a move away from the traditional concept of VaD as the criteria for diagnosis were largely based on those used for AD, which does not take into consideration cognitive deficits more commonly associated with cerebral vascular disease. Furthermore, the criteria are insensitive to detect early cases of cognitive impairment. Consequently, the term VCI was introduced in the 1990s by Hachinski and Bowler and is defined as a syndrome of cognitive impairment associated with either clinical stroke or subclinical vascular brain injury (confirmed by neuroimaging), and cognitive impairment affecting at least one cognitive domain (Hachinski & Bowler, 1993).

VCI is an umbrella term, acting to incorporate the broad spectrum of disease severity ranging from mild cognitive impairment to more severe VaD. It is characterized by a dysfunction in multiple cognitive domains, including learning and memory. However, impairments in attention and executive function have been suggested to be more pronounced, possibly supporting a neurocognitive distinction between VCI and AD (Hachinski et al., 2006; Selnes & Vinters, 2006). Vascular alterations contributing to VCI are diverse (Iadecola, 2013) (Table 11.2). For example, small vessel disease (atherosclerotic plaques, deposition of hyaline in the vascular wall (lipophylainosis), concentric fibrotic changes in the vascular wall causing stiffening and distortion (arteriosclerosis), or loss of vascular wall integrity (fibroid necrosis)) affecting arterioles <300 μm in diameter are frequently associated with VCI (Thal, Grinberg, & Attems, 2012), and result in the formation of white matter lesions (leukoaraiosis), lacunaes, micro-infarcts, hemorrhages, and microbleeds (Table 11.2). Global cerebral hypoperfusion (either transient or chronic), occurring as a result of cardiac arrest, cardiac failure, arrhythmias, hypotension, or extracranial large vessel atherosclerosis and vascular stiffening is another common cause of VCI (Iadecola, 2013). Arterial occlusions affecting specific areas of the brain associated with cognition ("strategic infarct" dementia) and multiple arterial occlusions occurring over a period of time resulting in multi-infarct dementia (Thal et al., 2012) also result in VCI (Table 11.2). Focal ischemic stroke, which results in severe and prolonged reductions

TABLE 11.2   Causes of VCI

Lesions/condition	Vascular pathology
White matter disease	SVD (atherosclerosis, lipohyalinosis, arteriosclerosis, fibrinoid necrosis)
White matter lesions (leukoaraiosis) and small infarcts (lacunaes)	CADASIL (a rare, genetic form of SVD)
Focal hypoperfusion	
Ischemic stroke (arterial occlusion)	Thrombosis/embolis
Arteriovenous malformation	Abnormal artery–vein transition
Global hypoperfusion	Cardiac dysfunction (cardiac arrest, cardiac failure, arrhythmias), hypotension, large vessel atherosclerosis (carotid occlusion)
Watershed or border zone infarcts	
"Strategic infarct" dementia	SVD, embolic events
Solitary or multiple small infarcts	
Multi-infarct dementia	Thromboembolism (large to medium arteries/arterioles), SVD, CAA
Multiple arterial occlusions	
Microinfarcts	SVD, CAA, thromboembolism
Small arterial occlusion	
Microbleeds and hemorrhages	SVD, CAA, CADASIL, vascular rupture
Mixed dementia	Atherosclerosis (large to medium arteries/arterioles), SVD, CAA
White matter lesions, hemorrhages, infarction, and AD pathology	

CAA, cerebral amyloid angiopathy; CADASIL, cerebral autosomal dominant arteriopathy with subcortical infarcts and leukoencephalopathy; SVD, small vessel disease; WML, white matter lesions.

in cerebral perfusion (Leys, Hénon, Mackowiak-Cordoliani, & Pasquier, 2005) is a major cause of VCI (discussed in detail in this chapter) and indeed, the term VCI has now replaced earlier descriptions such as post-stroke dementia for referring to the cognitive dysfunction associated with stroke (Gorelick et al., 2011). Unlike earlier classifications, a diagnosis of VCI does not require the presence of significant memory impairment. As memory impairment is neither the most pronounced nor consistently observed cognitive deficit observed after stroke (Cumming, Marshall, & Lazar, 2013), adoption of the term VCI and its associated diagnostic criteria may allow a greater sensitivity of detection of cognitive abnormalities in stroke sufferers and possibly facilitate the differential diagnosis of VCI and AD.

AD and VCI, and the pathogenic causes and mechanisms underlying these two conditions, have traditionally been considered separate entities (Iadecola, 2010). However, as discussed later in this chapter, several lines of evidence now indicate that there is a definite overlap between vascular and neurodegenerative causes of dementia. First, while pure cases of AD and VCI do exist, it is now widely recognized that in many cases, dementia is, in fact, of mixed pathology, characterized by the presence of both vascular lesions and AD pathology (Gold, Giannakopoulos, Herrmann, Bouras, & Kövari, 2007; Iadecola, 2010). Furthermore, while the degree of impairment may vary, there do not appear to be entirely discrete profiles of cognitive dysfunction for VCI and AD (Selnes & Vinters, 2006), further supporting overlap between the two conditions. Second, there is substantial overlap in associated vascular risk factors (e.g., hypertension, diabetes mellitus, high cholesterol) (Sahathevan, Brodtmann, & Donnan, 2011) for both conditions. Third, there is clinical and experimental evidence that AD is associated with both structural and functional alterations of the cerebral circulation, most likely due to the actions of Aβ (Iadecola, 2010). Finally, evidence suggests that cerebral vascular lesions can aggravate cognitive deficits induced by AD and, conversely, AD may increase the risk of ischemia. Thus, both VCI and AD are important causes and contributors to cognitive dysfunction associated with clinical stroke, and therapies aimed at treating their combined effects on the poststroke brain are likely to have the best chance at improving outcome.

THE ROLE OF THE NVU IN AD AND VCI

Strong evidence indicates that similar to stroke, disruption of the NVU is involved in the pathogenesis of VCI (Figure 11.3). Furthermore, there is an emerging view that vascular abnormalities are a feature of not only VCI, but also neurodegenerative condition, including AD. Thus, vascular dysfunction may contribute to cognitive impairment and dementias, irrespective of the underlying brain pathology.

There is evidence of marked alterations in the structure of cerebral vessels in AD, even in the absence of Aβ deposition (CAA). For example, several studies have reported that the incidence and severity of atherosclerotic narrowing of carotid arteries and of larger intracranial arteries is greater in patients with AD than in nondemented individuals (Beach et al., 2007; Hofman et al., 1997; Roher et al., 2011). At the microvascular level, vascular pathology is characterized by vessel thickening (Brown, Moody, Thore, Challa, & Anstrom, 2007; Zarow, Barron, Chui, & Perlmutter, 1997). Cerebral capillary degeneration is reported to be present in almost all AD brains at postmortem (la Torre de, 2002), and appear particularly prevalent in the hippocampus (Fischer, Siddiqi, & Yusufaly, 1990). In the

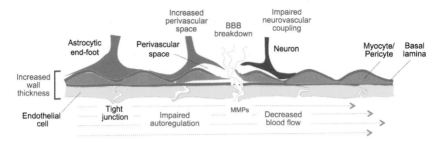

FIGURE 11.3 Dysfunction of the neurovascular unit in Alzheimer's disease (AD) and vascular cognitive impairment (VCI). Structural alterations in the neurovascular unit are induced by both AD and VCI. Matrix metalloproteinase (MMP) mediated tight junction degradation, astrocytic endfoot detachment, and increases in perivascular space and wall thickness are observed. The functional consequences of these structural deficits are increased blood–brain barrier (BBB) disruption, decreased cerebral blood flow, and impaired autoregulation and neurovascular coupling.

later stages of the disease, cells of the NVU including pericytes, endothelial, and smooth muscle cells, also exhibit degenerative changes (Kalaria, 1997). Furthermore, in CAA, accumulation of Aβ in the media of cerebral vessels leads to a series of pathological structural changes (see the section, "Contribution of Aβ to Neurovascular Dysfunction in AD"). Similarly, in the setting of VCI atherosclerotic narrowing of circle of Willis arteries has been demonstrated in people with both pure VCI and mixed dementia (Beach et al., 2007), consistent with an attenuation of cerebral perfusion. White matter microvascular density has also been shown to be decreased, both in areas associated with lesions and nonlesioned regions, in people with leukoaraiosis (Brown et al., 2007). Increased wall thickness and increases in perivascular space have been observed in white matter lesions (Fernando et al., 2006; van Swieten et al., 1991). Interestingly the increase in perivascular space of vessels was more pronounced in white matter lesions of demented individuals (Fernando et al., 2006), possibly suggesting that these changes contribute to both lesion development and cognitive dysfunction; however, the converse is also possible.

Consistent with the structural changes, there is clinical and experimental evidence of blood–brain barrier disruption in AD. Neuropathological studies have shown blood–brain barrier leakage in the brains of patients with AD (Zipser et al., 2007), and blood–brain barrier disruption occurs in transgenic APP mice (Hartz et al., 2012; Paul, Strickland, & Melchor, 2007). Pronounced opening of the blood–brain barrier has also been demonstrated in individuals with VCI using both biochemical approaches and with brain imaging (Farrall & Wardlaw, 2009; Taheri et al., 2011). These findings are supported by experimental studies in which reductions in blood–brain barrier disruption are associated with decreased cognitive dysfunction and vice versa (Lee et al., 2014; Zhang, Yeung, McAlonan, Chung, & Chung, 2013). Matrix metalloproteinases (MMPs)—a family of zinc-dependent endopeptidases that degrade extracellular matrix involved in NVU damage and blood–brain barrier breakdown following stroke (Yang & Rosenberg, 2011)—appear to be important in both VCI and AD. Activity of MMP-9 was shown to be elevated in the cerebrospinal fluid of patients with VCI, suggesting that this isoform may be important (Adair et al., 2004). However, in another study, activity of MMP-3 was increased while cerebrospinal fluid MMP-9 levels were unaffected (Candelario-Jalil et al., 2011). Interestingly, in this study they observed a reduction in MMP-2 activity that was negatively correlated with blood–brain barrier permeability. Thus, while there appears to be a relationship, this interaction requires further investigation to explain how decreases in MMPs could correlate with increased blood–brain barrier dysfunction. Similar variability has been observed for AD, with MMP-9 increased in the plasma of AD patients in one study (Lorenzl et al., 2003), and unaffected in the cerebrospinal fluid in another (Adair et al., 2004). Thus, while the exact MMP isoforms involved in this process may vary depending on patient characteristics, and may also present differently in plasma versus cerebrospinal fluid, overall there appears to be evidence supporting a relationship between MMPs, blood–brain barrier disruption, and the pathogenesis of VCI and AD.

Experimental and clinical evidence indicates that cerebral blood flow and functional hyperemia are reduced in AD (Iadecola, 2010; Pimentel-Coelho & Rivest, 2012). Studies using transgenic APP mice suggest that cerebral autoregulation is severely impaired (Claassen & Zhang, 2011). However, research on cerebral autoregulation in human AD has only recently begun (Claassen & Zhang, 2011). Out of the handful of studies thus far (van Beek, Sijbesma, Jansen, Rikkert, & Claassen, 2010; van Beek, Sijbesma, Olde Rikkert, & Claassen, 2010), only one (Abeelen den, Lagro, van Beek, & Claassen, 2014) has found evidence of impaired cerebral autoregulation in Alzheimer's patients. In recent years, a large number of studies have suggested that cerebral vascular abnormalities may develop before the onset of dementia. Luckhaus et al. observed decreased cerebral blood flow in patients with mild cognitive impairment (MCI) (Luckhaus et al., 2008). Furthermore, a decrease in cerebral blood flow in specific brain regions can be used to identify patients with MCI that will rapidly progress to AD. For example, MCI patients with significant hypoperfusion of areas linked to memory function (e.g., hippocampal-amygdaloid complex) converted to AD in up to 3 years of follow-up, whereas patients with normal cerebral blood flow in these brain areas did not (Johnson & Albert, 2000; Johnson et al., 1998). Experimental studies also support the notion that cerebral vascular dysfunction may be an early event in AD. Indeed, some studies have shown that abnormalities in functional hyperemia, neurovascular coupling, and reductions in cerebral blood flow occur in APP transgenic mice at an early age when amyloid plaques and cognitive deficits are not yet present (Niwa et al., 2000, 2002; Niwa, Kazama, Younkin, Carlson, & Iadecola, 2002; Park et al., 2005). Consistent with these findings, cerebral blood flow is reduced in areas corresponding to white matter lesions (Kobari, Meyer, Ichijo, & Oravez, 1990; Markus, Lythgoe, Ostegaard, O'Sullivan, & Williams, 2000; Marstrand et al., 2002; O'Sullivan et al., 2002) and decreased perfusion has also been observed in nearby normal-appearing white matter (O'Sullivan et al., 2002), suggesting that a generalized reduction in white matter perfusion in these subjects could contribute to cognitive dysfunction. Vascular reactivity and cerebral autoregulation are also attenuated in individuals with white matter lesions (Marstrand et al., 2002; Matsushita et al., 1994), as is neurovascular coupling, evaluated using experimental models of VCI (Jackman & Iadecola, 2013; Kunz & Iadecola, 2009).

Mechanistically, oxidative stress and inflammation are likely to contribute to the neurovascular dysfunction observed in the setting of both AD and VCI (Iadecola & Davisson, 2008; Iadecola, 2013). Oxidative stress and inflammation are induced within brain tissue associated with white matter lesions (Back et al., 2011; Fernando et al., 2006). Furthermore, circulating markers of oxidative stress are elevated, antioxidant enzymes decreased (Casado, Encarnación López-Fernández, Concepción Casado, & La Torre de, 2008; Ihara et al., 1997), and markers of inflammation increased (Rouhl et al., 2012) in patients with VCI and AD. Experimental studies of VCI have demonstrated that high doses of antioxidants (Liu et al., 2007; Xu et al., 2010) and the NADPH-oxidase inhibitor apocynin (Shen et al., 2011) are effective at decreasing cognitive impairment following induction of chronic hypoperfusion. Similarly, interventions aimed at decreasing inflammation, including mucosal tolerance to E-selectin, or treatment with huperzine A (Wakita et al., 2008; Wang, Zhang, & Tang, 2010), are also effective. Taken together, there is strong evidence supporting a key role of inflammation and oxidative stress in the pathogenesis of VCI and AD.

Contribution of Aβ to Neurovascular Dysfunction in AD

Deposition of Aβ peptide can occur in the walls of cerebral leptomeningeal and intracortical arteries, arterioles, and capillaries, which is the hallmark of CAA. $A\beta_{1-42}$ is principally found in parenchymal plaques, whereas in CAA, Aβ is predominately composed of the shorter $A\beta_{1-40}$ species, and to a lesser extent, $A\beta_{1-42}$ (Attems, Jellinger, Thal, & Van Nostrand, 2011). CAA is frequently observed in the elderly with or without AD, appearing in ~10–30% of brain autopsies and >80% when in the presence of accompanying AD (Jellinger & Attems, 2010). The exact source of Aβ in CAA is not known; however, studies of experimental models of CAA suggest that it may be primarily generated by nearby neurons and subsequently becomes deposited in the vessel wall (Herzig et al., 2004; Herzig, Nostrand, & Jucker, 2006). Ineffective transport of Aβ out of the brain or peripheral Aβ (e.g., derived from the liver) may also contribute to vascular deposition (Eisele et al., 2010; Sutcliffe, Hedlund, Thomas, Bloom, & Hilbush, 2011; Weller, Subash, Preston, Mazanti, & Carare, 2008). Furthermore, accelerated Aβ deposition may occur as result of downregulation of the transvascular transport system (Bell et al., 2009).

CAA is most commonly recognized as a cause of spontaneous ICH. In advanced cases, CAA can cause a series of structural changes in the vessel wall, which include loss of smooth muscle cells, concentric splitting of the vessel wall, development of microaneurysms, and fibrinoid necrosis. These CAA-related vascular changes are often accompanied by perivascular leakage of blood products and appear to be a major cause of symptomatic CAA-related ICH. Evidence indicates that CAA is also an important cause of cognitive impairment, and is associated with a high prevalence of vascular lesions including watershed white matter lesions and cortical microinfarcts, and microhemorrhages (Attems et al., 2011; Viswanathan & Greenberg, 2011). Importantly, accumulating evidence suggests that CAA may influence cerebral vascular function, even at very early stages of disease. Using APP mice, Christie et al. were the first to show that deposition of Aβ causes impaired vasodilator responses of cerebral vessels at an age predating loss of smooth muscle cells (Christie, Yamada, Moskowitz, & Hyman, 2001). More recently, other studies have also shown age-dependent deficits in vasodilator capacity in cerebral vessels from APP transgenic mice (Nicolakakis et al., 2008; Park et al., 2008). Deleterious effects of CAA on blood–brain barrier function have also been reported (Hartz et al., 2012).

Of potential importance, recent evidence links elevations in soluble Aβ with impaired cerebral vascular function even in the absence of vascular deposition. For instance, physiologically relevant levels of exogenous Aβ peptides constrict cerebral vessels of humans and rodents (Chisari, Merlo, Sortino, & Salomone, 2010; Paris et al., 2003). Furthermore, some (Niwa et al., 2000, 2002; Niwa, Kazama, et al., 2002; Park et al., 2005) but not all (Shin et al., 2007) studies have reported that APP transgenic mice exhibit cerebral vascular dysfunction such as impaired nitric oxide (NO·)-dependent relaxation responses, reductions in cerebral blood flow, abnormal cerebrovascular autoregulation, and impaired functional hyperaemia at an early age when amyloid plaques and behavioral deficits are not yet present. Studies using cultured cells suggest that soluble Aβ may also promote blood–brain barrier disruption. For example, treatment of cultured rodent or human brain endothelial cells with $Aβ_{1-40}$ triggers tight junction disorganization and increases transendothelial cell resistance (Marco & Skaper, 2006; Velasquez & Kotarek, 2008). In addition to its effects on endothelial cells, soluble Aβ might be deleterious to other cells of the NVU. For example, exogenous Aβ can induce degeneration of cultured human brain pericytes (Verbeek, de Waal, Schipper, & Van Nostrand, 1997). Importantly, studies of pericyte-deficient mice indicate that a loss of pericytes leads to chronic cerebral hypoperfusion and blood–brain barrier disruption (Bell et al., 2010). Thus, Aβ-induced changes in pericyte function may contribute to vascular dysfunction in AD.

The mechanisms underlying Aβ-related vascular dysfunction appear to depend on the stage of disease progression. Several studies suggest that similar to neurons, Aβ is toxic to vascular smooth muscle cells, pericytes, and cerebral endothelial cells. Thus, at later stages of AD, cerebral vascular dysfunction may result from alterations in the vasomotor apparatus, including loss of smooth muscle integrity and pericytes (Park et al., 2014). In the case of CAA, Aβ may also present a mechanical obstacle to vessel dilatation, or by interfering with intracellular signaling pathways (Christie et al., 2001).

In the case of soluble Aβ, strong experimental evidence implicates oxidative stress as a key mechanism. First, exogenous treatment of cerebral microvessels with $A\beta_{1-40}$ augments the production of ROS (Park et al., 2004). Second, APP transgenic mice exhibit signs of vascular oxidative stress (Park et al., 2004, 2008; Tong et al., 2009). Third, overexpression of superoxide dismutase (scavenger of superoxide) or treatment with ROS scavengers improves vascular function in APP transgenic mice (Iadecola et al., 1999; Nicolakakis et al., 2008; Niwa, Carlson, & Iadecola, 2000; Park et al., 2004, 2008). Several lines of evidence point toward the Nox oxidases, a family of prooxidant enzymes, as the enzymatic source of the ROS responsible for the vascular effects of Aβ peptides. The Nox oxidase cytosolic subunits, p47phox and p67phox, are translocated to the membrane in brain tissue from AD patients (Shimohama et al., 2000), suggesting that Nox oxidases are activated during AD. Park et al. (2005) found that either inhibition of Nox oxidases or genetic deletion of the catalytic subunit for Nox2 oxidase isoform counteracts the oxidative stress and vascular dysfunction induced by exogenous $A\beta_{1-40}$. Moreover, genetic deletion of Nox2 abrogates the cerebral vascular dysfunction that occurs in young APP transgenic mice prior to the development of cognitive deficits and plaque deposition (Park et al., 2005). In the case of CAA, the role of Nox oxidases and oxidative stress in cerebral vascular dysfunction is less clear. Early studies showed that Nox oxidase inhibition or Nox2 deletion restored vascular function in cerebral arteries from older APP mice that display frequent CAA, suggesting that in more advanced stages of pathology, Nox oxidase-derived ROS remain a key mediator of Aβ-induced cerebral vascular dysfunction (Hamel, Nicolakakis, Aboulkassim, Ongali, & Tong,

2008; Park et al., 2008). In contrast, however, a more recent study reported that oxidative stress contributed to cerebral vascular dysfunction in young but not older APP mice (Park et al., 2014).

Neurovascular Dysfunction: Cause or Consequence of Neurodegeneration?

There is ample evidence that neurovascular dysfunction contributes to the pathogenesis of stroke-related brain injury and VCI (del Zoppo, 2010; Iadecola, 2013). Moreover, there are many examples that vascular deficits lead to secondary neurodegeneration, including CADASIL (cerebral autosomal dominant arteriopathy with subcortical infarcts and leukoencephalopathy) (Chabriat, Joutel, Dichgans, Tournier-Lasserve, & Bousser, 2009). It is important to note, however, that direct evidence showing a cause–effect relationship between cerebral hypoperfusion and the initiation/development of Alzheimer's neurodegeneration is still lacking, and if so, in which direction it operates. Indeed, it is plausible that reduced cerebral blood flow may at least partially occur secondarily to neurodegenerative disease, rather than being its cause, whereby neuronal death from a nonvascular cause results in decreased metabolic requirement, and consequently less perfusion of that brain region (Kim et al., 2012). Furthermore, the hippocampus has a rich blood supply and is relatively unaffected by reductions in cerebral blood flow (Phan, Donnan, Srikanth, Chen, & Reutens, 2009). Thus clinically, we might argue that cerebral hypoperfusion may be relatively unlikely to cause localized hippocampal injury in AD, suggesting hypoperfusion in AD is more likely to occur secondarily to an initial decrease in local brain metabolism. On the other hand, evidence favoring a hypoperfusion-to-neurodegenerative of casuality is compelling: (i) brain hypoperfusion may precede neurodegenerative changes in AD; (ii) Aβ causes cerebral vascular abnormalities, often before the development of plaque formation; and (iii) a broad spectrum of risk factors for AD have a vascular component that reduces cerebral perfusion (e.g., hypertension, diabetes).

VCI AFTER STROKE

Incidence and Prevalence

At the severe end of the cognitive spectrum, poststroke dementia is present in 10% of people before first stroke, newly developed in 10% soon after first stroke, and present in more than a third of people after recurrent stroke (Pendlebury & Rothwell, 2009). Prevalence rates for poststroke cognitive impairment are substantially higher. In the Canadian Study of

Health and Aging, 64% of stroke survivors had cognitive impairment, compared to only 21% of stroke-free controls (Jin, Di Legge, Ostbye, Feightner, & Hachinski, 2006). Direct comparison was made in a Swedish study: dementia was more common in stroke than controls (28% versus 7%), as was cognitive impairment (72% versus 36%) (Linden, Skoog, Fagerberg, Steen, & Blomstrand, 2004). There is evidence that stroke is related to cognitive impairment–no dementia at 2-year follow-up,—and that recurrent stroke is related to incident dementia at this time point (Srikanth, Quinn, Donnan, Saling, & Thrift, 2006).

Clinical Presentation of VCI

Given the heterogeneity in the location and extent of damage caused by stroke, it is not surprising that the clinical presentation of cognitive impairment is highly variable. In the 1990s, large studies in stroke populations revealed generalized cognitive deficits across multiple cognitive domains, including attention, memory, language, and visuo-spatial ability (Hochstenbach, Mulder, van Limbeek, Donders, & Schoonderwaldt, 1998; Tatemichi et al., 1994b). More recent work indicates that stroke tends to have a greater deleterious impact on attention and executive function than on memory. In a community-based comparison of stroke patients with population controls, patients had greater deficits in visuo-spatial ability, executive function, attention, and language, but not in memory (Srikanth et al., 2003). Other population-based evidence suggests that stroke is associated with a higher risk of nonamnestic (odds ratio (OR) = 2.85) than amnestic (OR = 1.77) cognitive impairment, and is associated with every cognitive domain except memory (Knopman et al., 2009). Longitudinal studies are consistent with this pattern. At 5 years post-stroke, deficits were more pronounced in processing speed ($z = -2.16$) and executive function ($z = -1.92$) than in visual memory ($z = -0.43$) and verbal memory ($z = -0.32$) (Barker-Collo, Feigin, Parag, Lawes, & Senior, 2010). Contributing evidence comes not only from stroke but from studies of other vascular changes. The presence of at least one silent infarct identified on autopsy was related most strongly with perceptual speed and least strongly with episodic memory (Schneider et al., 2003). MRI data from elderly participants indicate that appearance of new lacunes is associated with decline in executive function and processing speed, but not with memory (Jokinen et al., 2011).

Direct comparisons between patient groups give some support to the relative sparing of memory performance in cerebrovascular disease. VaD patients tend to have less long-term memory impairment but more frontal-executive impairment than Alzheimer's patients (Looi & Sachdev, 1999). Memory performance of patients with cerebrovascular

TABLE 11.4 Experimental Evaluation of Poststroke VCI

Cognitive domain	Test	Details	Evaluates	Reference
LEARNING AND MEMORY				
Nonspatial	Novel object recognition test	Animals familiarized with two objects (initial exposure) and then one object is subsequently changed (preference test)	Primary visual learning, short- and long-term memory	Carmo et al. (2013)
Nonspatial	Nonspatial water maze	Animals learn to escape water by swimming to a hidden platform paired with salient intramaze cue	Associative learning and memory	Truong et al. (2012)
Spatial/contextual	Morris water maze	Animals learn to escape from water by swimming to a hidden platform	Working memory	Bingham et al. (2012); Bouet et al. (2007); Carmo et al. (2013); Karhunen et al. (2003); Li et al. (2009), 2013; Roof et al. (2001); Truong et al. (2012); Winter et al. (2004); Zanier et al. (2013); Zhang et al. (2012)
Spatial/contextual	Radial arm maze	Central hub from which eight arms radiate with food rewards placed strategically at particular arms	Reference and working memory, problem solving and executive function	Chu et al. (2010); Sakai et al. (1996)
Spatial/working	Y maze	Animals placed in one arm and exploration into other arms/novel arm evaluated	Spatial and working memory	Carmo et al. (2013); Jin et al. (2010); Zhang et al. (2012)

(*Continued*)

TABLE 11.4 (Continued)

Cognitive domain	Test	Details	Evaluates	Reference
Associative learning	Passive (or active) avoidance test	Suppress normal tendency to enter darkened chamber (passive); proactive response to escape chamber; e.g., in response to foot shock (active)	Memory acquisition, short-term, long-term, and working memory and memory consolidation	Bouet et al. (2007); Hattori et al. (2000)
Learned behavior	Operant conditions (partial reinforcement)	Operant conditioning chamber (Skinner box) with retractable levers for reward/food	Assessment of time perception, behavioral flexibility	Ferrara et al. (2009); Linden et al. (2014)
	Fear conditioning (contextual and cued)	Tone paired with foot shock	Ability of emotional memory to influence behavior	Chin et al. (2013)
		Contextual fear evaluated by freezing behavior and cued testing freezing in response to without shock		
INFORMATION PROCESSING				
Reaction time	Modified operant conditioning	Modified operant conditioning chamber (Skinner box)	Latency to respond	Hoff et al. (2006)
Auditory processing	Modified prepulse inhibition/acoustic startle reflex	A large amplitude involuntary response following a startle eliciting stimulus	Acoustic discrimination	Truong et al. (2012)

both greater path length to reach the hidden platform during acquisition, indicating impaired learning, and decreased time in the quadrant previously containing the platform during the probe trial, consistent with impaired memory retention (Roof et al., 2001). Consistent with sustained impairments in cognitive function after stroke, rats subjected to transient ischemia had impaired performance during both the acquisition and probe trial phases of the MWM at 30 days (Li et al., 2013) and 12 months after injury (Karhunen et al., 2003).

Interestingly, not all studies in rats have supported striking poststroke deficits in learning and memory using the MWM (Andersen, Zimmer, & Sams-Dodd, 1999; Markgraf, Johnson, Braun, & Bickers, 1997), suggesting that there is some variability. There also appears to be a great degree of variability in findings with the MWM after stroke in mice. For example, early poststroke deficits in MWM performance have been demonstrated 5 days after transient MCAO (Truong, Venna, McCullough, & Fitch, 2012) and at 15–20 days after stroke (Gibson & Murphy, 2004), as have more sustained deficits (up to 8–12 weeks after transient MCAO (Truong et al., 2012; Zanier et al., 2013). Yet, a number of other studies have in fact failed to detect consistent poststroke cognitive impairments in mice using the same test (Bouet et al., 2007; Ji, Kronenberg, Balkaya, Färber, & Gertz, 2009; Klapdor & van der Staay, 1998; Winter, Bert, Fink, Dirnagl, & Endres, 2004; Zhang et al., 2013). Whether this variability is due to the inability of the MWM to reliably detect them remains to be established. In support of the latter, there is a great degree of variability in the protocols of MWM used by different researchers, including testing environments, indices of performance, and whether test-specific motor function is evaluated (discussed below), which are all likely to contribute. In addition, interstudy variability in the model of experimental stroke utilized, such as rodent gender or strain (e.g., Sprague-Dawley, Lister-hooded, Wistar rats; C57Bl/6J, Swiss, 129S6/Sv mice), the protocol of MCAO surgery (e.g., stroke intensity, anesthesia) or the timing of poststroke cognitive evaluation, is also likely to be central to the inconsistencies and should also be taken into consideration.

In addition to the evaluation of learning and memory, operant conditioning is becoming increasingly popular for evaluating behavioral reinforcement after experimental stroke (Chin et al., 2013; Ferrara, Bejaoui El, Seyen, Tirelli, & Plumier, 2009; Linden, Fassotte, Tirelli, Plumier, & Ferrara, 2014), and indeed different schedules of partial reinforcement, including fixed-interval and fixed ratio, have been utilized to demonstrate deficits 5 weeks after stroke in rodents (Ferrara et al., 2009; Linden et al., 2014). Deficits in reaction times have also been demonstrated. Using a modified operant conditioning chamber, rats subjected to unilateral cortical photothrombotic infarction had decreased reaction speeds that became most pronounced 3–4 weeks after injury (Hoff et al., 2006). Interestingly, auditory processing was recently evaluated following transient cerebral ischemia in mice. Truong and colleagues demonstrated "aphasia-like" deficits in rapid auditory processing 23 days after ischemia using a modified prepulse inhibition paradigm (Truong et al., 2012), supporting the presence of language-related cognitive loss after experimental stroke. At least with regard to the published literature, newly developed touchscreen technology for evaluating cognitive function (Bussey et al., 2012) is yet to be applied to experimental stroke research. This approach to cognitive testing has several advantages over existing techniques, namely its automated apparatus, the ability to test multiple cognitive domains, and, as discussed below, the potential to control for sensory-motor deficits, and will undoubtedly enhance our understanding of poststroke cognitive dysfunction.

Due to the absence of language and verbal skills in animals, evaluation of cognition is largely dependent on motor function. This is a concern for experimental studies evaluating poststroke cognition, particularly at early time points after surgery, as motor impairment is one of the major hallmarks of cerebral ischemia. To limit the influence of motor dysfunction on cognitive testing, experimenters should wait until motor function has normalized before evaluating cognition, and demonstrate this by performing appropriate controls, such as the visible platform test for the MWM. However, this may not always be possible in practice and for these reasons is not always strictly adhered to. Furthermore, it is not a particularly clinically relevant scenario to evaluate cognition in the absence of motor dysfunction, as stroke survivors frequently have coexisting motor and cognitive impairments.

The importance of performing the appropriate controls and demonstrating normal test-specific motor function for experimental poststroke cognitive testing was highlighted by a recent study. They demonstrated that while induction of MCAO increased path length and escape latency, consistent with previously defined deficits in learning and memory with the MWM, rats in fact displayed deficits in sensory motor function using the visible platform test, which could account for the changes (Bingham et al., 2012). Interestingly, the authors then went on to demonstrate that with training, both control and stroked mice demonstrated a similar rate of learning and furthermore, after achieving a set level of performance, had similar long-term memory retention providing further evidence for the importance of controlling for motor dysfunction. The ability to account for motor impairments is one of the major advantages of touchscreen cognitive testing (mentioned above). Motor dysfunction can be monitored by recording control response latencies; that is, the time taken to respond to a stimulus displayed on the screen or to retrieve a reward after a correct response, therefore allowing the cognitive component of the test to be evaluated independently. Thus, touchscreen technology may represent a particularly attractive approach for the evaluation of poststroke cognition.

AD AND STROKE

Clinical Findings

It is now widely recognized that there is an overlap of vascular pathology and AD in a large proportion of dementias. Indeed, neuropathological changes characteristic of AD (amyloid plaques and neurofibrillary tangles) and cerebral vascular lesions (lacunes, subcortical white matter lesions, microbleeds, microinfarcts) are observed together in up to 50% of cases of dementia (Jellinger, 2013). Furthermore, ischemic lesions in cortical watershed zones have also been reported in AD (Suter et al., 2002). Emerging evidence suggests that ischemia influences the expression and severity of dementia in AD. In the Rochester study, the risk of dementia, as well as the risk of AD, doubled in stroke patients over a 25-year follow-up period (Kokmen, Whisnant, O'Fallon, Chu, & Beard, 1996), indicating that the association between stroke and dementia could not be accounted for by VCI alone. Honig et al. (2003) also reported a positive association between stroke and AD, however that was strongest in the presence of known vascular risk factors including hypertension and diabetes. Moreover, disturbances of episodic memory, considered the hallmark of AD, have been linked with infarcts even after accounting for AD neuropathology (Schneider et al., 2007).

There is also evidence that ischemia may accelerate the tempo of dementia. Indeed, a history of stroke is associated with greater cognitive decline in Alzheimer's patients carrying the apolipoprotein E epsilon4 (APOE-epsilon4) gene (Helzner et al., 2009), and the presence of deep subcortical white matter lesions accelerates the rate of conversion from MCI to AD (Prasad, Wiryasaputra, Ng, & Kandiah, 2011). Even silent brain infarcts, which may be identified by neuroimaging even in patients with no history of transient ischemic attack or stroke, are reported to increase the risk of dementia, including AD (Vermeer et al., 2003). These findings may be all the more crucial when we consider the high prevalence of silent brain infarcts in the general population (Vermeer, Longstreth, & Koudstaal, 2007). Several autopsy studies of people with probable or possible dementia also show a potential link between stroke and AD. For example, in the Nun Study, the prevalence of dementia and cognitive impairment among cases that met the neuropathologic criteria for AD was much greater in those who also exhibited lacunar infarcts in the basal ganglia, thalamus, or deep white matter, than in those without vascular lesions (Snowdon et al., 1997). Troncoso et al. reported findings of the Baltimore Study of Aging, which found that symptomatic or asymptomatic macro- and microinfarcts increased the odds of dementia (Troncoso et al., 2008). Strozyk et al. found that the presence of macroscopic vascular lesions (lacunar infarcts, white matter lesions) was associated with dementia and AD (Strozyk et al., 2010).

On the other hand, some but not all (Troncoso et al., 2008) evidence indicates that AD may increase the risk of stroke (Kalaria, 2000). For example, the incidence of acute ischemic infarcts, and particularly of acute cerebral hemorrhage, is reported to be higher in patients with AD than in controls at autopsy, which may indicate a greater frequency of stroke-related deaths in the AD population (Jellinger & Mitter-Ferstl, 2003). More recently, Tolppanen et al. reported that people with AD, with no history of previous strokes, have a higher risk of hemorrhagic strokes (Tolppanen et al., 2013). Furthermore, two separate studies found that elderly individuals without previous stroke but with cognitive dysfunction, which may represent a clinical diagnosis of AD, were at increased risk of stroke (Ferrucci et al., 1996; Gale, Martyn, & Cooper, 1996).

Interactions Between Stroke and AD

The finding that most cases of dementia exhibit a mixture of vascular and Alzheimer's pathology may be coincidental, as both conditions are common in aged individuals (Hachinski, 2011). Furthermore, both processes may influence dementia expression independently, and therefore their cumulative effects on cognitive deficits would be additive. This relationship appears to be true for advanced cases of mixed dementia, with studies showing that vascular pathology influences cognition independently of Alzheimer's-associated neuropathology (Chui et al., 2006; Schneider, Wilson, Bienias, Evans, & Bennett, 2004). Notwithstanding, in early stages of disease some evidence suggests a synergistic interaction between vascular and neurodegenerative pathology. For example, moderate Alzheimer's pathology has a much greater impact on cognitive function in patients who also have ischemic lesions (e.g., lacunar infarcts, ischemic white matter lesions) (Petrovitch et al., 2005; Snowdon et al., 1997). Evidence also suggests that this effect of vascular pathology is even more prominent during earlier stages of AD (Esiri, Nagy, Smith, Barnetson, & Smith, 1999; Schneider et al., 2004). Thus, vascular brain injury may interact synergistically with Alzheimer's pathology to cause more severe cognitive dysfunction than either process alone, especially in the early stages of disease progression. Importantly, AD is associated with cerebral vascular abnormalities and, as discussed above, AD is potentially linked to an increased propensity to ischemic brain injury. Thus, it is likely that a bidirectional relationship exists. Specifically, it has been proposed that ischemic brain injury aggravates Alzheimer's-associated pathology, which in turn may make the brain more susceptible to ischemia/hypoxia (Iadecola, 2010). Indeed, as discussed below, experimental evidence suggests that hypoxia and/ischemia accelerates the accumulation of Aβ in the brain (Garcia-Alloza et al., 2011; Kitaguchi et al., 2009; Koike, Green, Blurton-Jones, & LaFerla, 2010; Li et al., 2009; Sun et al., 2006; Weller et al., 2008; Wen, Onyewuchi, Yang, Liu, & Simpkins, 2004), and there is strong evidence Aβ causes substantial cerebral vascular impairment (Iadecola, 2010).

Mechanistic Links

It is likely that numerous vital mechanisms link ischemia and AD. For example, several experimental studies have shown that cerebral ischemia (e.g., MCAO) or chronic cerebral hypoperfusion (e.g., permanent bilateral carotid artery occlusion) are associated with increased APP expression, β-secretase and γ-secretase activity, and Aβ deposition (Liu, Xing, Wang, Liu, & Li, 2012; Okamoto et al., 2012; Pimentel-Coelho & Rivest, 2012). Furthermore, elevations in Aβ production are reported to persist for many months and positively correlate with cognitive impairment (Liu et al., 2012; Zhiyou et al., 2009). Several mechanisms for the removal of Aβ from the brain have been described, however the vascular pathway is thought to be a major route (Weller et al., 2008). Experimental evidence indicates that these vascular clearance pathways are impaired in ischemic stroke (Arbel-Ornath et al., 2013), raising the possibility that stroke may reduce the clearance of Aβ. Indeed a recent study using real-time imaging in live mice demonstrated that ischemic lesions led to a rapid but transient deposition of Aβ, independent of changes in Aβ metabolism (Garcia-Alloza et al., 2011). Similarly, chronic cerebral hypoperfusion in rats impairs the clearance of Aβ from the brain (Liu et al., 2012). Thus, cerebral hypoperfusion and ischemia induced by stroke (and vascular risk factors) may accelerate and potentially initiate the neurodegeneration cascade in AD by increasing Aβ production and reducing its clearance from the brain. Studies of humans, however, have produced mixed results. Indeed, some studies (Huang et al., 2012; Jendroska et al., 1995; Ly et al., 2012), but not others (Aho, Jolkkonen, & Alafuzoff, 2006; Garcia-Alloza et al., 2011; Marchant et al., 2013; Mastaglia, Byrnes, & Johnsen, 2003), have found evidence of altered amyloid deposition in stroke patients or in demented patients with reduced cerebral blood flow, suggesting that cerebral vascular injury is at most just one of the many contributors to Aβ deposition. This latter point notwithstanding, it has been proposed that whereas stroke may promote Aβ deposition, Aβ in turn, could threaten brain perfusion and increase the propensity to ischemic damage by causing cerebral vascular dysfunction (Iadecola, 2010; Viswanathan & Greenberg, 2011). Consistent with this hypothesis, there is some evidence that AD and CAA may increase the risk of ischemic brain injury (Gurol et al., 2013; Holland et al., 2008; Kalaria, 2000; Suter et al., 2002). Furthermore, several experimental studies have also shown that mouse models of AD (transgenic mice overexpressing a mutated human APP gene) exhibit greater brain injury after focal or global cerebral ischemia (Garcia-Alloza et al., 2011; Heikkinen et al., 2014; Koistinaho et al., 2002; Zhang, Eckman, Younkin, Hsiao, & Iadecola, 1997), an effect that is in part related to vascular dysfunction (Zhang et al., 1997). Abnormalities in the white matter have been reported in AD patients and APP transgenic mice (Horsburgh et al., 2011). Oligodendrocytes, which are critical for maintaining and regulating myelination of axons, are particularly vulnerable to Aβ-induced toxicity (Horsburgh et al., 2011). Thus, the vascular dysfunction induced by Aβ, as well as direct toxic effects on oligodendrocytes, may contribute to white matter injury (Horsburgh et al., 2011).

Another vital mechanism by which ischemia may influence the expression of dementia in AD is by worsening cholinergic deficits, which are known to contribute to cognitive decline in AD (Francis, Palmer, Snape, & Wilcock, 1999). Indeed, ischemic injury is known to lead to widespread disconnection of cholinergic innervations from the basal forebrain to other parts of the brain including the neocortex (Kalaria, Akinyemi, & Ihara, 2012). Importantly, the cholinergic deficit in AD not only involves deficits in cholinergic neurotransmission but also leads to a loss of innervation of cortical blood vessels, which is an important regulator of cortical cerebral blood flow (Tong & Hamel, 1999). Thus, vascular cholinergic deficit may cause cerebral hypoperfusion and an increased propensity to ischemic injury.

Several mechanisms that may lower the threshold of neuronal death in AD, or conversely sensitize the brain to ischemic injury, have been described. Among these, oxidative stress and neuroinflammation appear to play an important role. Indeed, activated microglia, reactive astrocytes, and other markers of inflammation including complement factors and cytokines characterize the development of Alzheimer's neuropathology and ischemic lesions (Koistinaho & Koistinaho, 2005). Likewise, excess levels of toxic free radicals are prominent pathological features of both conditions (Honjo, Black, & Verhoeff, 2012).

PROSPECTS FOR TREATMENT AND PREVENTION

Despite advances in our understanding of stroke-related VCI and AD, treatment options remain limited or nonexistent. Thus, in the absence of interventions targeting the specific mechanisms of vascular and neurodegenerative cognitive impairment, therapeutic strategies are currently focused on lifestyle and risk factor control, in addition to cognitive rehabilitative or compensatory approaches.

Given the heterogeneity in vascular-related cognitive impairment, it is not surprising that the most promising interventions have been identified for focal cortical deficits. A review of evidence-based cognitive rehabilitation identified evidence for cognitive-linguistic therapies to treat aphasia and for visuo-spatial rehabilitation to treat neglect after stroke, but effective treatments in other cognitive domains were lacking (Cicerone et al., 2005). Constraint-induced treatment protocols can improve functional communication in chronic aphasia after stroke (Meinzer, Rodriguez, & Gonzalez Rothi, 2012). There is also evidence that low-frequency repetitive transcranial magnetic stimulation (rTMS) can improve language abilities in patients with chronic nonfluent aphasia. The stimulation is thought to modulate and inhibit overactivity in the right hemisphere homologous language sites, such as the inferior frontal gyrus. Significant improvements following rTMS have been reported in picture naming (Naeser et al., 2005), expressive language and auditory comprehension (Barwood et al., 2011), and semantic fluency (Szaflarski et al., 2011) of stroke patients with chronic aphasia. In stroke patients with hemispatial neglect, using prism glasses to create an optical shift of the visual field to the right has been successful. The original 1998 study included patients in the first year after stroke and revealed improvements in sensorimotor and spatial function after a single prism adaptation (Rossetti et al., 1998). Positive effects of prism adaptation have also been demonstrated in chronic stroke. In patients who were 1–7 years poststroke and had persistent neglect, an 8-week intervention with prism glasses produced improved eye movements on the neglected side and improved standing center of gravity (Shiraishi, Yamakawa, Itou, Muraki, & Asada, 2008). It should be noted, however, that these promising treatments for aphasia and neglect are yet to be tested in large randomized controlled trials, and have not been widely adopted in clinical practice. Cochrane reviews have revealed the paucity of studies in poststroke cognitive rehabilitation for both memory deficits (Nair das & Lincoln, 2007) and attention deficits (Loetscher & Lincoln, 2013). There has been one randomized control that reported benefits of "attention process training" in 78 stroke patients with attention deficits, with the positive effects maintained at 6-month follow-up (Barker-Collo et al., 2009).

Treatment attempts aimed at reducing generalized VCI have had limited success. In the PROGRESS trial, active blood pressure lowering reduced the risk of cognitive decline (Tzourio et al., 2003), but this positive result was not reproduced in the PRoFESS trial (Diener et al., 2008). There is some evidence that escitalopram (normally used to treat depression) and rivastigmine (typically used in AD) can benefit cognitive function in stroke patients (Jorge, Acion, Moser, Adams, & Robinson, 2010; Narasimhalu et al., 2010). Yet these studies featured improvement in one or two cognitive measures in the context of multiple other cognitive outcomes that were not affected, and baseline performance was not balanced between the groups. Increasing physical activity is another potential treatment; it has been shown to improve cognitive performance in the cognitively impaired (Heyn, Abreu, & Ottenbacher, 2004) and those at risk of AD (Lautenschlager et al., 2008). The relationship between physical and mental activity may also apply to stroke patients, but there is little empirical evidence to date. In a systematic review, only 12 studies were identified that employed physical activity interventions in stroke and had cognitive outcomes (Cumming, Tyedin, Churilov, Morris, & Bernhardt, 2012). One of the few that had cognition as a primary focus found stroke patients exposed to an 8-week exercise program had improved information processing speed relative to control patients (Quaney et al., 2009).

Another approach to addressing VCI is to compensate for reduced cognitive abilities rather than attempting to restore them. One example is the use of an electronic paging system, which was found to be effective in compensating for everyday memory and planning problems after brain injury (Wilson, Emslie, Quirk, & Evans, 2001) and after stroke (Fish, Manly, Emslie, & Evans, 2008). Compensatory strategies can also be internally generated. Rather than attempt to restore reaction time to a normal speed, Winkens and colleagues (Winkens, Van Heugten, Wade, Habets, & Fasotti, 2009) introduced a "time pressure management" strategy: stroke patients were taught to compensate for mental slowness in real-life tasks by reorganizing the execution of subtasks that had a time pressure component. The treatment group significantly outperformed controls on speed of performance on everyday tasks at 3-month follow-up.

The age-old assertion that prevention is better than a cure is particularly relevant to VCI and AD, where improving cognitive deficits is so challenging. The revelation that vascular risk factors, particularly in mid-life, confer higher risk for AD (Kivipelto et al., 2001) opens the way for modification of these risk factors and reduction in AD prevalence. Potential prevention strategies include control of blood pressure, cholesterol, diabetes, and obesity; increased cognitive, physical and social activity; and greater intake of antioxidants and polyunsaturated fatty acids (Middleton & Yaffe, 2009). Barnes and Yaffe (Barnes & Yaffe, 2011) calculated that a 25% reduction in seven potentially modifiable risk factors (diabetes, mid-life hypertension, mid-life obesity, smoking, depression, cognitive inactivity, low education) could prevent three million AD cases worldwide. Such preventive measures can be applied to the whole population, but people with stroke and VCI should be a strong target given their elevated vascular risk profile.

CONCLUSIONS

Stroke and dementia, and their pathophysiological mechanisms, have been known for many years. However, we have only recently discovered the close links that exist between them. Stroke often leads to cognitive impairments and frequently also to dementia, particularly in recurrent stroke. But the relationship is bigger than that—vascular risk factors have also been shown to increase the risk of AD. Now that we have epidemiological evidence supporting the vascular contribution to degenerative dementia, we now need to work toward increasing our understanding of the mechanisms mediating this relationship. Animal studies have provided useful insights, and future work will undoubtedly increase our understanding of this complex interaction. Ultimately, as rates of stroke and dementia continue to climb, the development of more targeted strategies will work toward alleviating the cognitive fallout of this devastating disease.

References

Abeelen den, A. S. S. M. -V., Lagro, J., van Beek, A. H. E. A., & Claassen, J. A. H. R. (2014). Impaired cerebral autoregulation and vasomotor reactivity in sporadic Alzheimer's disease. *Current Alzheimer Research, 11*, 11–17.

Adair, J. C., et al. (2004). Measurement of gelatinase B (MMP-9) in the cerebrospinal fluid of patients with vascular dementia and Alzheimer disease. *Stroke, 35*, e159–e162.

Aho, K., et al. (1980). Cerebrovascular disease in the community: Results of a WHO collaborative study. *Bulletin of the World Health Organization, 58*, 113–130.

Aho, L., Jolkkonen, J., & Alafuzoff, I. (2006). Beta-amyloid aggregation in human brains with cerebrovascular lesions. *Stroke, 37*, 2940–2945.

Allen, C. L., & Bayraktutan, U. (2009). Oxidative stress and its role in the pathogenesis of ischaemic stroke. *International Journal of Stroke, 4*, 461–470.

Andersen, M. B., Zimmer, J., & Sams-Dodd, F. (1999). Specific behavioral effects related to age and cerebral ischemia in rats. *Pharmacology, Biochemistry, and Behavior, 62*, 673–682.

Arbel-Ornath, M., et al. (2013). Interstitial fluid drainage is impaired in ischemic stroke and Alzheimer's disease mouse models. *Acta Neuropathologica, 126*, 353–364.

Astrup, J., Siesjo, B. K., & Symon, L. (1981). Thresholds in cerebral ischemia—The ischemic penumbra. *Stroke, 12*, 723–725.

Attems, J., Jellinger, K., Thal, D. R., & Van Nostrand, W. (2011). Review: Sporadic cerebral amyloid angiopathy. *Neuropathology and Applied Neurobiology, 37*, 75–93.

Back, S. A., et al. (2011). White matter lesions defined by diffusion tensor imaging in older adults. *Annals of Neurology, 70*, 465–476.

Barker-Collo, S., Feigin, V. L., Parag, V., Lawes, C. M. M., & Senior, H. (2010). Auckland stroke outcomes study. Part 2: Cognition and functional outcomes 5 years poststroke. *Neurology, 75*, 1608–1616.

Barker-Collo, S. L., et al. (2009). Reducing attention deficits after stroke using attention process training: A randomized controlled trial. *Stroke, 40*, 3293–3298.

Barnes, D. E., & Yaffe, K. (2011). The projected effect of risk factor reduction on Alzheimer's disease prevalence. *Lancet Neurology, 10*, 819–828.

Barwood, C. H. S., et al. (2011). Improved language performance subsequent to low-frequency rTMS in patients with chronic non-fluent aphasia post-stroke. *European Journal of Neurology, 18*, 935–943.

Beach, T. G., et al. (2007). Circle of Willis atherosclerosis: Association with Alzheimer's disease, neuritic plaques and neurofibrillary tangles. *Acta Neuropathologica, 113*, 13–21.

Bell, R. D., et al. (2009). SRF and myocardin regulate LRP-mediated amyloid-beta clearance in brain vascular cells. *Nature Cell Biology, 11*, 143–153.

Bell, R. D., et al. (2010). Pericytes control key neurovascular functions and neuronal phenotype in the adult brain and during brain aging. *Neuron, 68*, 409–427.

Bingham, D., Martin, S. J., Macrae, I. M., & Carswell, H. V. O. (2012). Watermaze performance after middle cerebral artery occlusion in the rat: The role of sensorimotor versus memory impairments. *Journal of Cerebral Blood Flow and Metabolism, 32*, 989–999.

Blasi, F., et al. (2014). Recognition memory impairments after subcortical white matter stroke in mice. *Stroke, 45*, 1468–1473.

Bouet, V., et al. (2007). Sensorimotor and cognitive deficits after transient middle cerebral artery occlusion in the mouse. *Experimental Neurology, 203*, 555–567.

Brown, W. R., Moody, D. M., Thore, C. R., Challa, V. R., & Anstrom, J. A. (2007). Vascular dementia in leukoaraiosis may be a consequence of capillary loss not only in the lesions, but in normal-appearing white matter and cortex as well. *Journal of the Neurological Sciences, 257*, 62–66.

Bussey, T. J., et al. (2012). New translational assays for preclinical modelling of cognition in schizophrenia: The touchscreen testing method for mice and rats. *Neuropharmacology, 62*, 1191–1203.

Candelario-Jalil, E., et al. (2011). Matrix metalloproteinases are associated with increased blood-brain barrier opening in vascular cognitive impairment. *Stroke, 42*, 1345–1350.

Carmo, M. R. S., et al. (2013). ATP P2Y1 receptors control cognitive deficits and neurotoxicity but not glial modifications induced by brain ischemia in mice. *European Journal of Neuroscience, 39*, 614–622.

Casado, A., Encarnación López-Fernández, M., Concepción Casado, M., & La Torre de, R. (2008). Lipid peroxidation and antioxidant enzyme activities in vascular and Alzheimer dementias. *Neurochemical Research, 33*, 450–458.

Chabriat, H., Joutel, A., Dichgans, M., Tournier-Lasserve, E., & Bousser, M.-G. (2009). CADASIL. *The Lancet Neurology, 8*, 643–653.

Chin, Y., et al. (2013). Involvement of glial P2Y$_1$ receptors in cognitive deficit after focal cerebral stroke in a rodent model. *Journal of Neuroinflammation, 10*, 95.

Chisari, M., Merlo, S., Sortino, M. A., & Salomone, S. (2010). Long-term incubation with beta-amyloid peptides impairs endothelium-dependent vasodilatation in isolated rat basilar artery. *Pharmacology Research, 61*, 157–161.

Chrissobolis, S., Miller, A. A., Drummond, G. R., Kemp-Harper, B. K., & Sobey, C. G. (2011). Oxidative stress and endothelial dysfunction in cerebrovascular disease. *Frontiers in Bioscience, 16*, 1733–1745.

Christie, R., Yamada, M., Moskowitz, M., & Hyman, B. (2001). Structural and functional disruption of vascular smooth muscle cells in a transgenic mouse model of amyloid angiopathy. *The American Journal of Pathology, 158*, 1065–1071.

Chu, L. -S., et al. (2010). Minocycline inhibits 5-lipoxygenase expression and accelerates functional recovery in chronic phase of focal cerebral ischemia in rats. *Life Sciences, 86*, 170–177.

Chui, H. C., et al. (2006). Cognitive impact of subcortical vascular and Alzheimer's disease pathology. *Annals of Neurology, 60*, 677–687.

Cicerone, K. D., et al. (2005). Evidence-based cognitive rehabilitation: Updated review of the literature from 1998 through 2002. *Archives of Physical Medicine and Rehabilitation, 86*, 1681–1692.

Claassen, J. A., & Zhang, R. (2011). Cerebral autoregulation in Alzheimer's disease. *Journal of Cerebral Blood Flow and Metabolism, 31,* 1572–1577.

Claesson, L., Lindén, T., Skoog, I., & Blomstrand, C. (2005). Cognitive impairment after stroke – impact on activities of daily living and costs of care for elderly people. *Cerebrovascular Diseases, 19,* 102–109.

Cordova, C. A., Jackson, D., Langdon, K. D., Hewlett, K. A., & Corbett, D. (2014). Impaired executive function following ischemic stroke in the rat medial prefrontal cortex. *Behavioural Brain Research, 258,* 106–111.

Cumming, T. B., Marshall, R. S., & Lazar, R. M. (2013). Stroke, cognitive deficits, and rehabilitation: Still an incomplete picture. *International Journal of Stroke, 8,* 38–45.

Cumming, T. B., Tyedin, K., Churilov, L., Morris, M. E., & Bernhardt, J. (2012). The effect of physical activity on cognitive function after stroke: A systematic review. *International Psychogeriatrics, 24,* 557–567.

del Zoppo, G. J. (2010). The neurovascular unit in the setting of stroke. *Journal of Internal Medicine, 267,* 156–171.

DeVries, A. C., Nelson, R. J., Traystman, R. J., & Hurn, P. D. (2001). Cognitive and behavioral assessment in experimental stroke research: Will it prove useful? *Neuroscience and Biobehavioral Reviews, 25,* 325–342.

Diener, H. -C., et al. (2008). Effects of aspirin plus extended-release dipyridamole versus clopidogrel and telmisartan on disability and cognitive function after recurrent stroke in patients with ischaemic stroke in the Prevention Regimen for Effectively Avoiding Second Strokes (PRoFESS) trial: A double-blind, active and placebo-controlled study. *Lancet Neurology, 7,* 875–884.

Dirnagl, U. (2012). Pathobiology of injury after stroke: The neurovascular unit and beyond. *Annals of the New York Academy of Sciences, 1268,* 21–25.

Dirnagl, U., Iadecola, C., & Moskowitz, M. A. (1999). Pathobiology of ischaemic stroke: An integrated view. *Trends in Neurosciences, 22,* 391–397.

Douiri, A., Rudd, A. G., & Wolfe, C. D. A. (2013). Prevalence of poststroke cognitive impairment South London stroke register 1995–2010. *Stroke, 44,* 138–145.

Dubois, B., et al. (2007). Research criteria for the diagnosis of Alzheimer's disease: Revising the NINCDS-ADRDA criteria. *Lancet Neurology, 6,* 734–746.

Durukan, A., & Tatlisumak, T. (2007). Acute ischemic stroke: Overview of major experimental rodent models, pathophysiology, and therapy of focal cerebral ischemia. *Pharmacology Biochemistry and Behavior, 87,* 179–197.

Eisele, Y. S., et al. (2010). Peripherally applied Abeta-containing inoculates induce cerebral beta-amyloidosis. *Science, 330,* 980–982.

Esiri, M. M., Nagy, Z., Smith, M. Z., Barnetson, L., & Smith, A. D. (1999). Cerebrovascular disease and threshold for dementia in the early stages of Alzheimer's disease. *Lancet, 354,* 919–920.

Farrall, A. J., & Wardlaw, J. M. (2009). Blood–brain barrier: Ageing and microvascular disease—Systematic review and meta-analysis. *Neurobiology of Aging, 30,* 337–352.

Fernando, M. S., et al. (2006). White matter lesions in an unselected cohort of the elderly: Molecular pathology suggests origin from chronic hypoperfusion injury. *Stroke, 37,* 1391–1398.

Ferrara, A., Bejaoui El, S., Seyen, S., Tirelli, E., & Plumier, J. -C. (2009). The usefulness of operant conditioning procedures to assess long-lasting deficits following transient focal ischemia in mice. *Behavioural Brain Research, 205,* 525–534.

Ferri, C. P., et al. (2005). Global prevalence of dementia: A Delphi consensus study. *Lancet, 366,* 2112–2117.

Ferrucci, L., et al. (1996). Cognitive impairment and risk of stroke in the older population. *Journal of the American Geriatrics Society, 44,* 237–241.

Fischer, V. W., Siddiqi, A., & Yusufaly, Y. (1990). Altered angioarchitecture in selected areas of brains with Alzheimer's disease. *Acta Neuropathologica, 79,* 672–679.

Fish, J., Manly, T., Emslie, H., & Evans, J. J. (2008). Compensatory strategies for acquired disorders of memory and planning: Differential effects of a paging system for patients with brain injury of traumatic versus cerebrovascular aetiology. *Journal of Neurology, Neurosurgery, and Psychiatry, 79*, 930–935.

Francis, P. T., Palmer, A. M., Snape, M., & Wilcock, G. K. (1999). The cholinergic hypothesis of Alzheimer's disease: A review of progress. *Journal of Neurology, Neurosurgery, and Psychiatry, 66*, 137–147.

Gale, C. R., Martyn, C. N., & Cooper, C. (1996). Cognitive impairment and mortality in a cohort of elderly people. *BMJ, 312*, 608–611.

Garcia-Alloza, M., et al. (2011). Cerebrovascular lesions induce transient β-amyloid deposition. *Brain, 134*, 3697–3707.

Gibson, C. L., & Murphy, S. P. (2004). Progesterone enhances functional recovery after middle cerebral artery occlusion in male mice. *Journal of Cerebral Blood Flow & Metabolism, 24*, 805–813.

Girouard, H., & Iadecola, C. (2006). Neurovascular coupling in the normal brain and in hypertension, stroke, and Alzheimer disease. *Journal of Applied Physiology, 100*, 328–335.

Go, A. S., Mozaffarian, D., Roger, V. L., & Benjamin, E. J. (2014). Heart disease and stroke statistics—2014 update: A report from the American Heart Association. *Circulation, 129*, e28–e292.

Gold, G., Giannakopoulos, P., Herrmann, F. R., Bouras, C., & Kövari, E. (2007). Identification of Alzheimer and vascular lesion thresholds for mixed dementia. *Brain, 130*, 2830–2836.

Gorelick, P. B., et al. (2011). Vascular contributions to cognitive impairment and dementia: A statement for healthcare professionals from the American Heart Association/American Stroke Association. *Stroke, 42*, 2672–2713.

Gurol, M. E., et al. (2013). Cerebral amyloid angiopathy burden associated with leukoaraiosis: A positron emission tomography/magnetic resonance imaging study. *Annals of Neurology, 73*, 529–536.

Hachinski, V. (2011). Stroke and Alzheimer disease: Fellow travelers or partners in crime? *Archives of Neurology, 68*, 797–798.

Hachinski, V., et al. (2006). Welcome harmonizations. *Stroke, 37* 2197–2197.

Hachinski, V. C., & Bowler, J. V. (1993). Vascular dementia. *Neurology, 43*, 2159–2160 – Author reply 2160–2161.

Hamel, E., Nicolakakis, N., Aboulkassim, T., Ongali, B., & Tong, X. -K. (2008). Oxidative stress and cerebrovascular dysfunction in mouse models of Alzheimer's disease. *Experimental Physiology, 93*, 116–120.

Hartz, A. M. S., et al. (2012). Amyloid-β contributes to blood–brain barrier leakage in transgenic human amyloid precursor protein mice and in humans with cerebral amyloid angiopathy. *Stroke, 43*, 514–523.

Hattori, K., et al. (2000). Cognitive deficits after focal cerebral ischemia in mice. *Stroke, 31*, 1939–1944.

Heikkinen, R., et al. (2014). Susceptibility to focal and global brain ischemia of Alzheimer mice displaying aβ deposits: Effect of immunoglobulin. *Aging and Disease, 5*, 76–87.

Helzner, E. P., et al. (2009). Contribution of vascular risk factors to the progression in Alzheimer disease. *Archives of Neurology, 66*, 343–348.

Herzig, M. C., Nostrand, W. E., & Jucker, M. (2006). Mechanism of cerebral β-amyloid angiopathy: Murine and cellular models. *Brain Pathology, 16*, 40–54.

Herzig, M. C., et al. (2004). Abeta is targeted to the vasculature in a mouse model of hereditary cerebral hemorrhage with amyloidosis. *Nature Neuroscience, 7*, 954–960.

Heyn, P., Abreu, B. C., & Ottenbacher, K. J. (2004). The effects of exercise training on elderly persons with cognitive impairment and dementia: A meta-analysis. *Archives of Physical Medicine and Rehabilitation, 85*, 1694–1704.

Hochstenbach, J., Mulder, T., van Limbeek, J., Donders, R., & Schoonderwaldt, H. (1998). Cognitive decline following stroke: A comprehensive study of cognitive decline following stroke. *Journal of Clinical and Experimental Neuropsychology, 20,* 503–517.

Hoff, E. I., Blokland, A., Rutten, K., Steinbusch, H. W. M., & van Oostenbrugge, R. J. (2006). Dissociable effects in reaction time performance after unilateral cerebral infarction: A comparison between the left and right frontal cortices in rats. *Brain Research, 1069,* 182–189.

Hofman, A., et al. (1997). Atherosclerosis, apolipoprotein E, and prevalence of dementia and Alzheimer's disease in the Rotterdam study. *The Lancet, 349,* 151–154.

Holland, C. M., et al. (2008). Spatial distribution of white-matter hyperintensities in Alzheimer disease, cerebral amyloid angiopathy, and healthy aging. *Stroke, 39,* 1127–1133.

Honig, L. S., et al. (2003). Stroke and the risk of Alzheimer disease. *Archives of Neurology, 60,* 1707–1712.

Honjo, K., Black, S. E., & Verhoeff, N. P. L. G. (2012). Alzheimer's disease, cerebrovascular disease, and the β-amyloid cascade. *The Canadian Journal of Neurological Sciences, 39,* 712–728.

Horsburgh, K., et al. (2011). Axon-glial disruption: The link between vascular disease and Alzheimer's disease? *Biochemical Society Transactions, 39,* 881–885.

Hossmann, K.-A. (2008). Cerebral ischemia: Models, methods and outcomes. *Neuropharmacology, 55,* 257–270.

Howells, D. W., et al. (2010). Different strokes for different folks: The rich diversity of animal models of focal cerebral ischemia. *Journal of Cerebral Blood Flow and Metabolism, 30,* 1412–1431.

Huang, K.-L., et al. (2012). Amyloid deposition after cerebral hypoperfusion: Evidenced on [(18)F]AV-45 positron emission tomography. *Journal of the Neurological Sciences, 319,* 124–129.

Hunter, A. J., Mackay, K. B., & Rogers, D. C. (1998). To what extent have functional studies of ischaemia in animals been useful in the assessment of potential neuroprotective agents? *Trends in Pharmacological Sciences, 19,* 59–66.

Iadecola, C. (2010). The overlap between neurodegenerative and vascular factors in the pathogenesis of dementia. *Acta Neuropathologica, 120,* 287–296.

Iadecola, C. (2013). The pathobiology of vascular dementia. *Neuron, 80,* 844–866.

Iadecola, C., & Anrather, J. (2011). The immunology of stroke: From mechanisms to translation. *Nature Medicine, 17,* 796–808.

Iadecola, C., & Davisson, R. L. (2008). Hypertension and cerebrovascular dysfunction. *Cell Metabolism, 7,* 476–484.

Iadecola, C., et al. (1999). SOD1 rescues cerebral endothelial dysfunction in mice overexpressing amyloid precursor protein. *Nature Neuroscience, 2,* 157–161.

Ihara, M., & Tomimoto, H. (2011). Lessons from a mouse model characterizing features of vascular cognitive impairment with white matter changes. *Journal of Aging Research, 2011,* 978761.

Ihara, Y., et al. (1997). Free radicals and superoxide dismutase in blood of patients with Alzheimer's disease and vascular dementia. *Journal of the Neurological Sciences, 153,* 76–81.

Ingles, J. L., Wentzel, C., Fisk, J. D., & Rockwood, K. (2002). Neuropsychological predictors of incident dementia in patients with vascular cognitive impairment, without dementia. *Stroke, 33,* 1999–2002.

Jackman, K., & Iadecola, C. (2013). Neurovascular regulation in the ischemic brain. *Antioxidants & Redox Signaling,* in press.

Jackman, K. A., et al. (2009). Reduction of cerebral infarct volume by apocynin requires pretreatment and is absent in Nox2-deficient mice. *British Journal of Pharmacology, 156,* 680–688.

Jackman, K., Kunz, A., & Iadecola, C. (2011). Modeling focal cerebral ischemia *in vivo*. *Methods in Molecular Biology, 793*, 195–209.

Jellinger, K. A. (2013). Pathology and pathogenesis of vascular cognitive impairment-a critical update. *Frontiers in Aging Neuroscience, 5*, 17.

Jellinger, K. A., & Attems, J. (2010). Prevalence of dementia disorders in the oldest-old: An autopsy study. *Acta Neuropathologica, 119*, 421–433.

Jellinger, K. A., & Mitter-Ferstl, E. (2003). The impact of cerebrovascular lesions in Alzheimer disease—A comparative autopsy study. *Journal of Neurology, 250*, 1050–1055.

Jendroska, K., et al. (1995). Ischemic stress induces deposition of amyloid beta immunoreactivity in human brain. *Acta Neuropathologica, 90*, 461–466.

Ji, S., Kronenberg, G., Balkaya, M., Färber, K., & Gertz, K. (2009). Acute neuroprotection by pioglitazone after mild brain ischemia without effect on long-term outcome. *Experimental Neurology, 216*, 321–328.

Jin, K., et al. (2010). Transplantation of human neural precursor cells in Matrigel scaffolding improves outcome from focal cerebral ischemia after delayed postischemic treatment in rats. *Journal of Cerebral Blood Flow and Metabolism, 30*, 534–544.

Jin, Y. -P., Di Legge, S., Ostbye, T., Feightner, J. W., & Hachinski, V. (2006). The reciprocal risks of stroke and cognitive impairment in an elderly population. *Alzheimer's & Dementia, 2*, 171–178.

Jiwa, N. S., Garrard, P., & Hainsworth, A. H. (2010). Experimental models of vascular dementia and vascular cognitive impairment: A systematic review. *Journal of Neurochemistry, 115*, 814–828.

Johnson, K. A., & Albert, M. S. (2000). Perfusion abnormalities in prodromal AD. *Neurobiology of Aging, 21*, 289–292.

Johnson, K. A., et al. (1998). Preclinical prediction of Alzheimer's disease using SPECT. *Neurology, 50*, 1563–1571.

Jokinen, H., et al. (2011). Incident lacunes influence cognitive decline: The LADIS study. *Neurology, 76*, 1872–1878.

Jorge, R. E., Acion, L., Moser, D., Adams, H. P., & Robinson, R. G. (2010). Escitalopram and enhancement of cognitive recovery following stroke. *Archives of General Psychiatry, 67*, 187–196.

Joutel, A., et al. (2010). Cerebrovascular dysfunction and microcirculation rarefaction precede white matter lesions in a mouse genetic model of cerebral ischemic small vessel disease. *Journal of Clinical Investigation, 120*, 433–445.

Kahles, T., & Brandes, R. P. (2012). NADPH oxidases as therapeutic targets in ischemic stroke. *Cellular and Molecular Life Sciences, 69*, 2345–2363.

Kalaria, R. N. (1997). Cerebrovascular degeneration is related to amyloid-beta protein deposition in Alzheimer's disease. *Annals of the New York Academy of Sciences, 826*, 263–271.

Kalaria, R. N. (2000). The role of cerebral ischemia in Alzheimer's disease. *Neurobiology of Aging, 21*, 321–330.

Kalaria, R. N., Akinyemi, R., & Ihara, M. (2012). Does vascular pathology contribute to Alzheimer changes? *Journal of the Neurological Sciences, 322*, 141–147.

Karhunen, H., et al. (2003). Long-term functional consequences of transient occlusion of the middle cerebral artery in rats: A 1-year follow-up of the development of epileptogenesis and memory impairment in relation to sensorimotor deficits. *Epilepsy Research, 54*, 1–10.

Kim, H. A., et al. (2012). Vascular cognitive impairment and Alzheimer's disease: Role of cerebral hypoperfusion and oxidative stress. *Naunyn-Schmiedeberg's Archives of Pharmacology, 385*, 953–959.

Kitaguchi, H., et al. (2009). Chronic cerebral hypoperfusion accelerates amyloid beta deposition in APPSwInd transgenic mice. *Brain Research, 1294*, 202–210.

Kivipelto, M., et al. (2001). Midlife vascular risk factors and Alzheimer's disease in later life: Longitudinal, population based study. *BMJ, 322*, 1447–1451.

Klapdor, K., & van der Staay, F. J. (1998). Repeated acquisition of a spatial navigation task in mice: Effects of spacing of trials and of unilateral middle cerebral artery occlusion. *Physiology & Behavior, 63*, 903–909.

Knopman, D. S., et al. (2009). Association of prior stroke with cognitive function and cognitive impairment: A population-based study. *Archives of Neurology, 66*, 614–619.

Kobari, M., Meyer, J. S., Ichijo, M., & Oravez, W. T. (1990). Leukoaraiosis: Correlation of MR and CT findings with blood flow, atrophy, and cognition. *AJNR American Journal of Neuroradiology, 11*, 273–281.

Koike, M. A., Green, K. N., Blurton-Jones, M., & LaFerla, F. M. (2010). Oligemic hypoperfusion differentially affects tau and amyloid-β. *The American Journal of Pathology, 177*, 300–310.

Koistinaho, M., & Koistinaho, J. (2005). Interactions between Alzheimer's disease and cerebral ischemia—Focus on inflammation. *Brain Research Reviews, 48*, 240–250.

Koistinaho, M., et al. (2002). Amyloid precursor protein transgenic mice that harbor diffuse a deposits but do not form plaques show increased ischemic vulnerability: Role of inflammation. *Proceedings of the National Academy of Sciences, 99*, 1610–1615.

Kokmen, E., Whisnant, J. P., O'Fallon, W. M., Chu, C. P., & Beard, C. M. (1996). Dementia after ischemic stroke: A population-based study in Rochester, Minnesota (1960–1984). *Neurology, 46*, 154–159.

Krafft, P. R., et al. (2012). Etiology of stroke and choice of models. *International Journal of Stroke, 7*, 398–406.

Kunz, A., Anrather, J., Zhou, P., Orio, M., & Iadecola, C. (2006). Cyclooxygenase-2 does not contribute to postischemic production of reactive oxygen species. *Journal of Cerebral Blood Flow & Metabolism, 27*, 545–551.

Kunz, A., & Iadecola, C. (2009). Chapter 14 Cerebral vascular dysregulation in the ischemic brain. *Handbook of Clinical Neurology, 92*, 283–305.

la Torre de, J. C. (2002). Alzheimer disease as a vascular disorder: Nosological evidence. *Stroke, 33*, 1152–1162.

Lautenschlager, N. T., et al. (2008). Effect of physical activity on cognitive function in older adults at risk for Alzheimer disease: A randomized trial. *JAMA, 300*, 1027–1037.

Lee, J. Y., et al. (2014). Fluoxetine inhibits transient global ischemia-induced hippocampal neuronal death and memory impairment by preventing blood-brain barrier disruption. *Neuropharmacology, 79*, 161–171.

Leys, D., Hénon, H., Mackowiak-Cordoliani, M. -A., & Pasquier, F. (2005). Poststroke dementia. *The Lancet Neurology, 4*, 752–759.

Li, L., et al. (2009). Hypoxia increases Abeta generation by altering beta- and gamma-cleavage of APP. *Neurobiology of Aging, 30*, 1091–1098.

Li, W., et al. (2013). Transient focal cerebral ischemia induces long-term cognitive function deficit in an experimental ischemic stroke model. *Neurobiology of Disease, 59*, 18–25.

Li, W. -L., et al. (2009). Chronic fluoxetine treatment improves ischemia-induced spatial cognitive deficits through increasing hippocampal neurogenesis after stroke. *Journal of Neuroscience Research, 87*, 112–122.

Linden, J., Fassotte, L., Tirelli, E., Plumier, J. -C., & Ferrara, A. (2014). Assessment of behavioral flexibility after middle cerebral artery occlusion in mice. *Behavioural Brain Research, 258*, 127–137.

Linden, T., Skoog, I., Fagerberg, B., Steen, B., & Blomstrand, C. (2004). Cognitive impairment and dementia 20 months after stroke. *Neuroepidemiology, 23*, 45–52.

Liu, C., et al. (2007). Baicalein improves cognitive deficits induced by chronic cerebral hypoperfusion in rats. *Pharmacology Biochemistry and Behavior, 86*, 423–430.

Liu, H., Xing, A., Wang, X., Liu, G., & Li, L. (2012). Regulation of β-amyloid level in the brain of rats with cerebrovascular hypoperfusion. *Neurobiology of Aging, 33* 826. e31–42.

Loetscher, T., & Lincoln, N. B. (2013). Cognitive rehabilitation for attention deficits following stroke. *Cochrane Database of Systematic Reviews, 5,* CD002842.

Looi, J. C., & Sachdev, P. S. (1999). Differentiation of vascular dementia from AD on neuropsychological tests. *Neurology, 53,* 670–678.

Lorenzl, S., et al. (2003). Increased plasma levels of matrix metalloproteinase-9 in patients with Alzheimer's disease. *Neurochemistry International, 43,* 191–196.

Luckhaus, C., et al. (2008). Detection of changed regional cerebral blood flow in mild cognitive impairment and early Alzheimer's dementia by perfusion-weighted magnetic resonance imaging. *NeuroImage, 40,* 495–503.

Ly, J. V., et al. (2012). Subacute ischemic stroke is associated with focal 11C PiB positron emission tomography retention but not with global neocortical Aβ deposition. *Stroke, 43,* 1341–1346.

Marchant, N. L., et al. (2013). The aging brain and cognition: Contribution of vascular injury and aβ to mild cognitive dysfunction. *JAMA Neurology, 70,* 488–495.

Marco, S., & Skaper, S. D. (2006). Amyloid β-peptide1–42 alters tight junction protein distribution and expression in brain microvessel endothelial cells. *Neuroscience Letters, 401,* 219–224.

Markgraf, C. G., Johnson, M. P., Braun, D. L., & Bickers, M. V. (1997). Behavioral recovery patterns in rats receiving the NMDA receptor antagonist MDL 100,453 immediately post-stroke. *Pharmacology, Biochemistry, and Behavior, 56,* 391–397.

Markus, H. S., Lythgoe, D. J., Ostegaard, L., O'Sullivan, M., & Williams, S. C. (2000). Reduced cerebral blood flow in white matter in ischaemic leukoaraiosis demonstrated using quantitative exogenous contrast based perfusion MRI. *Journal of Neurology, Neurosurgery, and Psychiatry, 69,* 48–53.

Marstrand, J. R., et al. (2002). Cerebral perfusion and cerebrovascular reactivity are reduced in white matter hyperintensities. *Stroke, 33,* 972–976.

Mastaglia, F. L., Byrnes, M. L., & Johnsen, R. D. (2003). Prevalence of cerebral vascular amyloid-β deposition and stroke in an aging Australian population: A postmortem study. *Journal of Clinical Neuroscience, 10,* 186–189.

Matsushita, K., et al. (1994). Periventricular white matter lucency and cerebral blood flow autoregulation in hypertensive patients. *Hypertension, 23,* 565–568.

Meinzer, M., Rodriguez, A. D., & Gonzalez Rothi, L. J. (2012). First decade of research on constrained-induced treatment approaches for aphasia rehabilitation. *Archives of Physical Medicine and Rehabilitation, 93,* S35–S45.

Middleton, L. E., & Yaffe, K. (2009). Promising strategies for the prevention of dementia. *Archives of Neurology, 66,* 1210–1215.

Moskowitz, M. A., Lo, E. H., & Iadecola, C. (2010). The science of stroke: Mechanisms in search of treatments. *Neuron, 67,* 181–198.

Naeser, M. A., et al. (2005). Improved picture naming in chronic aphasia after TMS to part of right Broca's area: An open-protocol study. *Brain and Language, 93,* 95–105.

Nair das, R., & Lincoln, N. (2007). Cognitive rehabilitation for memory deficits following stroke. *Cochrane Database of Systematic Reviews, 3,* CD002293.

Narasimhalu, K., et al. (2010). A randomized controlled trial of rivastigmine in patients with cognitive impairment no dementia because of cerebrovascular disease. *Acta Neurologica Scandinavica, 121,* 217–224.

Nicolakakis, N., et al. (2008). Complete rescue of cerebrovascular function in aged Alzheimer's disease transgenic mice by antioxidants and pioglitazone, a peroxisome proliferator-activated receptor gamma agonist. *Journal of Neuroscience, 28,* 9287–9296.

Niwa, K., Carlson, G. A., & Iadecola, C. (2000). Exogenous A beta1-40 reproduces cerebrovascular alterations resulting from amyloid precursor protein overexpression in mice. *Journal of Cerebral Blood Flow & Metabolism, 20,* 1659–1668.

Niwa, K., Kazama, K., Younkin, S. G., Carlson, G. A., & Iadecola, C. (2002). Alterations in cerebral blood flow and glucose utilization in mice overexpressing the amyloid precursor protein. *Neurobiology of Disease, 9,* 61–68.

Niwa, K., et al. (2000). Abeta 1-40-related reduction in functional hyperemia in mouse neo-cortex during somatosensory activation. *Proceedings of the National Academy of Sciences of the United States of America, 97*, 9735–9740.

Niwa, K., et al. (2002). Cerebrovascular autoregulation is profoundly impaired in mice over-expressing amyloid precursor protein. *American Journal of Physiology, 283*, H315–H323.

Okamoto, Y., et al. (2012). Cerebral hypoperfusion accelerates cerebral amyloid angiopathy and promotes cortical microinfarcts. *Acta Neuropathologica, 123*, 381–394.

O'Sullivan, M., et al. (2002). Patterns of cerebral blood flow reduction in patients with ischemic leukoaraiosis. *Neurology, 59*, 321–326.

Paris, D., et al. (2003). Vasoactive effects of A beta in isolated human cerebrovessels and in a transgenic mouse model of Alzheimer's disease: Role of inflammation. *Neurological Research, 25*, 642–651.

Park, L., et al. (2004). Abeta-induced vascular oxidative stress and attenuation of functional hyperemia in mouse somatosensory cortex. *Journal of Cerebral Blood Flow & Metabolism, 24*, 334–342.

Park, L., et al. (2005). NADPH-oxidase-derived reactive oxygen species mediate the cer-ebrovascular dysfunction induced by the amyloid beta peptide. *Journal of Neuroscience, 25*, 1769–1777.

Park, L., et al. (2008). Nox2-derived radicals contribute to neurovascular and behavioral dysfunction in mice overexpressing the amyloid precursor protein. *Proceedings of the National Academy of Sciences, 105*, 1347–1352.

Park, L., et al. (2014). Age-dependent neurovascular dysfunction and damage in a mouse model of cerebral amyloid angiopathy. *Stroke, 45*, 1815–1821.

Pasquini, M., Leys, D., Rousseaux, M., Pasquier, F., & Hénon, H. (2007). Influence of cogni-tive impairment on the institutionalisation rate 3 years after a stroke. *Journal of Neurology, Neurosurgery, and Psychiatry, 78*, 56–59.

Paul, J., Strickland, S., & Melchor, J. P. (2007). Fibrin deposition accelerates neurovascular damage and neuroinflammation in mouse models of Alzheimer's disease. *Journal of Experimental Medicine, 204*, 1999–2008.

Pendlebury, S. T., & Rothwell, P. M. (2009). Prevalence, incidence, and factors associated with pre-stroke and post-stroke dementia: A systematic review and meta-analysis. *Lancet Neurology, 8*, 1006–1018.

Petrovitch, H., et al. (2005). AD lesions and infarcts in demented and non-demented Japanese-American men. *Annals of Neurology, 57*, 98–103.

Phan, T. G., Donnan, G. A., Srikanth, V., Chen, J., & Reutens, D. C. (2009). Heterogeneity in infarct patterns and clinical outcomes following internal carotid artery occlusion. *Archives of Neurology, 66*, 1523–1528.

Pimentel-Coelho, P. M., & Rivest, S. (2012). The early contribution of cerebrovascular factors to the pathogenesis of Alzheimer's disease. *European Journal of Neuroscience, 35*, 1917–1937.

Prasad, K., Wiryasaputra, L., Ng, A., & Kandiah, N. (2011). White matter disease indepen-dently predicts progression from mild cognitive impairment to Alzheimer's disease in a clinic cohort. *Dementia and Geriatric Cognitive Disorders, 31*, 431–434.

Quaney, B. M., et al. (2009). Aerobic exercise improves cognition and motor function post-stroke. *Neurorehabilitation and Neural Repair, 23*, 879–885.

Querfurth, H. W., & LaFerla, F. M. (2010). Alzheimer's disease. *The New England Journal of Medicine, 362*, 329–344.

Reed, B. R., et al. (2007). Profiles of neuropsychological impairment in autopsy-defined Alzheimer's disease and cerebrovascular disease. *Brain, 130*, 731–739.

Rodriguiz, R. M., & Wetsel, W. C. (2006). In E. D. Levin & J. J. Buccafusco (Eds.), *Chapter 12: Assessments of cognitive deficits in mutant mice. Animal Models of Cognitive Impairment*. Boca Raton, FL: CRC Press.

Roher, A. E., et al. (2011). Intracranial atherosclerosis as a contributing factor to Alzheimer's disease dementia. *Alzheimer's & Dementia, 7*, 436–444.

Roof, R. L., Schielke, G. P., Ren, X., & Hall, E. D. (2001). A comparison of long-term functional outcome after 2 middle cerebral artery occlusion models in rats. *Stroke, 32,* 2648–2657.

Rossetti, Y., et al. (1998). Prism adaptation to a rightward optical deviation rehabilitates left hemispatial neglect. *Nature, 395,* 166–169.

Rouhl, R. P. W., et al. (2012). Vascular inflammation in cerebral small vessel disease. *Neurobiology of Aging, 33,* 1800–1806.

Sahathevan, R., Brodtmann, A., & Donnan, G. A. (2011). Dementia, stroke, and vascular risk factors; a review. *International Journal of Stroke, 7,* 61–73.

Sakai, N., et al. (1996). Behavioral studies on rats with transient cerebral ischemia induced by occlusion of the middle cerebral artery. *Behavioural Brain Research, 77,* 181–188.

Sarazin, M., de Souza, L. C., Lehéricy, S., & Dubois, B. (2012). Clinical and research diagnostic criteria for Alzheimer's disease. *Neuroimaging Clinics of North America, 22,* 23–32, viii.

Schneider, J. A., Boyle, P. A., Arvanitakis, Z., Bienias, J. L., & Bennett, D. A. (2007). Subcortical infarcts, Alzheimer's disease pathology, and memory function in older persons. *Annals of Neurology, 62,* 59–66.

Schneider, J. A., Wilson, R. S., Bienias, J. L., Evans, D. A., & Bennett, D. A. (2004). Cerebral infarctions and the likelihood of dementia from Alzheimer disease pathology. *Neurology, 62,* 1148–1155.

Schneider, J. A., et al. (2003). Relation of cerebral infarctions to dementia and cognitive function in older persons. *Neurology, 60,* 1082–1088.

Selnes, O. A., & Vinters, H. V. (2006). Vascular cognitive impairment. *Nature Clinical Practice Neurology, 2,* 538–547.

Shen, J., et al. (2011). Interrupted reperfusion reduces the activation of NADPH oxidase after cerebral I/R injury. *Free Radical Biology & Medicine, 50,* 1780–1786.

Shimohama, S., et al. (2000). Activation of NADPH oxidase in Alzheimer's disease brains. *Biochemical and Biophysical Research Communications, 273,* 5–9.

Shin, H. K., et al. (2007). Age-dependent cerebrovascular dysfunction in a transgenic mouse model of cerebral amyloid angiopathy. *Brain, 130,* 2310–2319.

Shiraishi, H., Yamakawa, Y., Itou, A., Muraki, T., & Asada, T. (2008). Long-term effects of prism adaptation on chronic neglect after stroke. *NeuroRehabilitation, 23,* 137–151.

Snowdon, D. A., et al. (1997). Brain infarction and the clinical expression of Alzheimer disease. The Nun study. *JAMA, 277,* 813–817.

Srikanth, V. K., Quinn, S. J., Donnan, G. A., Saling, M. M., & Thrift, A. G. (2006). Long-term cognitive transitions, rates of cognitive change, and predictors of incident dementia in a population-based first-ever stroke cohort. *Stroke, 37,* 2479–2483.

Srikanth, V. K., et al. (2003). Increased risk of cognitive impairment 3 months after mild to moderate first-ever stroke: A community-based prospective study of nonaphasic English-speaking survivors. *Stroke, 34,* 1136–1143.

Stanimirovic, D. B., & Friedman, A. (2012). Pathophysiology of the neurovascular unit: Disease cause or consequence? *Journal of Cerebral Blood Flow & Metabolism, 32,* 1207–1221.

Strozyk, D., et al. (2010). Contribution of vascular pathology to the clinical expression of dementia. *Neurobiology of Aging, 31,* 1710–1720.

Sun, X., et al. (2006). Hypoxia facilitates Alzheimer's disease pathogenesis by up-regulating BACE1 gene expression. *Proceedings of the National Academy of Sciences of the United States of America, 103,* 18727–18732.

Sutcliffe, J. G., Hedlund, P. B., Thomas, E. A., Bloom, F. E., & Hilbush, B. S. (2011). Peripheral reduction of β-amyloid is sufficient to reduce brain β-amyloid: Implications for Alzheimer's disease. *Journal of Neuroscience Research, 89,* 808–814.

Suter, O. -C., et al. (2002). Cerebral hypoperfusion generates cortical watershed microinfarcts in Alzheimer disease. *Stroke, 33,* 1986–1992.

Szaflarski, J. P., et al. (2011). Excitatory repetitive transcranial magnetic stimulation induces improvements in chronic post-stroke aphasia. *Medical Science Monitor, 17,* CR132–CR139.

Taheri, S., et al. (2011). Blood-brain barrier permeability abnormalities in vascular cognitive impairment. *Stroke, 42*, 2158–2163.

Tajiri, N., et al. (2013). *In vivo* animal stroke models: A rationale for rodent and non-human primate models. *Translational Stroke Research, 4*, 308–321.

Tatemichi, T. K., et al. (1994a). Dementia after stroke is a predictor of long-term survival. *Stroke, 25*, 1915–1919.

Tatemichi, T. K., et al. (1994b). Cognitive impairment after stroke: Frequency, patterns, and relationship to functional abilities. *Journal of Neurology, Neurosurgery, and Psychiatry, 57*, 202–207.

Taylor, T. N., et al. (1996). Lifetime cost of stroke in the United States. *Stroke, 27*, 1459–1466.

Thal, D. R., Grinberg, L. T., & Attems, J. (2012). Vascular dementia: Different forms of vessel disorders contribute to the development of dementia in the elderly brain. *Experimental Gerontology, 47*, 816–824.

Tolppanen, A. -M., et al. (2013). Incidence of stroke in people with Alzheimer disease: A national register-based approach. *Neurology, 80*, 353–358.

Tong, X. -K., & Hamel, E. (1999). Regional cholinergic denervation of cortical microvessels and nitric oxide synthase-containing neurons in Alzheimer's disease. *Neuroscience, 92*, 163–175.

Tong, X. -K., et al. (2009). Simvastatin improves cerebrovascular function and counters soluble amyloid-beta, inflammation and oxidative stress in aged APP mice. *Neurobiology of Disease, 35*, 406–414.

Troncoso, J. C., et al. (2008). Effect of infarcts on dementia in the Baltimore longitudinal study of aging. *Annals of Neurology, 64*, 168–176.

Truong, D. T., Venna, V. R., McCullough, L. D., & Fitch, R. H. (2012). Deficits in auditory, cognitive, and motor processing following reversible middle cerebral artery occlusion in mice. *Experimental Neurology, 238*, 114–121.

Tzourio, C., et al. (2003). Effects of blood pressure lowering with perindopril and indapamide therapy on dementia and cognitive decline in patients with cerebrovascular disease. *Archives of Internal Medicine, 163*, 1069–1075.

Ueno, K. -I., et al. (2002). Behavioural and pharmacological relevance of stroke-prone spontaneously hypertensive rats as an animal model of a developmental disorder. *Behavioural Pharmacology, 13*, 1–13.

van Beek, A. H. E. A., Sijbesma, J. C., Jansen, R. W. M. M., Rikkert, M. G. M. O., & Claassen, J. A. H. R. (2010). Cortical oxygen supply during postural hypotension is further decreased in Alzheimer's disease, but unrelated to cholinesterase-inhibitor use. *Journal of Alzheimer's Disease, 21*, 519–526.

van Beek, A. H. E. A., Sijbesma, J. C., Olde Rikkert, M. G. M., & Claassen, J. A. H. R. (2010). Galantamine does not cause aggravated orthostatic hypotension in people with Alzheimer's disease. *Journal of the American Geriatrics Society, 58*, 409–410.

van Swieten, J. C., et al. (1991). Periventricular lesions in the white matter on magnetic resonance imaging in the elderly. A morphometric correlation with arteriolosclerosis and dilated perivascular spaces. *Brain, 114*(Pt 2), 761–774.

Velasquez, F. G., & Kotarek, J. A. (2008). Soluble aggregates of the amyloid-β protein selectively stimulate permeability in human brain microvascular endothelial monolayers. *Journal of Neurochemistry, 107*, 466–477.

Verbeek, M. M., de Waal, R. M., Schipper, J. J., & Van Nostrand, W. E. (1997). Rapid degeneration of cultured human brain pericytes by amyloid beta protein. *Journal of Neurochemistry, 68*, 1135–1141.

Vermeer, S. E., Longstreth, W. T., Jr, & Koudstaal, P. J. (2007). Silent brain infarcts: A systematic review. *The Lancet Neurology, 6*, 611–619.

Vermeer, S. E., et al. (2003). Silent brain infarcts and the risk of dementia and cognitive decline. *The New England Journal of Medicine, 348*, 1215–1222.

Viswanathan, A., & Greenberg, S. M. (2011). Cerebral amyloid angiopathy in the elderly. *Annals of Neurology, 70,* 871–880.

Wakita, H., et al. (2008). Mucosal tolerization to E-selectin protects against memory dysfunction and white matter damage in a vascular cognitive impairment model. *Journal of Cerebral Blood Flow & Metabolism, 28,* 341–353.

Walder, C. E., et al. (1997). Ischemic stroke injury is reduced in mice lacking a functional NADPH oxidase. *Stroke, 28,* 2252–2258.

Wang, J., Zhang, H. Y., & Tang, X. C. (2010). Huperzine a improves chronic inflammation and cognitive decline in rats with cerebral hypoperfusion. *Journal of Neuroscience Research, 88,* 807–815.

Wang, M., et al. (2012). Cognitive deficits and delayed neuronal loss in a mouse model of multiple microinfarcts. *Journal of Neuroscience, 32,* 17948–17960.

Warlow, C., Sudlow, C., Dennis, M., Wardlaw, J., & Sandercock, P. (2003). Stroke. *Lancet, 362,* 1211–1224.

Washida, K., et al. (2010). Nonhypotensive dose of telmisartan attenuates cognitive impairment partially due to peroxisome proliferator-activated receptor-gamma activation in mice with chronic cerebral hypoperfusion. *Stroke, 41,* 1798–1806.

Weller, R. O., Subash, M., Preston, S. D., Mazanti, I., & Carare, R. O. (2008). Perivascular drainage of amyloid-beta peptides from the brain and its failure in cerebral amyloid angiopathy and Alzheimer's disease. *Brain Pathology, 18,* 253–266.

Wen, Y., Onyewuchi, O., Yang, S., Liu, R., & Simpkins, J. W. (2004). Increased beta-secretase activity and expression in rats following transient cerebral ischemia. *Brain Research, 1009,* 1–8.

Wilson, B. A., Emslie, H. C., Quirk, K., & Evans, J. J. (2001). Reducing everyday memory and planning problems by means of a paging system: A randomised control crossover study. *Journal of Neurology, Neurosurgery, and Psychiatry, 70,* 477–482.

Winkens, I., Van Heugten, C. M., Wade, D. T., Habets, E. J., & Fasotti, L. (2009). Efficacy of time pressure management in stroke patients with slowed information processing: A randomized controlled trial. *Archives of Physical Medicine and Rehabilitation, 90,* 1672–1679.

Winter, B., Bert, B., Fink, H., Dirnagl, U., & Endres, M. (2004). Dysexecutive syndrome after mild cerebral ischemia? Mice learn normally but have deficits in strategy switching. *Stroke, 35,* 191–195.

Xu, Y., et al. (2010). Green tea polyphenols inhibit cognitive impairment induced by chronic cerebral hypoperfusion via modulating oxidative stress. *The Journal of Nutritional Biochemistry, 21,* 741–748.

Yamamoto, Y., et al. (2009). Nobiletin improves brain ischemia-induced learning and memory deficits through stimulation of CaMKII and CREB phosphorylation. *Brain Research, 1295,* 218–229.

Yang, S. -H., et al. (2006). Endovascular middle cerebral artery occlusion in rats as a model for studying vascular dementia. *Age (Dordr), 28,* 297–307.

Yang, Y., & Rosenberg, G. A. (2011). Blood-brain barrier breakdown in acute and chronic cerebrovascular disease. *Stroke, 42,* 3323–3328.

Zanier, E. R., et al. (2013). Six-month ischemic mice show sensorimotor and cognitive deficits associated with brain atrophy and axonal disorganization. *CNS Neuroscience & Therapeutics, 19,* 695–704.

Zarow, C., Barron, E., Chui, H. C., & Perlmutter, L. S. (1997). Vascular basement membrane pathology and Alzheimer's disease. *Annals of the New York Academy of Sciences, 826,* 147–159.

Zhang, F., Eckman, C., Younkin, S., Hsiao, K. K., & Iadecola, C. (1997). Increased susceptibility to ischemic brain damage in transgenic mice overexpressing the amyloid precursor protein. *Journal of Neuroscience, 17,* 7655–7661.

Zhang, X., Huang, G., Liu, H., Chang, H., & Wilson, J. X. (2012). Folic acid enhances Notch signaling, hippocampal neurogenesis, and cognitive function in a rat model of cerebral ischemia. *Nutritional Neuroscience, 15,* 55–61.

Zhang, X., Yeung, P. K. K., McAlonan, G. M., Chung, S. S. M., & Chung, S. K. (2013). Transgenic mice over-expressing endothelial endothelin-1 show cognitive deficit with blood–brain barrier breakdown after transient ischemia with long-term reperfusion. *Neurobiology of Learning and Memory, 101,* 46–54.

Zhiyou, C., et al. (2009). Upregulation of BACE1 and beta-amyloid protein mediated by chronic cerebral hypoperfusion contributes to cognitive impairment and pathogenesis of Alzheimer's disease. *Neurochemical Research, 34,* 1226–1235.

Zipser, B. D., et al. (2007). Microvascular injury and blood-brain barrier leakage in Alzheimer's disease. *Neurobiology of Aging, 28,* 977–986.

Zlokovic, B. V. (2008). The blood-brain barrier in health and chronic neurodegenerative disorders. *Neuron, 57,* 178–201.

Cerebral Innate Immunity: A New Conceptual Framework for Alzheimer's Disease

David Gate and Terrence Town

Department of Physiology and Biophysics, Zilkha Neurogenetic Institute, Keck School of Medicine, University of Southern California, Los Angeles, CA, United States

OUTLINE

Cerebral Innate Immunity in Alzheimer Pathoetiology   362

An Historical Perspective: Innate Immunity in AD   364

Aβ Immunotherapy and the Role of Mononuclear Phagocytes in Amyloid Plaque Clearance   365

Novel Strategies for Targeting Peripheral Macrophages Versus Brain-Resident Microglia   367

Targeting Cardinal Anti-Inflammatory Cytokines to Restrict Cerebral Amyloidosis   369

Chemokines Recruit Monocytes to Aβ Plaques and Dying Neurons   371

Blocking Inflammatory ILs 12 and 23 Prevents Plaque Buildup in Transgenic Mice   372

PPARγ Agonists Reduce Inflammation While Boosting Microglial Aβ Uptake   373

Inflammasome Activation in AD   375

Beclin 1 Regulates Microglial Phagocytosis and May Be Impaired in AD   375

Genes, Environment and Alzheimer's Disease.
DOI: http://dx.doi.org/10.1016/B978-0-12-802851-3.00012-7

361

A New Generation of AD Pharmacotherapeutics Targeting
Innate Immunity   376

Concluding Remarks   378

Acknowledgements   380

References   380

CEREBRAL INNATE IMMUNITY IN ALZHEIMER PATHOETIOLOGY

Typical pathological changes that occur in Alzheimer's disease (AD) include brain atrophy, particularly in the cerebral cortex and hippocampus, and large-scale neuronal cell loss; however, neuronal loss is neither specific to Alzheimer-type dementia nor indicative of it. AD is histopathologically defined on the basis of two hallmarks: deposition of amyloid-β (Aβ) peptides as β-amyloid plaques and formation of neurofibrillary tangles (NFTs; chiefly comprised of the hyperphosphorylated microtubule-associated protein *tau*) (Selkoe, 2001). In his original description of the histopathology of AD in 1907, Alois Alzheimer identified a third pathological feature: gliosis, or inflammation of the brain's support cells known as glia. For decades after Alzheimer's original case report, gliosis was viewed as an epiphenomenon with little relevance to AD etiology. Yet, it is now accepted that gliosis and brain inflammation are integral to AD pathoetiology, and that targeting inflammation and the immune system represent promising therapeutic approaches for AD (Wyss-Coray, 2006). Gliosis involves the proliferation or hypertrophy of several different types of glial cells, namely astrocytes and microglia. Astrocytosis is typically found surrounding amyloid plaques in the AD brain, perhaps in an attempt to "wall off" these aggregated structures. Microglia, the resident macrophages of the central nervous system (CNS), are found in and around amyloid plaques in an "activated" state; yet, they fail to effectively phagocytose and clear Aβ deposits (Akiyama et al., 2000; McGeer & McGeer, 2003; Patel et al., 2005). Microglia have been shown to secrete increased levels of pro-inflammatory cytokines such as tumor necrosis factor-α (TNF-α), granulocyte-macrophage colony stimulating factor, and interleukin-1β (IL-1β), which correlate with amyloid plaque load in transgenic mouse models of AD (Akiyama et al., 2000; Benzing et al., 1999; McGeer & McGeer, 2003; McGowan et al., 1999; Patel et al., 2005; Rogers et al., 1996). The complex interplay among inflammatory mediators in the context of AD is shown in Figure 12.1.

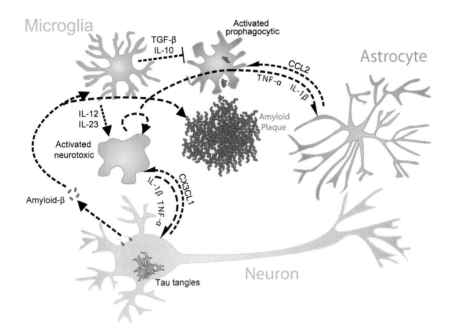

FIGURE 12.1   Inflammatory factors that play roles in Alzheimer's disease. Amyloid-β peptides aggregate, form dense β-amyloid plaques, and activate microglia. Microglia transition from quiescent to neurotoxic activation via IL-12/IL-23 signaling. On the other hand, phagocytic microglia are beneficial vis-à-vis amyloid clearance, but are inhibited by the cardinal anti-inflammatory cytokines, TGF-β and IL-10. Neurotoxic microglia secrete factors such as IL-1β and TNF-α that can damage neurons at supraphysiologic abundance. Factors secreted by neurons, such as CX3CL1, enlist activated microglia to perpetrate neurotoxicity. Cytokines and chemokines secreted by microglia stimulate a response by astrocytes that can act on microglia, driving recruitment and modulating neuroinflammation. Neurons, in response to intracellular injury from tau tangles or neurotoxic cytokines, can die and thereby influence the inflammatory phenotypes of microglia and astrocytes. IL, interleukin; TGF, transforming growth factor; TNF, tumor necrosis factor. *Source: Adapted from Doty, Guillot-Sestier, & Town (2015).*

The context of microglial activation is key for determining whether these cells play pathological or beneficial roles in AD evolution (Gate, Rezai-Zadeh, Jodry, Rentsendorj, & Town, 2010; Rezai-Zadeh, Gate, Gowing, & Town, 2011). Although the relationship between brain inflammatory responses (including microgliosis) and AD-like pathology is not so straightforward, immunization of transgenic mouse models of AD or AD patients with Aβ peptide results in mitigation of cerebral amyloidosis (McGeer & McGeer, 2004; Nicoll et al., 2003; Schenk et al., 1999), a beneficial effect that has been attributed to phagocytic activation of microglia (Bard et al., 2000, 2003; Nicoll et al., 2003). Increased expression of inflammatory mediators in postmortem AD patient brains has also been reported, supporting the notion that neuroinflammation may serve to exacerbate AD pathogenesis. Further, epidemiologic findings have shown that nonsteroidal anti-inflammatory drugs (NSAIDs) reduce AD risk by up to 50% (in t' Veld et al., 2001; Szekely et al., 2004; Zandi et al., 2002). Additionally, treatment of transgenic mouse models of cerebral amyloidosis with NSAIDs reduces microglial activation, which strongly correlates with diminished β-amyloid deposition (Lim et al., 2000; Lim, Chu, et al., 2001; Lim, Yang, et al., 2001). However, the causal relationship between brain inflammation and AD pathology is not so straightforward, as NSAIDs for AD primary prevention failed to produce a positive signal, and may even hasten conversion of mild cognitively impaired individuals to AD (Group et al., 2008; Leoutsakos, Muthen, Breitner, Lyketsos, & Team, 2012). Furthermore, NSAID treatment effects differ at various stages of disease (Breitner et al., 2011).

Interestingly, recent AD genomewide association studies have implicated innate immune genes as major risk loci, including *TREM2*, a gene encoding a microglial receptor (Brouwers et al., 2012; Jonsson et al., 2013; Zhang et al., 2013). These findings have had a watershed effect on the field by calling our attention to the importance of cerebral innate immunity in AD. Defining a cerebral innate immune phenotype that is beneficial in the context of AD will be fundamental toward understanding this debilitating disease. This chapter will focus on various approaches taken by geneticists and cellular and molecular biologists utilizing human cases and animal models to interrogate the intricate relationship between innate immunity and AD pathology.

AN HISTORICAL PERSPECTIVE: INNATE IMMUNITY IN AD

Studies dating back to the late 1980s have reported on the intricate spatial association between microglia and senile plaques (Aβ deposits) in brains of AD patients (Dickson et al., 1988; Haga, Akai, & Ishii, 1989). These initial observations sparked the first of many debates concerning the role of microglia in AD pathology. The presence of microglia in close vicinity of senile plaques supports the notion that these cells actively phagocytose and clear amyloid (Itagaki, McGeer, Akiyama, Zhu, & Selkoe, 1989). Yet, some have challenged that microglial cells are actually the source of amyloid fibrils, and actually seed Aβ plaques (Roher, Gray, & Paula-Barbosa, 1988; Wegiel & Wisniewski, 1990; Wisniewski, Vorbrodt, Wegiel, Morys, & Lossinsky, 1990; Wisniewski, Wegiel, Wang, Kujawa, & Lach, 1989). With evidence building for neuroinflammatory signals being key players in AD pathology, the question of cause and effect is paramount

(Wyss-Coray, 2006). Does neuroinflammation drive, or is it driven by senile plaques and NFTs? Perhaps even more importantly, can "good" neuroinflammation be harnessed to attenuate AD pathology?

Neuropathologist Henryk Wisniewski offered a tantalizing explanation for beneficial versus detrimental roles of brain macrophages. He proposed dichotomous functions of brain-resident microglia and blood-borne macrophages in the context of AD pathology. Relying on immunoelectron microscopy, Wisniewski reported that, in the rare instance of comorbidity of stroke with AD, Aβ fibrils could be identified in lysosomal compartments of what he perceived to be brain-infiltrating macrophages (Wisniewski, Barcikowska, & Kida, 1991). By contrast, Wisniewski did not observe amyloid fibrils in lysosomes of microglia associated with classical amyloid plaques (Wisniewski et al., 1991). Wisniewski's findings were later corroborated by an additional study, which also noted that Aβ deposits were cleared in macrophage-dense brain regions of AD patients with comorbid stroke (Akiyama et al., 1996). These early experiments raised the possibility that infiltrating peripheral macrophages were more adept at phagocytosing and clearing Aβ deposits than brain-resident microglia. However, the qualitative nature of morphological assessments made in these studies left many open questions, including (i) do microglia and macrophages have differing Aβ phagocytic capabilities, (ii) could blood-derived macrophages restrict cerebral amyloidosis, and (iii) why were infiltrating macrophages excluded from the AD brain in the absence of stroke comorbidity? Answers to these questions are beginning to be revealed, as geneticists and cellular and molecular biologists have started to flesh out how the innate immune system interacts with AD pathology.

Aβ IMMUNOTHERAPY AND THE ROLE OF MONONUCLEAR PHAGOCYTES IN AMYLOID PLAQUE CLEARANCE

Although the studies by Wisniewski and others highlighted the potential of mononuclear phagocytes to clear cerebral amyloid, they did not address whether the immune system could be harnessed to militate against AD. Dale Schenk and colleagues brought the field significantly closer to definitively answering this question in the late 1990s. By actively immunizing AD model mice with $Aβ_{1-42}$ peptide plus adjuvant, these researchers were able to prevent cerebral amyloid accumulation as well as clear existing amyloid plaques (Schenk et al., 1999). These findings gave rise to the field of Aβ immunotherapy, and led to the Elan/Wyeth active Aβ vaccine early developmental clinical trials. Unfortunately, the phase IIa Aβ vaccine trial (AN-1792) was terminated when approximately

6% of vaccinated AD patients developed aseptic meningoencephalitis. This condition was thought to have arisen from brain-infiltrating Aβ-specific autoaggressive T-cells (Town, 2009; Town, Tan, Flavell, & Mullan, 2005). Shortly after the initial preclinical AD mouse model vaccination studies, Schenk's immunotherapeutic approach was advanced by others who passively transferred antibodies raised against the $Aβ_{1-42}$ peptide to transgenic AD model mice. Similar to Schenk's original active Aβ vaccine, these researchers found that passive immunization effectively attenuated AD-like pathology in transgenic mice (Bard et al., 2000). Following the suspension of AN-1792, passive Aβ immunotherapy became particularly attractive, because it circumvents an active Aβ immune response driven by T-cells. Additional studies have advanced the field of Aβ immunotherapy by employing Aβ peptide fragments that lack cytotoxic T-cell epitopes or by altering vaccine parameters including adjuvants, carrier proteins, and route of administration for active immunization and by other immunotherapy strategies, including DNA-based vaccines (Agadjanyan et al., 2005; Fu, Liu, Frost, & Lemere, 2010; Klyubin et al., 2005; Movsesyan et al., 2008; Nikolic et al., 2007; Obregon et al., 2008; Petrushina et al., 2007; Town, 2009; Vasilevko, Xu, Previti, Van Nostrand, & Cribbs, 2007).

Interestingly, active and passive Aβ immunization approaches are thought to work by the same mechanism of action: antibody-mediated Aβ phagocytosis. It has been demonstrated that Aβ antibodies stimulate microglia to phagocytose and clear amyloid present *ex vivo* in AD brain sections via engagement of IgG-recognizing microglial Fc receptors (Bard et al., 2000). Furthermore, *in vivo* multiphoton microscopic analysis of passively immunized AD transgenic mice has shown Aβ antibody-mediated disruption of amyloid plaques in real-time, an affect purportedly mediated by amyloid phagocytes (Bacskai et al., 2001, 2003). Despite restriction of cerebral amyloid after Aβ immunization, there appear to be negative side effects to this immunotherapeutic strategy. Subsequent studies demonstrated that AD model mice either actively or passively vaccinated with Aβ antibodies develop cerebrovascular microhemorrhage (Pfeifer et al., 2002; Wilcock et al., 2004). Additionally, a number of reports have linked Aβ immunotherapy to exacerbated deposition of β-amyloid in cerebrovessels, known as cerebral amyloid angiopathy (CAA), which is believed to result from overly exuberant antibody-mediated amyloid clearance from the brain parenchyma into the vasculature. Accordingly, while microglia seem to clear antibody-opsonized Aβ deposits, this mechanism does not operate in isolation, and may occur with detrimental side effects. Future Aβ vaccination strategies will need to be designed to increase amyloid clearance while circumventing potentially unsafe side effects such as cerebral microhemorrhage or CAA.

NOVEL STRATEGIES FOR TARGETING PERIPHERAL MACROPHAGES VERSUS BRAIN-RESIDENT MICROGLIA

Irradiation chimeras are a commonly used approach to study infiltration of bone marrow-derived cells into the healthy and diseased CNS. In this method, recipient animals are exposed to a lethal dose of irradiation, which effectively kills all bone marrow cells (but not microglia, which are radio-resistant), and then transplanted with bone marrow expressing a tracer such as green fluorescent protein (GFP). The advantage of this method is the ability to distinguish bone marrow-derived macrophages from resident CNS microglia, as activated microglia and macrophages share many of the same cell surface receptors and signaling proteins. Initial studies reported that unmanipulated, healthy animals demonstrated physiologic renewal of perivascular and meningeal macrophages by GFP[+] bone marrow-derived cells, but rarely showed GFP[+] resident parenchymal microglia (Bacskai et al., 2003; Kennedy & Abkowitz, 1997, 1998). However, it was later observed that quantitatively minor renewal of parenchymal microglia by peripherally-derived, GFP-expressing donor bone marrow cells occurred in the normal CNS (Priller et al., 2001). Using bone-marrow chimeras, it has been demonstrated that enhanced CNS engraftment of mononuclear cells of bone marrow origin occurs in acute models of brain injury (Priller et al., 2001). Succeeding studies utilized bone-marrow chimeras to investigate the possibility of mononuclear cell infiltration in mouse models of chronic neurodegenerative diseases including experimental autoimmune encephalomyelitis, amyotrophic lateral sclerosis, Parkinson's disease, prion-associated disease, and AD (Kang & Rivest, 2007; Lewis, Solomon, Rossi, & Krieger, 2009; Mildner et al., 2007; Ponomarev, Shriver, Maresz, & Dittel, 2005).

Two independent studies have confirmed infiltration of peripherally-derived mononuclear cells in the CNS of cerebral amyloidosis mouse models (Simard, Soulet, Gowing, Julien, & Rivest, 2006; Stalder et al., 2005). While both studies showed a clear spatial association between bone marrow-derived cells and amyloid plaques, electron microscopy analysis of GFP-labeled amoeboid cells with macrophage-like ultrastructural features did not disclose evidence of amyloid phagocytosis (Stalder et al., 2005). By contrast, others have shown presence of amyloid deposits in GFP[+] infiltrating mononuclear phagocytes (Simard et al., 2006). In the latter study, the authors crossed a doubly-transgenic mouse model of AD with CD11b-tyrosine kinase (TK[mut30]) transgenic mice. These animals have proliferating CD11b[+] cells that can be specifically ablated by administration of the antiviral drug ganciclovir. This next-generation model system has demonstrated the importance of proliferating major histocompatibility class II$^+$ peripheral mononuclear phagocytes in restricting β-amyloid plaques (Simard et al., 2006).

More recent results have shown a significant reduction in Aβ deposit accumulation in another double transgenic mouse model of AD transplanted with wild-type bone marrow. Reduction in cerebral Aβ load was further improved by grafting AD transgenic animals with mononuclear cells deficient in E prostanoid receptor subtype 2 (EP2) (Keene et al., 2010). Increased efficiency of Aβ clearance was attributed to the higher phagocytic capacity of EP2 knockout mononuclear cells (Keene et al., 2010; Shie, Montine, Breyer, & Montine, 2005). The beneficial effect of reconstituting the hematopoietic compartment of AD transgenic mice with wild-type bone marrow has recently been substantiated. In a similar study, engraftment of wild-type bone marrow mitigated cerebral amyloid load and correlated with decreased expression of the canonical pro-inflammatory cytokines TNF-α and IL-1β, and with increased abundance of the cardinal anti-inflammatory molecules IL-4 and IL-10 (Zhu et al., 2011).

While radiation chimerism has been widely used to address key questions related to immunity in the normal and diseased CNS, a number of technical concerns have been raised regarding this experimental paradigm (Ajami, Bennett, Krieger, Tetzlaff, & Rossi, 2007; Mildner et al., 2007). The use of alternative strategies for investigating the role of peripheral macrophages in Aβ clearance and AD pathology will undoubtedly lead to a clarification of the role of these cells. In fact, a novel study demonstrated that depletion of perivascular macrophages by liposome-encapsulated clodronate exacerbated CAA as evidenced by increased deposition of cerebral Aβ peptides in leptomeningeal and cortical blood vessels in a mouse model of cerebral amyloid deposition. Those authors demonstrated the converse effect on CAA when increasing turnover of perivascular macrophages following chitin administration (Hawkes & McLaurin, 2009), again indicative of a direct relationship between peripheral macrophages and cerebral amyloid.

As mentioned, the role of brain-resident microglia in clearing Aβ in AD patients and in mouse models of the disease is a controversial area. Work from the group of Mathias Jücker has suggested that brain-resident microglia do not play a significant role in Aβ clearance. These researchers utilized a ganciclovir-activated CD11b-TK suicide gene approach to selectively ablate microglia for 2–4 weeks in two different mouse models of cerebral amyloidosis (APP/PS1 and APP23). After selective ablation of microglia in these mice, no differences were detected between treated and untreated animals with respect to total Aβ burden, plaque morphology, or distribution of cerebral Aβ deposits (Grathwohl et al., 2009). However, it remains possible that ablation of brain macrophages for only 2–4 weeks is not long enough to observe altered Aβ plaque dynamics in these transgenic mice. Unfortunately, due to toxicity associated with administration of ganciclovir to CD11b-TK transgenic mice, ablating cerebral macro-phages for longer than 4 weeks is not feasible, thus limiting the extent of the conclusions that could be drawn.

TARGETING CARDINAL ANTI-INFLAMMATORY CYTOKINES TO RESTRICT CEREBRAL AMYLOIDOSIS

While myriad immune molecules exist to keep inflammation in check, combined activity of the key immunoregulatory cytokines transforming growth factor-β (TGF-β) and IL-10 ensures that inflammatory responses are carried out in an acute, controlled fashion without becoming chronic and inducing excessive tissue damage (Li & Flavell, 2008). The fact that prolonged, low-level activation of brain inflammatory processes occurs and is likely pathogenic in AD (Selkoe, 2001; Wyss-Coray et al., 1997) raises the question of whether these immunomodulatory pathways are dysregulated. Interestingly, when a transgenic mouse line overexpressing TGF-β1 (Wyss-Coray et al., 1995) was crossed with the PDAPP mouse model of brain amyloid deposition (Games et al., 1995), vascular β-amyloid deposits were markedly accelerated, suggesting an amyloidogenic role for TGF-β1 *in vivo* (Wyss-Coray et al., 1997). However, despite increases in vascular amyloid deposits, these bigenic animals had less parenchymal amyloid burden, suggesting opposing effects of TGF-β1 on vascular versus parenchymal amyloid deposition (Wyss-Coray et al., 2001). These results were supported by findings from the laboratories of Tony Wyss-Coray and Lennart Mücke in AD patients' brains. Those investigators found that TGF-β1 mRNA levels were approximately threefold increased in AD patient frontal cortex ver-sus nondemented controls, and a positive correlation was noted between TGF-β1 mRNA levels and vascular amyloid deposits (Wyss-Coray et al., 1997), while an inverse association was found between TGF-β1 mRNA and parenchymal Aβ deposits (Wyss-Coray et al., 2001). Despite increased TGF-β1 mRNA levels in AD frontal cortex, protein levels of TGFβRII in human AD brains are, on average, only half as abundant versus nondemented controls (Tesseur et al., 2006), suggesting that TGF-β signaling is ultimately reduced in AD patient brains. If this were the case, then abrogation of TGF-β pathway activity may represent a pathogenic disinhibition of neuroinflam-mation in the AD brain. When taken together, the above findings raise the interesting and important question of the cellular mechanistic underpin-nings of the TGF-β/AD pathology relationship.

While TGF-β signaling often functions to suppress inflammation, results show that inhibition of TGF-β signaling in peripheral macrophages leads to brain infiltration of these cells and resolution of cerebral amyloidosis in concert with increased brain IL-10 levels (Town et al., 2008). Specifically, our group genetically interrupted TGF-β, and downstream Smad 2/3 signaling in peripheral macrophages. In this model, peripheral monocytes (as opposed to brain-resident microglia) were specifically targeted by engineering a CD11c promoter-driven dominant-negative TGF-β type II receptor transgene in C57BL/6 mice (CD11c-DNR mice) (Laouar et al., 2008). We subsequently crossed CD11c-DNR mice with the Tg2576 AD mouse model, and evaluated behavioral impairment and AD-like pathology (Town et al., 2008). Notably, bigenic animals exhibited partial amelioration of cognitive impairment and presented with reduced astrocytosis, with concomitant 90% attenuation of brain parenchymal Aβ and reduction of CAA (Town et al., 2008). These preclinical therapeutic effects were associated with increased infiltration of Aβ-containing peripheral mononuclear phagocytes in and around cerebral vessels and amyloid plaques. Further, Aβ could be localized within the cytoplasm of these cells, suggesting a productive Aβ phagocytosis/clearance response. Based on these findings, it appears that removing the immunosuppressive TGF-β signal to peripheral monocytes allows these cells to gain access into the brain, while simultaneously maximizing their potential to phagocytose Aβ. As *ex vivo* validation of the latter, our group found approximately threefold increased phagocytosis of Aβ by CD11c-DNR versus wild-type macrophages (Town et al., 2008). These results suggest that inhibition of TGF-β-Smad 2/3 signaling promotes peripheral mononuclear phagocyte recruitment to brains of cerebral amyloid-depositing transgenic mice and reduces Aβ burden via phagocytic amyloid clearance.

In a similar manner to TGF-β, IL-10 suppresses overly exuberant inflammatory responses and inhibits effector function of myeloid cells by blocking pro-inflammatory cytokine pathways (Banchereau, Pascual, & O'Garra, 2012). Several lines of evidence implicate aberrant IL-10 signaling in AD. Notably, elevated IL-10 signaling was observed in reactive glia neighboring β-amyloid plaques in aged Tg2576 mice (Apelt & Schliebs, 2001). Additionally, a functional polymorphism within the *Il10* gene has been linked to increased risk of AD (Arosio et al., 2004; Lio et al., 2003; Vural et al., 2009). To further explore how anti-inflammatory IL-10 signaling affects Aβ pathology, we and others have investigated the effects of IL-10 modulation in the brains of APP transgenic mouse models. After finding that the IL-10 signaling pathway was elevated in AD patient brains, our group utilized the APP/PS1 mouse model of cerebral amyloidosis crossed with a mouse deficient for *Il10* (APP/PS1+*Il10*−/−) (Guillot-Sestier et al., 2015). We then relied on a novel method to quantify activated Aβ phagocytic microglia using an *in silico* 3D modeling technique. This technique revealed activated Aβ phagocytic microglia that restricted cerebral amyloidosis in APP/PS1+*Il10*−/− mice (Guillot-Sestier et al., 2015). Notably, genomewide RNA sequencing of APP/PS1+ brains showed modulation of innate immune genes that are known to promote neuroinflammation. Moreover, *Il10* deficiency preserved synaptic integrity and mitigated cognitive disturbance in APP/PS1 mice, suggesting a detrimental role of IL-10 in Alzheimer pathophysiology. To probe the mechanism for these findings, *in vitro* knock-down of microglial *Il10-Stat3* signaling was performed and led to augmented Aβ phagocytosis, while addition of exogenous IL-10 had the converse effect.

In a complimentary approach, Todd Golde's group investigated the effects of adeno-associated virus (AAV2/1) overexpression of IL-10 in brains of two APP transgenic mouse models. In this scenario, IL-10 expression resulted in increased Aβ accumulation and impaired learning and memory (Chakrabarty et al., 2015). Interestingly, *Il10* deficient animals had reduced apolipoprotein E (ApoE) mRNA levels in our report, as observed by RNA sequencing (RNAseq) and quantitative real-time PCR (qPCR) (Guillot-Sestier et al., 2015), while transcriptome analyses performed by Chakrabarty et al. (2015) demonstrated enhanced IL-10 signaling with concomitantly increased ApoE expression in IL-10 expressing APP mice. Importantly, in the Chakrabarty and colleagues work, ApoE protein was selectively increased in the plaque-associated insoluble cellular fraction, which they linked to direct interaction of ApoE with aggregated Aβ in IL-10 expressing APP transgenic mice (Chakrabarty et al., 2015). In both studies, *ex vivo* analyses showed that IL-10 and ApoE can separately impair microglial Aβ phagocytosis. Interestingly, *Il10* deficiency also partially overcame isoform-specific inhibition of human ApoE on microglial Aβ uptake (E4 > E3 > E2). Together, these results suggest two mechanisms by which *Il10* blockade enables cerebral Aβ clearance: (i) by decreasing microglial expression of ApoE and (ii) by opposing ApoE–Aβ binding and therefore endorsing Aβ phagocytosis (Guillot-Sestier et al., 2015). Collectively, these studies demonstrate a negative effect of IL-10 on Aβ proteostasis and cognition in APP mouse models and suggest that rebalancing innate immunity by blocking the IL-10 anti-inflammatory response may be therapeutically relevant for AD. Moreover, these independent and complimentary results (which utilize divergent methods) demonstrate the complex interplay between innate immunity and proteostasis in AD.

CHEMOKINES RECRUIT MONOCYTES TO Aβ PLAQUES AND DYING NEURONS

Recent evidence suggests that chemokines (chemotactic cytokines produced by both immune and nonimmune cells (Laing & Secombes, 2004)) are also critical for mononuclear phagocyte trafficking in AD. Interestingly, deficiency of the chemokine gene *Ccr2* has been shown to significantly impair trafficking of mononuclear cells to Aβ plaques in the Tg2576 mouse model of cerebral amyloidosis (El Khoury et al.,

2007). Perhaps more significantly, diminished recruitment of mononuclear phagocytes in this model was associated with increased mortality and higher Aβ plaque load. These data indicate that Ccr2 plays a nonredundant role in restricting cerebral amyloidosis by promoting accumulation of monocytes/microglia that are capable of phagocytosing amyloid.

Fractalkine (Cx3cl1) is a chemotactic agent for T-cells and monocytes. The membrane-bound form of fractalkine plays an important role in regulating leukocyte adhesion and migration at the endothelium. Expression of the fractalkine receptor (Cx3cr1) is reportedly restricted within the CNS to microglia (Savarin-Vuaillat & Ransohoff, 2007). A new perspective on the role of fractalkine-microglial signaling in AD was demonstrated by knocking out microglial Cx3cr1 (Fuhrmann et al., 2010). In order to observe the interaction between neurons and microglia, researchers crossed 3x Tg-AD mice with two additional transgenic lines—one in which subsets of layer III and V cortical neurons express yellow fluorescent protein (YFP) (Feng et al., 2000), and another carrying a GFP knockin substituting the endogenous murine *Cx3cr1* locus (Jung et al., 2000). Utilizing a sophisticated intravital two-photon imaging approach to analyze microglial interactions with neighboring neurons in brains of 3x Tg-AD mice enabled tracking of microglial Cx3cr1 expression. Importantly, ablation of the endogenous murine fractalkine receptor interfered with neuron-microglial crosstalk. Although neurodegeneration in 3x Tg-AD mice was meager (~1.8% of neurons are lost), two-photon imaging of 4- to 6-month-old YFP-expressing 3x Tg-AD mice revealed neuronal loss over a 2- to 4-week observation period. Time-lapse images using YFP to mark neurons and GFP as a surrogate for *Cx3cr1* deficient microglia in genetically manipulated 3x Tg-AD mice showed that YFP$^+$ neurons in the *Cx3cr1$^{-/-}$*-3x Tg-AD mice survived while neurons in *Cx3cr1*-sufficient YFP-expressing mice died. Further, *Cx3cr1*-sufficient microglia in 3x Tg-AD mice rapidly localized to neurons that were destined for death, while microglia mobilization to dying neurons in *Cx3cr1$^{-/-}$*-3x Tg-AD animals occurred with delayed kinetics (Fuhrmann et al., 2010). These data suggest that fractalkine/Cx3cl1 signaling promotes microglial homing to dying neurons. However, a critical question left unanswered in this scenario is whether microglia initiate the neuronal death cascade or simply serve as downstream effectors.

BLOCKING INFLAMMATORY ILS 12 AND 23 PREVENTS PLAQUE BUILDUP IN TRANSGENIC MICE

Recent studies suggest that blocking inflammatory cytokine pathways could slow development of AD-like pathology. Specifically, it has been shown that simultaneously mollifying two microglial cytokines, IL-12 and IL-23, prevents plaque buildup and improves cognition in transgenic mouse models of cerebral amyloid deposition (Vom Berg et al., 2012). Interestingly, both cytokines are present in copious quantity in the cerebrospinal fluid of human AD patients. Each cytokine is made up of two subunits, with the p40 subunit being the common chain. Activated microglia in APP/PS1 mice produce more mRNA for IL-12/23 p40 and IL-23 p19. When APP/PS1 mice were bred with mice lacking IL12/23 p40, IL-12 p35, or IL-23 p19, cortical plaque load was reduced in all knockouts relative to APP/PS1 controls. Interestingly, mice lacking the common p40 subunit showed the most marked effect. These results suggest that both IL-12 and IL-23 play a role in AD pathology, and that "reining in" those cytokines might suppress Aβ accumulation. To determine whether peripheral macrophages or brain-resident microglia mediated these effects, the authors irradiated APP/PS1 and APP/PS1/$p40^{-/-}$ mice to eliminate their peripheral macrophages (Vom Berg et al., 2012). They then replenished peripheral cells by injecting bone marrow from wild-type or *p40* knockouts, thereby creating mice with IL-12/23 p40 restricted to either CNS resident or peripheral immune cells. The latter accumulated plaque loads comparable to APP/PS1 mice, while the former had reduced cerebral amyloid load, suggesting that p40 action takes place within brain-resident innate immune cells, and not in peripheral myelomonocytes (Vom Berg et al., 2012). This work offers an intriguing treatment strategy, as Ustekinumab, a humanized monoclonal blocking antibody against IL-12/23, is already approved by the FDA for the treatment of skin autoimmune disease.

PPARγ AGONISTS REDUCE INFLAMMATION WHILE BOOSTING MICROGLIAL Aβ UPTAKE

The diabetes drugs rosiglitazone and pioglitazone were once promising AD treatments, but lost their luster when rosiglitazone failed in a phase III clinical trial (Gold et al., 2010). Rosiglitazone and pioglitazone belong to the thiazolidinedione class of compounds, which bind to the nuclear receptor peroxisome proliferator-activated receptor-γ (PPARγ). PPARγ forms a heterodimer with retinoid X receptor (RXR) to affect gene transcription. Because the PPARγ pathway curbs pro-inflammatory gene expression (Jiang, Ting, & Seed, 1998; Ricote, Li, Willson, Kelly, & Glass, 1998), focus has been directed toward testing the effects of PPARγ agonists such as DSP-8658 on microglial Aβ phagocytosis (Yamanaka et al., 2012). Over 4 h, DSP-8658 treated microglia engulfed approximately twice as much Aβ as control cells and knocking down PPARγ abolished this effect. Intriguingly, adding PPARγ and RXR agonists together produced an additive effect, allowing microglia to engulf up to four times the amount of

Aβ as unstimulated cells. In determining the mechanism behind improved phagocytosis, researchers have uncovered an essential role for the receptor CD36, a scavenger receptor that microglia nonspecifically rely upon to bind Aβ. In this model, PPARγ activation promotes CD36 activity, suggesting that pharmacological stimulation of PPARγ enables CD36-dependent microglial engulfment of Aβ (Yamanaka et al., 2012). Importantly, treatment of APP/PS1 mice with DSP-8658 promoted microglial-mediated Aβ uptake *in vivo* (Yamanaka et al., 2012).

It is important to note that ApoE transcription is stimulated by RXR heterodimers formed with either PPARγ or liver X receptors (LXRs). Previous studies have shown that PPARγ and LXR agonists mitigate Aβ pathology and improve symptoms in AD mice (Donkin et al., 2010; Heneka & Landreth, 2007). It was recently reported that the RXR agonist bexarotene, which targets both RXR-PPARγ and RXR-LXR pathways, promotes rapid clearance of brain amyloid in mouse models of cerebral amyloidosis (Cramer et al., 2012). Gary Landreth's group delivered bexarotene orally to APP/PS1 mice at 2, 6, or 11 months of age. Using a novel, *in vivo* microdialysis method, the researchers observed dramatic decreases in soluble Aβ levels in brain interstitial fluid of 2-month-old APP/PS1 mice as soon as 6h posttreatment (Cramer et al., 2012). Additionally, they witnessed more than 50% reduction in Aβ plaque area within just 72h of administration, which reportedly stimulated rapid reversal of cognitive, social, and olfactory deficits and improved neural circuit function (Cramer et al., 2012). The compound had no influence on interstitial fluid Aβ in ApoE-deficient animals, suggesting that ApoE is required for the Aβ-reducing effect. Furthermore, those authors observed abundant Aβ-containing microglia following bexarotene treatment, and suggested microglial involvement in the phagocytic removal of Aβ deposits (Cramer et al., 2012). These results were especially tantalizing, because bexarotene is approved by the FDA to treat certain types of skin cancer caused by T-cells, and could hypothetically be used in prevention trials for asymptomatic patients in whom antiamyloid therapies are most likely to be effective. However, other groups attempting to reproduce these findings have concluded that bexarotene has little to no impact on plaque burden in amyloidosis mouse models, and have called for extreme caution when considering this compound for use in AD patients (Price et al., 2013; Tesseur et al., 2013; Veeraraghavalu et al., 2013). Others were able to replicate the significant decrease of interstitial fluid Aβ, but not the effects on amyloid deposition in APP/PS1 mice expressing human ApoE3 and ApoE4 isoforms (Fitz, Cronican, Lefterov, & Koldamova, 2013). While it has been argued that drug formulation is critical to bexarotene-mediated effects (Landreth et al., 2013), the microglial-mediated reduction in plaque burden following bexarotene will require substantiation.

INFLAMMASOME ACTIVATION IN AD

The expression of multiple surface amyloid fibrils on evolutionarily ancient microorganisms—including bacteria and fungi—has led some to suggest that the recognition of fibrillar amyloid by the immune system may be a consequence of the evolution of antimicrobial defense mechanisms in the brain (Heneka, Kummer, & Latz, 2014). The NOD-, LRR-, and pyrin domain-containing 3 (NLRP3) inflammasome is particularly important in the development of acute and chronic inflammatory responses, as it can detect a wide range of aggregated molecules. Inflammasomes consist of a sensor molecule from the NOD-like receptor (NLR) family or the pyrin and HIN domain-containing protein family, the adaptor protein ASC and caspase 1. Sensing of β-amyloid aggregates by the germline encoded, evolutionarily conserved toll-like receptors is followed by activation of intracellular machinery that leads to the assembly of the NLRP3 inflammasome (Halle et al., 2008). The expression of active caspase 1 is increased in brains of patients with AD compared with those of age-matched controls (Heneka et al., 2013). In the APP/PS1 mouse model of cerebral amyloidosis, mice that are deficient for NLRP3 or the inflammasome adaptor ASC are mostly protected from accumulation of amyloid pathology (Halle et al., 2008; Heneka et al., 2013). Finally, NLRP3-deficient APP/PS1 mice demonstrate nearly restored cognitive function and are protected from Aβ-induced suppression of synaptic plasticity (Heneka et al., 2013). These results implicate the NLRP3/caspase-1 axis in AD pathogenesis, and suggest that NLRP3 inflammasome inhibition may be a new AD therapeutic modality.

These findings emphasize the role that inflammation and impaired Aβ phagocytosis play in sporadic AD, and suggest that this form of the disease is the result of inefficient clearance of Aβ, rather than enhanced production. This idea is substantiated by a study in which a neuroinflammatory stimulus was administered to young APP transgenic mice, which dramatically increased plaque production 11 months later (Krstic et al., 2012). Intriguingly, an immune stimulus given to wild-type mice early in life increases their susceptibility to AD-like pathology when given a second immune stimulus as adults (Krstic et al., 2012). This suggests that neuroinflammation provides fertile grounds for AD pathology to flourish, and implicates early infection as a major risk factor for disease later in life.

BECLIN 1 REGULATES MICROGLIAL PHAGOCYTOSIS AND MAY BE IMPAIRED IN AD

Beclin 1 has long been known to regulate autophagy, a cellular degradation pathway, through studies performed in yeast. Interestingly, recent work from Tony Wyss-Coray's group described a role for beclin 1 in recycling of phagocytic receptors in mammalian microglia, a function analogous to its yeast ortholog. Observing that levels of beclin 1 were decreased in cortical brain extracts of AD patients, these researchers utilized both the microglial BV2 cell line infected with control or beclin 1 shRNA lentivirus and primary microglia from heterozygous beclin 1 deficient mice (Lucin et al., 2013). These studies revealed that loss of beclin 1 decreased phagocytosis of latex beads by flow cytometry and live-cell imaging (Lucin et al., 2013). In an additional experiment, BV2 cells with decreased beclin 1 levels cleared less Aβ when cultured on brain slices from APP transgenic mice as measured by immunohistochemistry and enzyme-linked immunosorbent assay (Lucin et al., 2013). Finally, amyloid fibrils injected into mouse frontal cortex were cleared more efficiently in wild-type than in beclin 1-deficient mice (Lucin et al., 2013). These results are corroborated by a separate study in which heterozygous beclin 1 deficiency in APP transgenic mice resulted in synaptic loss and enhanced Aβ deposition in conjunction with increased microglial activation (Pickford et al., 2008). Together, these results suggest that beclin 1 regulates phagocytosis, and that impaired microglial phagocytosis may contribute to Aβ accumulation in AD. However, these studies did not explore the role of beclin 1 in receptor recycling. Future studies may go on to identify the repertoire of cell surface receptors that are regulated by beclin 1, which could potentially reveal therapeutic targets.

A NEW GENERATION OF AD PHARMACOTHERAPEUTICS TARGETING INNATE IMMUNITY

As mentioned, the amyloid cascade hypothesis purports that mismetabolism and subsequent deposition of Aβ peptides as amyloid plaques is the principal etiopathological event in AD, which sets into motion a cascade of disease-perpetrating events. For this reason, methods to reduce brain amyloid remain the primary AD pharmacotherapeutic strategy in current clinical trials. In this chapter, a variety of methods have been presented in which manipulation of the innate immune system allows phagocytes to restrict and clear cerebral amyloid plaques. In fact, it would take over 20 years and the advent of modern cellular and molecular biology for us to begin to understand this innate immune cell dichotomy (Figure 12.2). An immunotherapeutic strategy targeting peripheral innate immune cells is attractive, as it avoids the issue of getting drugs through the blood–brain barrier, which continues to be a major obstacle to AD pharmacotherapeutics. In theory, small-molecule drugs could be administered systemically to target and mobilize populations of phagocytes in the periphery, which could subsequently infiltrate the CNS and home to and clear amyloid deposits.

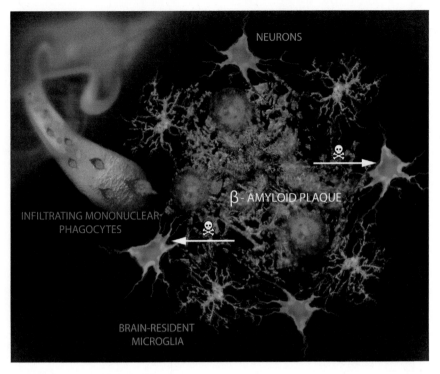

FIGURE 12.2   Innate immune cellular players in Alzheimer's disease. A penetrating cerebral artery is shown on the left, and a β-amyloid plaque within the brain parenchyma is shown to the right. Peripheral mononuclear phagocytes (macrophages) are infiltrating the brain parenchyma and homing to the β-amyloid plaque, where they phagocytose and clear deposited Aβ. Peripheral mononuclear phagocytes are depicted in green, brain-resident microglia are magenta, neurons are blue, and a β-amyloid plaque is shown in red. *Source: Reproduced from Gate et al. (2010).*

If innate immune cells are to be targeted as an AD therapeutic modality, then strategies for (i) increasing brain recruitment and (ii) augmenting amyloid clearance potential of these cells need to be developed. Given the exquisite complexity of the innate immune system and the intricate nature of the interplay with AD pathologies, a focused strategy targeting a specific cellular subset may ultimately prove most successful. Although there are limited data supporting or refuting this possibility, it stands to reason that endorsing a beneficial form of innate immunity could lead to amyloid clearance without exacerbating neuroinflammation and bystander injury in the context of AD neuropathology. Therefore, continued investigation is warranted into pharmacotherapeutics aimed at immunomodulatory pathways that target innate immunity as a therapeutic modality for AD.

CONCLUDING REMARKS

In this chapter, we have focused on preclinical strategies to promote innate immunity as an important immunotherapeutic approach for AD (summarized in Table 12.1). Seminal observations from Henryk Wisniewski over two decades ago led to the first evidence that peripheral macrophages possess Aβ clearance aptitude. These findings were groundbreaking, as they called our attention to the possibility that monocytes could function as amyloid phagocytes. We have since gained a greater understanding of the mechanisms governing Aβ phagocytosis and clearance by innate immune cells. Despite significant advances toward understanding the cellular and molecular underpinnings of this "good" form of neuroinflammation, additional important questions include (i) whether antiamyloid properties of phagocytes can be harnessed without incurring inflammation-induced bystander injury and (ii) which subset(s) of these cells will provide the greatest amyloid clearance potential. A deeper understanding of the molecular pathways leading to beneficial innate immune responses to clear pathogenic misfolded proteins will be the key to ushering in a new era of AD pharmacotherapeutics. Rather than shutting neuroinflammation off completely, rebalancing it toward a beneficial innate immune response may allow us to harness innate immunity in the fight against AD.

TABLE 12.1   Preclinical Innate Immunotherapy in Alzheimer's Disease

Study	Model system	Observations
Akiyama et al. (1996), Wisniewski et al. (1991)	Electron microscopy of AD patient brain samples with comorbid stroke	CNS infiltrating peripheral mononuclear cells containing lysosomal amyloid
Bard et al. (2000, 2003), Schenk et al. (1999)	Active Aβ vaccine; passive immunization of AD mice	Attenuated AD-like pathology in transgenic mice
Stalder et al. (2005)	APP23/GFP+ bone marrow chimeric mice	CNS infiltration of GFP+ mononuclear cells mediates reduced amyloid burden
Simard et al. (2006)	APP/PSI/ GFP+ bone marrow chimeric mice	Age/Aβ-dependent brain infiltration of GFP+ mononuclear cells; reduced amyloid burden
Zhu et al. (2011)	APP/PSI/wild-type bone marrow chimeric mice	Reduced amyloid burden
Keene et al. (2010)	APP/PSI/EP2$^{-/-}$ chimeric mice	Increased cortical mononuclear cells; reduced amyloid burden

(Continued)

TABLE 12.1 (Continued)

Study	Model system	Observations
Hawkes and McLaurin (2009)	Clodronate or chitin-treated TgCRND8 mice	Opposing effects of clodronate and chitin on perivascular macrophage turnover and CAA
Chakrabarty (2015), Guillot-Sestier et al. (2015), Town et al. (2008)	Tg2576/CD11c-DNR bigenic mice; APP/PSI$^+$$Il10^{-/-}$ mice; (AAV2/1) mediated over-expression of IL-10	Targeting cardinal anti-inflammatory cytokines TGF-β or IL-10 restricts cerebral amyloidosis by monocyte-mediate Aβ phagocytosis
El Khoury et al. (2007)	Tg2576/$Ccr2^{-/-}$ crossed mice	Reduced CD11b$^+$ cells in proximity to plaques; increased amyloid burden
Grathwohl et al. (2009)	APP23 and APP/PS1/$CD11b$-$HSVTK$ crossed mice	Reduced CD11b$^+$ cells in both periphery and CNS; amyloid deposits unaffected
Fuhrmann et al. (2010)	3x Tg-AD/$Cx3cr1^{-/-}$ crossed mice	Reduced mononuclear density in CNS; reduced neuronal loss; amyloid burden unaffected
Vom Berg et al. (2012)	IL-12/23 $p40^{-/-}$, IL-12 $p35^{-/-}$, or IL-23 $p19^{-/-}$ mice crossed with APP/PSI mice	IL-12 and IL-23 play a role in AD pathology, and blocking these cytokines suppresses Aβ accumulation
Cramer et al. (2012); Yamanaka et al. (2012)	PPARγ agonist (DSP-8658) treatment of microglia; bexarotene treatment of multiple mouse models of amyloidosis	PPARγ and LXR agonists mitigate Aβ pathology and improve cognitive deficits
Halle et al. (2008); Heneka et al. (2013)	$Nlrp3^{-/-}$ or $Casp1^{-/-}$ and APP/PS1 crossed mice	NLRP3 inflammasome deficiency skews microglial cells to a phagocytic phenotype results in decreased Aβ deposition
Lucin et al. (2013); Pickford et al. (2008)	Beclin1$^{+/-}$ mice; AD patient samples	Beclin 1 regulates retromer trafficking and phagocytosis and is impaired in AD

AD, Alzheimer's disease; CNS, central nervous system; GFP, green fluorescent protein; Aβ, amyloid-β; IL, interleukin.

Acknowledgements

We thank Dr Marie-Victoire Guillot-Sestier for helpful discussion and Dr Kevin Doty (USC Zilkha Neurgenetic Institute, Los Angeles, CA, USA) for assistance with the design and creation of Figure 12.1. D.G. is supported by an NIH National Research Service Award 1F31NS083339-01A1. This work was supported by the National Institute on Aging (5R00AG029726-04 and 3R00AG029726-04S1, to T.T.), the National Institute on Neurologic Disorders and Stroke (1R01NS076794-01, to T.T.), and startup funds from the Zilkha Neurogenetic Institute.

References

Agadjanyan, M. G., Ghochikyan, A., Petrushina, I., Vasilevko, V., Movsesyan, N., Mkrtichyan, M., et al. (2005). Prototype Alzheimer's disease vaccine using the immunodominant B cell epitope from beta-amyloid and promiscuous T cell epitope pan HLA DR-binding peptide. *Journal of Immunology, 174*(3), 1580–1586.

Ajami, B., Bennett, J. L., Krieger, C., Tetzlaff, W., & Rossi, F. M. V. (2007). Local self-renewal can sustain CNS microglia maintenance and function throughout adult life. *Nature Neuroscience, 10*(12), 1538–1543.

Akiyama, H., Barger, S., Barnum, S., Bradt, B., Bauer, J., Cole, G. M., et al. (2000). Inflammation and Alzheimer's disease. *Neurobiology of Aging, 21*(3), 383–421.

Akiyama, H., Kondo, H., Mori, H., Kametani, F., Nishimura, T., Ikeda, K., et al. (1996). The amino-terminally truncated forms of amyloid beta-protein in brain macrophages in the ischemic lesions of Alzheimer's disease patients. *Neuroscience Letters, 219*(2), 115–118.

Apelt, J., & Schliebs, R. (2001). Beta-amyloid-induced glial expression of both pro- and anti-inflammatory cytokines in cerebral cortex of aged transgenic Tg2576 mice with Alzheimer plaque pathology. *Brain Research, 894*(1), 21–30.

Arosio, B., Trabattoni, D., Galimberti, L., Bucciarelli, P., Fasano, F., Calabresi, C., et al. (2004). Interleukin-10 and interleukin-6 gene polymorphisms as risk factors for Alzheimer's disease. *Neurobiology of Aging, 25*(8), 1009–1015. http://dx.doi.org/10.1016/j.neurobiolaging. 2003.10.009.

Bacskai, B. J., Hickey, G. A., Skoch, J., Kajdasz, S. T., Wang, Y., Huang, G. -F., et al. (2003). Four-dimensional multiphoton imaging of brain entry, amyloid binding, and clearance of an amyloid-beta ligand in transgenic mice. *Proceedings of the National Academy of Sciences of the United States of America, 100*(21), 12462–12467.

Bacskai, B. J., Kajdasz, S. T., Christie, R. H., Carter, C., Games, D., Seubert, P., et al. (2001). Imaging of amyloid-beta deposits in brains of living mice permits direct observation of clearance of plaques with immunotherapy. *Nature Medicine, 7*(3), 369–372.

Banchereau, J., Pascual, V., & O'Garra, A. (2012). From IL-2 to IL-37: The expanding spectrum of anti-inflammatory cytokines. *Nature Immunology, 13*(10), 925–931. http://dx.doi. org/10.1038/ni.2406.

Bard, F., Barbour, R., Cannon, C., Carretto, R., Fox, M., Games, D., et al. (2003). Epitope and isotype specificities of antibodies to beta -amyloid peptide for protection against Alzheimer's disease-like neuropathology. *Proceedings of the National Academy of Sciences of the United States of America, 100*(4), 2023–2028. http://dx.doi.org/10.1073/pnas.0436286100.

Bard, F., Cannon, C., Barbour, R., Burke, R. L., Games, D., Grajeda, H., et al. (2000). Peripherally administered antibodies against amyloid beta-peptide enter the central nervous system and reduce pathology in a mouse model of Alzheimer disease. *Nature Medicine, 6*(8), 916–919.

Benzing, W. C., Wujek, J. R., Ward, E. K., Shaffer, D., Ashe, K. H., Younkin, S. G., et al. (1999). Evidence for glial-mediated inflammation in aged APP(SW) transgenic mice. *Neurobiology of Aging, 20*(6), 581–589.

Breitner, J. C., Baker, L. D., Montine, T. J., Meinert, C. L., Lyketsos, C. G., Ashe, K. H., et al. (2011). Extended results of the Alzheimer's disease anti-inflammatory prevention trial. *Alzheimer's & Dementia, 7*(4), 402–411. http://dx.doi.org/10.1016/j.jalz.2010.12.014.

Brouwers, N., Van Cauwenberghe, C., Engelborghs, S., Lambert, J. C., Bettens, K., Le Bastard, N., et al. (2012). Alzheimer risk associated with a copy number variation in the complement receptor 1 increasing C3b/C4b binding sites. *Molecular Psychiatry, 17*(2), 223–233. http://dx.doi.org/10.1038/mp.2011.24.

Chakrabarty, P., Li, A., Ceballos-Diaz, C., Eddy, J. A., Funk, C. C., Moore, B., et al. (2015). IL-10 alters immunoproteostasis in APP mice, increasing plaque burden and worsening cognitive behavior. *Neuron, 85*(3), 519–533. http://dx.doi.org/10.1016/j.neuron.2014.11.020.

Cramer, P. E., Cirrito, J. R., Wesson, D. W., Lee, C. Y., Karlo, J. C., Zinn, A. E., et al. (2012). ApoE-directed therapeutics rapidly clear beta-amyloid and reverse deficits in AD mouse models. *Science, 335*(6075), 1503–1506. http://dx.doi.org/10.1126/science.1217697.

Dickson, D. W., Farlo, J., Davies, P., Crystal, H., Fuld, P., & Yen, S. H. (1988). Alzheimer's disease. A double-labeling immunohistochemical study of senile plaques. *The American Journal of Pathology, 132*(1), 86–8101.

Donkin, J. J., Stukas, S., Hirsch-Reinshagen, V., Namjoshi, D., Wilkinson, A., May, S., et al. (2010). ATP-binding cassette transporter A1 mediates the beneficial effects of the liver X receptor agonist GW3965 on object recognition memory and amyloid burden in amyloid precursor protein/presenilin 1 mice. *The Journal of Biological Chemistry, 285*(44), 34144–34154. http://dx.doi.org/10.1074/jbc.M110.108100.

Doty, K. R., Guillot-Sestier, M. V., & Town, T. (2015). The role of the immune system in neurodegenerative disorders: Adaptive or maladaptive? *Brain Research, 1617*, 155–173. http://dx.doi.org/10.1016/j.brainres.2014.09.008.

El Khoury, J., Toft, M., Hickman, S. E., Means, T. K., Terada, K., Geula, C., et al. (2007). Ccr2 deficiency impairs microglial accumulation and accelerates progression of Alzheimer-like disease. *Nature Medicine, 13*(4), 432–438.

Feng, G., Mellor, R. H., Bernstein, M., Keller-Peck, C., Nguyen, Q. T., Wallace, M., et al. (2000). Imaging neuronal subsets in transgenic mice expressing multiple spectral variants of GFP. *Neuron, 28*(1), 41–51.

Fitz, N. F., Cronican, A. A., Lefterov, I., & Koldamova, R. (2013). Comment on "ApoE-directed therapeutics rapidly clear beta-amyloid and reverse deficits in AD mouse models". *Science, 340*(6135), 924-c. http://dx.doi.org/10.1126/science.1235809.

Fu, H. J., Liu, B., Frost, J. L., & Lemere, C. A. (2010). Amyloid-beta immunotherapy for Alzheimer's disease. *CNS & Neurological Disorders Drug Targets, 9*(2), 197–206.

Fuhrmann, M., Bittner, T., Jung, C. K. E., Burgold, S., Page, R. M., Mitteregger, G., et al. (2010). Microglial Cx3cr1 knockout prevents neuron loss in a mouse model of Alzheimer's disease. *Nature Neuroscience, 13*(4), 411–413.

Games, D., Adams, D., Alessandrini, R., Barbour, R., Berthelette, P., Blackwell, C., et al. (1995). Alzheimer-type neuropathology in transgenic mice overexpressing V717f beta-amyloid precursor protein. *Nature, 373*(6514), 523–527.

Gate, D., Rezai-Zadeh, K., Jodry, D., Rentsendorj, A., & Town, T. (2010). Macrophages in Alzheimer's disease: The blood-borne identity. *Journal of Neural Transmission, 117*(8), 961–970.

Gold, M., Alderton, C., Zvartau-Hind, M., Egginton, S., Saunders, A. M., Irizarry, M., et al. (2010). Rosiglitazone monotherapy in mild-to-moderate Alzheimer's disease: Results from a randomized, double-blind, placebo-controlled phase III study. *Dementia and Geriatric Cognitive Disorders, 30*(2), 131–146. http://dx.doi.org/10.1159/000318845.

Grathwohl, S. A., Kalin, R. E., Bolmont, T., Prokop, S., Winkelmann, G., Kaeser, S. A., et al. (2009). Formation and maintenance of Alzheimer's disease beta-amyloid plaques in the absence of microglia. *Nature Neuroscience, 12*(11), 1361–1363.

Group, A. R., Martin, B. K., Szekely, C., Brandt, J., Piantadosi, S., Breitner, J. C., et al. (2008). Cognitive function over time in the Alzheimer's Disease Anti-inflammatory Prevention

Trial (ADAPT): Results of a randomized, controlled trial of naproxen and celecoxib. *Archives of Neurology, 65*(7), 896–905. http://dx.doi.org/10.1001/archneur.2008.65.7. nct70006.

Guillot-Sestier, M. V., Doty, K. R., Gate, D., Rodriguez, J., Jr., et al., Leung, B. P., Rezai-Zadeh, K., et al. (2015). Il10 deficiency rebalances innate immunity to mitigate Alzheimer-like pathology. *Neuron, 85*(3), 534–548. http://dx.doi.org/10.1016/j.neuron.2014.12.068.

Haga, S., Akai, K., & Ishii, T. (1989). Demonstration of microglial cells in and around senile (neuritic) plaques in the Alzheimer brain. An immunohistochemical study using a novel monoclonal antibody. *Acta Neuropathologica, 77*(6), 569–575.

Halle, A., Hornung, V., Petzold, G. C., Stewart, C. R., Monks, B. G., Reinheckel, T., et al. (2008). The NALP3 inflammasome is involved in the innate immune response to amyloid-beta. *Nature Immunology, 9*(8), 857–865. http://dx.doi.org/10.1038/ni.1636.

Hawkes, C. A., & McLaurin, J. (2009). Selective targeting of perivascular macrophages for clearance of beta-amyloid in cerebral amyloid angiopathy. *Proceedings of the National Academy of Sciences of the United States of America, 106*(4), 1261–1266.

Heneka, M. T., Kummer, M. P., & Latz, E. (2014). Innate immune activation in neurodegenerative disease. *Nature Reviews Immunology, 14*(7), 463–477. http://dx.doi.org/10.1038/nri3705.

Heneka, M. T., Kummer, M. P., Stutz, A., Delekate, A., Schwartz, S., Vieira-Saecker, A., et al. (2013). NLRP3 is activated in Alzheimer's disease and contributes to pathology in APP/PS1 mice. *Nature, 493*(7434), 674–678. http://dx.doi.org/10.1038/nature11729.

Heneka, M. T., & Landreth, G. E. (2007). PPARs in the brain. *Biochim Biophys Acta, 1771*(8), 1031–1045. http://dx.doi.org/10.1016/j.bbalip.2007.04.016.

in t' Veld, B. A., Ruitenberg, A., Hofman, A., Launer, L. J., van Duijn, C. M., Stijnen, T., et al. (2001). Nonsteroidal antiinflammatory drugs and the risk of Alzheimer's disease. *The New England Journal of Medicine, 345*(21), 1515–1521. http://dx.doi.org/10.1056/NEJMoa010178.

Itagaki, S., McGeer, P. L., Akiyama, H., Zhu, S., & Selkoe, D. (1989). Relationship of microglia and astrocytes to amyloid deposits of Alzheimer disease. *Journal of Neuroimmunology, 24*(3), 173–182.

Jiang, C., Ting, A. T., & Seed, B. (1998). PPAR-gamma agonists inhibit production of monocyte inflammatory cytokines. *Nature, 391*(6662), 82–86. http://dx.doi.org/10.1038/34184.

Jonsson, T., Stefansson, H., Steinberg, S., Jonsdottir, I., Jonsson, P. V., Snaedal, J., et al. (2013). Variant of TREM2 associated with the risk of Alzheimer's disease. *The New England Journal of Medicine, 368*(2), 107–116. http://dx.doi.org/10.1056/NEJMoa1211103.

Jung, S., Aliberti, J., Graemmel, P., Sunshine, M. J., Kreutzberg, G. W., Sher, A., et al. (2000). Analysis of fractalkine receptor CX(3)CR1 function by targeted deletion and green fluorescent protein reporter gene insertion. *Molecular and Cellular Biology, 20*(11), 4106–4114.

Kang, J., & Rivest, S. (2007). MyD88-deficient bone marrow cells accelerate onset and reduce survival in a mouse model of amyotrophic lateral sclerosis. *The Journal of Cell Biology, 179*(6), 1219–1230.

Keene, C. D., Chang, R. C., Lopez-Yglesias, A. H., Shalloway, B. R., Sokal, I., Li, X., et al. (2010). Suppressed accumulation of cerebral amyloid {beta} peptides in aged transgenic Alzheimer's disease mice by transplantation with wild-type or prostaglandin E2 receptor subtype 2-null bone marrow. *The American Journal of Pathology, 177*(1), 346–354.

Kennedy, D. W., & Abkowitz, J. L. (1997). Kinetics of central nervous system microglial and macrophage engraftment: Analysis using a transgenic bone marrow transplantation model. *Blood, 90*(3), 986–993.

Kennedy, D. W., & Abkowitz, J. L. (1998). Mature monocytic cells enter tissues and engraft. *Proceedings of the National Academy of Sciences of the United States of America, 95*(25), 14944–14949.

Klyubin, I., Walsh, D. M., Lemere, C. A., Cullen, W. K., Shankar, G. M., Betts, V., et al. (2005). Amyloid beta protein immunotherapy neutralizes Abeta oligomers that disrupt synaptic plasticity *in vivo*. *Nature Medicine, 11*(5), 556–561.

Krstic, D., Madhusudan, A., Doehner, J., Vogel, P., Notter, T., Imhof, C., et al. (2012). Systemic immune challenges trigger and drive Alzheimer-like neuropathology in mice. *Journal of Neuroinflammation, 9*, 151. http://dx.doi.org/10.1186/1742-2094-9-151.

Laing, K. J., & Secombes, C. J. (2004). Chemokines. *Developmental and Comparative Immunology, 28*(5), 443–460.

Landreth, G. E., Cramer, P. E., Lakner, M. M., Cirrito, J. R., Wesson, D. W., Brunden, K. R., et al. (2013). Response to comments on "ApoE-directed therapeutics rapidly clear beta-amyloid and reverse deficits in AD mouse models". *Science, 340*(6135), 924-g. http://dx.doi.org/10.1126/science.1234114.

Laouar, Y., Town, T., Jeng, D., Tran, E., Wan, Y., Kuchroo, V. K., et al. (2008). TGF-beta signaling in dendritic cells is a prerequisite for the control of autoimmune encephalomyelitis. *Proceedings of the National Academy of Sciences of the United States of America, 105*(31), 10865–10870.

Leoutsakos, J. M., Muthen, B. O., Breitner, J. C., Lyketsos, C. G., & Team, A. R. (2012). Effects of non-steroidal anti-inflammatory drug treatments on cognitive decline vary by phase of pre-clinical Alzheimer disease: Findings from the randomized controlled Alzheimer's disease anti-inflammatory prevention trial. *International Journal of Geriatric Psychiatry, 27*(4), 364–374. http://dx.doi.org/10.1002/gps.2723.

Lewis, C. -A. B., Solomon, J. N., Rossi, F. M., & Krieger, C. (2009). Bone marrow-derived cells in the central nervous system of a mouse model of amyotrophic lateral sclerosis are associated with blood vessels and express CX(3)CR1. *Glia, 57*(13), 1410–1419.

Li, M. O., & Flavell, R. A. (2008). Contextual regulation of inflammation: A duet by transforming growth factor-beta and interleukin-10. *Immunity, 28*(4), 468–476. http://dx.doi.org/10.1016/j.immuni.2008.03.003.

Lim, G. P., Chu, T., Yang, F. S., Beech, W., Frautschy, S. A., & Cole, G. M. (2001). The curry spice curcumin reduces oxidative damage and amyloid pathology in an Alzheimer transgenic mouse. *Journal of Neuroscience, 21*(21), 8370–8377.

Lim, G. P., Yang, F., Chu, T., Chen, P., Beech, W., Teter, B., et al. (2000). Ibuprofen suppresses plaque pathology and inflammation in a mouse model for Alzheimer's disease. *Journal of Neuroscience, 20*(15), 5709–5714.

Lim, G. P., Yang, F., Chu, T., Gahtan, E., Ubeda, O., Beech, W., et al. (2001). Ibuprofen effects on Alzheimer pathology and open field activity in APPsw transgenic mice. *Neurobiology of Aging, 22*(6), 983–991.

Lio, D., Licastro, F., Scola, L., Chiappelli, M., Grimaldi, L. M., Crivello, A., et al. (2003). Interleukin-10 promoter polymorphism in sporadic Alzheimer's disease. *Genes and Immunity, 4*(3), 234–238. http://dx.doi.org/10.1038/sj.gene.6363964.

Lucin, K. M., O'Brien, C. E., Bieri, G., Czirr, E., Mosher, K. I., Abbey, R. J., et al. (2013). Microglial beclin 1 regulates retromer trafficking and phagocytosis and is impaired in Alzheimer's disease. *Neuron, 79*(5), 873–886. http://dx.doi.org/10.1016/j.neuron.2013.06.046.

McGeer, E. G., & McGeer, P. L. (2003). Inflammatory processes in Alzheimer's disease. *Progress in Neuro-psychopharmacology & Biological Psychiatry, 27*(5), 741–749. http://dx.doi.org/10.1016/S0278-5846(03)00124-6.

McGeer, P. L., & McGeer, E. (2004). Immunotherapy for Alzheimer's disease. *Science of Aging Knowledge Environment, 2004*(27), pe29. http://dx.doi.org/10.1126/sageke.2004.27.pe29.

McGowan, E., Sanders, S., Iwatsubo, T., Takeuchi, A., Saido, T., Zehr, C., et al. (1999). Amyloid phenotype characterization of transgenic mice overexpressing both mutant amyloid precursor protein and mutant presenilin 1 transgenes. *Neurobiology of Disease, 6*(4), 231–244.

Mildner, A., Schmidt, H., Nitsche, M., Merkler, D., Hanisch, U. K., Mack, M., et al. (2007). Microglia in the adult brain arise from Ly-6ChiCCR2+ monocytes only under defined host conditions. *Nature Neuroscience*, 10(12), 1544–1553. http://dx.doi.org/10.1038/nn2015.

Movsesyan, N., Ghochikyan, A., Mkrtichyan, M., Petrushina, I., Davtyan, H., Olkhanud, P. B., et al. (2008). Reducing AD-like pathology in 3xTg-AD mouse model by DNA epitope vaccine—A novel immunotherapeutic strategy. *PLoS One*, 3(5). http//dx.doi.org/10.1371/journal.pone.0002124.

Nicoll, J. A., Wilkinson, D., Holmes, C., Steart, P., Markham, H., & Weller, R. O. (2003). Neuropathology of human Alzheimer disease after immunization with amyloid-beta peptide: A case report. *Nature Medicine*, 9(4), 448–452. http://dx.doi.org/10.1038/nm840.

Nikolic, W. V., Bai, Y., Obregon, D., Hou, H., Mori, T., Zeng, J., et al. (2007). Transcutaneous beta-amyloid immunization reduces cerebral beta-amyloid deposits without T cell infiltration and microhemorrhage. *Proceedings of the National Academy of Sciences of the United States of America*, 104(7), 2507–2512.

Obregon, D., Hou, H., Bai, Y., Nikolic, W. V., Mori, T., Luo, D., et al. (2008). CD40L disruption enhances Abeta vaccine-mediated reduction of cerebral amyloidosis while minimizing cerebral amyloid angiopathy and inflammation. *Neurobiology of Disease*, 29(2), 336–353.

Patel, N. S., Paris, D., Mathura, V., Quadros, A. N., Crawford, F. C., & Mullan, M. J. (2005). Inflammatory cytokine levels correlate with amyloid load in transgenic mouse models of Alzheimer's disease. *Journal of Neuroinflammation*, 2(1), 9. http://dx.doi.org/10.1186/1742-2094-2-9.

Petrushina, I., Ghochikyan, A., Mktrichyan, M., Mamikonyan, G., Movsesyan, N., Davtyan, H., et al. (2007). Alzheimer's disease peptide epitope vaccine reduces insoluble but not soluble/oligomeric Abeta species in amyloid precursor protein transgenic mice. *The Journal of Neuroscience*, 27(46), 12721–12731.

Pfeifer, M., Boncristiano, S., Bondolfi, L., Stalder, A., Deller, T., Staufenbiel, M., et al. (2002). Cerebral hemorrhage after passive anti-Abeta immunotherapy. *Science*, 298(5597) 1379–1379.

Pickford, F., Masliah, E., Britschgi, M., Lucin, K., Narasimhan, R., Jaeger, P. A., et al. (2008). The autophagy-related protein beclin 1 shows reduced expression in early Alzheimer disease and regulates amyloid beta accumulation in mice. *The Journal of Clinical Investigation*, 118(6), 2190–2199. http://dx.doi.org/10.1172/JCI33585.

Ponomarev, E. D., Shriver, L. P., Maresz, K., & Dittel, B. N. (2005). Microglial cell activation and proliferation precedes the onset of CNS autoimmunity. *Journal of Neuroscience Research*, 81(3), 374–389.

Price, A. R., Xu, G., Siemienski, Z. B., Smithson, L. A., Borchelt, D. R., Golde, T. E., et al. (2013). Comment on "ApoE-directed therapeutics rapidly clear beta-amyloid and reverse deficits in AD mouse models". *Science*, 340(6135), 924-d. http://dx.doi.org/10.1126/science.1234089.

Priller, J., Flugel, A., Wehner, T., Boentert, M., Haas, C. A., Prinz, M., et al. (2001). Targeting gene-modified hematopoietic cells to the central nervous system: Use of green fluorescent protein uncovers microglial engraftment. *Nature Medicine*, 7(12), 1356–1361.

Rezai-Zadeh, K., Gate, D., Gowing, G., & Town, T. (2011). How to get from here to there: Macrophage recruitment in Alzheimer's disease. *Current Alzheimer Research*, 8(2), 156–163.

Ricote, M., Li, A. C., Willson, T. M., Kelly, C. J., & Glass, C. K. (1998). The peroxisome proliferator-activated receptor-gamma is a negative regulator of macrophage activation. *Nature*, 391(6662), 79–82. http://dx.doi.org/10.1038/34178.

Rogers, J., Webster, S., Lue, L. F., Brachova, L., Civin, W. H., Emmerling, M., et al. (1996). Inflammation and Alzheimers disease pathogenesis. *Neurobiology of Aging*, 17(5), 681–686.

Roher, A., Gray, E. G., & Paula-Barbosa, M. (1988). Alzheimer's disease: Coated vesicles, coated pits and the amyloid-related cell. *Proceedings of the Royal Society of London Series B, Biological Sciences*, 232(1269), 367–373.

Savarin-Vuaillat, C., & Ransohoff, R. M. (2007). Chemokines and chemokine receptors in neurological disease: Raise, retain, or reduce? *Neurotherapeutics, 4*(4), 590–601.

Schenk, D., Barbour, R., Dunn, W., Gordon, G., Grajeda, H., Guido, T., et al. (1999). Immunization with amyloid-beta attenuates Alzheimer-disease-like pathology in the PDAPP mouse. *Nature, 400*(6740), 173–177. http://dx.doi.org/10.1038/22124.

Selkoe, D. J. (2001). Alzheimer's disease: Genes, proteins, and therapy. *Physiological Reviews, 81*(2), 741–766.

Shie, F. -S., Montine, K. S., Breyer, R. M., & Montine, T. J. (2005). Microglial EP2 is critical to neurotoxicity from activated cerebral innate immunity. *Glia, 52*(1), 70–77.

Simard, A. R., Soulet, D., Gowing, G., Julien, J. P., & Rivest, S. (2006). Bone marrow-derived microglia play a critical role in restricting senile plaque formation in Alzheimer's disease. *Neuron, 49*(4), 489–502. http://dx.doi.org/10.1016/j.neuron.2006.01.022.

Stalder, A. K., Ermini, F., Bondolfi, L., Krenger, W., Burbach, G. J., Deller, T., et al. (2005). Invasion of hematopoietic cells into the brain of amyloid precursor protein transgenic mice. *The Journal of Neuroscience, 25*(48), 11125–11132.

Szekely, C. A., Thorne, J. E., Zandi, P. P., Ek, M., Messias, E., Breitner, J. C., et al. (2004). Nonsteroidal anti-inflammatory drugs for the prevention of Alzheimer's disease: A systematic review. *Neuroepidemiology, 23*(4), 159–169. http://dx.doi.org/10.1159/000078501.

Tesseur, I., Lo, A. C., Roberfroid, A., Dietvorst, S., Van Broeck, B., Borgers, M., et al. (2013). Comment on "ApoE-directed therapeutics rapidly clear beta-amyloid and reverse deficits in AD mouse models". *Science, 340*(6135), 924-e. http://dx.doi.org/10.1126/science.1233937.

Tesseur, I., Zou, K., Esposito, L., Bard, F., Berber, E., Can, J. V., et al. (2006). Deficiency in neuronal TGF-beta signaling promotes neurodegeneration and Alzheimer's pathology. *The Journal of Clinical Investigation, 116*(11), 3060–3069.

Town, T. (2009). Alternative Abeta immunotherapy approaches for Alzheimer's disease. *CNS & Neurological Disorders Drug Targets, 8*(2), 114–127.

Town, T., Laouar, Y., Pittenger, C., Mori, T., Szekely, C. A., Tan, J., et al. (2008). Blocking TGF-beta-Smad2/3 innate immune signaling mitigates Alzheimer-like pathology. *Nature Medicine, 14*(6), 681–687.

Town, T., Tan, J., Flavell, R. A., & Mullan, M. (2005). T-cells in Alzheimer's disease. *Neuromolecular Medicine, 7*(3), 255–264.

Vasilevko, V., Xu, F., Previti, M. L., Van Nostrand, W. E., & Cribbs, D. H. (2007). Experimental investigation of antibody-mediated clearance mechanisms of amyloid-beta in CNS of Tg-SwDI transgenic mice. *The Journal of Neuroscience, 27*(49), 13376–13383.

Veeraraghavalu, K., Zhang, C., Miller, S., Hefendehl, J. K., Rajapaksha, T. W., Ulrich, J., et al. (2013). Comment on "ApoE-directed therapeutics rapidly clear beta-amyloid and reverse deficits in AD mouse models". *Science, 340*(6135), 924-f. http://dx.doi.org/10.1126/science.1235505.

Vom Berg, J., Prokop, S., Miller, K. R., Obst, J., Kalin, R. E., Lopategui-Cabezas, I., et al. (2012). Inhibition of IL-12/IL-23 signaling reduces Alzheimer's disease-like pathology and cognitive decline. *Nature Medicine, 18*(12), 1812–1819. http://dx.doi.org/10.1038/nm.2965.

Vural, P., Degirmencioglu, S., Parildar-Karpuzoglu, H., Dogru-Abbasoglu, S., Hanagasi, H. A., Karadag, B., et al. (2009). The combinations of TNFalpha-308 and IL-6 174 or IL-10 1082 genes polymorphisms suggest an association with susceptibility to sporadic late-onset Alzheimer's disease. *Acta Neurologica Scandinavica, 120*(6), 396–401. http://dx.doi.org/10.1111/j.1600-0404.2009.01230.x.

Wegiel, J., & Wisniewski, H. M. (1990). The complex of microglial cells and amyloid star in three-dimensional reconstruction. *Acta Neuropathologica, 81*(2), 116–124.

Wilcock, D. M., Rojiani, A., Rosenthal, A., Subbarao, S., Freeman, M. J., Gordon, M. N., et al. (2004). Passive immunotherapy against Abeta in aged APP-transgenic mice reverses cognitive deficits and depletes parenchymal amyloid deposits in spite of increased vascular amyloid and microhemorrhage. *Journal of Neuroinflammation, 1*(1), 24.

Wisniewski, H. M., Barcikowska, M., & Kida, E. (1991). Phagocytosis of beta/A4 amyloid fibrils of the neuritic neocortical plaques. *Acta Neuropathologica, 81*(5), 588–590.

Wisniewski, H. M., Vorbrodt, A. W., Wegiel, J., Morys, J., & Lossinsky, A. S. (1990). Ultrastructure of the cells forming amyloid fibers in Alzheimer disease and scrapie. *American Journal of Medical Genetics Supplement, 7,* 287–297.

Wisniewski, H. M., Wegiel, J., Wang, K. C., Kujawa, M., & Lach, B. (1989). Ultrastructural studies of the cells forming amyloid fibers in classical plaques. *The Canadian Journal of Neurological Sciences, 16*(4 Suppl), 535–542.

Wyss-Coray, T. (2006). Inflammation in Alzheimer disease: Driving force, bystander or beneficial response? *Nature Medicine, 12*(9), 1005–1015. http://dx.doi.org/10.1038/nm1484.

Wyss-Coray, T., Feng, L., Masliah, E., Ruppe, M. D., Lee, H. S., Toggas, S. M., et al. (1995). Increased central nervous system production of extracellular matrix components and development of hydrocephalus in transgenic mice overexpressing transforming growth factor-beta 1. *The American Journal of Pathology, 147*(1), 53–67.

Wyss-Coray, T., Lin, C., Yan, F., Yu, G. Q., Rohde, M., McConlogue, L., et al. (2001). TGF-beta1 promotes microglial amyloid-beta clearance and reduces plaque burden in transgenic mice. *Nature Medicine, 7*(5), 612–618.

Wyss-Coray, T., Masliah, E., Mallory, M., McConlogue, L., Johnson-Wood, K., Lin, C., et al. (1997). Amyloidogenic role of cytokine TGF-beta1 in transgenic mice and in Alzheimer's disease. *Nature, 389*(6651), 603–606.

Yamanaka, M., Ishikawa, T., Griep, A., Axt, D., Kummer, M. P., & Heneka, M. T. (2012). PPARgamma/RXRalpha-induced and CD36-mediated microglial amyloid-beta phagocytosis results in cognitive improvement in amyloid precursor protein/presenilin 1 mice. *The Journal of Neuroscience, 32*(48), 17321–17331. http://dx.doi.org/10.1523/JNEUROSCI.1569-12.2012.

Zandi, P. P., Anthony, J. C., Hayden, K. M., Mehta, K., Mayer, L., Breitner, J. C., et al. (2002). Reduced incidence of AD with NSAID but not H2 receptor antagonists: The cache county study. *Neurology, 59*(6), 880–886.

Zhang, B., Gaiteri, C., Bodea, L. G., Wang, Z., McElwee, J., Podtelezhnikov, A. A., et al. (2013). Integrated systems approach identifies genetic nodes and networks in late-onset Alzheimer's disease. *Cell, 153*(3), 707–720. http://dx.doi.org/10.1016/j.cell.2013.03.030.

Zhu, Y., Obregon, D., Hou, H., Giunta, B., Ehrhart, J., Fernandez, F., et al. (2011). Mutant presenilin-1 deregulated peripheral immunity exacerbates Alzheimer-like pathology. *Journal of Cellular and Molecular Medicine, 15*(2), 327–338. http://dx.doi.org/10.1111/j.1582-4934.2009.00962.x.

Type 2 Diabetes Mellitus as a Risk Factor for Alzheimer's Disease

Jacqueline A. Bonds[1], Peter C. Hart[2],
Richard D. Minshall[3,4], Orly Lazarov[5],
Jacob M. Haus[6] and Marcelo G. Bonini[2,7]

[1]Graduate Program in Neuroscience and Department of Anatomy and Cell Biology, College of Medicine, University of Illinois at Chicago, Chicago, IL, United States [2]Department of Pathology, University of Illinois at Chicago, Chicago, IL, United States [3]Department of Anesthesiology, University of Illinois at Chicago, Chicago, IL, United States [4]Department of Pharmacology, University of Illinois at Chicago, Chicago, IL, United States [5]Department of Anatomy and Cell Biology, College of Medicine, University of Illinois at Chicago, Chicago, IL, United States [6]Department of Kinesiology and Nutrition, University of Illinois at Chicago, Chicago, IL, United States [7]Department of Medicine, University of Illinois at Chicago, Chicago, IL, United States

OUTLINE

Epidemiology	388
T2DM: Clinical Description	390
Mouse Models of Type 2 Diabetes	391
IR Knockout Mouse (IR$^{-/-}$)	*391*
Tissue-Specific Deletion of IR	*391*
IRS Knockout Mice	*392*
GLUT4 Knockout (Global and Tissue Specific)	*393*
MKR Mouse Model of Insulin Resistance	*394*

Genes, Environment and Alzheimer's Disease.
DOI: http://dx.doi.org/10.1016/B978-0-12-802851-3.00013-9

Mechanism and Pathways   394
  Mitochondrial Dysfunction in Diabetes   394
  Oxidative Stress and Inflammation   396
  Advanced Glycated Endproducts and the Receptor for AGEs   397
  Diabetes and Endothelial Dysfunction   398

Cerebrovascular Complications in T2DM: Implications for AD
  Development   400
  Dysfunctional Cerebral Neovascularization in T2DM   400
  Hyperglycemia and Alzheimer's Disease   402
  Vascular Inflammation in T2DM   402

Alzheimer's Disease: When Inflammation and Vascular
  Dysfunction Get to the Brain   404
  Endothelial Cell Caveolin-1, Caveolae and the Access of Insulin
    to the Brain   404
  Metabolic Flexibility Enables Cellular Plasticity, Tissue Repair,
    Regeneration, and Function   406

Conclusion   407

References   407

EPIDEMIOLOGY

Type 2 diabetes mellitus (T2DM) is a metabolic disorder that results from reduced or impaired responsiveness to insulin signaling (e.g., insulin resistance). The reduced glucose uptake by insulin-target organs, along with impaired fatty acid metabolism, results in increased free fatty acid in the circulation, causes chronic hyperglycemia, and sustained inflammation, which can evolve to cardiovascular and neurological dysfunction (CDC, 2014). According to the Centers for Disease Control and Prevention (CDC), the number of newly-diagnosed adults with diabetes in the United States reached 1.7 million in 2010, nearly three times that observed in 1990 (CDC, 2014). Diabetes currently affects over 29 million US citizens (9.3% of the population) and the healthcare costs associated with treating these patients total $245 billion per year, including $176 billion in direct costs and $69 billion in indirect costs (disability, work loss, premature mortality) (CDC, 2014).

The alarming increase in prevalence of T2DM has warranted several large-scale prospective studies to determine socioeconomic and environmental risk factors that have led to its increased incidence. Many of these studies have demonstrated lifestyle as a prevalent factor in contributing to the early onset and increased prevalence of T2DM, with overnutrition and decreased physical exercise leading to obesity as the major contributing factors to the pathology (Meigs, 2010). T2DM is associated with chronic inflammatory states, as cytokine and acute-phase protein production are typically elevated in diabetic patients (Pickup & Crook, 1998; Spranger et al., 2003). For example, the European Prospective Investigation into Cancer and Nutrition–Potsdam study identified increased interleukin 6 (IL-6) (odds ratio (OR) = 2.57, 95% confidence interval (CI) 1.24–5.47) and the interaction of both high tumor necrosis factor alpha (TNF-α) and IL-1β (OR = 2.3, 95% CI 1.1–4.9) as factors that may predict T2DM pathogenesis (Spranger et al., 2003). In a separate study, patients with T2DM were also shown to exhibit low-grade inflammation as indicated by elevated plasma levels of C-reactive protein (CRP) and IL-6 compared to healthy subjects (Herder et al., 2005). The concept that chronic inflammation may be an important contributing factor in the etiology of T2DM is becoming widely accepted. If this is the case, then chronic inflammation associated with other systemic pathophysiological conditions such as cardiovascular disease and neuropathy may be linked.

In line with these notions, patients with T2DM exhibit significantly increased incidence of cardiovascular disease with an associated increase in plasma level of von Willebrand Factor, a procoagulant required for normal hemostasis that can predispose individuals to increased risk of microvascular thrombosis when levels become elevated (Meigs, 2010). Further, it was also shown that subjects suffering from diabetes have an increased risk of stroke-associated dementia (hazard ratio (HR) = 4.20, 95% CI 2.18–8.25) (Luchsinger, Tang, Stern, Shea, & Mayeux, 2001), as well as stoke-associated nondementia cognitive impairments (HR = 2.30, 95% CI 0.88–5.91), indicating that T2DM is tightly linked to several neuropathies through similar vascular complications.

Recently, T2DM has emerged as a risk factor for Alzheimer's disease (AD). This possibility has been further investigated in several prospective cohort and case-control studies. For example, one study conducted using a cohort of the Rochester Epidemiology Project demonstrated that T2DM increased the relative risk of developing AD over twofold in men (relative risk (RR) = 2.27, 95% CI 1.55–3.31), whereas there was a nonsignificant trend of an increase in women (Leibson et al., 1997). Interestingly, there was no significant interaction between T2DM and AD when stratified by age despite the fact that 25.9%, or 11.8 million, of individuals age 65 years or older have diabetes (CDC, 2014). However, it was found that 35% of AD patients also present with T2DM (Janson et al., 2004). These data suggest that T2DM independently confers risk in populations regardless of age and thereby may promote early onset of AD. Interestingly, a subset of patients in the Cardiovascular Health Study showed only a slight increase in risk of dementia based on diabetes status alone (HR 1.45, 95% CI 0.89–2.37), whereas there was a marked increase in relative risk when stratified by both diabetes and a variant of apolipoprotein E (ApoEε4), known to be an associated risk factor for early onset of sporadic AD (HR 4.53, 95% CI 2.47–8.30) (Irie et al., 2008).

Taken together, epidemiological data indicate that T2DM and other cardiovascular disease risk factors increase the incidence and reduce the age of onset of AD in human populations.

T2DM: CLINICAL DESCRIPTION

T2DM is a chronic disease characterized by a state of persistent hyperglycemia and reduced responsiveness to insulin (i.e., insulin resistance). It is associated with a number of pathologies of diverse etiology collectively referred to as diabetic complications. These include renal failure, blindness, coagulopathies, impaired wound healing, enhanced cardiovascular disease risk, and neurodegeneration. Insulin resistance is the most significant prognostic indicator of T2DM. In fact, T2DM is preceded by a phase of hyperinsulinemia and normal plasma glucose levels that indicate that there is a compensatory response by the pancreas to increase insulin levels in order to enhance glucose uptake by target organs. Eventually, pancreatic beta cells fail to meet the body's need for increased insulin levels in order to maintain glucose within the normal range, which leads to a gradual and sustained increase in plasma glucose and T2DM.

T2DM is typically diagnosed based upon plasma glucose criteria obtained during fasting (FPG) or a 2-h plasma glucose value following a 75-g oral glucose tolerance test (OGTT) where FPG ≥126 mg/dL (7.0 mmol/L after at least an 8-h fast, or 2-h plasma glucose ≥200 mg/dL (11.1 mmol/L) during an OGTT. More recently, hemoglobin A1C (A1C) values ≥6.5% have been added as a third option in the diagnosis of T2DM. Finally, if a patient presents with classic symptoms of hyperglycemia (i.e., polydipsia, polyfagia, polyuria, fatigue, malaise) or a hyperglycemic crisis, in those with a random plasma glucose of ≥200 mg/dL (11.1 mmol/L), diabetes as an underlying cause should be considered and follow-up testing performed. As with most diagnostic tests, a positive test result should be repeated when feasible to rule out laboratory or other sources of error (Standards of Medical Care in Diabetes, 2014).

Once it develops, T2DM has a dramatic impact on whole body metabolism well beyond the handling and utilization of sugars; it also affects lipid processing, protein synthesis, and modifies hormonal and cytokine balance that impact every tissue in the body. Most commonly, T2DM presents as numerous associated pathologies that can vary in severity, including hypertension, ketoacidosis, nephropathy, ocular and retinal diseases, and increased risk of stroke. Importantly, T2DM has been shown to increase morbidity due to complications from cardiovascular disease and diabetic neuropathy in large part because of its adverse effects on the vasculature (CDC, 2014).

MOUSE MODELS OF TYPE 2 DIABETES

In addition to diet-induced models of T2DM, several knockout and transgenic mouse models have been generated that mimic one or more aspects of the very complex T2DM condition. Here we will highlight some of the many models that were created based on their utility in generating insights regarding basic mechanisms of insulin resistance, insulin signaling, and metabolic syndrome associated with the development of T2DM. It is important to note that in target organs, insulin signaling is mediated by a family of receptor tyrosine kinases, the two most prominent being the insulin receptor (IR) and the insulin-like growth factor 1 receptor (IGF-1R). Both receptors operate through the activation of the same downstream effector kinases, although these signaling pathways exert different biological actions. Many of the animal models are based on manipulation of these receptors, namely genetic deletion of the IR, insulin receptor substrate molecule (IRS), and IGF-1R. Models were also created in which the major glucose transporter activated by insulin (GLUT4) was deleted globally or in a tissue-specific manner. Particularities and major insights provided by each of these models are presented below.

IR Knockout Mouse (IR$^{-/-}$)

Mice lacking the IR are born resistant to the signaling effects of insulin and therefore rapidly develop diabetes and hyperinsulinemia. These animals are not viable and die within days after birth, most likely because of severe diabetic ketoacidosis (Accili et al., 1996). Importantly, it was observed that deletion of the IR leads to rapid failure of pancreatic β-cells to secrete insulin due to chronic hyperinsulinemia-induced negative feedback. Thus, there are few studies conducted using the global IR$^{-/-}$ mouse due to the lethal phenotype of this knockout strain.

Tissue-Specific Deletion of IR

In an attempt to overcome the severe phenotype of whole body IR deletion, tissue-specific IR$^{-/-}$ mice were created using Cre-lox/P technology. Muscle (Bruning et al., 1998), adipose (both white and brown fat tissue) (Bluher, Kahn, & Kahn, 2003), liver (Cohen et al., 2007; Michael et al., 2000), and β-cell (Kulkarni et al., 1999) specific IR$^{-/-}$ mice were generated that revealed varied physiologic and pathologic responses due to the lack of IR-dependent signaling in each cell type. The muscle-specific $IR^{-/-}$ mouse exhibited a mild phenotype despite evidence of reduced insulin-dependent glucose uptake by muscle. Contrary to expectations, these mice did not develop either insulin resistance or diabetes because of compensatory signaling by the IGF-1R and by noninsulin dependent AMP-activated kinase (AMPK)-driven glucose uptake. IR deletion in adipose cells also did not alter insulin sensitivity, although there was an obvious reduction in fat mass and plasma triglyceride levels observed in these mice. Interestingly, mice lacking IR expression in fat cells lived longer.

Deletion of the IR in β-cells generated an unexpected and rather interesting phenotype: impaired glucose-driven insulin release and an increase in appetite. The study by Kulkarni et al. (1999) showed that $IR^{-/-}$ β-cells responded to glucose abnormally and failed to release insulin into the bloodstream immediately following an increase in plasma glucose level, implicating the involvement of the IR in regulation of insulin release.

Interestingly, liver-specific $IR^{-/-}$ knockout mice showed the most detrimental phenotype. Severe insulin resistance and hyperglycemia most likely are related to failure to suppress hepatic gluconeogenesis. The IR was also specifically knocked out in the heart, vascular endothelium (Kondo et al., 2003), and central nervous system (CNS) (Fisher, Bruning, Lannon, & Kahn, 2005). In cardiac tissue, IR deficiency led to reduced heart size and contractility, while in endothelial cells, IR deficiency promoted impaired hypoxic-driven neovascularization. IR deficiency in the CNS led to increased weight gain (Bruning et al., 2000) and defective counter-regulatory responses to hypoglycemia, suggesting that insulin signaling in the CNS promotes satiety.

Taken together, these models provide support for the concept that diabetes results from abnormal insulin signaling in canonical and noncanonical target organs, and that ablation of the IR in a particular cell type is useful for obtaining a better understanding of insulin signaling in specific tissue types. Nevertheless, none of the models completely mimicked the systemic metabolic features of diabetes.

IRS Knockout Mice

Many of the most prominent insulin-activated signaling mechanisms are the result of IRS-mediated signaling downstream of insulin binding to the IR or IGF-1R. Mouse models in which one or a combination of IRS-target genes were genetically inactivated revealed a dramatic retardation in development associated with insulin resistance (Cho et al., 2001; Fantin, Wang, Lienhard, & Keller, 2000; Ogata et al., 2000). However, knockout of IRS1 did not result in the development of diabetes (Araki et al.,

1994). Deletion of IRS2 was shown to lead to diabetes both by impairing insulin action and by promoting insulin deficiency. It was noted that in IRS2 knockout animals, there was significant β-cell death in parallel with a reduction in responsiveness of peripheral tissue to insulin signaling (Withers et al., 1998). IRS3 knockouts presented no discernible phenotype. The reason why the IRS3 knockout showed a lack of phenotypic consequences is not yet clear, although compensation by IRS1 has been suggested (Liu, Wang, Lienhard, & Keller, 1999). Knockout of IRS4 resulted in mild phenotypic consequences resembling those observed with IRS1 deficiency. Notably, IRS4 knockout mice were characterized as having reproductive abnormalities (Fantin et al., 2000). In double knockouts of IRS1 and another IRS subtype, a variety of phenotypes were observed, ranging from embryonic lethality (in the case of IRS1/IRS2 dKO) to lack of additive effectives in IRS1/IRS4 dKO animals whose phenotype was identical to that of IRS1 knockouts. In the case of IRS1/IRS3 knockouts, insulin resistance with lipoatrophy and reduced intrahepatic and intramuscular deposits of triglycerides was observed.

GLUT4 Knockout (Global and Tissue Specific)

Insulin-stimulated glucose uptake requires the GLUT4 glucose transporter and therefore genetic mouse models were developed to abolish glucose transport by this mechanism. GLUT4 knockouts exhibited a severe phenotype characterized by growth retardation, underdeveloped adipose tissue, cardiac hypertrophy, and insulin resistance (Katz, Stenbit, Hatton, DePinho, & Charron, 1995). Since GLUT4 is a major facilitator of glucose uptake in skeletal muscle, muscle-specific deletion of GLUT4 phenocopied many of the features of the global knockout including severe insulin resistance and glucose intolerance (Zisman et al., 2000). Interestingly, GLUT4 knockout mice also showed phenotypic gender disparity with males developing hyperglycemia in the postprandial state whereas female knockout mice did not develop hyperglycemia.

Because of the association between insulin resistance and cardiovascular disease, cardiac-specific GLUT4 deficient mice were developed. In these animals, cardiac hypertrophy associated with increased cardiomyocyte size was observed. Although GLUT4 deletion in the heart did not lead to impaired contractile function, recovery from ischemic insult was impaired. This indicates that GLUT4 is essential for insulin-independent glucose uptake in the heart under hypoxic conditions wherein glycolysis compensates for deficits in the oxidative metabolism.

Adipose tissue depletion of GLUT4 generated complex and variable phenotypic consequences in different mouse specimens. Hyperinsulinemia, most likely due to hepatic insulin resistance and a reduced capacity to uptake glucose, indicate that the lack of GLUT4 activity in adipocytes in this model results in prominent and disproportionate effects in distant tissues. Interestingly, despite the hyperinsulinemia and desensitization to insulin signaling in muscle and liver, diabetes does not develop in these mice.

MKR Mouse Model of Insulin Resistance

The functional disruption of IGF1R in transgenic MKR mice was shown to cause T2DM (Fernandez et al., 2001). A mutant IGF1R driven by a creatinine kinase promoter (Ckm) encodes for a functionally disabled dominant negative receptor where lysine 1003 is replaced by an arginine residue (L1003R). The expression of mutant receptor (primarily in skeletal muscle and to a lesser extent (~10%) in the heart) allows for insulin binding but disables downstream signaling resulting in insulin resistance. This is because upon binding to insulin, L1003R-IGF1R dimerizes with the endogenous IR and produces a functionally inactive receptor complex. As such, mice expressing mutant IGF1R rapidly develop hyperglycemia and insulin resistance that evolves to β-cell dysfunction and diabetes. Interestingly, MKR mice are not obese but rather show a slight decrease in body weight throughout life as compared to wild-type controls. With its close resemblance to the human diabetic state, including loss of β-cell function due to progressive hyperinsulinemia and hyperglycemia, the MKR mouse model has emerged as a powerful and quite unique tool for investigating mechanisms that connect insulin insensitivity to diabetes.

MECHANISM AND PATHWAYS

The profound modification of fundamental metabolic and regulatory pathways in diabetes imposes severe adaptive and compensatory responses that impact a broad array of cellular signaling effectors and their interactions with coregulators and repressors. In this section, an overview of major systems affected by diabetes that have a direct impact on vascular health will be reviewed. For detailed information, we refer the reader to the many excellent reviews that have focused on each specific area.

Mitochondrial Dysfunction in Diabetes

Diabetes is a state of profound metabolic dysfunction where glucose is highly abundant but insulin-sensitive cells have impaired access to it. These cells adapt to utilize alternative mechanisms for energy production such as enhanced lipid utilization and the production and uptake of ketogenic substrates. The various tissues adapt differently to diabetes depending on their biosynthetic, metabolic, and energetic demands. In the diabetic state, mitochondrial dysfunction ensues at a degree that depends on the severity of hyperglycemia, hyperinsulinemia, inflammation, and duration of the metabolic syndrome. Mitochondria are dynamic organelles that undergo restructuring based on rates of fusion and fission. Fused mitochondria constitute networks of interconnected organelles that during states of optimal homeostatic conditions exhibit a high yield of adenosine triphosphate (ATP) production per unit of glucose and low rate of reactive oxygen species generation. Physiologically, mitochondria fission is believed to be part of an intricate mechanism of quality control where damaged, old, and dysfunctional mitochondria are separated from the network and eliminated by the process of mitophagy. In metabolic syndrome and diabetes, mitochondria become fragmented and fission rates considerably increase, although the pathophysiologic significance of this observation is still unclear. One possibility is that diabetes causes accelerated mitochondrial dysfunction resulting from inflammation, oxidative stress, and the activation of alternative metabolic pathways (such as lipid mobilization) that further stress cells and tissues by enhancing the level of reactive by-products such as oxidized lipids and aldehydes. It is also possible that diabetes inhibits fusion by directly by altering fusion-protein expression or enhancing the expression of elements of the fission machinery. In any case, the observed increase in mitochondrial fragmentation in the diabetic state is associated with mitochondrial dysfunction, reduced bioenergetic efficiency and capacity, and increased reactive oxygen species (ROS) production. The reduction in ATP production promotes the activation of adaptive signaling in an attempt to bolster alternative energy generating pathways, such as glycolysis, autophagy, and mitochondrial biogenesis.

Bioenergetic processes are enhanced by AMPK, a master sensor in the cell that promotes the activation of energy-producing processes while inhibiting energy-consuming ones. AMPK activation enhances recycling of proteins and entire organelles via autophagy in parallel with the inhibition of protein synthesis and cellular proliferation. Intuitively, the persistence of this state would impair healing and tissue repair. ATP is not the only energetically significant metabolite impacted by mitochondrial dysfunction. Importantly, NADH is oxidized in mitochondria and when complex I fails NADH accumulates in its reduced form, which promotes conversion of pyruvate into lactate. This activates glycolysis, resulting in ATP generation. Furthermore, the inactivation of sirtuins and deacetylases, which normally have key epigenetic and regulatory roles, directly or indirectly promote detrimental pathologic effects as exemplified by the known role of SIRT1, SIRT3, SIRT4, and SIRT6 as tumor suppressors. SIRT3, for instance, activates the mitochondrial antioxidant protein

MnSOD to scavenge and detoxify superoxide radicals (Tao et al., 2010; Zhu et al., 2012), preventing damage to integral components of the mitochondria metabolic network, such as aconitase (Hausladen & Fridovich, 1994). Therefore, by promoting mitochondrial dysfunction, diabetes is associated with homeostatic imbalance that negatively impacts tissue renewal and repair. Even cells that are predominantly glycolytic (such as the vascular endothelium) are affected by mitochondrial dysfunction due to the central role of mitochondria as cellular signaling hubs.

Oxidative Stress and Inflammation

Hyperglycemia promotes the glycation of many proteins resulting in enhanced antigenic responses. Associated with hyperinsulinemia and dyslipidemia, glycation also potentiates inflammatory responses by enhancing the production of pro-inflammatory cytokines such as IL-6 and TNFα and repressing the production of anti-inflammatory signals such as IL-10 and adiponectin by adipose tissue and macrophages. Therefore, diabetes produces a state of inflammation that contributes to vascular and neurovascular disease directly and by activating tissue-specific inflammation. For example, it is now well established that chronic or exacerbated systemic inflammation eventually activates microglia, the resident macrophages of the brain. Once activated microglia-driven inflammation persists for prolonged periods of time. Studies devoted to understanding this mechanism have shown that while peripheral inflammation resolves within days, microgliosis takes weeks to months to fully resolve, thus imposing lasting modifications to the neural landscape. Indirectly, inflammation is associated with enhanced oxidative stress via increased production of ROS and diminishing antioxidant responses, with the net result being enhancement of steady state ROS levels. Reactive oxygen species are important participants in the regulation of cell signaling and cell communication, but when produced chronically at high levels, they alter the chemical composition of proteins, lipids, and DNA resulting in their disassembly, dysfunction, or fragmentation. Intuitively, the accumulation of dysfunctional or highly altered biomolecules leads to dysfunction at the cellular level, and in the case of endothelial cells, to endothelial dysfunction, which is thought to impair the function of blood vessels and alter endothelial cell interactions with other cell types (as detailed below in the section 'Diabetes and Endothelial Dysfunction'). Therefore, chronic inflammation and progression of oxidative stress are inseparable from and intimately related to the pathogenesis of diabetes. Directly or through their effect on the vasculature, chronic inflammation and oxidative stress also contribute to the many complications associated with diabetes including retinopathy, nephropathy, cardiopathy, and neuropathy.

Advanced Glycated Endproducts and the Receptor for AGEs

No organ system is immune from microvascular damage caused by hyperglycemia. Heart disease is increased by up to tenfold in people with diabetes compared to the general age-matched population. In addition, individuals with diabetes have a life span that is reduced by up to 15 years (Troiani, Draicchio, Bonci, & Zannoni, 1993; Walks, Lavau, Presta, Yang, & Bjorntorp, 1983; Welle, Thornton, Statt, & McHenry, 1994). Advanced glycation end-products (AGEs) are widely believed to modulate the development and progression of diabetic complications through Receptor for AGE (RAGE)-dependent and independent interactions. AGEs form as a result of nonenzymatic glycoxidation. This process is greatly accelerated in conditions such as T2DM and impaired glucose tolerance where hyperglycemia is evident. AGEs play a critical role in the pathogenesis of chronic diabetic complications by altering cellular structure and function. Particularly damaging effects of AGEs are produced by the covalent crosslinking of proteins, which confers resistance to proteolysis (DeGroot et al., 2001). Formation of AGEs on proteins causes structural distortion, loss of side chain charge, and functional impairment (Ahmed, Dobler, Dean, & Thornalley, 2005). Protein modification is also damaging when amino acid residues are located within sites critical for protein–protein interactions, enzyme–substrate interactions, or protein–DNA interactions. AGE deposits are commonly found in atherosclerotic plaques, vascular smooth muscle, and in myocardial tissues; serum AGE levels are elevated in diabetic patients with coronary heart disease (CHD) and correlate strongly with CHD severity (Sakata, Meng, Jimi, & Takebayashi, 1995). Similarly, both AGEs and RAGE have been found in the brain of patients with AD (Smith et al., 1994; Yan et al., 1996).

RAGE is a promiscuous pattern recognition receptor that spans the cell membrane by having distinct intracellular, transmembrane, and extracellular domains. It is part of the immunoglobulin superfamily of proteins and plays a role in mediating immune, inflammatory, and mitogenic responses. The ligand binding extracellular domain consists of three Ig-like domains: one V-type and two C-type domains. AGEs and other ligands, such as the S100 calgranulins, high mobility group box 1, amyloid-β (Aβ), and β-sheet peptides can bind to RAGE and initiate persistent activation of pro-inflammatory mediators such as nuclear factor-κB (Ramasamy, Yan, & Schmidt, 2012). Sturchler et al. (Sturchler, Galichet, Weibel, Leclerc, & Heizmann, 2008) demonstrated that oligomers of Aβ bind to the V-type domain, whereas Aβ aggregates associate with C-type domains. Under normal physiological conditions, RAGE expression is low until induced by pro-inflammatory signals. That said, the RAGE gene is transcriptionally regulated by factors such as IL-6 and NF-κβ and thus RAGE is capable of self-sustaining NF-κβ activation. RAGE–ligand interactions also result in an increased level of p65 and thus is capable of amplifying the inflammatory response (Bierhaus et al., 2005; Bierhaus & Nawroth, 2009). Furthermore, important clues are emerging on how the RAGE pathway is intertwined with cytokine, lipopolysaccharide, oxidized low-density lipoprotein, and hyperglycemia-initiated signaling cascades that exacerbate the inflammatory sequela (Cassese et al., 2008; Howard, McNeil, Xiong, Xiong, & McNeil, 2011; Lee et al., 2013; Piperi, Adamopoulos, Dalagiorgou, Diamanti-Kandarakis, & Papavassiliou, 2012; Remor et al., 2011; Yao & Brownlee, 2010).

Interruption of RAGE signaling is thought to be achieved by receptor ectodomain shedding via action of disintegrin and metalloproteinase 10 (ADAM10), which cleaves the RAGE ectodomain thereby creating a soluble form of RAGE (sRAGE) that is released into the interstitial space (Metz, Kojro, Rat, & Postina, 2012). Soluble forms of RAGE are decreased in inflammatory conditions such as insulin resistance and coronary artery disease (CAD) in humans, and treatment with recombinant sRAGE suppresses the development of atherosclerosis in animal models of accelerated diabetic CAD (Basta et al., 2006; Falcone et al., 2005; Park et al., 1998). Despite evidence that treatment with sRAGE suppresses diabetic atherosclerosis in preclinical models (Park et al., 1998) and that a decrease in the circulating concentration of RAGE is associated with increased severity of diabetic microvascular complications (Grossin et al., 2008), it is currently unknown whether RAGE ectodomain shedding can actually downregulate RAGE expression in the plasma membrane. Of note, mutant AβPP mice (which exhibit enhanced production of Aβ and its subsequent deposition in the brain) treated with sRAGE resulted in a decrease in total Aβ content of the brain and prevented additional parenchyme accumulation of Aβ oligomers in the brain (Hsia et al., 1999).

Diabetes and Endothelial Dysfunction

Oxidative stress, accumulation of oxidized and modified lipids in the blood plasma, elevation of inflammatory cytokines, and the generation of electrophiles such as aldehydes, nitroalkenes, and reactive organic compounds impose significant challenges to the endothelium. Reactive species modify electron rich thiol residues and thereby modify proteins with little specificity. They gain access to the cytosol and organelles though either active transport or passive diffusion through membranes. Of the many nonspecific targets of such species, phosphatase and tensin homolog (PTEN) and caveolin-1 (cav-1) are notorious because the modification of their activities promote significant changes in cell signaling and can even change endothelial cell identity. PTEN is generally considered to be a tumor suppressor. It opposes Akt signaling by limiting the availability of the Akt activating signal 3,4,5-phosphatydyl-triphosphate (3,4,5 PiP_3).

PTEN is also linked to the prevention of endothelial and epithelial to mesenchymal transition (EndoMT, EMT) and is generally believed to be a fundamental player in maintenance of the differentiated cellular phenotype. PTEN's activity depends on active-site residue Cys 124, a low pKa thiol particularly vulnerable to chemical modification and oxidation. When modified, Cys124 is rendered inactive and it can no longer hydrolyze 3,4,5-PiP$_3$. cav-1 is a scaffold protein present in cholesterol-enriched microdomains (CEM) of the plasma membrane. It is required for formation of caveolae, the small plasmalemmal vesicles produced upon invagination of plasma membrane CEMs. Caveolae are responsible for the transcellular transport (transcytosis) of macromolecules from the plasma through the endothelium. By directly associating with many client proteins, cav-1 regulates their function and maintains endothelial cell homeostasis. For example, a decrease in cav-1 expression results in dysregulated (uncoupled) endothelial nitric oxide synthase (eNOS) in that it becomes a primary source of oxidative stress. We and others are currently testing the hypothesis that an increase in oxidative stress dysregulates central pathways involved in maintaining endothelial cell homeostasis and the fully-differentiated, contact-inhibited, and quiescent state. Dysfunctional endothelium accelerates atherosclerosis, exhibits defective vasodilatory responses upon agonist stimulation, and fails to maintain barrier integrity resulting in the formation of edema and vascular cuffing. PTEN and cav-1 appear to be tightly linked in that the loss of PTEN activity can lead to cav-1 degradation, and vice versa, by mechanisms that are still under investigation but likely involve oxidative stress. Therefore, loss of function of critical signaling regulators in endothelial cells is thought to contribute to the etiology of cardiovascular disease characterized by vascular hyperpermeability, chronic inflammation, and hypertension while also compromising the supportive role endothelial cells play in maintaining vascular muscle, circulating immune cell, and neural progenitor cell functions dependent on the endothelial cell niche. Therefore, disruption of the ability of the endothelial niche to maintain homeostasis of partner cells may be linked to neuropathologies such as AD as reviewed below in the section 'Cerebrovascular Complications in T2DM: Implications for AD Development'.

Despite the fact that there are several converging pathways contributing to T2DM and AD pathogenesis, the complexity of the relationships between pathways that may promote their onset make it difficult to elucidate clinically relevant similarities between T2DM and AD. For instance, excess circulating neuropeptide-Y (NPY), a factor widely expressed in the CNSs that regulates many physiological processes including food intake, has been shown to facilitate metabolic syndrome (Kuo et al., 2007) and is elevated in T2DM patients (Rasul, Ilhan, Wagner, Luger, & Kautzky-Willer, 2012). However, NPY levels were found not to differ between T2DM and polyneuropathy (Rasul et al., 2012) or in patients with AD when compared to control subjects (Foster et al., 1986). While steady-state circulating levels of NPY may differ between patients of varying disease states, the contribution of NPY signaling to pro-inflammatory disorders such as obesity and metabolic syndrome may be relevant in the etiology of neuropathies in conjunction with other developing pathologies.

Taken together, the complex systemic effects of T2DM on neuropathies such as AD require further investigation. In addition to insulin resistance, chronic inflammation is a critical factor involved in the etiology of both T2DM and AD and may therefore be important to analyze T2DM as risk factor for AD pathogenesis.

CEREBROVASCULAR COMPLICATIONS IN T2DM: IMPLICATIONS FOR AD DEVELOPMENT

Microvessels support their surrounding environment via paracrine signaling mediated by trophic factors, such as VEGF, which induces endothelial cell proliferation, stimulates brain plasticity and remodeling, and reduces neuronal degeneration. In the injured brain, newly formed vessels create a niche that can support the migration and neuronal differentiation of neuroblasts (Dobkin, 2004; Zhang & Chopp, 2009). In turn, neuroblasts reinforce angiogenesis via release of vascular endothelial growth factor (Johansson, 2007). Under hypoxic conditions, there is an upregulation of hypoxia-inducible factor alpha, which promotes the formation of new blood vessels (angiogenesis) via activation of other pro-angiogenic factors such as VEGF, angiopoietins (Ang), and bFGF, to name a few (Ergul, Abdelsaid, Fouda, & Fagan, 2014). This same process occurs in areas of the uninjured brain and is an important response to the changing metabolic demands of the brain. This is of particular importance in the context of metabolic disorders such as T2DM, which has a significant impact on microvascular integrity.

Dysfunctional Cerebral Neovascularization in T2DM

Several studies have shown that there is an increase in cerebral neovascularization in T2DM (Li et al., 2010; Prakash, Johnson, Fagan, & Ergul, 2013; Silvestre & Levy, 2006). However, these new vessels do not mature properly, resulting in an increase of unperfusable vasculature and blood–brain barrier (BBB) hyperpermeability (Prakash et al., 2012). In turn, this creates a hypoxic environment where the metabolic demands of the surrounding tissue cannot be met. It is important here to take into account that these studies focused on neovascularization in the cerebral cortex and did not focus on the neurogenic areas of the brain that are associated with learning and memory: the subgranular layer (SGL) of the hippocampus and the subventricular zone (SVZ). The vasculature is a critical regulator of the proliferative neural stem cell (NSC) microenvironment. Neurogenesis plays a role in the restorative attempts following an insult to the brain. NSCs are intimately associated with vascular endothelial cells within the niches and their self-renewal, proliferation, and early differentiation tightly depend on homeostatic endothelial functions (Eichmann & Thomas, 2013). In the SVZ, there is direct access to factors in blood and cerebrospinal fluid, and thus the established vasculature network of the adult SVZ niche is a key component of the specialized microenvironment that supports NSCs and their progeny (Tavazoie et al., 2008). In the SGL, neurogenesis is closely associated with a process of active vascular recruitment and subsequent remodeling (Palmer, Willhoite, & Gage, 2000). Up to 37% of the cells proliferating in the SGZ are endothelial precursors (Abrous, Koehl, & Le Moal, 2005). Furthermore, NPCs and angioblasts proliferate together in clusters associated with the microvasculature of the SGL, and cells within these clusters express the VEGF-receptor Flk-1 (Palmer et al., 2000). The clustering of neural and endothelial precursors suggests that neurogenesis involves cross-talk with endothelial precursors.

In the context of T2DM, Beauquis et al. (2010) showed that vascularization of the hippocampus is significantly decreased. This same study also shows that while there is a significant increase in the number of proliferating neuroblasts, these cells do not survive to incorporate into the granular cell layer. Their findings imply that the survival of new neurons depends heavily on the ability of the brain to provide functional vasculature. The formation of new neurons in both the SVZ and the SGL are crucial to brain plasticity, an important function for recovery after trauma (Lazarov, Mattson, Peterson, Pimplikar, & van Praag, 2010). This process of neurogenesis has been shown to be defective in AD (Demars, Hu, Gadadhar, & Lazarov, 2010; Lazarov et al., 2010).

Interestingly, these findings are similar to what has been shown following an ischemic stroke, another risk factor associated with T2DM. It has been observed that depletion of neurogenesis exacerbates stroke outcome leading to increased infarct volume and enhanced neurological deficits (Wang et al., 2012). It has also been shown that following stroke, there is an increase in angiogenesis and the number of NPCs 72h and 7 days after injury (Font, Arboix, & Krupinski, 2010). However, most NPCs die and neural replacement does not take place. Taken together, the compromised vascular function observed in T2DM not only increases the risk of an ischemic event but can also undermine the regenerative capacity of the brain, which can cause and/or accelerate the onset of AD.

Hyperglycemia and Alzheimer's Disease

There are several risk factors that contribute to the development of AD, including insulin resistance and T2DM. According to a recent study published in the journal *Diabetes*, more than 80% of AD cases presented with either T2DM or an impaired glucose metabolism disorder (data collected from the Mayo Clinic Alzheimer Disease Patient Registry) (Janson et al., 2004). Focusing on the patients with T2DM, there was a significant positive correlation between the duration of diabetes and the density of Aβ plaques (adjusting for age) (Janson et al., 2004). One possible explanation for this observation is the presence of hyperglycemia, a common complication in T2DM. Hyperglycemia has been observed to result in increased expression of the RAGE and RAGE ligands either by exposure to elevated levels of blood glucose, or through increased ROS production (Smith, Sayre, & Perry, 1996; Yan, Stern, & Schmidt, 1997; Yao & Brownlee, 2010). These, as well as other advanced glycation end products, are also observed in Aβ plaques and neurofibrillary tangles present in AD.

Diabetes induced hyperglycemia causes severe deficits in cerebrovascular structure and function. Increased vascular tone/rigidity, decreased cerebral blood flow, and increased expression of matrix metalloproteinases (MMPs) causes a worsening of BBB function hyperglycemic patients (Ergul, Li, Elgebaly, Bruno, & Fagan, 2009). These symptoms are accompanied by degeneration of endothelial cells and pericytes, as well as increased aggregation and adhesion of platelets to the endothelium (Cheng et al., 2015; Lorenzi, Cagliero, & Toledo, 1985; Vinik, Erbas, Park, Nolan, & Pittenger, 2001; Williams et al., 1998). As a result of compromised vascular function, hyperglycemia is a major contributor to neuronal injury, particularly following stroke (Capes, Hunt, Malmberg, Pathak, & Gerstein, 2001; Ergul, Kelly-Cobbs, Abdalla, & Fagan, 2012; Sasaki et al., 2001; Seners, Turc, Oppenheim, & Baron, 2015; Seners et al., 2014). Together, these cerebrovascular complications contribute to the increased risk of developing AD in T2DM.

Vascular Inflammation in T2DM

A common complication in T2DM is the formation of atherosclerotic plaques and impaired endothelium-dependent vasodilation. Increased levels of CRP, as well as the modified lipoproteins within these plaques, potentiate the release and activation of inflammatory molecules, namely IL-1, IL-6, interferon gamma (IFN-γ), and TNF-α. Elevated levels of CRP have been shown to decrease the expression of eNOS, thereby disrupting NO signaling and interfering with the ability of endothelial cells to respond to changes in blood flow (Venugopal, Devaraj, Yuhanna, Shaul, &

Jialal, 2002; Verma et al., 2002). Endothelial cell dysfunction combined with elevated levels of pro-inflammatory cytokines induce the expression of endothelial adhesion molecules, thereby increasing the influx of inflammatory cells across the endothelium and vessel wall hardening (Blake & Ridker, 2001; Davies et al., 1993; De Vriese, Verbeuren, Van de Voorde, Lameire, & Vanhoutte, 2000).

IL-6 and TNF-α activate endothelial cells to synthesize cellular adhesion molecules (CAMs) (Etter et al., 1998). The initial rolling of inflammatory cells along endothelial cells is mediated by selectins. E-selectin is expressed by endothelial cells and P-selectin is also expressed by platelets, as well as endothelial cells. Upregulation of both selectins has been associated with T2DM and increased serum levels of soluble E-selectin has been shown to predict an ischemic event (Matsumoto, Fujishima, Moriuchi, Saishoji, & Ueki, 2010; Neubauer et al., 2010). The immunoglobulin family of CAMs mediates the attachment and transendothelial migration of leukocytes. These include the intracellular adhesion molecule-1 and vascular adhesion molecule (Davies et al., 1993).

As the plaque continues to build within the vessel wall, monocytes attach and differentiate into macrophages. Micropinocytosis of modified LDLs by macrophages induces formation of foam cells. Formation of these cells increases ROS production and macrophage recruitment contributing further to the impairment of NO signaling and endothelial dysfunction (Rajagopalan, Meng, Ramasamy, Harrison, & Galis, 1996; Yorek, 2003). In response to inflammatory cell recruitment, the vessel wall attempts to repair itself by inducing the proliferation of smooth muscle cells. These cells surround the foam cells, thereby creating a fibrous cap, which further contributes to impaired blood flow. The plaques within the cap become destabilized through the production of ROS and activation of MMPs (Galis, Sukhova, Lark, & Libby, 1994; Rajagopalan et al., 1996; Shah et al., 1995). The presence of IFN-γ inhibits the production of collagen, thereby contributing to weakening of the fibrous cap. Formations of cholesterol crystals erode the endothelium causing encroachment of the plaque into the vessel lumen leading to obstructed blood flow. Rupture of the plaque releases the hardened lipids into the blood stream where they can travel to smaller arterioles to potentially induce an ischemic event and neurovascular uncoupling.

Impairments in cerebral blood flow causes significant injury to neurons and the connections that are made between them, which is a contributing factor to the cognitive decline observed in AD (Marchant et al., 2012). In an attempt to isolate the effects of cerebrovascular alterations in cognition in the context of AD, Marchant et al. discovered that cerebrovascular disease and Aβ aggregation were independent contributors to cognitive impairment. This study, as well as several others, suggests that the cognitive decline observed in cerebrovascular disease and AD may share common pathways (altered endothelial regulation of cerebral blood flow, pathological changes in BBB integrity and function, and neurovascular uncoupling) (Girouard & Iadecola, 2006; Girouard, Park, Anrather, Zhou, & Iadecola, 2007; Marchant et al., 2012; Mooradian, 1988).

ALZHEIMER'S DISEASE: WHEN INFLAMMATION AND VASCULAR DYSFUNCTION GET TO THE BRAIN

Endothelial Cell Caveolin-1, Caveolae and the Access of Insulin to the Brain

As presented above, many are the pathways that contribute to change the vasculature in sustained and deleterious ways. How then do functional changes in the vascular network affect the viability of neural tissue impacting cognition, motion, and memory? Recent findings indicate that sustained inflammation affects the capacity of endothelium to transport peptides from the circulation to underlying tissues. The endothelium, in fact, plays a critical role in actively transporting a number of macromolecular compounds through the blood vessel wall. This transport (generally referred to as "transcytosis") is mediated by endocytic structures called caveolae that reside on the plasma membrane and internalize macromolecules such as insulin and albumin via c-Src and dynamin-dependent vesicle fission (Sverdlov, Shajahan, & Minshall, 2007), which then move across the cell body where they fuse with the basal membrane and exocytose stored contents. Formation of caveolae is dependent on the expression of cav-1, a 22 kDa protein that assembles into oligomeric membrane-associated chains that promote the invagination of cholesterol-enriched domains (John, Vogel, Tiruppathi, Malik, & Minshall, 2003; Minshall et al., 2000; Vogel et al., 2001). Caveolae also contain receptors, for example for insulin and albumin, which upon activation promote the transcytosis of bound and fluid phase macromolecules from the circulation to the underlying cells (Wang, Wang, Aylor, & Barrett, 2015; Wang, Wang, & Barrett, 2011) (Figure 13.1). Recently, it was demonstrated that cav-1 is susceptible to nitric oxide and oxidative stress with the chemical modification of a specific cysteine residue (Cys 156) being sufficient to trigger protein degradation and depletion (Bakhshi et al., 2013). Oxidative stress accompanies inflammation and is significantly enhanced in diabetic patients (Brouwers et al., 2010; Giacco & Brownlee, 2010; Kohen Avramoglu et al., 2013; Pi et al., 2009; Stadler, 2012), providing a link between chronic inflammation and alterations to the endothelium that impact cerebrovascular homeostasis. Depletion of cav-1 disrupts caveolae formation and caveolae-dependent transport. Since this mechanism is

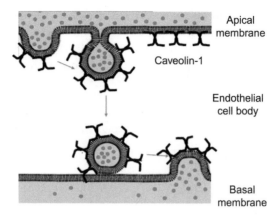

FIGURE 13.1   Caveolae-mediated transendothelial transport. Caveolin-1 is required for caveolae formation and transcellular transport (transcytosis) of plasma macromolecules.

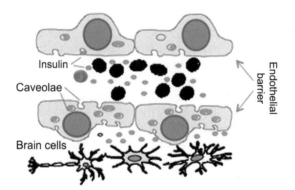

FIGURE 13.2   Schematic representation of transendothelial insulin transport via caveolae. Endocytosis and transcytosis of insulin through the endothelial barrier requires caveolae. Inflammation, degradation of caveolin-1, and depletion of caveolae associated with insulin resistance and reduced signaling in the brain may be the common features of T2DM and AD.

largely responsible for the uptake and transport of insulin, a 51 amino acid polypeptide (~5.8 kDa) impairment of caveolae formation at the level of the endothelium is likely to negatively impact the access of insulin to the brain (Wang, Liu, Li, & Barrett, 2006; Wang et al., 2011, 2015) (Figure 13.2).

In the brain, a variety of glucose transporters (GLUT1, GLUT2, GLUT3, GLUT4, GLUT5, and GLUT8) are expressed at different levels in the various cells of the brain (McEwen & Reagan, 2004). GLUT5 for instance is prevalent in microglia whereas GLUT1 predominates in astrocytes. In the cerebellum, neural subpopulations in the cortex, and the hippocampus, which largely controls memory, the insulin-activated transporter GLUT4 is particularly important as it controls a large fraction of the glucose uptake by the neurons in these regions. It is therefore not only possible but likely that diminished access of insulin to these areas will produce significant deficits in the ability of these cells to nourish themselves, function normally, and survive stress. The role of cav-1 and the ability of the neurovascular endothelium to maintain insulin signaling in these areas of the brain is an active area of research that will likely provide a clearer picture regarding the link between diabetes and risk of developing AD.

Metabolic Flexibility Enables Cellular Plasticity, Tissue Repair, Regeneration, and Function

Different cellular functional states and the interactions with the microenvironment require a defined metabolic program that supports activities such as plasticity, repair, and regeneration by providing energy and building blocks that enable protein and membrane synthesis, DNA duplication and transcription, resistance to stress, and proliferation. Metabolism is fueled by glucose, glutamine, lipids, ketone bodies, and to a lesser extent, amino acids. Glucose and glutamine are by far the most significant sources of energy due to their indispensable roles in supporting glycolysis, the tricarboxylic acid cycle, and indirectly mitochondrial respiration, the three pillars of cellular energy metabolism. The different catabolic and anabolic needs of proliferation, quiescence, differentiation, stemness, as well as the requirements imposed by transitions between these states requires that a combination of substrates be utilized by cells reacting to microenvironmental stimulation or entrance into a specific program. For instance, transition between quiescence and proliferation requires the activation of mitochondrial respiration (OXPHOS) to provide large amounts of ATP needed for cellular division. Reentry into quiescence activates a switch back to glycolysis as the primary means of sustaining this stationary state. The same way differentiation is marked by an upregulation of aerobic respiration, glycolysis is not only a feature but as many recent studies indicate, a requirement of stem-like states and pluripotency. As such, failure to finely regulate and effectively execute these metabolic programs can lead to premature differentiation with the exhaustion of the stem cell pool, incomplete differentiation (with loss of specialized functions), and vulnerability to stress and cell cycle arrest with tissue degeneration the likely outcome.

It is immediately obvious how a deficit in insulin signaling in specific areas of the brain might contribute to tissue injury and disease. As mentioned above in the section 'Endothelial Cell Caveolin-1, Caveolae and the Access of Insulin to the Brain', hippocampal, cerebellar, and some subpopulations of cortical neurons depend on insulin to capture and utilize glucose. Lack of insulin signaling in these cells imposes an obligatory switch in fuel utilization with a negative impact on cells that normally depend on glucose. These include cells that are directly involved in processes like learning, memory, and cognition (hippocampus), and cells responsible for cerebral regeneration and plasticity (neural stem and progenitor cells).

CONCLUSION

In summary, many different mechanisms are likely to contribute to the onset, progression, and severity of neural degeneration and AD as they relate to vascular disease and diabetes. The mechanisms proposed here are viewed as central etiologic components and likely useful therapeutic targets. Currently, enhancement of insulin signaling in the brain is being pursued as a potential therapeutic for AD, and both pharmacologic and lifestyle changes that have already been shown to have a positive effect in the management of diabetes itself include healthier diets, exercise, and lower caloric intake. These not only reinforce the epidemiologic connection between diabetes, vascular dysfunction, and AD, but also reiterate the fact that effective treatments for AD need to take into consideration vascular health and endocrine interventions.

References

Abrous, D. N., Koehl, M., & Le Moal, M. (2005). Adult neurogenesis: From precursors to network and physiology. *Physiological Reviews, 85,* 523–569.

Accili, D., Drago, J., Lee, E. J., Johnson, M. D., Cool, M. H., Salvatore, P., et al. (1996). Early neonatal death in mice homozygous for a null allele of the insulin receptor gene. *Nature Genetics, 12,* 106–109.

Ahmed, N., Dobler, D., Dean, M., & Thornalley, P. J. (2005). Peptide mapping identifies hotspot site of modification in human serum albumin by methylglyoxal involved in ligand binding and esterase activity. *The Journal of Biological Chemistry, 280,* 5724–5732.

Araki, E., Lipes, M. A., Patti, M. E., Bruning, J. C., Haag, B., III, Johnson, R. S., et al. (1994). Alternative pathway of insulin signalling in mice with targeted disruption of the IRS-1 gene. *Nature, 372,* 186–190.

Bakhshi, F. R., Mao, M., Shajahan, A. N., Piegeler, T., Chen, Z., Chernaya, O., et al. (2013). Nitrosation-dependent caveolin 1 phosphorylation, ubiquitination, and degradation and its association with idiopathic pulmonary arterial hypertension. *Pulmonary Circulation, 3,* 816–830.

Basta, G., Sironi, A. M., Lazzerini, G., Del Turco, S., Buzzigoli, E., Casolaro, A., et al. (2006). Circulating soluble receptor for advanced glycation end products is inversely associated with glycemic control and S100A12 protein. *The Journal of Clinical Endocrinology and Metabolism, 91,* 4628–4634.

Beauquis, J., Homo-Delarche, F., Giroix, M. H., Ehses, J., Coulaud, J., Roig, P., et al. (2010). Hippocampal neurovascular and hypothalamic-pituitary-adrenal axis alterations in spontaneously type 2 diabetic GK rats. *Experimental Neurology, 222,* 125–134.

Bierhaus, A., Humpert, P. M., Morcos, M., Wendt, T., Chavakis, T., Arnold, B., et al. (2005). Understanding RAGE, the receptor for advanced glycation end products. *Journal of Molecular Medicine, 83*, 876–886.

Bierhaus, A., & Nawroth, P. P. (2009). Multiple levels of regulation determine the role of the receptor for AGE (RAGE) as common soil in inflammation, immune responses and diabetes mellitus and its complications. *Diabetologia, 52*, 2251–2263.

Blake, G. J., & Ridker, P. M. (2001). Novel clinical markers of vascular wall inflammation. *Circulation Research, 89*, 763–771.

Bluher, M., Kahn, B. B., & Kahn, C. R. (2003). Extended longevity in mice lacking the insulin receptor in adipose tissue. *Science, 299*, 572–574.

Brouwers, O., Niessen, P. M., Haenen, G., Miyata, T., Brownlee, M., Stehouwer, C. D., et al. (2010). Hyperglycaemia-induced impairment of endothelium-dependent vasorelaxation in rat mesenteric arteries is mediated by intracellular methylglyoxal levels in a pathway dependent on oxidative stress. *Diabetologia, 53*, 989–1000.

Bruning, J. C., Gautam, D., Burks, D. J., Gillette, J., Schubert, M., Orban, P. C., et al. (2000). Role of brain insulin receptor in control of body weight and reproduction. *Science, 289*, 2122–2125.

Bruning, J. C., Michael, M. D., Winnay, J. N., Hayashi, T., Horsch, D., Accili, D., et al. (1998). A muscle-specific insulin receptor knockout exhibits features of the metabolic syndrome of NIDDM without altering glucose tolerance. *Molecular Cell, 2*, 559–569.

Capes, S. E., Hunt, D., Malmberg, K., Pathak, P., & Gerstein, H. C. (2001). Stress hyperglycemia and prognosis of stroke in nondiabetic and diabetic patients: A systematic overview. *Stroke, 32*, 2426–2432.

Cassese, A., Esposito, I., Fiory, F., Barbagallo, A. P., Paturzo, F., Mirra, P., et al. (2008). In skeletal muscle advanced glycation end products (AGEs) inhibit insulin action and induce the formation of multimolecular complexes including the receptor for AGEs. *The Journal of Biological Chemistry, 283*, 36088–36099.

Centers for Disease Control and Prevention (CDC), (2014). *National diabetes statistics report: Estimates of diabetes and its burden in the United States, 2014.* Atlanta, GA: US Department of Health and Human Services.

Cheng, Z., Jiang, X., Pansuria, M., Fang, P., Mai, J., Mallilankaraman, K., et al. (2015). Hyperhomocysteinemia and hyperglycemia induce and potentiate endothelial dysfunction via mu-calpain activation. *Diabetes, 64*(3), 947–959.

Cho, H., Mu, J., Kim, J. K., Thorvaldsen, J. L., Chu, Q., Crenshaw, E. B., III, et al. (2001). Insulin resistance and a diabetes mellitus-like syndrome in mice lacking the protein kinase Akt2 (PKB beta). *Science, 292*, 1728–1731.

Cohen, S. E., Kokkotou, E., Biddinger, S. B., Kondo, T., Gebhardt, R., Kratzsch, J., et al. (2007). High circulating leptin receptors with normal leptin sensitivity in liver-specific insulin receptor knock-out (LIRKO) mice. *The Journal of Biological Chemistry, 282*, 23672–23678.

Davies, M. J., Gordon, J. L., Gearing, A. J., Pigott, R., Woolf, N., Katz, D., et al. (1993). The expression of the adhesion molecules ICAM-1, VCAM-1, PECAM, and E-selectin in human atherosclerosis. *The Journal of Pathology, 171*, 223–229.

DeGroot, J., Verzijl, N., Wenting-Van Wijk, M. J., Bank, R. A., Lafeber, F. P., Bijlsma, J. W., et al. (2001). Age-related decrease in susceptibility of human articular cartilage to matrix metalloproteinase-mediated degradation: The role of advanced glycation end products. *Arthritis and Rheumatism, 44*, 2562–2571.

Demars, M., Hu, Y. S., Gadadhar, A., & Lazarov, O. (2010). Impaired neurogenesis is an early event in the etiology of familial Alzheimer's disease in transgenic mice. *Journal of Neuroscience Research, 88*, 2103–2117.

De Vriese, A. S., Verbeuren, T. J., Van de Voorde, J., Lameire, N. H., & Vanhoutte, P. M. (2000). Endothelial dysfunction in diabetes. *British Journal of Pharmacology, 130*, 963–974.

Dobkin, B. H. (2004). Strategies for stroke rehabilitation. *Lancet Neurology, 3,* 528–536.

Eichmann, A., & Thomas, J. L. (2013). Molecular parallels between neural and vascular development. *Cold Spring Harbor Perspectives in Medicine, 3,* a006551.

Ergul, A., Abdelsaid, M., Fouda, A. Y., & Fagan, S. C. (2014). Cerebral neovascularization in diabetes: Implications for stroke recovery and beyond. *Journal of Cerebral Blood Flow and Metabolism, 34,* 553–563.

Ergul, A., Kelly-Cobbs, A., Abdalla, M., & Fagan, S. C. (2012). Cerebrovascular complications of diabetes: Focus on stroke. *Endocrine, Metabolic & Immune Disorders Drug Targets, 12,* 148–158.

Ergul, A., Li, W., Elgebaly, M. M., Bruno, A., & Fagan, S. C. (2009). Hyperglycemia, diabetes and stroke: Focus on the cerebrovasculature. *Vascular Pharmacology, 51,* 44–49.

Etter, H., Althaus, R., Eugster, H. P., Santamaria-Babi, L. F., Weber, L., & Moser, R. (1998). IL-4 and IL-13 downregulate rolling adhesion of leukocytes to IL-1 or TNF-alpha-activated endothelial cells by limiting the interval of E-selectin expression. *Cytokine, 10,* 395–403.

Falcone, C., Emanuele, E., D'Angelo, A., Buzzi, M. P., Belvito, C., Cuccia, M., et al. (2005). Plasma levels of soluble receptor for advanced glycation end products and coronary artery disease in nondiabetic men. *Arteriosclerosis, Thrombosis, and Vascular Biology, 25,* 1032–1037.

Fantin, V. R., Wang, Q., Lienhard, G. E., & Keller, S. R. (2000). Mice lacking insulin receptor substrate 4 exhibit mild defects in growth, reproduction, and glucose homeostasis. *American Journal of Physiology Endocrinology and Metabolism, 278,* E127–E133.

Fernandez, A. M., Kim, J. K., Yakar, S., Dupont, J., Hernandez-Sanchez, C., Castle, A. L., et al. (2001). Functional inactivation of the IGF-I and insulin receptors in skeletal muscle causes type 2 diabetes. *Genes & Development, 15,* 1926–1934.

Fisher, S. J., Bruning, J. C., Lannon, S., & Kahn, C. R. (2005). Insulin signaling in the central nervous system is critical for the normal sympathoadrenal response to hypoglycemia. *Diabetes, 54,* 1447–1451.

Font, M. A., Arboix, A., & Krupinski, J. (2010). Angiogenesis, neurogenesis and neuroplasticity in ischemic stroke. *Current Cardiology Reviews, 6,* 238–244.

Foster, N. L., Tamminga, C. A., O'Donohue, T. L., Tanimoto, K., Bird, E. D., & Chase, T. N. (1986). Brain choline acetyltransferase activity and neuropeptide Y concentrations in Alzheimer's disease. *Neuroscience Letters, 63,* 71–75.

Galis, Z. S., Sukhova, G. K., Lark, M. W., & Libby, P. (1994). Increased expression of matrix metalloproteinases and matrix degrading activity in vulnerable regions of human atherosclerotic plaques. *The Journal of Clinical Investigation, 94,* 2493–2503.

Giacco, F., & Brownlee, M. (2010). Oxidative stress and diabetic complications. *Circulation Research, 107,* 1058–1070.

Girouard, H., & Iadecola, C. (2006). Neurovascular coupling in the normal brain and in hypertension, stroke, and Alzheimer disease. *Journal of Applied Physiology (1985), 100,* 328–335.

Girouard, H., Park, L., Anrather, J., Zhou, P., & Iadecola, C. (2007). Cerebrovascular nitrosative stress mediates neurovascular and endothelial dysfunction induced by angiotensin II. *Arteriosclerosis, Thrombosis, and Vascular Biology, 27,* 303–309.

Grossin, N., Wautier, M. P., Meas, T., Guillausseau, P. J., Massin, P., & Wautier, J. L. (2008). Severity of diabetic microvascular complications is associated with a low soluble RAGE level. *Diabetes & Metabolism, 34,* 392–395.

Hausladen, A., & Fridovich, I. (1994). Superoxide and peroxynitrite inactivate aconitases, but nitric oxide does not. *The Journal of Biological Chemistry, 269,* 29405–29408.

Herder, C., Illig, T., Rathmann, W., Martin, S., Haastert, B., Muller-Scholze, S., et al. (2005). Inflammation and type 2 diabetes: Results from KORA Augsburg. *Gesundheitswesen, 67*(Suppl 1), S115–S121.

II. GENETIC AND ENVIRONMENTAL RISK FACTORS

Howard, A. C., McNeil, A. K., Xiong, F., Xiong, W. C., & McNeil, P. L. (2011). A novel cellular defect in diabetes: Membrane repair failure. *Diabetes, 60*, 3034–3043.

Hsia, A. Y., Masliah, E., McConlogue, L., Yu, G. Q., Tatsuno, G., Hu, K., et al. (1999). Plaque-independent disruption of neural circuits in Alzheimer's disease mouse models. *Proceedings of the National Academy of Sciences of the United States of America, 96*, 3228–3233.

Irie, F., Fitzpatrick, A. L., Lopez, O. L., Kuller, L. H., Peila, R., Newman, A. B., et al. (2008). Enhanced risk for Alzheimer disease in persons with type 2 diabetes and APOE epsilon4: The Cardiovascular Health Study Cognition Study. *Archives of Neurology, 65*, 89–93.

Janson, J., Laedtke, T., Parisi, J. E., O'Brien, P., Petersen, R. C., & Butler, P. C. (2004). Increased risk of type 2 diabetes in Alzheimer disease. *Diabetes, 53*, 474–481.

Johansson, B. B. (2007). Regeneration and plasticity in the brain and spinal cord. *Journal of Cerebral Blood Flow and Metabolism, 27*, 1417–1430.

John, T. A., Vogel, S. M., Tiruppathi, C., Malik, A. B., & Minshall, R. D. (2003). Quantitative analysis of albumin uptake and transport in the rat microvessel endothelial monolayer. *American Journal of Physiology Lung Cellular and Molecular Physiology, 284*, L187–L196.

Katz, E. B., Stenbit, A. E., Hatton, K., DePinho, R., & Charron, M. J. (1995). Cardiac and adipose tissue abnormalities but not diabetes in mice deficient in GLUT4. *Nature, 377*, 151–155.

Kohen Avramoglu, R., Laplante, M. A., Le Quang, K., Deshaies, Y., Despres, J. P., Larose, E., et al. (2013). The genetic and metabolic determinants of cardiovascular complications in type 2 diabetes: Recent insights from animal models and clinical investigations. *Canadian Journal of Diabetes, 37*, 351–358.

Kondo, T., Vicent, D., Suzuma, K., Yanagisawa, M., King, G. L., Holzenberger, M., et al. (2003). Knockout of insulin and IGF-1 receptors on vascular endothelial cells protects against retinal neovascularization. *The Journal of Clinical Investigation, 111*, 1835–1842.

Kulkarni, R. N., Bruning, J. C., Winnay, J. N., Postic, C., Magnuson, M. A., & Kahn, C. R. (1999). Tissue-specific knockout of the insulin receptor in pancreatic beta cells creates an insulin secretory defect similar to that in type 2 diabetes. *Cell, 96*, 329–339.

Kuo, L. E., Kitlinska, J. B., Tilan, J. U., Li, L., Baker, S. B., Johnson, M. D., et al. (2007). Neuropeptide Y acts directly in the periphery on fat tissue and mediates stress-induced obesity and metabolic syndrome. *Nature Medicine, 13*, 803–811.

Lazarov, O., Mattson, M. P., Peterson, D. A., Pimplikar, S. W., & van Praag, H. (2010). When neurogenesis encounters aging and disease. *Trends in Neurosciences, 33*, 569–579.

Lee, D., Lee, K. H., Park, H., Kim, S. H., Jin, T., Cho, S., et al. (2013). The effect of soluble RAGE on inhibition of angiotensin II-mediated atherosclerosis in apolipoprotein E deficient mice. *PLoS One, 8*, e69669.

Leibson, C. L., Rocca, W. A., Hanson, V. A., Cha, R., Kokmen, E., O'Brien, P. C., et al. (1997). The risk of dementia among persons with diabetes mellitus: A population-based cohort study. *Annals of the New York Academy of Sciences, 826*, 422–427.

Li, W., Prakash, R., Kelly-Cobbs, A. I., Ogbi, S., Kozak, A., El-Remessy, A. B., et al. (2010). Adaptive cerebral neovascularization in a model of type 2 diabetes: Relevance to focal cerebral ischemia. *Diabetes, 59*, 228–235.

Liu, S. C., Wang, Q., Lienhard, G. E., & Keller, S. R. (1999). Insulin receptor substrate 3 is not essential for growth or glucose homeostasis. *The Journal of Biological Chemistry, 274*, 18093–18099.

Lorenzi, M., Cagliero, E., & Toledo, S. (1985). Glucose toxicity for human endothelial cells in culture. Delayed replication, disturbed cell cycle, and accelerated death. *Diabetes, 34*, 621–627.

Luchsinger, J. A., Tang, M. X., Stern, Y., Shea, S., & Mayeux, R. (2001). Diabetes mellitus and risk of Alzheimer's disease and dementia with stroke in a multiethnic cohort. *American Journal of Epidemiology, 154*, 635–641.

Marchant, N. L., Reed, B. R., DeCarli, C. S., Madison, C. M., Weiner, M. W., Chui, H. C., et al. (2012). Cerebrovascular disease, beta-amyloid, and cognition in aging. *Neurobiology of Aging, 33* 1006 e25–1006 e36.

Matsumoto, K., Fujishima, K., Moriuchi, A., Saishoji, H., & Ueki, Y. (2010). Soluble adhesion molecule E-selectin predicts cardiovascular events in Japanese patients with type 2 diabetes mellitus. *Metabolism, 59,* 320–324.

McEwen, B. S., & Reagan, L. P. (2004). Glucose transporter expression in the central nervous system: Relationship to synaptic function. *European Journal of Pharmacology, 490,* 13–24.

Meigs, J. B. (2010). Epidemiology of type 2 diabetes and cardiovascular disease: Translation from population to prevention: The Kelly West award lecture 2009. *Diabetes Care, 33,* 1865–1871.

Metz, V. V., Kojro, E., Rat, D., & Postina, R. (2012). Induction of RAGE shedding by activation of G protein-coupled receptors. *PLoS One, 7,* e41823.

Michael, M. D., Kulkarni, R. N., Postic, C., Previs, S. F., Shulman, G. I., Magnuson, M. A., et al. (2000). Loss of insulin signaling in hepatocytes leads to severe insulin resistance and progressive hepatic dysfunction. *Molecular Cell, 6,* 87–97.

Minshall, R. D., Tiruppathi, C., Vogel, S. M., Niles, W. D., Gilchrist, A., Hamm, H. E., et al. (2000). Endothelial cell-surface GP60 activates vesicle formation and trafficking via G(i)-coupled Src kinase signaling pathway. *The Journal of Cell Biology, 150,* 1057–1070.

Mooradian, A. D. (1988). Diabetic complications of the central nervous system. *Endocrine Reviews, 9,* 346–356.

Neubauer, H., Setiadi, P., Gunesdogan, B., Pinto, A., Borgel, J., & Mugge, A. (2010). Influence of glycaemic control on platelet bound CD40-CD40L system, P-selectin and soluble CD40 ligand in type 2 diabetes. *Diabetic Medicine, 27,* 384–390.

Ogata, N., Chikazu, D., Kubota, N., Terauchi, Y., Tobe, K., Azuma, Y. (2000). Insulin receptor substrate-1 in osteoblast is indispensable for maintaining bone turnover. *The Journal of Clinical Investigation, 105,* 935–943.

Palmer, T. D., Willhoite, A. R., & Gage, F. H. (2000). Vascular niche for adult hippocampal neurogenesis. *The Journal of Comparative Neurology, 425,* 479–494.

Park, L., Raman, K. G., Lee, K. J., Lu, Y., Ferran, L. J., Jr., Chow, W. S., et al. (1998). Suppression of accelerated diabetic atherosclerosis by the soluble receptor for advanced glycation endproducts. *Nature Medicine, 4,* 1025–1031.

Pi, J., Bai, Y., Daniel, K. W., Liu, D., Lyght, O., Edelstein, D., et al. (2009). Persistent oxidative stress due to absence of uncoupling protein 2 associated with impaired pancreatic beta-cell function. *Endocrinology, 150,* 3040–3048.

Pickup, J. C., & Crook, M. A. (1998). Is type II diabetes mellitus a disease of the innate immune system? *Diabetologia, 41,* 1241–1248.

Piperi, C., Adamopoulos, C., Dalagiorgou, G., Diamanti-Kandarakis, E., & Papavassiliou, A. G. (2012). Crosstalk between advanced glycation and endoplasmic reticulum stress: Emerging therapeutic targeting for metabolic diseases. *The Journal of Clinical Endocrinology and Metabolism, 97,* 2231–2242.

Prakash, R., Johnson, M., Fagan, S. C., & Ergul, A. (2013). Cerebral neovascularization and remodeling patterns in two different models of type 2 diabetes. *PLoS One, 8,* e56264.

Prakash, R., Somanath, P. R., El-Remessy, A. B., Kelly-Cobbs, A., Stern, J. E., Dore-Duffy, P., et al. (2012). Enhanced cerebral but not peripheral angiogenesis in the Goto-Kakizaki model of type 2 diabetes involves VEGF and peroxynitrite signaling. *Diabetes, 61,* 1533–1542.

Rajagopalan, S., Meng, X. P., Ramasamy, S., Harrison, D. G., & Galis, Z. S. (1996). Reactive oxygen species produced by macrophage-derived foam cells regulate the activity of vascular matrix metalloproteinases in vitro. Implications for atherosclerotic plaque stability. *The Journal of Clinical Investigation, 98,* 2572–2579.

Ramasamy, R., Yan, S. F., & Schmidt, A. M. (2012). The diverse ligand repertoire of the receptor for advanced glycation endproducts and pathways to the complications of diabetes. *Vascular Pharmacology, 57,* 160–167.

II. GENETIC AND ENVIRONMENTAL RISK FACTORS

Rasul, S., Ilhan, A., Wagner, L., Luger, A., & Kautzky-Willer, A. (2012). Diabetic polyneuropathy relates to bone metabolism and markers of bone turnover in elderly patients with type 2 diabetes: Greater effects in male patients. *Gender Medicine, 9*, 187–196.

Remor, A. P., de Matos, F. J., Ghisoni, K., da Silva, T. L., Eidt, G., Burigo, M., et al. (2011). Differential effects of insulin on peripheral diabetes-related changes in mitochondrial bioenergetics: Involvement of advanced glycosylated end products. *Biochimica et Biophysica Acta, 1812*, 1460–1471.

Sakata, N., Meng, J., Jimi, S., & Takebayashi, S. (1995). Nonenzymatic glycation and extractability of collage in human atherosclerotic plaques. *Atherosclerosis, 116*, 63–75.

Sasaki, N., Toki, S., Chowei, H., Saito, T., Nakano, N., Hayashi, Y., et al. (2001). Immunohistochemical distribution of the receptor for advanced glycation end products in neurons and astrocytes in Alzheimer's disease. *Brain Research, 888*, 256–262.

Seners, P., Turc, G., Oppenheim, C., & Baron, J. C. (2015). Incidence, causes and predictors of neurological deterioration occurring within 24 h following acute ischaemic stroke: A systematic review with pathophysiological implications. *Journal of Neurology, Neurosurgery, and Psychiatry, 86*(1), 87–94.

Seners, P., Turc, G., Tisserand, M., Legrand, L., Labeyrie, M. A., Calvet, D., et al. (2014). Unexplained early neurological deterioration after intravenous thrombolysis: Incidence, predictors, and associated factors. *Stroke, 45*, 2004–2009.

Shah, P. K., Falk, E., Badimon, J. J., Fernandez-Ortiz, A., Mailhac, A., Villareal-Levy, G., et al. (1995). Human monocyte-derived macrophages induce collagen breakdown in fibrous caps of atherosclerotic plaques. Potential role of matrix-degrading metalloproteinases and implications for plaque rupture. *Circulation, 92*, 1565–1569.

Silvestre, J. S., & Levy, B. I. (2006). Molecular basis of angiopathy in diabetes mellitus. *Circulation Research, 98*, 4–6.

Smith, M. A., Sayre, L. M., & Perry, G. (1996). Diabetes mellitus and Alzheimer's disease: Glycation as a biochemical link. *Diabetologia, 39*, 247.

Smith, M. A., Taneda, S., Richey, P. L., Miyata, S., Yan, S. D., Stern, D., et al. (1994). Advanced maillard reaction end products are associated with Alzheimer disease pathology. *Proceedings of the National Academy of Sciences of the United States of America, 91*, 5710–5714.

Spranger, J., Kroke, A., Mohlig, M., Hoffmann, K., Bergmann, M. M., Ristow, M., et al. (2003). Inflammatory cytokines and the risk to develop type 2 diabetes: Results of the prospective population-based European Prospective Investigation into Cancer and Nutrition (EPIC)—Potsdam Study. *Diabetes, 52*, 812–817.

Stadler, K. (2012). Oxidative stress in diabetes. *Advances in Experimental Medicine and Biology, 771*, 272–287.

Standards of Medical Care in Diabetes (2014). Diabetes Care 37(Suppl. 1), S14–S79.

Sturchler, E., Galichet, A., Weibel, M., Leclerc, E., & Heizmann, C. W. (2008). Site-specific blockade of RAGE-Vd prevents amyloid-beta oligomer neurotoxicity. *The Journal of Neuroscience, 28*, 5149–5158.

Sverdlov, M., Shajahan, A. N., & Minshall, R. D. (2007). Tyrosine phosphorylation-dependence of caveolae-mediated endocytosis. *Journal of Cellular and Molecular Medicine, 11*, 1239–1250.

Tao, R., Coleman, M. C., Pennington, J. D., Ozden, O., Park, S. H., Jiang, H., et al. (2010). SIRT3-mediated deacetylation of evolutionarily conserved lysine 122 regulates MnSOD activity in response to stress. *Molecular Cell, 40*, 893–904.

Tavazoie, M., Van der Veken, L., Silva-Vargas, V., Louissaint, M., Colonna, L., Zaidi, B., et al. (2008). A specialized vascular niche for adult neural stem cells. *Cell Stem Cell, 3*, 279–288.

Troiani, D., Draicchio, F., Bonci, A., & Zannoni, B. (1993). Responses of vestibular neurons to stimulation of cortical sensorimotor areas in the cat. *Archives Italiennes de Biologie, 131*, 137–146.

Venugopal, S. K., Devaraj, S., Yuhanna, I., Shaul, P., & Jialal, I. (2002). Demonstration that C-reactive protein decreases eNOS expression and bioactivity in human aortic endothelial cells. *Circulation, 106,* 1439–1441.

Verma, S., Wang, C. H., Li, S. H., Dumont, A. S., Fedak, P. W., Badiwala, M. V., et al. (2002). A self-fulfilling prophecy: C-reactive protein attenuates nitric oxide production and inhibits angiogenesis. *Circulation, 106,* 913–919.

Vinik, A. I., Erbas, T., Park, T. S., Nolan, R., & Pittenger, G. L. (2001). Platelet dysfunction in type 2 diabetes. *Diabetes Care, 24,* 1476–1485.

Vogel, S. M., Easington, C. R., Minshall, R. D., Niles, W. D., Tiruppathi, C., Hollenberg, S. M., et al. (2001). Evidence of transcellular permeability pathway in microvessels. *Microvascular Research, 61,* 87–101.

Walks, D., Lavau, M., Presta, E., Yang, M. -U., & Bjorntorp, P. (1983). Refeeding after fasting in the rat: Effects of dietary-induced obesity on energy balance regulation. *The American Journal of Clinical Nutrition, 37,* 387–395.

Wang, H., Liu, Z., Li, G., & Barrett, E. J. (2006). The vascular endothelial cell mediates insulin transport into skeletal muscle. *American Journal of Physiology Endocrinology and Metabolism, 291,* E323–E332.

Wang, H., Wang, A. X., Aylor, K., & Barrett, E. J. (2015). Caveolin-1 phosphorylation regulates vascular endothelial insulin uptake and is impaired by insulin resistance in rats. *Diabetologia, 58,* 1344–1353.

Wang, H., Wang, A. X., & Barrett, E. J. (2011). Caveolin-1 is required for vascular endothelial insulin uptake. *American Journal of Physiology Endocrinology and Metabolism, 300,* E134–E144.

Wang, X., Mao, X., Xie, L., Sun, F., Greenberg, D. A., & Jin, K. (2012). Conditional depletion of neurogenesis inhibits long-term recovery after experimental stroke in mice. *PLoS One, 7,* e38932.

Welle, S., Thornton, C., Statt, M., & McHenry, B. (1994). Postprandial myofibrillar and whole body protein synthesis in young and old human subjects. *The American Journal of Physiology, 267,* E599–E604.

Williams, S. B., Goldfine, A. B., Timimi, F. K., Ting, H. H., Roddy, M. A., Simonson, D. C., et al. (1998). Acute hyperglycemia attenuates endothelium-dependent vasodilation in humans *in vivo. Circulation, 97,* 1695–1701.

Withers, D. J., Gutierrez, J. S., Towery, H., Burks, D. J., Ren, J. M., Previs, S., et al. (1998). Disruption of IRS-2 causes type 2 diabetes in mice. *Nature, 391,* 900–904.

Yan, S. D., Chen, X., Fu, J., Chen, M., Zhu, H., Roher, A., et al. (1996). RAGE and amyloid-beta peptide neurotoxicity in Alzheimer's disease. *Nature, 382,* 685–691.

Yan, S. D., Stern, D., & Schmidt, A. M. (1997). What's the RAGE? The receptor for advanced glycation end products (RAGE) and the dark side of glucose. *European Journal of Clinical Investigation, 27,* 179–181.

Yao, D., & Brownlee, M. (2010). Hyperglycemia-induced reactive oxygen species increase expression of the receptor for advanced glycation end products (RAGE) and RAGE ligands. *Diabetes, 59,* 249–255.

Yorek, M. A. (2003). The role of oxidative stress in diabetic vascular and neural disease. *Free Radical Research, 37,* 471–480.

Zhang, Z. G., & Chopp, M. (2009). Neurorestorative therapies for stroke: Underlying mechanisms and translation to the clinic. *Lancet Neurology, 8,* 491–500.

Zhu, Y., Park, S. H., Ozden, O., Kim, H. S., Jiang, H., Vassilopoulos, A., et al. (2012). Exploring the electrostatic repulsion model in the role of SIRT3 in directing MnSOD acetylation status and enzymatic activity. *Free Radical Biology & Medicine, 53,* 828–833.

Zisman, A., Peroni, O. D., Abel, E. D., Michael, M. D., Mauvais-Jarvis, F., Lowell, B. B., et al. (2000). Targeted disruption of the glucose transporter 4 selectively in muscle causes insulin resistance and glucose intolerance. *Nature Medicine, 6,* 924–928.

Index

Note: Page numbers followed by "*f*" and "*t*" refer to figures and tables, respectively.

A
"A disintegrin and metalloprotease domain" (ADAM) family, 100–101
AADvac-1, 156
ACI-35, 156
Adenosine triphosphate (ATP), 153, 277, 394–395
Adrenocorticotropic hormone (ACTH) secretion, 61
Advanced glycation end-products (AGEs), 397–398
Age-dependent lifestyle changes, 198
Age-related cognitive decline, 30
Age-related "disconnection" syndrome, 31
Aging, 62–64, 199, 204
Air pollution, effects of, 213–214
Allostasis, 61–62
Almorexant, 304–305
α-secretases, 100–103, 240–241, 241*f*, 301–302
Amnestic mild cognitive impairment (aMCI), 37, 39, 41–42, 300–301
  associative memory deficit in, 37
  sleep–wake cycle, 300–301
AMPA receptors, 7, 299
Amyloid cascade hypothesis, 136, 138, 376
Amyloid plaque clearance
  Aβ immunotherapy and mononuclear phagocytes' role in, 365–366
Amyloid plaques, 136, 203, 240–241, 249, 376
Amyloid precursor protein (APP), 95, 138, 240–241, 244, 296
  and AD, 113–117
    altered APP processing/translocation, 115–116
    altered signal transduction, 116–117
    disrupted axonal transport, 113–114
    disrupted neurogenesis, 114–115
  amyloid-beta (Aβ) peptide, 98, 99*f*, 111–113
  APP gene and its homologs, 99–100

APP knockout mice, 113
APP protein, its domains, and its proteolytic fragments, 100–102
  cholesterol and APP processing, 103–104
  functions of, 106–111
    axonal transport, 108–109
    metal binding/redox, 111
    neurite outgrowth, 107–108
    neurogenesis, 106–107
    synaptic formation and function, 109
    transcription factor, 109–110
  -interacting proteins, 104–106
    low-density lipoprotein receptors, 104–105
    YENPTY motif, interactors with, 104
  intracellular trafficking and processing of, 102–103
  processing, 250
  proposed functions of, 106*t*
  proteolysis of, 241*f*
Amyloid precursor-like protein (APLP), 242–243
Amyloid-beta (Aβ) immunization approaches, 366
Amyloid-beta (Aβ) immunotherapy and mononuclear phagocytes' role in amyloid plaque clearance, 365–366
Amyloid-beta (Aβ) independent mechanisms, 169
Amyloid-beta (Aβ) pathology, sleep and, 301–308
Amyloid-beta (Aβ) peptide, 30, 98, 99*f*, 111–113, 240–241, 296, 330–333, 363*f*
  and neurovascular dysfunction in AD, 330–333
  synthesis, 256–257
Amyloidogenic processing enzymes, 103–104
Antidepressants and mood-stabilizing treatments, 65–66
Antioxidant consumption, 215–216
Aplysia sensory neurons, 9–10

APOE4, 172–178, 182–187, 206, 216–217
Apolipoprotein E (apoE), 169, 296, 389–390
  APOE genotypes and cognitive functions in nondemented individuals, 173–174
  APOE receptors, 179–181
  and brain glucose metabolism, 176–177
  and cerebrovascular functions, 181–183
  and cholesterol metabolism in synaptic functions, 174–176
  ε2 allele of, 172
  and inflammatory response, 183–185
  and insulin signaling, 176–177
  isoforms and AD risk, 172–173, 173*f*
  isoform-specific pathogenic pathways for AD, 186*f*
  mitochondria dysfunction, 177–179
  perspectives, 185–187
  synaptic functions, 179–181
  tau phosphorylation, 177–179
Apolipoprotein E receptor 2 (apoER2), 104–105, 179–180
APP intracellular domain (AICD), 101–102, 104, 107, 109–110
APP/PS1 mouse model
  treadmill running in, 210
APP[V717I] mice, 249
APP-like protein-1 (APLP1), 99–100
APP-like protein-2 (APLP2), 99–100
arg/arg3.1 gene, 8
Arrhythmias, 325–326, 326*t*
Arterial dissection, 320–321
Arterial occlusions, 325–326, 326*t*
Arteriosclerosis, 325–326, 326*t*
Associative memory deficit, in amnestic MCI, 37
Astrocytes, 60
Astrocytosis, 362
Asymmetric cognitive decline in preclinical AD, 42–43
Atherosclerotic plaques, 325–326
Atkinson–Shiffrin memory model, 3*f*
Attention process training, 345
Autophagy-lysosome pathway, 147–148
Axonal transport, APP in, 108–109

B

BACE1 KO mice, 242–243
  phenotypes detected in different lines of, 243*t*
Bacterial artificial chromosome (BAC) clone, 249–250

Baddeley's model of working memory, 3*f*
Bapineuzumab, 172
Bax gene, 76–77
BDNF gene, 8, 16
Beclin 1, 375–376
Beta-amyloid, 240–241
β-secretase, 240–242, 241*f*, 247–248, 253–254, 256–257, 301–302
  cleavage by, 102–103
Beta-site amyloid precursor protein cleaving enzyme 1 (BACE1) protein, 101–103, 241–258
  in AD, 247–253
  BACE1 null mice, 242–243
  carboxyl-terminal fragment (CTF), 241–242
  in cognitive function, 239
  elevation in AD and acute brain injuries, 253–256
  experimental TBI, 256
  GGA3-dependent regulation, 253–256
  human BACE1 (hBACE1), 245–247
  inhibition, 256–258
  substrates, 244–245
  therapeutic approach for AD, 256–258
  trafficking, 241–242
  transgenic and knockin models of, 246–247, 247*t*
  transgenic models, 245–247
  validated substrates of, 246*t*
Bilingualism
  and AD risk, 202–203
  benefits of, 206
Biomarkers, of AD, 31, 43
Block Design Test, 42–43
Blood–brain barrier (BBB), 181, 256–257, 328*f*, 376, 400–401
BMS-241027, 155
Boston Naming Test, 42–43
Brain glucose metabolism, APOE4 and, 176–177
Brain reserve, 203–205
Brain structures underlying memory processes, 5
Brain-derived neurotrophic factor (BDNF) signalling, 214, 218
Brain-resident microglia, peripheral macrophages versus, 367–369
BrdU, 76–78

C

^{14}C levels, 77–78
C99 peptide, 101–102

Ca²⁺-calmodulin-dependent protein kinases (CaMK), 10–11
CaMKI, 10–11
CaMKII, 10–11
CaMKIV, 10–11
Ca²⁺/calmodulin-sensitive NO synthases (NOS), 11
CA3 pyramidal cells, 5, 56f, 67–68
CADASIL (cerebral autosomal dominant arteriopathy with subcortical infarcts and leukoencephalopathy), 333
Caenorhabditis elegans, 99–100
Caffeine, 216–217
California Verbal Learning Test-II (CVLT), 31–34, 34f
Calmodulin-sensitive NO synthases (NOS), 11
Calpain, 146
Cam Kinase II (CaMKII) promoter, 245–246
CaMK IV activity, 11
cAMP responsive element (CRE), 8–9
cAMP responsive element-binding protein (CREB)
    pathways activating, 9–13
    pathways activating CREB phosphorylation, 14f
    role of, in memory, 8–9
cAMP/PKA pathway, 9–11
cAMP-dependent protein kinase A (PKA), 148
Cancer, 155, 320, 388–389
Cardiac arrest, 325–326, 326t
Cardiac failure, 325–326, 326t
Cardinal anti-inflammatory cytokines, targeting, 369–371
Cardioembolism, 320–321
Casein kinase 1 (CK1), 148
Caspases, 146
Cathepsins, 146
Caveolae, 404–406
Caveolin-1, 404–406
Cellular adhesion molecules (CAMs), 403
Cellular basis of memory, 6–8
Central nervous system (CNS), 99–100, 185, 305, 362, 392
Cerebral amyloid angiopathy (CAA), 324–325, 331, 366
Cerebral amyloidosis, targeting cardinal anti-inflammatory cytokines to restrict, 369–371
Cerebral β-amyloidosis, 306–307
Cerebral blood flow (CBF), 181
Cerebral capillary degeneration, 327–328

Cerebral innate immunity, 361
    Aβ immunotherapy and mononuclear phagocytes' role
    in amyloid plaque clearance, 365–366
    AD pharmacotherapeutics targeting innate immunity, 376–377
    in Alzheimer pathoetiology, 362–364
    beclin 1, 375–376
    cardinal anti-inflammatory cytokines, targeting, 369–371
    chemokines, 371–372
    historical perspective, 364–365
    inflammasome activation in AD, 375
    inflammatory ILS 12 and 23, blocking, 372–373
    peripheral macrophages versus brain-resident microglia, 367–369
    PPARγ agonists, 373–374
Cerebral ischemia, 322
Cerebral microhemorrhage, 366
Cerebral spinal fluid (CSF)
    examining Aβ and tau in, 203
Cerebrospinal fluid (CSF), 182–183, 301
Cerebrovascular complications in T2DM, 400–404
Cerebrovascular functions, apoE and, 181–183
c-fos gene, 8
cGMP signaling, 11–12
Chaperone-mediated autophagy, 147
Chemical exposure and AD Risk, 211–212
Chemokines, 321–322, 371–372
CHIP (C-terminus of heat shock cognate 70 interacting protein), 146–147
Cholesterol, 103–104, 175
    and AD risk, 215–217
    and APP processing, 103–104
Cholesterol metabolism in synaptic functions, 174–176
Cholesterol-enriched microdomains (CEM), 398–399
Chromatin structure remodeling, 17
Chronic ablation of neurogenesis, 70–71
Chronic lower respiratory disease, 320
Chronic traumatic encephalopathy (CTE), 138, 281–282
Cognition, 52–53, 66
Cognitive decline, models of
    in healthy aging, 30–31
Cognitive decline, sleep disturbances and, 300–301

Cognitive decline
  AD-associated
    vs. normal aging, 31–34
    APOE4 and, 173–174, 181
    C-reactive protein (CRP) and, 183–184
    nonsteroidal anti-inflammatory drug
      (NSAID) treatment, 183–184
Cognitive disturbance, mechanisms of
  following TBI, 273
Cognitive dysfunction, 268, 271–272, 320
  TBI-related, 271–273, 278–279
Cognitive function, 324–327
  BACE1 in, 239
Cognitive impairment and dementia
  AD, 324–325
  vascular cognitive impairment (VCI),
    325–326
    causes of, 326t
  vascular dementia (VaD), 324–325
Cognitive processes, 52, 59–60, 275
Cognitive reserve, 203–207
  air pollution and tobacco smoke, 213–214
  benefits of cognitive complexity
    following onset of dementia,
    207–209
  chemical exposure and AD Risk, 211–212
  and education, 201–204
  metals, 212–213
  microbiome, 217–218
  nutrition, 214–218
  physical activity and exercise, 209–211
  sleep and circadian rhythm, 218–219
  socialization, 219
Cognitive stimulation therapy (CST),
  207–209
Cognitive training, 204–205, 207–208
Computed tomography (CT), 269
Contextual memory, 70–71
Controlled cortical impact (CCI), 112, 255
Copper, 111, 212–213
Coronary artery disease (CAD), 183–184,
  398
Coronary heart disease (CHD), 172, 397
Corticotropin releasing hormone (CRH), 61
C-reactive protein (CRP), 183–184, 388–389
CREB binding protein (CBP), 8–9, 14f, 16
CREB phosphorylation at Ser133, 10–11
CREB-regulated transcriptional co-activator
  (CRTC), 8–9
Cushing's syndrome, 61–62
Cyclin-dependent kinase 5 (Cdk5), 148
Cytochrome oxidase activity, 177–178
Cytokine, 372–373, 388–389, 397–398

D
Dab-1 protein, 104, 115–116
Davunetide, 155
Declarative memory, 4–5
Default-mode network, 303–304
Dementia, 324–327
  cognitive complexity following the onset
    of, 207–209
Dentate gyrus (DG), 52–60, 66–71, 75, 77
Depth of coma, 270–271
Diabetes. See also Type 2 diabetes mellitus
  (T2DM)
  and endothelial dysfunction, 398–400
  and mitochondrial dysfunction, 394–396
Dichlorodiphenyldichloroethane (DDE),
  exposure to, 211
Dichlorodiphenyltrichloroethane (DDT),
  exposure to, 211
Diet, 64–65, 216, 219–220
  copper, 212–213
  fish and vegetables, 215–216
  fruits and vegetables, 215, 219–220
  high cholesterol, 216–217
  high-fat, 216–217
  zinc, 212–213
Diffuse axonal injury (DAI), 268, 274
  as mechanism of cognitive dysfunction
    following TBI, 274–276
Diffuse plaques, 136
Di-leucine sorting signal, 241–242
Disability Rating Scale (DRS), 277–278
Disabled-1 (Dab-1), 104
Disconnection syndrome, age-related, 31
Disrupted axonal transport, 113–114
Dividing cells, labeling, 76
DNMT inhibitor, 15
Drosophila melanogaster, 99–100, 297–298
Dysfunctional cerebral neovascularization
  in T2DM, 400–401

E
E prostanoid receptor subtype 2 (EP2), 368
E1, E2, and E3 enzymes, 146–147
Education
  cognitive reserve and, 201–204, 206
  and risk for AD, 203
Encoding, 2
Endoplasmic reticulum, 102–103
Endothelial dysfunction, diabetes and,
  398–400
Endothelial nitric oxide synthase (eNOS),
  321–322, 398–399
Energy crisis, 277

Enriched environment, 62, 199*f*
Enrichment, environmental, 64–65
Entorhinal cortex, 5, 56*f*, 58–59
Environmental enrichment (EE), 64–65, 198–199
Environmental factors in disease development. *See* Lifestyle and AD
Epidermal growth factor receptor (EGFR), 107
Epigallocatechin gallate (EGCG), 214–215
Epigenetics, 13
Episodic buffer, 2–3
Episodic memory, 4–5, 32–34, 36–37, 41, 297–298
Epothilone D, 155
ERK signaling, 12
ERK1/2, 12
E-selectin, 403
Euchromatin, 15
Event-related potentials (ERPs), normal aging with preclinical AD vs., 35–36
Executive dysfunction, 270*f*, 272
  TBI-related, 276
Exercise and AD risk modulation, 209–211
Explicit (declarative) system, 280
Extracranial large vessel atherosclerosis, 325–326

F

Familial Alzheimer's disease (FAD), proteins linked to, 74–75
Familial form of AD (FAD), 247–248
Fat, diet high in, 216–217
FE65, 104, 115–116
Fear conditioning (FC), 11
Fetal 3R Tau isoform (0N3R), 151
Fibroid necrosis, 325–326
Fish and vegetables, 215–216
5XFAD transgenic mouse line, 251–253
Fluid-Attenuated Inversion Recovery (FLAIR)-MRI, 181
Fluorodeoxyglucose (FDG)-PET hypometabolism, 31–32
Focal injuries, 279
  as mechanism of cognitive dysfunction following TBI, 279–281
Focal ischemic stroke, 325–326
Fractalkine (Cx3cl1), 372
Frontal dysfunction, 280
Frontal lobe function, 279–280
Frontal syndrome, 279–280
Fruits and vegetables, 215, 219–220

FTD with parkinsonism linked to chromosome 17 (FTDP-17), 138–139
Fuld Object Memory Test, 32–33
Full scale IQ (FSIQ), 278

G

Gamma-aminobutyric acid (GABA), 57
γ-secretase, 100–101, 240–241, 241*f*, 256–258, 301–302
  modulation, 257–258
GGA1, 2, and 3, 241–242
Glasgow Coma Scale (GCS), 270–271
Glasgow Outcome Scale-extended (GOSe), 277–278
Glial-fibrillary acidic protein (GFAP), 55–56
Gliosis, 362
Global cerebral hypoperfusion, 325–326
Glucocorticoid receptors (GR), 61–62
Glucose, 176, 406
Glucose metabolism, cerebral, 176–177
GLUT4 knockout, 393–394
Glutamine, 406
Glycogen synthase kinase-3β (GSK-3β), 148
G-protein coupled receptor, 9–10
GRB2, 104
Green fluorescent protein (GFP), 367
GSK3-β, 216
Gut microbiota, 217–218

H

H2B acetylation, 15–16
H4K12 acetylation, 15–16
HAPP Swedish mutation (hAPPswe), 249
HDAC2, 16–17
HDAC3, 16–17
HDAC4, 16–17
Heart disease, 320
Hemorrhagic stroke, 320–323
Hilar interneurons, 67
Hippocampal circuitry, adult mammalian, 58–59
Hippocampal dysfunction and AD, 73–75
Hippocampal trisynaptic circuit, 6*f*
Hippocampus, 5, 52–53, 58–61, 68–69, 73–74, 204
Histone acetylation, 13–17
Histone acetyltransferase (HAT), 8–9, 14–16
  in gene transcription, 13–17
Histone deacetylases (HDACs), 14*f*, 15–17
  in gene transcription, 13–17
Hsc70, 147
Hsp90, 147, 155

Human BACE1 (hBACE1), 245–247, 249–250
Hyperglycemia, 390, 396
  and AD, 402
Hyperinsulinemia, 391, 393–394
Hypotension, 325–326
Hypothalamic–pituitary–adrenal (HPA) axis, 61, 71–73

I

Ibuprofen, 183–184
Immunoelectron microscopy, 365
Implicit (procedural) memory system, 280
Inflammasome activation in AD, 375
Inflammatory ILS 12 and 23, blocking, 372–373
Inflammatory response, apoE and, 183–185
Innate immunity, AD pharmacotherapeutics targeting, 376–377
Insulin, 177, 390
Insulin access to brain, 404–406
Insulin receptor (IR), 391
Insulin receptor substrate molecule (IRS), 391
Insulin signaling, APOE4 and, 176–177
Insulin-like growth factor 1 receptor (IGF-1R), 391
Interferon gamma (IFN-γ), 402–403
Interstitial fluid (ISF), 302–303
  diurnal fluctuation of, 304–305
Intracellular trafficking and processing of amyloid precursor protein, 102–103
Intracerebral hemorrhage (ICH), 320–321
IPN007, 156
IR knockout mouse (IR$^{-/-}$), 391
Iron neurotoxicity, 112–113
IRS knockout mice, 392–393
Ischemic stroke, 320–321, 401
  experimental models of, 326t
  pathogenesis of, 321f

J

JIP1, 104

K

Ketamine/xylazine anesthesia, 306
α-Ketoglutarate dehydrogenase complex (KGDHC), 177–178
KFERQ, 147
Knockin (KI) model of BACE1, 246–247

L

Lacunar stroke, 320–321
Large-artery atherothrombosis, 320–321
Late-onset form of AD (LOAD), 296
LDL receptor family, 104–105
LDL-receptor related protein-1 (LRP1), 104–105
  LRP1b, 104–105
Learning and memory, 1
  brain structures underlying memory processes, 5
  cAMP responsive element-binding protein (CREB)
    pathways activating, 9–13
    role of, in memory, 8–9
  cellular basis of memory, 6–8
  histone acetyltransferase (HATs) in gene transcription, 13–17
  histone deacetylases (HDACs) in gene transcription, 13–17
  memory processes, 2–5
  from Plato's wax tablet to genes, 2
  sleep for, 297–300
Leukoaraiosis, 325–328, 326t
Lifestyle and AD, 197
  APPswe/PS1ΔE9 mouse model of, 199
  brain reserve, 204–205
  cognitive reserve, 205–207
    air pollution and tobacco smoke, 213–214
    benefits of cognitive complexity following onset of dementia, 207–209
    chemical exposure and AD Risk, 211–212
    metals, 212–213
    microbiome, 217–218
    nutrition and microbiome, 214–218
    physical activity and exercise, 209–211
    sleep and circadian rhythm, 218–219
    socialization, 219
  environmental factors and their influence on AD risk, 200t–201t
  epidemiological studies, 201–207
    cognitive reserve and education, 201–204
Lipophylainosis, 325–326
Lipoprotein receptors, low-density, 104–105
Liver X receptors (LXRs), 374
LOAD, 198, 296
Logical Memory subtest, 33–34, 34f
Long-term depression (LTD), 6–7

Long-term memory, 2–5, 280
  different types of, 4f
Long-term potentiation (LTP), 6–7, 64, 111, 175
Loss of consciousness (LOC), 270–271
Low-density lipoprotein receptor (LDLR), 104–105, 179–181
Low-density lipoprotein receptor-related protein 1 (LRP1), 179–180

M

Macroautophagy, 147
Magnetic resonance imaging (MRI), 38, 43, 182, 269
MAPK/ERK kinase (MEK), 12
MAPT, 136–138
Matrix metalloproteinase 9 (MMP9), 182–183
Matrix metalloproteinases (MMPs), 328–329, 328f, 402
Maturing cells, labeling, 76
Medial diencephalic structures, 5
Medial temporal lobes, 5, 36, 274
Mediterranean diet
  and AD risk, 215
MEK inhibitors, 12
Memory deficits, 16, 272
Memory deficits, in preclinical AD, 37–42
  in associative memory, 37
  in pattern separation, 38
  in prospective memory, 39–40
  in recollection and familiarity, 38–39
  in remote memory, 40–41
  in working memory, 41–42
Memory processes, 2–5
  brain structures underlying, 5
Metabolic flexibility, 406–407
Metal binding sites, of APP, 111
Metals, exposure to, 212–213
Microbiome, 217–218
  nutrition and, 214–218
Microglia, 60, 184–185, 362, 363f, 364–365, 396, 405–406
Micropinocytosis, 403
Microtubule-associated proteins (MAPs), 137
Middle cerebral artery (MCA), 322–323
Middle cerebral artery occlusion (MCAO), 322–323
Mild cognitive impairment (MCI), 31, 36f, 149, 173–174, 206–207, 282, 325–326, 329–330

Mild TBI (mTBI), 268
Mineralocorticoid receptors (MR), 62
Mitochondrial dysfunction in diabetes, 394–396
Mitogen-activated protein kinases (MAPKs), 12
MKR mouse model of insulin resistance, 394
Modal model, 2–3
Molecular pathways in AD and cognitive function, 135
  normal production and function of tau, 136–137
  pathogenic tau, 138–139
  tau-mediated neuronal deficits, 143–154
    loss-of-function versus gain-of-function, 143
    posttranslational modifications of tau, 148–151
    regulation of tau homeostasis, 146–148
    tau aggregation, 143–146
    tau missorting, 151–154
  therapeutic implications, 154–156
  transgenic animal models for tauopathies, 144t
Mononuclear phagocytes, 365–368, 371–372
Mood-stabilizing treatments, antidepressants and, 65–66
Morris water maze, 113, 297–298, 336
Mouse models of type 2 diabetes mellitus, 391–394
  GLUT4 knockout, 393–394
  IR knockout mouse, 391
  IRS knockout mice, 392–393
  MKR mouse model of insulin resistance, 394
  tissue-specific deletion of IR, 391–392
Multi-infarct dementia, 325–326, 326t, 335–336, 335t
Multiple arterial occlusions, 325–326, 326t
Multiple neurotransmitter receptors, 7

N

N400 ERP, 35
NADPH oxidase, 321–322
Neocortex, 5, 298–299, 344
Nestin, 55–56
Neural progenitor cells (NPCs), 53–54, 107
Neural progenitors, 52–53, 57, 66, 107
Neural stem cells, 53–56, 63, 114–115
Neuregulin 1 (NRG1), 240, 244–245
Neurite outgrowth, APP and, 107–108

Neuritic dystrophy, 301–302
Neuritic plaques, 136
Neurodegeneration, 30, 136, 138, 150, 205, 243t
Neurodegenerative processes, 281–282
    as mechanism of cognitive dysfunction following TBI, 282–283
Neurofibrillary tangles (NFTs), 98, 136, 143–146, 152, 281–282, 301–302, 362
Neurogenesis, 400–401
    disrupted, 114–115
    embryonic and adult, 106–107
    manipulating, in rodents, 76–77
Neurogenesis, adult, 51
    adult-born neurons, development and maturation of, 55–58
    adult mammalian hippocampal circuitry, 58–59
    developmental stages in, 59f
    factors affecting, 59–67
        aging, 62–64
        antidepressants and mood-stabilizing treatments, 65–66
        environmental enrichment, physical activity, and diet, 64–65
        neurogenic niche, 60
        stress, 61–62
    identification and characterization of, 54–55
    methodologies for studying, 75–78
        human technologies, 77–78
        labeling dividing and maturing cells, 76
        manipulating neurogenesis in rodents, 76–77
    neurogenesis, hippocampal dysfunction, and AD, 73–75
        proteins linked to FAD regulating neurogenesis, 74–75
    proposed functions of, 67–73
        other aspects of spatial and contextual memory, 70–71
        pattern separation, 68–70
        regulation of mood and HPA-axis, 71–73
Neurogenesis, hippocampal dysfunction, and AD, 73–75
    proteins linked to FAD regulate neurogenesis, 74–75
Neurogenic niche, 60
Neurogenic theory of depression, 65
Neuroinflammation, 184–185, 301–302

Neurologically based models, 31
Neurometabolic changes, 277
    as mechanism of cognitive dysfunction following TBI, 277–279
Neuronal acetylation levels, dysregulation of, 15
Neuronal deficits, tau-mediated, 143–154
    aggregation of tau, 143–146
    loss-of-function versus gain-of-function, 143
    missorting of tau, 151–154
    posttranslational modifications, 148–151
    regulation of tau homeostasis, 146–148
        autophagy-lysosome pathway, 147–148
        ubiquitin-proteasome pathway, 146–147
Neuron-enriched endosomal protein of 21 kDa (NEEP21), 105
Neurons, adult-born
    development and maturation of, 55–58
Neuropeptide-Y (NPY), 399–400
Neuroplasticity, 52, 67, 198–199
Neuropsychology, 269
Neurotransmitter release, 11, 109
Neurovascular dysfunction in AD, 330–333
    Aβ contribution to, 330–333
    neurodegenerative changes, 333
Neurovascular unit (NVU), 322, 323f
    role of, in AD and vascular cognitive impairment (VCI), 327–333
Nitric oxide (NO), 11–12, 14f
Nitrosative stress, 321–322
NMDA receptor (NMDAR), 7
NO/cGMP/protein kinase G (PKG) pathway, 11–12
NOD-like receptor (NLR), 375
Nonautobiographical semantic memory, 41
Nondeclarative memory, 4–5
Nondemented individuals
    APOE genotypes and cognitive functions in, 173–174
Nonenzymatic glycoxidation, 397
Nonsteroidal anti-inflammatory drugs (NSAIDs), 363–364
    long-term use, 183–184
Normal aging, AD-associated cognitive decline vs., 31–34
NOTCH intracellular domain (NICD), 109–110
Notch-Jagged 1 (Jag1) signaling pathway, 244–245
Nutrition, and AD risk, 214–218

O

One-Back working memory test, 31–32
Oral glucose tolerance test (OGTT), 390
Orexin, 304–305
Oxidative stress, 321–322, 321*f*, 330,
 332–333, 398–399, 404–405
 and inflammation, 396

P

p38 MAPK, 12–13
P600 ERP, 35
Passive avoidance, 11
Pattern separation, 68–70
 deficits in, 38
PDAPP mouse line 109, 251
Peripheral macrophages versus brain-
 resident microglia, 367–369
Peroxisome proliferator-activated
 receptor-γ (PPARγ), 373–374
Pesticides, exposure to, 211
Phonological loop, 2–3
Phosphatidylinositol clathrin assembly
 lymphoid-myeloid leukemia
 (PICALM) gene, 116
Physical activity, 64–65
 and AD risk modulation, 209–211
PIB-PET, 181, 186–187
Pin1, 104
Pittsburgh Compound B (PiB), 31, 203
PKA, 9–10, 148–149
PLD3, 115–116
Pleiotropic polypeptide cytokine, 369–370
Positron emission tomography (PET)
 imaging, 31, 278
Posttranslation modifications (PTMs), 13
Posttraumatic amnesia (PTA), 270–271
PP2A, 148, 153–154
PPARγ agonists, 373–374
Preamyloid deposits, 136
Preclinical AD
 asymmetric cognitive decline in, 42–43
 memory deficits in, 37–42
 associative memory, deficits in, 37
 pattern separation, deficits in, 38
 prospective memory, deficits in, 39–40
 recollection and familiarity, deficits in,
  38–39
 remote memory, deficits in, 40–41
 working memory, deficits in, 41–42
 vs. normal aging with event-related
  potentials (ERPs), 35–36
Preclinical innate immunotherapy in AD,
 378*t*–379*t*

Presenilin, 212
Presenilin-1 (PS1), 98, 138, 256–257
Presenilin-2 (PS2), 98, 138
Processing speed, 30–31, 275
Progenitors, neural, 52–53
Pro-inflammatory cytokines, 321–322
Pro-inflammatory enzymes, 321–322
Prospective memory, deficits in, 39–40
Proteases, 100–101, 274, 321–322
Proteins, APP-interacting, 104–106
 low-density lipoprotein receptors,
  104–105
 YENPTY motif, interactors with, 104
PS1/γ-secretase, 74–75
P-selectin, 403
Punch drunk syndrome, 281–282

R

Rapid eye movement (REM), 297–298
Reactive oxygen species (ROS), 321–322,
 394–396
Receptor for AGE (RAGE), 397
Recollection and familiarity, deficits in,
 38–39
Reelin gene, 116
Regional cerebral metabolism (rCM), 278
Registration memory, 280
Religious orders and AD risk, 201–202
Remote memory, deficits in, 40–41
Repetitive transcranial magnetic
 stimulation (rTMS), 345
Resveratol derivatives, 214
Retinoid X receptor (RXR), 175–176,
 373–374
Retrieval, 2, 41
Rodents, manipulating neurogenesis in,
 76–77
rTg4510, 143–146

S

sAPPα, 75, 106–107
sAPPβ, 102–103
Selective Reminding Test, 32–33
Selective serotonin reuptake inhibitors
 (SSRIs), 60
Semantic memory, 4–5, 41
Ser133, CREB phosphorylation at, 10–11
Ser400, phosphorylation of, 148–149
Serotonin, 9–10, 60
Shc, 104
ShcA, 104
Short-term memory, 2–3, 280

Single Photon Emission Computed Tomography, 181–182
SIRT1, 16–17
Slave systems, 2–3
Sleep
  and Aβ pathology, 301–308
  and circadian rhythm, 218–219
  deprivation, 297–298
  disturbances, and cognitive decline, 300–301
  for learning and memory, 297–300
Sleep–wake cycle and AD, 295
Social recognition, 11
Socialization, 219
SorLA, 102–103, 115, 180f
Sox2, 55–56
Spatial memory, 16, 70–71, 251
Spike-timing dependent plasticity, 6–7
Src homology-2 (SH2), 104
Stealth pathology, 284
Storage, 2
Strategic infarct dementia, 325–326
Stress, 61–62
Stress-induced coping mechanisms, 72
Stroke, 320–323
  AD and, 341–344
    clinical findings, 341–342
    interactions between, 342–343
    mechanistic links, 343–344
  pathogenesis of, 321–322
  treatment and prevention, 344–346
  vascular cognitive impairment (VCI) after, 333–341
    animal models of, 335t
    clinical presentation, 334–335
    experimental evaluation of poststroke, 337t–338t
    experimental studies, 335–341
    incidence and prevalence, 333–334
Stroop effect, 42
Subarachnoid hemorrhage, 320–321
Subgranular layer (SGL), 52–53, 107, 400–401
Subventricular zone (SVZ), 52–55, 62–63
Synaptic degeneration, 174–175
Synaptic dysfunction, 30, 35, 172, 174–175
Synaptic formation and function, 109
Synaptic function, 109, 111
Synaptic plasticity, 7–13, 15–17
Synaptic tagging, 8
Synaptic transmission, efficacy of, 6–7

T
TAG1 protein, 107
Tau
  Aβ pathogenesis, 138
  aggregation, 143–146
  homeostasis, regulation of, 146–148
  isoforms and disease-causing mutations, 137f
  loss-of-function versus gain-of-function, 143
  missorting, 151–154
    consequence of, 153–154
    under pathological conditions and disease models, 152–153
  mutations associated with tauopathies, 140t–142t
  neuronal deficits mediated by, 143–154
  normal production and function of, 136–137
  pathogenic, 138–139
  posttranslational modifications of, 148–151
    acetylation, 151
    hyperphosphorylation, 148–151
  targeting, in AD, 154–156
Tau phosphorylation
  apoE, mitochondria dysfunction, and, 177–179
Tauopathies, 138
Thy1 promoter, 249
Tissue-specific deletion of IR, 391–392
Tobacco smoke, exposure to, 213–214
Toll-like receptor (TLR) ligands, 184
Transcranial Doppler echo, 181
Transcytosis, 398–399, 404–405, 405f
Transforming growth factor-β (TGF-β), 369
Transgenic mice, 245–247
Trans-Golgi network (TGN), 102–103
Transthyretin, 109–110
Traumatic brain injury (TBI), 111, 239, 296–297
  cognitive deficits, 271–273
  diffuse axonal injury (DAI), 274
    as mechanism of cognitive dysfunction following TBI, 274–276
  focal injuries, 279
    as mechanism of cognitive dysfunction following TBI, 279–281
  mechanisms of cognitive disturbance following, 273
  mild, 268
  neurodegenerative processes, 281–282

as mechanism of cognitive dysfunction following TBI, 282–283
neurometabolic changes, 277
   as mechanism of cognitive dysfunction following TBI, 277–279
Trisynaptic circuit, 56*f*, 58–59
Tumor necrosis factor-α (TNF-α), 362, 388–389
Tumor necrosis factor-α converting enzyme (TACE), 240–241
Twins with greater involvement in cognitive leisure activities, 202–203
Type I neural stem cells, 56–57
Type 2 diabetes mellitus (T2DM)
   cerebrovascular complications in, 400–404
   clinical description, 390–391
   dysfunctional cerebral neovascularization, 400–401
   epidemiology, 388–390
   hyperglycemia and AD, 402
   inflammation and vascular dysfunction, 404–407
     caveolae, 404–405
     insulin access to brain, 404–406
     metabolic flexibility, 406–407
   mechanism and pathways, 394–400
     advanced glycation endproducts (AGEs), 397–398
     diabetes and endothelial dysfunction, 398–400
     mitochondrial dysfunction in diabetes, 394–396
     oxidative stress and inflammation, 396
   mouse models of, 391–394
     GLUT4 knockout, 393–394
     IR knockout mouse, 391
     IRS knockout mice, 392–393
     MKR mouse model of insulin resistance, 394
     tissue-specific deletion of IR, 391–392
   as risk factor for AD, 387
   vascular inflammation in, 402–404
Type II intermediate NPCs, 56–57
Type III neuroblasts, 56–57

U

Ubiquitin-proteasome pathway, 146–147

V

Vacuolar protein sorting 10 protein (Vps10p) domain receptors, 115
Vascular cognitive impairment (VCI), 320, 328*f*, 333–341
   animal models of, 335*t*
   causes of, 326*t*
   clinical presentation of, 334–335
   experimental evaluation of poststroke, 337*t*–338*t*
   experimental studies, 335–341
   incidence and prevalence, 333–334
   PRoFESS trial, 345–346
   PROGRESS trial, 345–346
   "time pressure management" strategy, 346
Vascular dementia (VaD), 324–325
Vascular endothelial growth factor receptor 1 (VEGFR1), 244–245
Vascular inflammation in T2DM, 402–404
Vascular stiffening, 325–326
Vasculitis, 320–321
Very low-density lipoprotein receptor (VLDLR), 179–180
Visuo-spatial sketchpad, 2–3
von Willebrand Factor, 389

W

Wechsler Memory Scale-Revised, 33–34
WMS-R Logical Memory Test, 34*f*
Wnt signaling, 110, 116–117
Wnt/β-catenin signaling pathway, 74–75
Working memory, 2–4, 280
   Baddeley's model of, 3*f*
   deficits in, 41–42

X

X11 binding, 104

Y

Yellow fluorescent protein (YFP), 372
YENPTY motif, 100
   interactors with, 104

Z

zif268 gene, 8, 16
Zinc, in AD brain, 212–213